T0296597

UNCERTAINTIES IN NUMERICAL WEATHER PREDICTION

UNCERTAINTIES IN NUMERICAL WEATHER PREDICTION

Edited by

HARALDUR ÓLAFSSON

JIAN-WEN BAO

ELSEVIER

Elsevier
Radarweg 29, PO Box 211, 1000 AE Amsterdam, Netherlands
The Boulevard, Langford Lane, Kidlington, Oxford OX5 1GB, United Kingdom
50 Hampshire Street, 5th Floor, Cambridge, MA 02139, United States

Library of Congress Cataloging-in-Publication Data
A catalog record for this book is available from the Library of Congress

British Library Cataloguing-in-Publication Data
A catalogue record for this book is available from the British Library

ISBN: 978-0-12-815491-5

For information on all Elsevier publications
visit our website at https://www.elsevier.com/books-and-journals

Publisher: Candice Janco
Acquisitions Editor: Amy Shapiro
Editorial Project Manager: Andrea Dulberger
Production Project Manager: Surya Narayanan Jayachandran
Cover Designer: Matthew Limbert

Typeset by SPi Global, India

Contents

6. Ensemble data assimilation for estimating analysis uncertainty

Roland Potthast

7. Subgrid turbulence mixing

Xu Zhang

8. Uncertainties in the surface layer physics parameterizations

Haiqin Li and Jian-Wen Bao

9. Radiation

Kristian Pagh Nielsen, Laura Rontu, and Emily Gleeson

10. Uncertainties in the parameterization of cloud microphysics: An illustration of the problem

Jian-Wen Bao, Sara Michelson, and Evelyn Grell

11. Mesoscale orographic flows

Haraldur Ólafsson and Hálfdán Ágústsson

12. Numerical methods to identify model uncertainty

Harald Sodemann and Hanna Joos

13. Dynamic identification and tracking of errors in numerical simulations of the atmosphere

Haraldur Ólafsson

Contributors

Hálfdán Ágústsson Kjeller vindteknikk Ltd., Lillestrøm, Norway; Belgingur Ltd.; School of Atmospheric Sciences, Science Institute, University of Iceland, Reykjavík, Iceland

Jian-Wen Bao NOAA/Physical Sciences Laboratory, Boulder, CO, United States

Peter Bechtold European Centre of Medium-Range Weather Forecast, Reading, United Kingdom

Roberto Buizza Scuola Superiore Sant'Anna; Centre for Climate Change studies and Sustainable Actions (3CSA), Pisa, Italy

Emily Gleeson Climate, Research and Applications Division, Met Eireann, Dublin, Ireland

Evelyn Grell NOAA/Physical Sciences Laboratory; CIRES, University of Colorado Boulder, Boulder, CO, United States

Hanna Joos Institute for Atmospheric and Climate Science, ETH Zürich, Zürich, Switzerland

Haiqin Li Cooperative Institute for Research in Environmental Sciences, University of Colorado Boulder; NOAA/Global Systems Laboratory, Boulder, CO, United States

Sara Michelson NOAA/Physical Sciences Laboratory; CIRES, University of Colorado Boulder, Boulder, CO, United States

Kristian Pagh Nielsen Department of Research and Development, Danish Meteorological Institute, Copenhagen, Denmark

Haraldur Ólafsson School of Atmospheric Sciences, Science Institute, and Faculty of Physical Sciences, University of Iceland, Veðurfélagið and the Icelandic Meteorological Office, Reykjavík, Iceland

Roland Potthast Deutscher Wetterdienst (DWD), Offenbach, Germany; Applied Mathematics, University of Reading, Reading, United Kingdom

Laura Rontu Meteorological Research, Finnish Meteorological Institute, Helsinki, Finland

Harald Sodemann Geophysical Institute and Bjerknes Centre for Climate Research, University of Bergen, Bergen, Norway

Ning Wang Cooperative Institute for Research in the Atmosphere, Colorado State University, and Earth System Research Laboratory, Boulder, CO, United States

Nigel Wood Met Office, Dynamics Research, Exeter, United Kingdom

Xu Zhang Shanghai Typhoon Institute, China Meteorological Administration, Key Laboratory of Numerical Modeling for Tropical Cyclone, China Meteorological Administration, Shanghai, China

Preface

This book project was motivated by our observation that on the one hand it has become widely accepted that numerical weather prediction (NWP) has uncertainties, but on the other hand, the main-steam textbooks rarely go deep to discuss in detail where the uncertainties lie. Most freshly-graduated students may even feel reluctant to accept the notion that there is a degree of empiricism in the development of NWP models. Such empiricism is required in NWP model development because there remain questions unanswerable using current theories and observations about the actual dynamical and physical processes in nature that govern the evolution of weather-related atmospheric circulations. There is a gap between what is being taught in schools about the uncertainty in NWP associated with empiricism and the reality in the operational field of numerical weather prediction. This book is an experimental effort to fill the gap by inviting actual practitioners of NWP model development and applications to contribute in-depth overviews of the real uncertainties in NWP models.

The book aims to present a comprehensive overview of some key issues in numerical weather prediction with emphasis on uncertainties and aspects of predictability of elements of the atmosphere. The book starts with an introduction to the governing equations and to numerical methods applied to solve these equations by discretization in time and space. The perspective of improvements in solving the dynamic equations is discussed in general and in particular concerning the development of supercomputing techniques. As a continuation of Chapter 1, Chapter 2 discusses uncertainties arising from the discretization of the shallow-water equations and addresses numerical techniques to maintain stability and increasing calculation efficiency. In Chapter 3, the governing equations are revisited, and the elements of a numerical weather prediction system are explained. There is a discussion of errors and chaos and the concept of probabilistic weather prediction is introduced. The perturbation of initial conditions is explained, leading the reader to ensemble forecasting and presentation of products of probabilistic forecasts. Chapter 4 continues in the line of probability. It addresses the sources and the growth of errors, limits of prediction, and the dependence of predictability on the scales at which the atmospheric processes occur. One of the large uncertainties of numerical weather prediction stems from water in the atmosphere. Chapter 5 presents perspectives of the parameterization of unresolved moist processes and interaction of physics and dynamics, waves at different scales, turbulent mixing, convection, and radiation.

Observations used in the initialization of NWP models have both instrumentational and representational uncertainties. Chapter 6 provides a theoretical overview of how to deal with these uncertainties in the process of data assimilation used to initialize NWP models. Chapter 7 provides an overview of the treatment of subgrid turbulent mixing in NWP models and the associated uncertainty in the treatment due to the very fact that the modeling of turbulence remains semiempirical. The uncertainty in the lower boundary conditions of NWP models at the earth's surface is concisely but pointedly discussed in Chapter 8. The ultimate energy source of the atmospheric motion is solar radiation. Chapter 9 describes the processes governing the radiation-atmosphere interaction and their uncertainty related to the uncertain aspect of the cloud modeling in NWP models. Chapter 10 further discusses the uncertainty in the process modeling of cloud microphysics that is widely used in NWP models.

Chapter 11 presents the main patterns of orographic flows, with emphasis on mesoscale structures, referring to spatial scales and the governing momentum equation. The patterns are presented concerning a mountain wind diagram, based on dimensionless parameters. The challenge of forecasting the individual patterns is addressed together with an assessment of flow pattern related uncertainties.

NWP practitioners need to understand the degree of uncertainty in the product of NWP. Chapter 12 introduces a few numerical methods for quantifying the uncertainty.

The book concludes with Chapter 13, presenting a dynamically based method of analyzing and tracing errors in numerical models of the atmosphere. The method is applied to various cases of erroneous forecasts of mean sea level pressure, surface winds, and precipitation.

We wholeheartedly thank all the contributors to this book, and it is our sincere hope that it will be of good use and enjoyment to research students and scientists in the flourishing field of atmospheric sciences.

Haraldur Ólafsson
Jian-Wen Bao

CHAPTER

1

Dynamical cores for NWP: An uncertain landscape*

Nigel Wood

Met Office, Dynamics Research, Exeter, United Kingdom

1 Introduction

The dynamical core of a numerical weather prediction model is that part of the model that is responsible for the fluid-dynamical processes that can be represented (with whatever degree of accuracy) by the spatial and temporal discretization of the governing equations. This is distinct from the "physics" packages, which represent the effects of nonfluid-dynamical processes, such as radiation, and of fluid-dynamical processes that occur at scales that cannot be resolved by the smallest scales represented by the discretized equations, typically such as boundary-layer turbulence.

The first operational numerical weather forecast was created in the summer of 1954 at the Swedish Hydrometeorological Institute, followed less than a year later by the US Joint Numerical Weather Prediction Unit (e.g., Persson, 2005). Creating those forecasts required enormous amounts of insight into the underpinning physics of the problem and, using that insight, ingenuity as to how to derive equations that could be successfully solved using the limited and emerging computer resources available. Those first models were based on manipulations to, and approximations of, the inviscid, Euler equations in a rotating frame of reference. Those first models neglected the effects of unresolved processes and so represent what will be referred to as an integration of a dynamical core, that is, the inviscid, energy-conserving aspects of the complete system. Over the last 60 years, the development of the dynamical core parts of numerical weather (and climate) prediction models can be regarded as the progressive relaxation of the approximations made in those early days. This has been

Uncertainties in Numerical Weather Prediction
https://doi.org/10.1016/B978-0-12-815491-5.00001-X

made possible principally by the exponential increase in the speed of computers as well as improved methods of deriving the initial conditions for those models; in a sense the cunning of numerical modelers has been replaced by the cunning of computational scientists! However, we are perhaps at the end of that journey and need again to explore how to squeeze the most out of a limited resource.

Initially, of course there were only a very small number of different models. As time progressed that number increased as differences in both approach (numerical scheme and level of approximation) and application (weather, climate, global, and regional) increased. But as the number of approximations applied has reduced and the understanding of the numerical methods available has increased, one might have anticipated that the number of different models would now be reducing. That is not the case. An indication of this is given by Tables 1 and 2 of Marras et al. (2016) in which are listed, respectively, 26 existing NWP systems and 28 that are under development. In fact, in the early decades of the 21st century, we seem to be in an era where the number of approaches and hence different models appear to be increasing. The following sections are aimed at giving a potential architect of a new dynamical core an idea of the range of choices that are available and that need to be made. From this, it will perhaps become clear why we appear to not yet be converging on one approach to modeling of the dynamical core.

The discussion covers a range of topics each of which is covered in a mostly qualitative, nonanalytical manner, with the intention of giving a broad overview of the issues involved rather than specific detail and derivation. The interested reader is referred to any of the excellent text books available that cover each of the topics in much more analytic detail. Two of particular relevance to this topic are Durran (2010) and Lauritzen et al. (2011).

2 Governing equations

The unapproximated Euler equations can be written symbolically in terms of the physical terms. Starting with the wind field:

- Horizontal winds:

$$\text{Tendency} + \text{Transport} = \text{Coriolis} + \text{Pressure Gradient}. \tag{1}$$

- Vertical winds:

$$\text{Tendency} + \text{Transport} = \text{Coriolis} + \text{Pressure Gradient} + \text{Gravity}. \tag{2}$$

The earliest dynamical cores applied four key approximations to these equations. The first is the geostrophic approximation. This assumes that the Coriolis terms are in exact balance with the pressure gradient terms and, therefore, neglect the tendency and transport of the horizontal winds (the left-hand side of Eq. 1). Subsequent models relaxed this by allowing for some aspects of the transport term: the quasigeostrophic models allow for transport of the geostrophic wind by the geostrophic wind and semigeostrophic models allow for transport of the geostrophic wind by the full wind. But no contemporary NWP models make any of these approximations; they solve the full Eq. (1).

The second is the shallow-atmosphere approximation. This assumes that the depth of the atmosphere is small compared to the radius of the Earth. This allows certain simplifications in the equations to be made. Many models still make this approximation (together with the traditional approximation discussed later) though there are exceptions such as the series of dynamical cores used in the Met Office's Unified Model (White and Bromley, 1995; Davies et al., 2005; Wood et al., 2014).

The third is the hydrostatic approximation. This neglects the tendency and transport of the vertical component of the wind. Models employing this approximation often, but not always, also apply the shallow-atmosphere approximation (and the resulting equations are termed as the primitive equations). In this case, it is assumed that the vertical pressure gradient balances the acceleration due to gravity. When the shallow-approximation is not made, the hydrostatic approximation is referred to as the quasihydrostatic approximation. The hydrostatic approximation is only valid for horizontal scales that are significantly longer than vertical scales (see, e.g., Holton, 1992; Davies et al., 2003). Therefore, as horizontal resolution continues to increase, most new model formulations are not hydrostatic (but it is notable that the European Center's model remains hydrostatic despite running very successfully at the highest, global, operational resolution).

The fourth is the traditional approximation. White et al. (2005) discuss the application (or not) of the shallow and hydrostatic approximations and associated issues of consistency of the equation sets with various physical conservation principles (namely conservation of mass, energy, axial angular momentum, and potential vorticity). In particular, the authors found that when the shallow approximation is made then the Coriolis terms also need to be approximated in order for the equations to retain the principle of conservation of axial angular momentum. Specifically, axial angular momentum is conserved if the horizontal component of the Earth's rotation vector is neglected, meaning that the Coriolis term in Eq. (2) vanishes completely and the term in Eq. (1) is modified. This approximation is referred to as the traditional approximation and has generally been made in conjunction with the shallow-atmosphere approximation. It followed that there were four choices of consistent equation sets, determined by whether each of the hydrostatic and shallow approximations are or are not made independently (with the traditional approximation being made when the shallow approximation is made).

However, there has been something of a revival of interest in this area in the recent years with a number of recent papers revisiting and generalizing this work such as Charron and Zadra (2014) and Staniforth (2014a). In particular, Tort and Dubos (2014) showed that consistency can in fact be achieved while retaining a horizontal component of the Earth's rotation vector, albeit in an approximate form. This means that there are then six consistent equation sets, determined by whether the hydrostatic approximation is or is not made, together with one of: deep atmosphere and full Coriolis terms; shallow-atmosphere and traditional approximation; and shallow-atmosphere and approximated full Coriolis.

For thermodynamic variables, the equations are simpler, all having the form:

$$\text{Tendency} + \text{Transport} = \text{Divergence}. \tag{3}$$

This equation (and its simplified form in which the divergence term is not present) applies to any of the thermodynamic variables such as temperature, T, pressure, p, and density, ρ and any function of any of these, for example, $\Pi, \theta, \ln\theta, \ln p$, etc., where θ denotes the potential temperature and Π is the Exner function. These are related to each other by the equation of state

$$p = RT\rho, \tag{4}$$

where R is the specific gas constant, and the definitions

$$\Pi \equiv \left(\frac{p}{p_0}\right)^{\kappa}, \tag{5}$$

where p_0 is a reference pressure, typically taken to be 1000 hPa and $\kappa \equiv R/c_p$ with c_p the specific heat capacity of air at constant pressure, and

$$\theta \equiv \frac{T}{\Pi}. \tag{6}$$

The equations have been given symbolically at this stage to avoid having to make any specific choices about exactly which variables are used in the numerical implementation of the equations. For the analyst working with the continuous equations, the choice of variables is one of convenience, the solution is independent of how the equations are written. However, numerical modelers and discrete analysts do not have this luxury; when the equations are discretized only a (small) finite number of the properties of the continuous equations are preserved by any particular discretization. And, the properties that are preserved are strongly related to the choice of how the discrete equations are written. So, to tie down exactly the design of the dynamical core, a choice has to be made for how the winds will be represented and which thermodynamic variables will be used as prognostic variables.

For the winds (and for the moment assuming the horizontal surface is flat), there are broadly two choices:

- *The components of the velocity*. Let these be denoted by (u,v,w), where (u,v) represent the horizontal components (perpendicular to gravity) and w represents the vertical component parallel to gravity.
- *The components of the momentum*. These are simply the components of the velocity multiplied by the density, ρ.

The advantage of using the momentum is that it simplifies the pressure gradient term (it is a linear term involving only the pressure) and it allows the transport terms to be written in a form that allows exact numerical conservation of linear momentum. The latter property is useful for local regional models for which linear momentum is a natural quantity. This form is used, for example, in the Weather Research and Forecasting model (see Skamarock et al., 2019 and also the WRF web-based resources at https://doi.org/10.5065/D6MK6B4K). However, on the sphere, it is axial angular momentum that is the conserved quantity. Although some consideration to using angular momentum has been given, the author is not aware of any model that uses this for the wind prognostic variables. The disadvantage of using momentum is that the velocity needs to be diagnosed from it to evaluate the transport term. How much of an issue this is depends on the specific discretization used.

The advantage of using velocity is that this is directly the quantity needed to evaluate the transport. This enables the use of the semi-Lagrangian scheme, of which more later, and it also enables the equations to be written directly in vector form (often referred to as the vector invariant form), which avoids the need for the explicit inclusion of potentially complicated terms that are specific to the specific choice of coordinates used. The disadvantages of the velocity form are a nonlinear pressure gradient term, that is, one that involves the product of the gradient of the pressure variable (e.g., p or Π) with another thermodynamic quantity (e.g., $1/\rho$ or θ), and the lack of a straightforward way of obtaining discrete conservation of linear, and angular, momentum.

In the case that the coordinate lines are perpendicular to each other (an orthogonal coordinate system), then there is no ambiguity in what the components (u,v,w) represent (other than choices of scaling factors). When the coordinates are not orthogonal, though, there are further choices to be made about what wind components to use. The two natural ones are: the contravariant components in which the wind vector is expressed in terms of base vectors that are parallel to the coordinate lines and the covariant components in which the base vector of one coordinate direction is perpendicular to the plane defined by the other coordinate directions. However, in meteorology, a third alternative is often used. When the Earth's surface is not flat because of orography (hills and mountains), the vertical coordinate is often chosen such that, near the surface at least, surfaces of constant vertical coordinate are parallel to the orography. Such a coordinate is termed a terrain-following coordinate, for example, that of Gal-Chen and Somerville (1975). Although either the contra- or covariant components could be used, Clark (1977) found that good numerical conservation of linear momentum and energy was much easier to achieve if a set of orthogonal base vectors are used to describe the wind field, with the vertical base vector parallel to the gravity vector, and this is a common practice in many contemporary NWP models (e.g., Wood et al., 2014).

It is worth noting that, for hydrostatic models (for which the vertical velocity is not a prognostic variable), there is a third choice for the wind variables. In place of the two horizontal components of velocity, this approach uses the vertical component of vorticity and the divergence of the horizontal wind (i.e., the horizontal wind is split into its rotational and nonrotational components). This is a particularly attractive choice for models that are focused on the large-scale dynamics. Due to the thermal stratification of the atmosphere, the large-scale dynamics are dominated by the effects of the planetary rotation which are captured by the evolution of the vertical vorticity, and in particular the potential vorticity which judiciously combines the vorticity with the potential temperature (the interested reader is referred to any books on dynamical meteorology such as Holton, 1992 and Vallis, 2006). The divergence of the horizontal wind plays an essential role in the dynamics of the gravity waves (and neglecting the tendency of the divergence is a way of filtering out the gravity waves). However, in such models, the wind components are still required in order, for example, to evaluate how the vorticity is transported. In a vorticity-divergence-based model, these are obtained by inverting the relations between the winds and the vorticity and divergence. These operations are expensive for all but spectral models (for which the cost is effectively already paid for in the Fourier and Legendre transformations). For nonhydrostatic models, for which the interest includes much smaller scales than just planetary scales, the vertical velocity is a prognostic variable and all three components of the vorticity become relevant. The split into the vertical vorticity and divergence becomes a much less natural choice.

For the thermodynamic variables, there is a large range of choices. Essentially, any two independent choices out of T, p, ρ, and any function of any of these, for example, Π, θ, $\ln\theta$, $\ln p$, etc.

In discussing the pros and cons of various choices of thermodynamic variables, it is useful to introduce the isothermal profile. In the US Standard Atmosphere, the value of temperature departs from a value of 238.5 K by only ±20% between the surface and 100 km. Between 10 and 50 km, the variation from 243.5 K is only ±11%. A reasonable first estimate of how thermodynamic quantities vary in the atmosphere is, therefore, to assume that $T(z) = T_0 =$ constant, where z denotes height above the surface. Together with the equation of state and the assumption that the atmosphere is in hydrostatic balance the vertical variation of all the other thermodynamic variables can then be obtained. It is found that

$$p,\rho \propto \exp\left(-\frac{gz}{RT_0}\right),$$

and

$$\Pi \propto \theta^{-1} \propto \exp\left(-\frac{\kappa gz}{RT_0}\right).$$

These all, therefore, decay exponentially (except for θ which increases exponentially) with the exponent of Π and θ being rather smaller (slower decay/increase) than that for p and ρ by the factor κ, which has the value 2/7 for ideal dry air. The scale height over which p and ρ decay is given by RT_0/g which has a typical value of 7 km.

What then of the choices for thermodynamic variable?

An important question is whether the dynamical core should preserve the mass of the atmosphere. If it is required to do so, then one of the prognostic thermodynamic variables needs to be density *and* the density needs to be evolved using a numerically conservative scheme. If a quantity other than a linear function of density is used, or if a nonconservative numerical scheme is used, then mass conservation is only achieved by some form of global fixer that adds or removes the requisite amount of mass to restore the total mass. Such schemes inevitably lead to varying degrees of nonphysical transport of mass. An early example of a scheme that tried to localize that transport as much as possible is the scheme of Priestley (1993) with recent variants on that by, for example, Zerroukat and Allen (2015).

If density is not chosen as a prognostic variable, then a natural choice for one of the thermodynamic variables is pressure, either directly as p or as Π or, to avoid any numerical inaccuracies with its exponential decay, $\ln p$.

There seem to be two alternative routes to follow for which other variable to choose. Numerical schemes are most accurate when applied to quantities that vary smoothly. It has already been noted that temperature does not vary dramatically with height so this might be a natural choice that avoids the exponential changes. The other is potential temperature, θ. The appeal of this is that because $\ln\theta$ is proportional (in a dry atmosphere) to the entropy of the system, it is preserved in the absence of any diabatic processes, that is, the divergence term in Eq. (3) vanishes for θ and it is purely transported. The disadvantage is the exponential variation of θ but this can be avoided by using $\ln\theta$, which varies in a manner closer to temperature.

It is worth noting that a tempting choice is to use $\rho\theta$ as the transported variable since the equation for its evolution can be written in conservative form (i.e., a form that a numerical scheme can straightforwardly ensure the global integral of $\rho\theta$ is exactly conserved). However, it is important to note that from Eqs. (4)–(6) $\rho\theta$ is a function only of p (or equivalently Π). Therefore, the only remaining choice of independent thermodynamic variable would be a function only of T.

These choices have to be considered also in a broader sense, in terms of what impact they have on the design as a whole. A particular aspect is the form of the pressure gradient term in Eqs. (1), (2). This depends on the choice made for the wind components. If momentum is used, then Pressure Gradient $=\mathrm{Grad}(p)$ directly (where Grad denotes evaluation of the spatial gradient of its argument). Therefore, use of p might be more natural as one of the thermodynamic variables rather than using a derived quantity such as Π or $\ln p$, both of which would reintroduce an unnecessary nonlinearity into the scheme. If velocity is used, then Pressure Gradient $=\rho^{-1}\mathrm{Grad}(p)$. Then, perhaps ρ and p are natural choices. However, using the equation of state this can also be written as either $RT\mathrm{Grad}(\ln p)$ or $c_p\theta\mathrm{Grad}(\Pi)$. The choices seem to be increasing not reducing! However, help is at hand.

It will be seen below that an important role of the dynamical core is the propagation of a variety of different oscillatory modes, or waves. The seminal work of Thuburn and Woollings (2005) showed that the choice of thermodynamic variables (and how to represent them spatially) plays a critical role in how well the dynamical core represents those waves, and hence how accurate the scheme is. In particular, the authors identified (for a given choice of vertical coordinate) a single choice of thermodynamic variable that gave what they described as the optimal configuration. In a follow-up piece of work, Thuburn (2006) showed how, by using specific numerical forms of the pressure gradient, some of the suboptimal configurations can in fact be made to be optimal. Depending on one's perspective, this is either unfortunate (it does not help narrow the possibilities!) or fortunate (it leaves open more possibilities!).

It is clear that before one even gets close to defining a numerical scheme for a dynamical core, there is a very large range of options for the choice of continuous equation set and its constituent variables. To close this section, it is worth giving some examples of the range of choices that have been made in widely used and widely known models:

- The operational dynamical core of ECMWF's IFS model is a global NWP model (see ECMWF's web-based documentation at https://www.ecmwf.int/en/publications/ifs-documentation). It makes the traditional, shallow-atmosphere, and hydrostatic approximations. It uses a pressure-based vertical coordinate. The wind is represented using velocity components. Its two thermodynamic variables are temperature and geopotential height (a measure of the height associated with a given value of pressure).
- The WRF dynamical core (Skamarock et al., 2019) is principally used as a regional NWP model (but it does have a global capability). It makes a form of the shallow approximation but not the traditional nor hydrostatic approximations. It uses a pressure-based vertical coordinate. The wind is represented using momentum components. It predicts three thermodynamic variables: density and density-weighted potential temperature (both in a conservative form), together with the geopotential height which is used in the pressure gradient term. (The use of three prognostic thermodynamic variables means that instead of the equation of state being used to eliminate one of the thermodynamic variables from the

equation set, or being used as a constraint on the evolution of the thermodynamic variables, its derivative in time is used to create an additional prognostic equation.)

- The dynamical core of the Met Office's Unified Model (e.g., Walters et al., 2017) is both an operational global and operational regional NWP model. It makes neither the shallow-atmosphere, the traditional, nor the hydrostatic approximations. It uses a height-based vertical coordinate. The wind is represented using velocity components. Its two thermodynamic variables are density (but not in a conservative form) and potential temperature. However, the pressure gradient term is written in terms of the Exner function, which is derived from density and potential temperature using the equation of state.

3 Some physical properties

Before considering what numerical schemes to apply to the governing equations, it is important to have a clear idea of the nature of the physical phenomenon represented by those equations. There are perhaps three key properties of the equations represented by Eqs. (1)–(3). Two of them are related: the equations are energy preserving (they can be derived from Hamilton's principle, e.g., Salmon, 1988; Shepherd, 1990; Staniforth, 2014a); and they constitute a hyperbolic system of equations which means that they admit wave-like solutions. The third property, and one that sets them apart from many other similar equation sets, is that the presence of rotation means that the equations admit nontrivial steady states, in particular steady states with nonzero wind fields. The large-scale states of the atmosphere can often be regarded as the evolution from one near-steady state to another. Further, the transition from one near-steady state to another is achieved by the wave-like solutions to the equations. Therefore, it is essential for the accuracy of a dynamical core (a) that those steady states are well represented and maintained by the numerical schemes employed and (b) that the adjustment by the waves is well captured by the numerical schemes.

The leading order steady states have already been discussed. In the horizontal, it is geostrophic balance which can symbolically be represented as

$$0 = \text{Coriolis} + \text{Pressure Gradient}. \tag{7}$$

Since the Coriolis terms act perpendicular to the flow direction, geostrophic balance requires the wind field to be perpendicular to the pressure gradient term so that flow is parallel to lines of constant pressure.

In the vertical, it is hydrostatic balance which can be represented as

$$0 = \text{Pressure Gradient} + \text{Gravity}, \tag{8}$$

where the traditional approximation has been assumed so the Coriolis effect on vertical motion has been neglected. If the flow is in both geostrophic and hydrostatic balance, then there is also thermal wind balance which requires that any vertical shear in the horizontal wind must be balanced by a horizontal gradient of temperature perpendicular to the wind direction.

Wave motions form when the inertia of an air parcel is opposed by a restoring force. By considering the governing equations in the presence of perturbations to a geostrophically and hydrostatically balanced state, it is found (see, e.g., the comprehensive text books of Gill, 1982, Holton, 1992, and Vallis, 2006) that the five equations (three for the wind components and two for each of the chosen thermodynamic variables) support five modes of oscillations, as three different types of wave:

- *One Rossby wave*. This is due to the variation of the Coriolis effect with latitude (there needs to be a horizontal gradient in the background potential vorticity field). Rossby waves have a propagation speed, relative to the mean wind, that is close to zero. Indeed, on an *f*-plane, for which the Coriolis term is constant, the Rossby wave is purely transported by the mean wind, that is, it is stationary relative to the mean wind. Rossby waves are the most important waves for determining the large-scale evolution of the weather and are predominantly responsible for the large-scale patterns seen at mid-latitudes in satellite images.
- *Two gravity waves*. These are due to the buoyancy created by the effects of gravity (there needs to be a vertical gradient in the background potential temperature field). They propagate anisotropically and come in pairs that propagate in opposite directions. They are important for the adjustment of flow due to the presence of orography and strong convective disturbances. They have a propagation speed determined by the stratification of the atmosphere. A typical speed might be $\pm 50\ \mathrm{ms}^{-1}$ (relative to the wind speed). However, for the deepest modes that extend through a significant proportion of the troposphere, the maximum propagation speed is around $320\ \mathrm{ms}^{-1}$, similar to the speed of sound.
- *Two acoustic waves*. These are due to the compressibility of the atmosphere. They propagate isotropically and come in pairs that propagate in opposite directions. The direct effect of acoustic waves on weather forecast models is negligible. Indeed, the hydrostatic approximation filters out all acoustic modes except for one pair of modes that propagate purely horizontally. They appear in NWP models effectively as a by-product of not wanting to make the hydrostatic approximation because of how that approximation distorts the propagation of the gravity waves. (Acoustic waves can also be filtered by removing, or at least approximating, the compressibility of the atmosphere by making the anelastic approximation. However, this is of questionable accuracy for the largest scales. See Davies et al. (2003) for further discussion.) The speed of sound varies with height and temperature but according to the US Standard Atmosphere, it varies between a maximum of $340\ \mathrm{ms}^{-1}$ (relative to the wind speed) near the surface and a minimum of around $275\ \mathrm{ms}^{-1}$ at heights around 80–90 km.

Hence, a dynamical core has to handle transport by the mean wind as accurately as possible, while representing both accurately and stably the propagation of gravity waves. Unless the equation set has been appropriately approximated, the dynamical core also has to handle stably the propagation of the acoustic waves but the accuracy with which it does this is of secondary importance.

A further important aspect to consider for any system before trying to model it is the energy spectrum of its motions, that is, how much energy there is in a given range of spatial (usually horizontal) scales. Various efforts have been made to observe the atmosphere's spectrum, the most famous being the work of Nastrom and Gage (1985). The spectrum is important in deciding what range of scales it is important for a model to resolve. It is clearly important to capture accurately those scales where there is the most energy as those are likely to be the most important. In the atmosphere, these scales are very large, of the order of thousands of kilometers. There will also generally be a length and/or a time scale below which the flow has virtually no energy. These scales then place a lower bound on the space and/or time scales that it is sensible for a model to resolve. In the atmosphere, it is the molecular viscosity that ultimately puts a lower bound on the scales of motion; kinematic molecular viscosity of air has a value that is of order 10^{-5} m^2 s^{-1}. As a result, a complete representation of the atmosphere would have to represent all scales from the molecular (at most of order millimeters) to the planetary (of order 10,000 km in the horizontal), that is, scales covering at least eight orders of magnitude. This is, and will for a long time, remain beyond the reach of supercomputers. Therefore, simulations have to parameterize the effects of motions below some nominally small scale. It is, therefore, important to understand how energy flows from one range of scales to another. For example, if energy only flows from large scales to small scales (called a downward cascade), then the small scales are said to be slaved to the larger ones and parameterizing the effects of the small scales in terms of the larger ones is sensible. It turns out that to a large extent this is what happens in the atmosphere and it is this that has allowed accurate weather predictions with relatively limited computational resource. But as ever more accurate forecasts are sought the fact that there is some upward energy cascade from small scales to larger scales starts to become important. It also turns out (Holdaway et al., 2008) that the slope of the spectrum can influence, and even determine, the rate at which numerical methods converge to the correct answer; if the slope of the spectrum is in some sense shallow then increasing the accuracy of the numerical scheme will not necessarily improve the accuracy of the results—it is, in that case, more effective to improve the resolution of the model.

4 Discretizing in time

This section gives a very simple introduction to some basic time-stepping schemes. Much more detail can be found elsewhere such as the text book Durran (2010). Once we have the basic schemes their suitability as building blocks for a dynamical core will be examined.

First, we take a step toward a more mathematical description of the equations. Eqs. (1)–(3) can be represented as

$$\frac{dF}{dt} = G, \tag{9}$$

where dF/dt represents the rate of change of F with time, F represents a "state vector," that is, a vector whose components are the set of all prognostic variables (which themselves might be a vector of a number of discrete variables, e.g., the temperature at each discrete point in space), and G is the vector of all the terms that contribute to the tendency of F. The terms comprising G are sometimes referred to as the source terms or forcing terms. However, this can be misleading since most of the elements of G are functions of F and so are not external to the problem.

The role of the dynamical core is to step forward in time from some given state at time t to a new state at a discrete time later, say $t + \Delta t$, where Δt is the time interval, or time step. Let $t = n\Delta t$ for some integer n, the number of time steps already taken from some initial condition at time $t=0$, and let F^n denote the state of the model at time t. Then, the state at time $t+\Delta t$, after $n + 1$ time steps, can be obtained by integrating Eq. (9) from time t to time $t + \Delta t$, that is

$$F^{n+1} = F^n + \int_t^{t+\Delta t} G dt. \tag{10}$$

In this context, the challenge of temporal discretizations is then to estimate the integral of G over a time interval and some different approaches are now explored.

4.1 Some different approaches

4.1.1 *Single-stage, single-step schemes*

The simplest time-stepping scheme is obtained by approximating the integral on the right-hand side of Eq. (10) as simply $\Delta t G(t)$, which can be denoted as $\Delta t G^n$. Eq. (10) then becomes

$$F^{n+1} = F^n + \Delta t G^n. \tag{11}$$

In this scheme, the tendency of F is estimated entirely from quantities at time level n. It is, therefore, called an explicit scheme (since all contributions to the tendency of F are known explicitly) and it is also called a forward-in-time scheme or forward Euler.

An alternative is to instead estimate the integral as $\Delta t G(t+\Delta t)$, that is, $\Delta t G^{n+1}$. The expression for F^{n+1}, that is,

$$F^{n+1} = F^n + \Delta t G^{n+1}, \tag{12}$$

is an implicit expression for F^{n+1} since in general G^{n+1} depends on F^{n+1}, which is unknown. This scheme is, therefore, called an implicit scheme and is also known as a backward-in-time scheme or backward Euler.

Both of these schemes are only first-order accurate, which means that the accuracy of their solutions improves linearly as the time-step Δt is reduced.

A second-order scheme is one whose accuracy improves quadratically as Δt is reduced, that is, halving the time step reduces the error of the scheme by a factor of 4. An example of such a scheme is obtained by approximating the integral on the right-hand side of Eq. (10) using the trapezoidal rule, that is, approximating the integral as $(\Delta t/2)[G(t+\Delta t)+G(t)]$. The resulting scheme,

$$F^{n+1} = F^n + \frac{\Delta t}{2}\left(G^{n+1} + G^n\right), \tag{13}$$

is the Crank-Nicolson scheme.

These schemes are all single-step schemes since only one time level, F^n, is needed to advance to the next time level.

4.1.2 *Single-stage, multistep schemes*

The Crank-Nicolson scheme can alternatively be thought of as first approximating the integral using the mid-point rule, that is, approximating the integral as $\Delta t G(t+\Delta t/2)$, and second as approximating $G(t+\Delta t/2) \approx [G(t+\Delta t)+G(t)]/2$. With this view in mind, another scheme is obtained if, instead of integrating Eq. (9) over one time step, it is integrated over two time steps, that is, over a time interval of $2\Delta t$. Then,

$$F^{n+2} = F^n + \int_t^{t+2\Delta t} G dt. \tag{14}$$

Applying the midpoint rule to this gives

$$F^{n+2} = F^n + 2\Delta t G(t+\Delta t) \equiv F^n + 2\Delta t G^{n+1}. \tag{15}$$

This is now an explicit scheme for F^{n+2} since, once F^{n+1} has been evaluated, the right-hand side term is explicitly known. It is called the leap-frog scheme since the scheme advances from F^n to F^{n+2} leap-frogging over F^{n+1} (though F^{n+1} is needed to evaluate the tendency). The scheme is termed a multistep scheme since two time levels, F^n and F^{n+1}, are needed to advance to the next time level.

An alternative family of schemes can be achieved if, instead of integrating the time derivative over multiple time steps (two in the case of Eq. 14), multiple time-step values are used to estimate the integral of G. The two-step Adams-Bashforth scheme is one example. In this scheme, the values of G at time $t^n - \Delta t$ (denoted by G^{n-1}) and time t^n (G^n) are used to create a linear extrapolation of G valid over the interval $t \in [t, t + \Delta t]$, namely:

$$G(t) \approx \frac{(t^n - t)G^{n-1} + (t - t^{n-1})G^n}{\Delta t}. \tag{16}$$

Using this expression, the integral on the right-hand side of Eq. (10) can be evaluated as

$$\int_{t^n}^{t^n+\Delta t} G dt = \Delta t \left(\frac{3}{2} G^n - \frac{1}{2} G^{n-1} \right),$$

so that the two-step Adams-Bashforth scheme is

$$F^{n+1} = F^n + \Delta t \left(\frac{3}{2} G^n - \frac{1}{2} G^{n-1} \right). \tag{17}$$

4.1.3 Multistage single-step schemes

The forward-Euler scheme is only first-order accurate but it is a very simple and, therefore, cheap scheme. The idea of the next scheme is that this estimate for F^{n+1} can be used as a first predictor of F^{n+1} in the Crank-Nicolson scheme to create an explicit scheme but one that is more accurate than the forward-Euler scheme. The result is a two-stage scheme:

$$F^{(1)} = F^n + \Delta t G^n, \tag{18}$$

$$F^{n+1} = F^n + \frac{\Delta t}{2} \left(G^{(1)} + G^n \right), \tag{19}$$

where $F^{(1)}$ is the (first) predictor for F^{n+1} and $G^{(1)} \equiv G(F^{(1)})$. This scheme can be regarded alternatively as the first two stages of a multistage iterative scheme in which each iteration is given by

$$F^{(k)} = F^n + \frac{\Delta t}{2} \left(G^{(k-1)} + G^n \right), \tag{20}$$

where $F^{(k)}$ is the kth predictor for F^{n+1} and $G^{(k-1)} \equiv G\left(F^{(k-1)}\right)$ and $F^{(0)} \equiv F^n$. It can also be regarded as a predictor-corrector scheme in which $F^{(1)}$ is the predictor to which a corrector $\Delta t\left(G^{(1)} - G^n\right)/2$ is applied. Indeed, the iterative scheme can also be regarded as a multiple predictor-corrector scheme where, for $k > 1$:

$$F^{(k)} = F^{(k-1)} + \frac{\Delta t}{2}\left(G^{(k-1)} - G^{(k-2)}\right). \tag{21}$$

This scheme is limited to second-order accuracy no matter how many iterations, or correctors, are applied. At convergence, it converges to the second-order Crank-Nicolson scheme.

An important class of time-stepping schemes can be introduced by taking an alternative route to Eq. (19). Having evaluated an estimate, $F^{(1)}$, for F at time $t + \Delta t$, this estimate can be used with the forward-Euler scheme to generate an estimate, $\widetilde{F}^{n+2}$, for F at time $t + 2\Delta t$ (the tilde is used to distinguish this from the actual solution at time-step $n + 2$):

$$\widetilde{F}^{n+2} = F^{(1)} + \Delta t G^{(1)}. \tag{22}$$

Then, F^{n+1} can be obtained as the linear interpolation between F^n and $\widetilde{F}^{n+2}$, that is,

$$F^{n+1} = \frac{1}{2}F^n + \frac{1}{2}\left(F^{(1)} + \Delta t G^{(1)}\right). \tag{23}$$

On substituting for $F^{(1)}$ from Eq. (18), this can be seen to be exactly the same, second-order scheme as Eq. (19). But following a similar principle, a higher-order scheme can be obtained.

Again, the same estimate for $\widetilde{F}^{n+2} \equiv F(t + 2\Delta t)$ is obtained but this time it is used to linearly interpolate to an estimate, $F^{(2)}$, for F at time $t + \Delta t/2$:

$$F^{(2)} = \frac{3}{4}F^n + \frac{1}{4}\left(F^{(1)} + \Delta t G^{(1)}\right). \tag{24}$$

The third and final predictor is obtained in a similar manner by: first using $F^{(2)}$ and the forward-Euler method to obtain an estimate, $F^{n+3/2}$, for F at time $t + 3\Delta t/2$, namely

$$F^{n+3/2} = F^{(2)} + \Delta t G^{(2)}, \tag{25}$$

and then linearly interpolating between this and F^n to obtain the estimate for F^{n+1}:

$$F^{n+1} = \frac{1}{3}F^n + \frac{2}{3}\left(F^{(2)} + \Delta t G^{(2)}\right). \tag{26}$$

This scheme, known as the Shu-Osher method, is a particular form of a class of schemes known as explicit Runge-Kutta schemes. There is a very large literature on such schemes but for discussion in the context of NWP see, for example, Baldauf (2008), Weller et al. (2013), and Lock et al. (2014). The Shu-Osher method has the property that each stage is built exclusively from application of the same, forward-Euler, scheme. This permits the development of what are termed strong stability preserving schemes (Durran, 2010). Further, the seemingly arbitrary choice of steps has been made deliberately so that the scheme has the desirable property of being third-order accurate after three stages even when G is a nonlinear function of F (Baldauf, 2008). A more natural choice might be the iterative scheme

$$F^{(k)} = \frac{1}{k}\sum_{l=0}^{k-2} F^{(l)} + \frac{1}{k}\left(F^{(k-1)} + \Delta t G^{(k-1)}\right), \tag{27}$$

with $F^{(0)} \equiv F^n$. The first three stages of this scheme (which can be compared, respectively, with Eqs. 18, 24, 26) are

$$F^{(1)} = F^n + \Delta t G^n, \tag{28}$$

$$\begin{aligned} F^{(2)} &= \frac{1}{2}\left(F^n + F^{(1)} + \Delta t G^{(1)}\right) \\ &= F^n + \frac{1}{2}[\Delta t G^n + \Delta t G(F^n + \Delta t G^n)], \end{aligned} \tag{29}$$

$$\begin{aligned} F^{(3)} &= \frac{1}{3}\left(F^n + F^{(l)} + F^{(2)} + \Delta t G^{(2)}\right) \\ &= F^n + \frac{1}{2}\Delta t G^n + \frac{1}{6}\Delta t G(F^n + \Delta t G^n) \\ &\quad + \frac{1}{3}\Delta t G\left\{F^n + \frac{\Delta t}{2}[G^n + G(F^n + \Delta t G^n)]\right\}. \end{aligned} \tag{30}$$

This would appear to be a more natural scheme because if G is a linear function of F then each iteration of this scheme adds the next term in the series expansion for the operator $\exp(\Delta t G)$ applied to F^n, for example, in that case, Eqs. (29), (30) become

$$\begin{aligned} F^{(2)} &= F^n + \frac{1}{2}\left[\Delta t G^n + \Delta t G(F^n) + \Delta t^2 G(G(F^n))\right] \\ &= F^n + \Delta t G^n + \frac{\Delta t^2}{2} G(G(F^n)), \end{aligned} \tag{31}$$

and

$$F^{(3)} = F^n + \Delta t G^n + \frac{\Delta t^2}{2} G(G^n) + \frac{\Delta t^3}{6} G[G(G^n)]. \tag{32}$$

However, in contrast to the Shu-Osher method, the kth stage of this scheme is kth-order accurate only when G is a linear function of F.

4.2 Some numerical properties

A natural question to ask is how good or bad are each of these schemes?

It has been seen that the important aspects of any scheme for a dynamical core is how it handles transport and also wave-like motion. The essential temporal aspects of both of these processes are captured by Eq. (9) with the choice $G(F) = i\Omega F$, where F can now be complex with i denoting the imaginary unit and Ω is the frequency of the motion. For transport of a wave with wave number k (i.e., wavelength $2\pi/k$) by a wind of speed U, $\Omega = -Uk$, while for the propagation of a sound wave of the same wave number $\Omega = \pm c_s k$, where c_s is the speed of sound.

Eq. (9) is then

$$\frac{dF}{dt} = i\Omega F, \tag{33}$$

and its analytical evolution over a time step of length Δt is given by

$$F_{ana}^{n+1} \equiv F_{ana}(t+\Delta t) = F_{ana}(t)\exp(i\Omega\Delta t) \equiv F_{ana}^{n}\exp(i\Omega\Delta t). \quad (34)$$

This expresses the fact that over the time step the wave is transported, or propagated, without a change in amplitude and its phase is shifted by the amount $\Omega\Delta t$. If $\Omega\Delta t$ is a small quantity (i.e., if the time step is much less than the time scale implied by the frequency of the wave), then a Taylor series expansion in $\Omega\Delta t$ shows that

$$F_{ana}^{n+1} = \left[1 + i\Omega\Delta t - \frac{(\Omega\Delta t)^2}{2} - i\frac{(\Omega\Delta t)^3}{6} + O(\Omega\Delta t)^4\right] F_{ana}^{n}, \quad (35)$$

where $O(\Omega\Delta t)^4$ denotes all terms involving at least fourth powers of $\Omega\Delta t$.

4.2.1 *Single-stage, single-step schemes*

Applying the forward-Euler scheme to Eq. (33), the discrete solution is

$$F_{FE}^{n+1} = (1 + i\Omega\Delta t)F_{FE}^{n}. \quad (36)$$

Comparing Eq. (36) with Eq. (35) it is seen, as previously stated, that the forward-Euler scheme is first-order accurate; only the terms in the first two powers of $\Omega\Delta t$, that is, $(\Omega\Delta t)^0$ and $(\Omega\Delta t)^1$, match those terms in the analytic expression.

Further, Eq. (36) can be written as

$$F_{FE}^{n+1} = A_{FE}\exp(i\omega_{FE}\Delta t)F_{FE}^{n}, \quad (37)$$

where, what is termed the amplification factor of the scheme, $A_{FE} = |1 + i\Omega\Delta t|$ (defined to be positive), is given by

$$A_{FE} = +\sqrt{1 + (\Omega\Delta t)^2}, \quad (38)$$

and, what is termed the phase shift of the scheme, $\omega_{FE}\Delta t = \arg(1 + i\Omega\Delta t)$, is given by

$$\tan(\omega_{FE}\Delta t) = \Omega\Delta t. \quad (39)$$

From Eq. (34), the ideal value of A_{FE} is $A_{FE} = |\exp(i\Omega\Delta t)| = 1$ and the ideal value of $\omega_{FE}\Delta t$ is $\omega_{FE}\Delta t = \arg[\exp(i\Omega\Delta t)] = \Omega\Delta t$. In particular, it is seen that $A_{FE} > 1$ independently of how small the time step is. This means that at each time step the amplitude of F_{FE} increases. This is in contrast to the analytic solution for which the amplitude is unchanged. The forward-Euler scheme is, therefore, said to be unconditionally unstable for this problem (there is no condition that the time step can satisfy, other than being zero, that allows the scheme to be stable); it generates unrealistic solutions that become arbitrarily large after sufficient time. For sufficiently small time steps, such that the magnitude of $\Omega\Delta t$ is small, the phase shift of the wave is reasonably well captured by the forward-Euler scheme (since $\tan\alpha \to \alpha$ as $\alpha \to 0$) but for magnitudes of $\Omega\Delta t$ larger than 1 then the phase shift of the numerical scheme is reduced. It is constrained to always be less than $\pi/2$ no matter how large the actual shift $\Omega\Delta t$ is.

For the backward-Euler scheme, the solution to Eq. (33) is

$$F_{BE}^{n+1} = F_{BE}^{n} + i\Omega\Delta t F_{BE}^{n+1}. \quad (40)$$

Gathering together the two terms involving F_{BE}^{n+1} gives

$$(1 - i\Omega\Delta t)F_{BE}^{n+1} = F_{BE}^{n}, \tag{41}$$

and the solution is finally obtained by applying the inverse of $(1 - i\Omega\Delta t)$ to both sides to give

$$F_{BE}^{n+1} = (1 - i\Omega\Delta t)^{-1}F_{BE}^{n}. \tag{42}$$

When $\Omega\Delta t$ is small, the right-hand side can be written as

$$F_{BE}^{n+1} = \left[1 + i\Omega\Delta t - (\Omega\Delta t)^2 + O(\Omega\Delta t)^3\right]F_{BE}^{n}, \tag{43}$$

and by comparing Eq. (43) with Eq. (35) it is seen that the backward-Euler scheme is also first-order accurate.

Eq. (42) can be written as

$$F_{BE}^{n+1} = A_{BE}\exp(i\omega_{BE}\Delta t)F_{BE}^{n}, \tag{44}$$

where the amplification factor, A_{BE}, is given by

$$A_{BE} = +\frac{1}{\sqrt{1 + (\Omega\Delta t)^2}}, \tag{45}$$

and the phase shift, ω_{BE}, is again given by

$$\tan(\omega_{BE}\Delta t) = \Omega\Delta t. \tag{46}$$

The numerical phase of the backward-Euler scheme is, therefore, the same as that for the forward-Euler scheme. But, importantly, the amplitude is always less than 1, no matter how large a time step is used. This means that with each successive time step the amplitude of the wave will be reduced, tending after sufficiently many time steps, to zero. The scheme is, therefore, said to be unconditionally stable. But it is also said to be damping; in contrast to the physical system which preserves the amplitude of the wave, the numerical scheme adds damping to the system and eventually the wave will be eliminated.

The unconditional stability of the scheme comes with a cost. In general, G is an operator, usually one involving a spatial derivative, whose discrete representation will be a matrix. Then, multiplication of F_{BE}^{n+1} by $i\Omega\Delta t$ represents application of that operator, or multiplication by the matrix. Solution of the implicit scheme then involves evaluating the inverse of that operator. This is often a global problem, solving for one point in space involves the values at all, or at least a large number of, other points. This is a more expensive operation than solving for each value locally, particularly on massively parallel computers in which the problem is split across a large number of processors.

Applying the Crank-Nicolson scheme to Eq. (33), and using the same methodology as earlier, results in

$$F_{CN}^{n+1} = \left(\frac{1 + i\Omega\Delta t/2}{1 - i\Omega\Delta t/2}\right)F_{CN}^{n}, \tag{47}$$

or

$$F_{CN}^{n+1} = A_{CN}\exp(i\omega_{CN}\Delta t)F_{CN}^{n}, \tag{48}$$

where A_{CN} is given by

$$A_{CN} = +\sqrt{\frac{1+(\Omega\Delta t/2)^2}{1+(\Omega\Delta t/2)^2}} = 1, \tag{49}$$

and ω_{CN} is given by

$$\tan(\omega_{CN}\Delta t) = \frac{\Omega\Delta t}{1-(\Omega\Delta t/2)^2}. \tag{50}$$

Using the half-angle trigonometric relationships, the singularity at $(\Omega\Delta t/2)^2 = 1$ can be avoided by rewriting this as

$$\tan\left(\frac{\omega_{CN}\Delta t}{2}\right) \equiv \frac{1-\cos(\omega_{CN}\Delta t)}{\sin(\omega_{CN}\Delta t)} = \frac{\Omega\Delta t}{2}. \tag{51}$$

For small $\Omega\Delta t$ (Eq. 47) can be expanded as

$$F_{CN}^{n+1} = \left[1 + i\Omega\Delta t - \frac{(\Omega\Delta t)^2}{2} - i\frac{(\Omega\Delta t)^3}{4} + O(\Omega\Delta t)^4\right] F_{CN}^n. \tag{52}$$

Since Eq. (52) matches the first three terms in the expansion of the analytic solution (35), the scheme is seen to be second-order accurate. Also, the amplification factor is 1 independent of the size of the time step. The scheme, therefore, retains the analytic amplitude without growth or decay; the scheme is said to be neutrally stable. These two properties make it a very attractive scheme for numerical solution of wave-like processes. If it is important to get the phase properties of the wave correct (i.e., its propagation speed), then the time step must be chosen small enough to resolve the wave's frequency, that is, chosen to ensure that the magnitude of $\Omega\Delta t$ is less than 1 so that, from Eq. (51), $\omega_{CN} \approx \Omega$. However, by comparing Eq. (51) with Eqs. (46), (39), it is seen that the Crank-Nicolson scheme matches the phase error of the backward- or forward-Euler schemes but with a time step that is twice as large.

4.2.2 Single-stage, multistep schemes

Applying the leap-frog scheme to Eq. (33) gives

$$F_{LF}^{n+2} = F_{LF}^n + 2i\Omega\Delta t F_{LF}^{n+1}. \tag{53}$$

The previous methodology is now applied to two successive time steps. A solution is sought that has the form

$$F_{LF}^{n+1} = A_{LF}\exp(i\omega_{LF}\Delta t)F_{LF}^n, \tag{54}$$

and

$$F_{LF}^{n+2} = A_{LF}\exp(i\omega_{LF}\Delta t)F_{LF}^{n+1} = [A_{LF}\exp(i\omega_{LF}\Delta t)]^2 F_{LF}^n. \tag{55}$$

Substituting these into Eq. (53) leads to the requirement

$$[A_{LF}\exp(i\omega_{LF}\Delta t)]^2 = 1 + 2i\Omega\Delta t[A_{LF}\exp(i\omega_{LF}\Delta t)]. \tag{56}$$

Solving this quadratic for $A_{LF}\exp(i\omega_{LF}\Delta t)$ gives

$$F_{LF}^{n+1} = A_{LF}\exp(i\omega_{LF}\Delta t)F_{LF}^{n} = \left[i\Omega\Delta t \pm \sqrt{1-(\Omega\Delta t)^2}\right]F_{LF}^{n}. \tag{57}$$

From this it follows that, provided

$$(\Omega\Delta t)^2 \le 1, \tag{58}$$

then

$$A_{LF} = +\sqrt{(\Omega\Delta t)^2 + \left[1-(\Omega\Delta t)^2\right]} = 1, \tag{59}$$

(i.e., the scheme is neutrally stable) and

$$\tan(\omega_{LF}\Delta t) = \frac{\Omega\Delta t}{\pm\sqrt{1-(\Omega\Delta t)^2}}. \tag{60}$$

However, if

$$(\Omega\Delta t)^2 > 1,$$

then

$$A_{LF} = \left|\Omega\Delta t \pm \sqrt{(\Omega\Delta t)^2 - 1}\right|. \tag{61}$$

If $\Omega\Delta t > +1$, then the solution with the positive root has amplitude greater than 1; if $\Omega\Delta t < -1$, then the solution with the negative root has amplitude greater than 1. Therefore, whatever the sign of $\Omega\Delta t$, there is a root for which $A_{LF} > |\Omega\Delta t| > 1$ and the scheme is unstable. The scheme is said to be conditionally stable: it is stable if $(\Omega\Delta t)^2 \le 1$ but unstable otherwise.

What is additionally different about this scheme is that there are two solutions. By expanding the right-hand side of Eq. (57) for small $\Omega\Delta t$, it is found that the solution with the positive root corresponds to the physical solution (and is second-order accurate in time), while the solution with the negative root is a purely computational solution that arises entirely due to the chosen method of solution. Further, because of Eq. (59), this solution is not damped at all, it propagates without change of amplitude. It is characteristic of all multistep methods that, in addition to the physical solution, they support additional, spurious, or computational solutions. Such solutions often manifest themselves as high-frequency oscillations, referred to as noise. However, many such schemes are designed so that the computational modes are strongly damped in time, $A < 1$. Then, their presence might be acceptable provided that there are no source terms or forcing of the equations that would act to continually excite such solutions.

Applying the two-step Adams-Bashforth scheme to Eq. (33) gives

$$F_{AB}^{n+1} = F_{AB}^{n} + i\Omega\Delta t\left(\frac{3}{2}F_{AB}^{n} - \frac{1}{2}F_{AB}^{n-1}\right). \tag{62}$$

As previously, a solution is sought that has the form

$$F_{AB}^{n} = A_{AB} \exp(i\omega_{AB}\Delta t) F_{AB}^{n-1}, \tag{63}$$

and

$$F_{AB}^{n+1} = A_{AB} \exp(i\omega_{AB}\Delta t) F_{AB}^{n} = [A_{AB} \exp(i\omega_{AB}\Delta t)]^2 F_{AB}^{n-1}. \tag{64}$$

Substituting these into Eq. (62) leads to the requirement

$$[A_{AB} \exp(i\omega_{AB}\Delta t)]^2 = A_{AB} \exp(i\omega_{AB}\Delta t) + i\Omega\Delta t \left[\frac{3}{2} A_{AB} \exp(i\omega_{AB}\Delta t) - \frac{1}{2}\right]. \tag{65}$$

Solving this quadratic for $A_{AB} \exp(i\omega_{AB}\Delta t)$ gives

$$F_{AB}^{n+1} = A_{AB} \exp(i\omega_{AB}\Delta t) F_{AB}^{n} = \frac{1}{2}\left[1 + \frac{3}{2} i\Omega\Delta t \pm \sqrt{1 - \frac{9}{4}(\Omega\Delta t)^2 + i\Omega\Delta t}\right] F_{AB}^{n}. \tag{66}$$

As for the multistep leap-frog scheme, this scheme has two solutions. By expanding the right-hand side of Eq. (66) for small $\Omega\Delta t$ it is found that the two solutions evolve as

$$\left(F_{AB}^{n+1}\right)^{+} = \left[1 + 2i\left(\frac{\Omega\Delta t}{2}\right) - 2\left(\frac{\Omega\Delta t}{2}\right)^2 + 2i\left(\frac{\Omega\Delta t}{2}\right)^3 - 2\left(\frac{\Omega\Delta t}{2}\right)^4 + O\left[(\Omega\Delta t)^5\right]\right]\left(F_{AB}^{n}\right)^{+}, \tag{67}$$

and

$$\left(F_{AB}^{n+1}\right)^{-} = \left[i\left(\frac{\Omega\Delta t}{2}\right) + 2\left(\frac{\Omega\Delta t}{2}\right)^2 - 2i\left(\frac{\Omega\Delta t}{2}\right)^3 + 2\left(\frac{\Omega\Delta t}{2}\right)^4 + O\left[(\Omega\Delta t)^5\right]\right]\left(F_{AB}^{n}\right)^{-}. \tag{68}$$

Therefore, the solution with the positive root corresponds to the physical solution and it is seen to be second-order accurate in time. This is to be expected since $\left(3F_{AB}^{n} - F_{AB}^{n-1}\right)/2$ is a second-order estimate of $F_{AB}^{n-1/2}$. The amplification factor for this solution is found by: gathering all the real terms in Eq. (67) together; squaring the result; gathering all the imaginary terms together; squaring the result; summing these two squares, keeping only terms that are less than $O\left[(\Omega\Delta t)^5\right]$; and then approximating the square root of the result from a Taylor series expansion, again only keeping terms that are less than $O\left[(\Omega\Delta t)^5\right]$. The result is

$$A_{AB}^{+} = 1 + \frac{1}{4}(\Omega\Delta t)^4 + O\left[(\Omega\Delta t)^5\right], \tag{69}$$

and, for small $\Omega\Delta t$, it is seen to be unstable. Not surprisingly, it remains unstable for larger values of $\Omega\Delta t$ as is shown in Durran (2010).

The solution with the negative root is a purely computational solution that arises entirely due to the chosen method of solution. For small $\Omega\Delta t$, this evolves as

$$\left(F_{AB}^{n+1}\right)^{-} = \frac{1}{2} i(\Omega\Delta t)[1 - i(\Omega\Delta t) + H.O.T.] F_{AB}^{n}, \tag{70}$$

and has amplification factor

$$A_{AB}^{-} = \frac{1}{2}(\Omega\Delta t)\left[1 + O(\Omega\Delta t)^2\right]. \tag{71}$$

This is, therefore, an example of a multistep scheme for which the computational mode is strongly damped in time, at least for small $\Omega\Delta t$.

4.2.3 Multistage single-step schemes

The low-order accuracy of the forward-Euler scheme (and we can now say its lack of stability) motivated the multistage scheme (Eq. 20). Applying this scheme to Eq. (33) gives

$$F_{iter}^{(k)} = F_{iter}^{n} + i\left(\frac{\Omega\Delta t}{2}\right)\left(F_{iter}^{(k-1)} + F_{iter}^{n}\right). \tag{72}$$

The first four iterations of the scheme are

$$F_{iter}^{(1)} = [1 + i(\Omega\Delta t)]F_{iter}^{n}, \tag{73}$$

$$F_{iter}^{(2)} = \left[1 + i(\Omega\Delta t) - \frac{(\Omega\Delta t)^2}{2}\right]F_{iter}^{n}, \tag{74}$$

$$F_{iter}^{(3)} = \left[1 + i(\Omega\Delta t) - \frac{(\Omega\Delta t)^2}{2} - i\frac{(\Omega\Delta t)^3}{4}\right]F_{iter}^{n}, \tag{75}$$

$$F_{iter}^{(4)} = \left[1 + i(\Omega\Delta t) - \frac{(\Omega\Delta t)^2}{2} - i\frac{(\Omega\Delta t)^3}{4} + \frac{(\Omega\Delta t)^4}{8}\right]F_{iter}^{n}. \tag{76}$$

$F_{iter}^{(1)}$ is first-order accurate but all the others are all seen to be second-order accurate. Their amplification factors are found to be

$$A_{iter}^{(1)} = \sqrt{1 + (\Omega\Delta t)^2} > 1, \tag{77}$$

$$A_{iter}^{(2)} = \sqrt{1 + \frac{(\Omega\Delta t)^4}{4}} > 1, \tag{78}$$

$$A_{iter}^{(3)} = \sqrt{1 - \frac{(\Omega\Delta t)^4}{4}\left[1 - \frac{(\Omega\Delta t)^2}{4}\right]}, \tag{79}$$

$$A_{iter}^{(4)} = \sqrt{1 - \frac{(\Omega\Delta t)^6}{16}\left[1 - \frac{(\Omega\Delta t)^2}{4}\right]}. \tag{80}$$

Therefore, the first two iterations are unconditionally unstable, while the last two are stable provided that $(\Omega\Delta t/2)^2 \le 1$. By minimizing $A_{iter}^{(4)}$ as a function of $\Omega\Delta t$, it is found that the maximum damping occurs when $(\Omega\Delta t)^2 = 3$ for which $A_{iter}^{(4)}$ takes the value $\sqrt{37}/8 \approx 0.76$, that is, a reduction in amplitude of nearly a quarter per time step. For $\Omega\Delta t \le 1$, the maximum reduction is less than 2.5% per time step.

It is left to the reader to explore how the scheme behaves with further iterations for which an interesting pattern quickly emerges.

An important aspect of any iterative scheme is whether the scheme converges with increasing iterations. This can be examined by subtracting two successive iterates from each other, that is, subtracting Eq. (72) evaluated at iterate $k-1$ from Eq. (72) at iterate k. The result is

$$F_{iter}^{(k)} - F_{iter}^{(k-1)} = i\left(\frac{\Omega\Delta t}{2}\right)\left(F_{iter}^{(k-1)} - F_{iter}^{(k-2)}\right). \tag{81}$$

This shows that the difference between two successive iterations will only reduce in amplitude (i.e., the scheme will only converge as the number of iterations increases) if $(\Omega\Delta t/2)^2 \leq 1$.

The final example of a time-stepping scheme considered here is the three-stage explicit Runge-Kutta scheme defined by Eqs. (18), (24), (26). Applying those equations to Eq. (33) results in

$$F_{RK}^{(1)} = [1 + i(\Omega\Delta t)]F_{RK}^{n}, \tag{82}$$

$$F_{RK}^{(2)} = \frac{3}{4}F_{RK}^{n} + \frac{1}{4}[1 + i(\Omega\Delta t)]F_{RK}^{(1)}, \tag{83}$$

$$F_{RK}^{n+1} = \frac{1}{3}F_{RK}^{n} + \frac{2}{3}[1 + i(\Omega\Delta t)]F_{RK}^{(2)}. \tag{84}$$

Successively eliminating $F_{RK}^{(1)}$ and $F_{RK}^{(2)}$ from these equations and collecting terms together leads to

$$F_{RK}^{n+1} = \left[1 + i(\Omega\Delta t) - \frac{(\Omega\Delta t)^2}{2} - i\frac{(\Omega\Delta t)^3}{6}\right]F_{RK}^{n}. \tag{85}$$

Comparing this with Eq. (35) shows directly that the scheme is third-order accurate. Its amplification and phase shifts can be shown to be

$$A_{RK} = \sqrt{\left\{1 - \frac{(\Omega\Delta t)^4}{12}\left[1 - \frac{(\Omega\Delta t)^2}{3}\right]\right\}}, \tag{86}$$

and

$$\tan(\omega_{RK}\Delta t) = (\Omega\Delta t)\left[\frac{1 - (\Omega\Delta t)^2/6}{1 - (\Omega\Delta t)^2/2}\right]. \tag{87}$$

From Eq. (86), the scheme is seen to be conditionally stable, conditional on $(\Omega\Delta t)^2 \leq 3$. By minimizing A_{RK} as a function of $\Omega\Delta t$, it is found that the maximum damping occurs when $(\Omega\Delta t)^2 = 2$ when A_{RK} takes the value $2\sqrt{2}/3 \approx 0.94$, that is, a reduction in amplitude of 6% per time step. For $\Omega\Delta t \leq 1$, the maximum reduction is just under 3% per time step.

These examples only scratch the surface of all available time-stepping schemes but they do demonstrate some of the characteristics of the different types of scheme. Effective single-stage single-step schemes invariably involve an implicit aspect, which will in general require the inversion of an operator with the complications associated with that. Multistep schemes introduce computational solutions in addition to the desired physical solution. Care then needs to be taken to understand the behavior of those solutions to determine whether they are sufficiently controlled (i.e., they are sufficiently damped, without overly damping the physical solution) and that the mechanisms for forcing such solutions are understood. Explicit multistage (or iterative) methods come in many flavors. They require multiple evaluation of the operators but the single-step flavor of such schemes does not introduce computational solutions. They are conditionally stable. Implicit multistage methods do exist but their complexity precludes discussion here.

A number of the schemes discussed earlier are used in contemporary NWP models. For example, both ECMWF's IFS and the Met Office's Unified Model use an implicit scheme for the linear part of their equation set. For the nonlinear part, the IFS uses a forward-Euler scheme, while the Unified Model uses the iterative scheme. Because of its good energy conservation properties, the leap-frog scheme has sometimes been used in cloud-resolving models such as the MONC model (Brown et al., 2015). Both WRF and the KIAP's model KIM (e.g., Choi and Hong, 2016) use the Runge-Kutta scheme presented here as part of their time-stepping schemes.

5 Discretizing in space

To introduce a spatial aspect to the prototype (33) let $G \equiv \partial H/\partial x$, for some function H, so that it becomes

$$\frac{\partial F}{\partial t} = \frac{\partial H}{\partial x}, \tag{88}$$

where now, because F is a function of both time, t, and space, x, the derivative d/dt has been replaced by the partial derivative $\partial/\partial t$.

The domain of interest in the x-direction, $x \in [x_0, x_N]$, is considered to be partitioned into a number, N, of volumes or cells, $C_{i-1/2}$, $i = 1,\ldots,N$, such that the collection of such volumes spans the whole space and no two volumes overlap each other. Each cell is centered on a point $x = x_{i-1/2}$ with edges at $x = x_{i-1}$ and $x = x_i$. Although not essential, let each cell be of the same length Δx so that $x_i = x_{i-1} + \Delta x$.

5.1 Some different approaches

5.1.1 Finite-difference method

The derivation of the finite-difference scheme is to consider (Eq. 88) pointwise and make estimates of the derivative $(\partial H/\partial x)$ at $x = x_{i-1/2}$ by expanding H in a Taylor series about certain points. For example, if the points $x = x_i$ and x_{i-1} are used, then

$$H_i = H_{i-1/2} + \frac{\Delta x}{2}\left(\frac{\partial H}{\partial x}\right)\bigg|_{x_{i-1/2}} + \frac{(\Delta x/2)^2}{2}\left(\frac{\partial^2 H}{\partial x^2}\right)\bigg|_{x_{i-1/2}} + O(\Delta x^3), \tag{89}$$

and

$$H_{i-1} = H_{i-1/2} - \frac{\Delta x}{2}\left(\frac{\partial H}{\partial x}\right)\bigg|_{x_{i-1/2}} + \frac{(\Delta x/2)^2}{2}\left(\frac{\partial^2 H}{\partial x^2}\right)\bigg|_{x_{i-1/2}} + O(\Delta x^3). \tag{90}$$

Subtracting Eq. (90) from Eq. (89) leads to

$$\left(\frac{\partial H}{\partial x}\right)\bigg|_{x_{i-1/2}} = \frac{H_i - H_{i-1}}{\Delta x} + O(\Delta x^2), \tag{91}$$

so that Eq. (88) is approximated as

$$\frac{\partial F_{i-1/2}}{\partial t} = \frac{H_i - H_{i-1}}{\Delta x}. \tag{92}$$

This is a second-order accurate scheme and is termed a centered approximation for $\partial H/\partial x$ since the two values of H that are used to construct it are both the same distance away from $x_{i-1/2}$. More accurate schemes can be obtained by using more points. For example, a fourth-order accurate centered scheme can be obtained by using the four points H_{i+1}, H_i, H_{i-1}, and H_{i-2}.

The scheme (92) needs H to be known at the points $x = x_i$ (midway between $x_{i+1/2}$ and $x_{i-1/2}$) and x_{i-1} (midway between $x_{i-1/2}$ and $x_{i-3/2}$). If those values of H are known and stored, then the scheme is said to be a staggered scheme; H is staggered with respect to F. This leads to the scheme being very compact (having a "compact stencil"): $F_{i-1/2}$ is evaluated entirely in terms of its two neighboring values of H. If H is not staggered with respect to F, then the scheme must be constructed from at least $H_{i-3/2}$, $H_{i-1/2}$, and $H_{i+1/2}$. Expanding $H_{i+1/2}$ and $H_{i-3/2}$ about the point $x_{i-1/2}$ gives exactly the same equations as Eqs. (89), (90) but with Δx replaced by 2Δ, that is,

$$\left(\frac{\partial H}{\partial x}\right)\bigg|_{x_{i-1/2}} = \frac{H_{i+1/2} - H_{i-3/2}}{2\Delta x} + O(\Delta x^2), \tag{93}$$

(and similarly for $x_{i-3/2}$) so that Eq. (88) is discretized as

$$\frac{\partial F_{i-1/2}}{\partial t} = \frac{H_{i+1/2} - H_{i-3/2}}{2\Delta x}. \tag{94}$$

Note that this expression does not use the value $H_{i-1/2}$; by analogy with Eq. (53), $H_{i-1/2}$ has been leap-frogged over. It turns out that, in exactly the same way that the leap-frog time scheme supports a spurious computational solution in time, unstaggered centered spatial schemes, such as Eq. (94), support spurious computational solutions in space. In this case, it is straightforward to see that if $H_{i-1/2} = A(-1)^i$, for any arbitrary constant amplitude A, then the unstaggered scheme (93) will estimate $\partial H/\partial x|_{i-1/2}$ to be zero and F will remain unchanged despite H having arbitrarily large variation across each cell.

But in this unstaggered case, there are alternative possibilities. Suppose there is reason to believe that F is preferentially influenced by values of H to the left of $x_{i-1/2}$ (e.g., if H represents the effect of a wind blowing F from left to right). Then, from the Taylor series expansion for $H_{i-3/2}$,

$$H_{i-3/2} = H_{i-1/2} - \Delta x\left(\frac{\partial H}{\partial x}\right)\bigg|_{x_{i-1/2}} + O(\Delta x^2), \tag{95}$$

an estimate for $\partial H/\partial x|_{i-1/2}$ can be obtained using just $H_{i-1/2}$ and $H_{i-3/2}$, namely

$$\left(\frac{\partial H}{\partial x}\right)\bigg|_{x_{i-1/2}} = \frac{H_{i-1/2} - H_{i-3/2}}{\Delta x} + O(\Delta x). \tag{96}$$

Importantly, this is now only a first-order accurate method. It is by construction not centered and is termed an upwind scheme (though if the notional wind were to be blowing from the right then it would be a downwind scheme). In this case, Eq. (88) is discretized as

$$\frac{\partial F_{i-1/2}}{\partial t} = \frac{H_{i-1/2} - H_{i-3/2}}{\Delta x}. \tag{97}$$

If the four values $H_{i+1/2}$, $H_{i-1/2}$, $H_{i-3/2}$, and $H_{i-5/2}$ are used in a similar manner, then since

$$H_{i+1/2} = H_{i-1/2} + \Delta x\left(\frac{\partial H}{\partial x}\right)\bigg|_{x_{i-1/2}} + \frac{\Delta x^2}{2}\left(\frac{\partial^2 H}{\partial x^2}\right)\bigg|_{x_{i-1/2}} + \frac{\Delta x^3}{6}\left(\frac{\partial^3 H}{\partial x^3}\right)\bigg|_{x_{i-1/2}} + O(\Delta x^4), \tag{98}$$

$$H_{i-3/2} = H_{i-1/2} - \Delta x\left(\frac{\partial H}{\partial x}\right)\bigg|_{x_{i-1/2}} + \frac{\Delta x^2}{2}\left(\frac{\partial^2 H}{\partial x^2}\right)\bigg|_{x_{i-1/2}} - \frac{\Delta x^3}{6}\left(\frac{\partial^3 H}{\partial x^3}\right)\bigg|_{x_{i-1/2}} + O(\Delta x^4), \tag{99}$$

and

$$H_{i-5/2} = H_{i-1/2} - 2\Delta x\left(\frac{\partial H}{\partial x}\right)\bigg|_{x_{i-1/2}} + 4\frac{\Delta x^2}{2}\left(\frac{\partial^2 H}{\partial x^2}\right)\bigg|_{x_{i-1/2}} - 8\frac{\Delta x^3}{6}\left(\frac{\partial^3 H}{\partial x^3}\right)\bigg|_{x_{i-1/2}} + O(\Delta x^4), \tag{100}$$

it follows that

$$\left(\frac{\partial H}{\partial x}\right)\bigg|_{x_{i-1/2}} = \frac{2H_{i+1/2} + 3H_{i-1/2} - 6H_{i-3/2} + H_{i-5/2}}{6\Delta x} + O(\Delta x^3), \tag{101}$$

that is, a third-order accurate, upwind biased scheme. Applying this to Eq. (88), this gives

$$\frac{\partial F_{i-1/2}}{\partial t} = \frac{2H_{i+1/2} + 3H_{i-1/2} - 6H_{i-3/2} + H_{i-5/2}}{6\Delta x}. \tag{102}$$

5.1.2 Finite-volume method

The finite-volume discretization is achieved by integrating Eq. (88) in space over each control volume. This results in

$$\int_{x_{i-1}}^{x_i} \frac{\partial F}{\partial t} dx = \int_{x_{i-1}}^{x_i} \frac{\partial H}{\partial x} dx, \tag{103}$$

which can be written as

$$\Delta x \frac{\partial \hat{F}_{i-1/2}}{\partial t} = (H_i - H_{i-1}), \tag{104}$$

where $\hat{F}_{i-1/2} \equiv \int_{x_{i-1}}^{x_i} F dx$ and $H_i \equiv H(x_i)$. This equation is exact. However, the quantity H will in general be a function of F and not of $\hat{F}$. To derive F from $\hat{F}$, the finite-volume scheme makes the simple assumption that $F_{i-1/2} = \hat{F}_{i-1/2}$. In this case, Eq. (104) is exactly the same as Eq. (92). If the model is not staggered, then the values H_i and H_{i-1} will not be known and so the details of the finite-volume scheme are then how to construct them from the $F_{i-1/2}$'s.

Provided the method of deriving H_i is the same whether the construction is done to advance $F_{i-1/2}$ or $F_{i+1/2}$ (i.e., that H_i is uniquely defined) then there is a conservation law that is obtained by summing Eq. (104) over all the N control volumes leading to

$$\Delta x \frac{\partial}{\partial t}\left(\sum_{i=1}^{N} \hat{F}_{i-1/2}\right) = (H_N - H_0). \tag{105}$$

This states that the discrete volume integral of F only changes due to fluxes of F through the boundaries of the domain. The numerical scheme, therefore, preserves the analytical property that there is no net internal source or sink of F. The presence of such a conservation principle is a key aspect of the finite-volume method.

5.1.3 Finite-element method

It might be useful to introduce the finite-element scheme by comparing and contrasting it with the finite-difference scheme. There are two key differences between finite-element schemes and finite-difference ones.

The first difference is that the latter work with the strong form of the equation, that is, directly with Eq. (88). In contrast, the finite-element scheme uses what is termed the weak form. This first notes that if Eq. (88) holds then it also holds when multiplied by any well-behaved function, $w(x)$ say. Then, it notes that it will also hold when that product is integrated over the some spatial domain. The result is the weak form of the equation, that is,

$$\int_x w(x)\left(\frac{\partial F}{\partial t} - \frac{\partial H}{\partial x}\right)dx = 0, \quad \text{for all } w. \tag{106}$$

It says that rather than requiring equation (88) to hold in a pointwise manner, it is only required to hold when integrated against any well-behaved test function w. The strong and weak forms are clearly very closely related to each other in continuous space: Eq. (106) is satisfied whenever the strong form of the equation, that is, Eq. (88), is satisfied; and the strong form can be recovered by setting w to be the delta function. However, the power of the weak form is that it is satisfied by a much larger set of solutions than the strong form. Specifically, the weak form only requires that the projection of the residual of the strong form of the equation (the amount by which it is not solved) onto the function w vanishes; in other words, the "error is orthogonalized to w."

The second difference is that the finite-difference method predicts directly the value of F at certain points and the finite-volume method essentially predicts the value of F averaged over a cell or volume. In contrast, the finite-element method (and the closely related spectral method) predicts the coefficients, $\tilde{F}_k(t)$, of the expansion of F in terms of a finite number of basis (or trial) functions, $f_k(x)$. Specifically

$$F(x,t) = \sum_i \tilde{F}_i(t) f_i(x). \tag{107}$$

Similarly, H is expanded as

$$H(x,t) = \sum_i \widetilde{H}_i(t) h_i(x). \tag{108}$$

Substituting these into Eq. (88), the equation can be written as

$$\sum_i \left(\frac{d\widetilde{F}_i}{dt} f_i - \widetilde{H}_i \frac{dh_i}{dx} \right) = 0. \tag{109}$$

This is the strong form of the equation. The finite-element weak form is obtained by requiring Eq. (106) to hold not for all functions w but for a finite number w_j. Therefore, Eq. (106) is replaced by

$$\int_x w_j(x) \sum_i \left(\frac{d\widetilde{F}_i}{dt} f_i - \widetilde{H}_i \frac{dh_i}{dx} \right) dx = 0, \quad \text{for all } j. \tag{110}$$

Noting that $\widetilde{F}_i$ and $\widetilde{H}_i$ are functions only of time (Eq. 110) can be rearranged as

$$\sum_i \left(M_{ji} \frac{d\widetilde{F}_i}{dt} - S_{ji} \widetilde{H}_i \right) = 0, \quad \text{for all } j, \tag{111}$$

where the matrix **M** (termed the mass matrix) is defined as

$$M_{ji} \equiv \int_x w_j f_i dx, \tag{112}$$

and the matrix **S** (termed the stiffness matrix) is defined as

$$S_{ji} \equiv \int_x w_j \frac{dh_i}{dx} dx. \tag{113}$$

The Galerkin flavor of the finite-element method restricts the choice of test functions w to be the same as the basis, or trial, functions for F, that is, for each j, $w_j = f_i$ for some i. This implies that the projection onto the trial space of F, of any error in solving Eq. (109) must vanish. In other words, any error in solving that equation cannot have any component in the trial space of F, the solution "orthogonalizes the error to the trial space of F." Or in yet other words, the solution is optimal in the sense that the error is not in that function space.

Examples of low-order trial and test functions are piecewise-linear functions,

$$w_j(x) \equiv \begin{cases} \dfrac{x - x_{j-1}}{\Delta x}, & x_{j-1} \le x \le x_j \\ \dfrac{x_{j+1} - x}{\Delta x}, & x_j \le x \le x_{j+1} \\ 0, & \text{otherwise,} \end{cases} \tag{114}$$

and piecewise-constant functions,

$$w_{j-1/2}(x) \equiv \begin{cases} 1, & x_{j-1} \le x \le x_j \\ 0, & \text{otherwise.} \end{cases} \tag{115}$$

These are examples of schemes whose basis functions have compact support, that is, they are only nonzero for a limited range of values of x, typically in one cell or at most a small number of adjoining cells. Spectral element methods similarly have compact support but are usually much higher-order functions (and use specific numerical approximations to evaluate the integrals (112) and (113), see, e.g., Melvin et al., 2012). In contrast, the spectral method uses global basis functions that, other than at isolated points, are nonzero over the whole domain (in Cartesian geometry they are the sine and cosine functions of the discrete Fourier transform).

A function expanded in the basis functions (114) has the property of being continuous across the domain. Galerkin schemes employing such basis functions are referred to as continuous Galerkin (CG) schemes. In contrast, functions expanded in terms of basis functions like Eq. (115) are in general discontinuous across the edges of the cells and such schemes are referred to as discontinuous Galerkin (DG) schemes. An exciting development is the introduction to the meteorological field of the mixed finite-element method. In this approach rather than all variables being assigned to be either all continuous or all discontinuous, some variables are assigned continuous basis functions (e.g., Eq. 114) while others are assigned discontinuous ones (e.g., Eq. 115). The choice of variables is determined using ideas from differential geometry (see, e.g., Cotter and Shipton, 2012 for an accessible introduction to these schemes and their advantages).

5.2 Some numerical properties

To explore some properties of these schemes in the context of a dynamical core, we can return to the prototype equation (33) but now recognize that for almost all the relevant problems (inertial oscillation being the only exception) $i\Omega F$ in fact represents in some way $-c\partial F/\partial x$, where c is an appropriate propagation speed (e.g., the wind speed for transport or the speed of sound, either positive or negative, for acoustic waves). Therefore, consider the equation

$$\frac{\partial F}{\partial t} = -c\frac{\partial F}{\partial x}. \tag{116}$$

This is sometimes termed as the one-way wave equation.

This time the temporal aspects will be kept as continuous and the spatial aspects will be discretized. Therefore, consider a solution with spatial variation $\exp(ikx)$. The analytical solution is obtained by seeking a solution of the form

$$F = \exp(i\omega t - ikx). \tag{117}$$

Substituting this into Eq. (116) results in $\omega_{ana} = ck$ giving $F_{ana} = \exp[ik(ct - x)]$. The solutions to the numerical schemes introduced in the previous section are now compared with this analytical solution.

In what follows the schemes of Section 5.1 are considered with H chosen to be $-cF$.

5.2.1 Finite difference

We start with the second-order, centered finite-difference scheme. Seeking a solution to Eq. (92) of the form (117) gives

$$\begin{aligned}\omega_{cent} &= ic\frac{\exp(-ik\Delta x/2) - \exp(ik\Delta x/2)}{\Delta x} \\ &= c\frac{\sin(k\Delta x/2)}{(\Delta x/2)} \\ &= \omega_{ana}\left[1 - \frac{1}{6}\left(\frac{k\Delta x}{2}\right)^2 + O\left(\frac{k\Delta x}{2}\right)^4\right],\end{aligned} \tag{118}$$

where to emulate a staggered scheme an exact spatial representation of F_i and F_{i-1} has been used. It is seen that, as by construction, the scheme has a second-order error.

For the unstaggered scheme, the same result is obtained but with $\Delta x \rightarrow 2\Delta x$ in Eq. (118) so that the error is four times larger than for the staggered case. In addition, this scheme supports the computational solution discussed in relation to Eq. (94). This solution has the spatial form $F_{i-1/2} = A(-1)^i = A\exp(-ik_C x)$ for $k_C\Delta x = \pi$. It is seen from Eq. (118) with $\Delta x \rightarrow 2\Delta x$ that when $k\Delta x = k_C\Delta x = \pi$ then $\omega_{cent} = 0$, that is, the computational solution is stationary, despite the true gradient of F being far from zero.

Next, consider the upwind scheme (97). Substituting Eq. (117) into Eq. (97) requires

$$\begin{aligned}\omega_{up} &= ic\frac{1 - \exp(ik\Delta x)}{\Delta x} \\ &= c\frac{\sin(k\Delta x)}{\Delta x} + ic\frac{1 - \cos(k\Delta x)}{\Delta x} \\ &= \omega_{ana}\left[1 + i\left(\frac{k\Delta x}{2}\right) + O(k\Delta x)^2\right].\end{aligned} \tag{119}$$

As expected, this scheme has a first-order error. However, the fact that the first-order term is imaginary means that the scheme will either grow or decay in time depending on the sign of the term. Substituting Eq. (119) into Eq. (117) shows that $F \approx F_{ana}\exp[-kct(k\Delta x/2)]$. The scheme is, therefore, stable (albeit damping) if $c > 0$, that is, provided the discretization is indeed an upwind biased one and not a downwind biased one. The rate of damping is determined by $k\Delta x$. This means that the well-resolved waves, for which $k\Delta x$ is small, are much less damped than those that are poorly resolved.

Applying the methodology to Eq. (102) gives

$$\begin{aligned}\omega_{3up} &= ic\frac{2\exp(-ik\Delta x) + 3 - 6\exp(ik\Delta x) + \exp(2ik\Delta x)}{6\Delta x} \\ &= ic\frac{3 - 4\cos(k\Delta x) + \cos(2k\Delta x) - i[8\sin(k\Delta x) - \sin(2k\Delta x)]}{6\Delta x} \\ &= \omega_{ana}\left[1 + i\frac{(k\Delta x)^3}{12} + O(k\Delta x)^4\right].\end{aligned} \tag{120}$$

The scheme is seen to have a third-order error. The scheme is again stable if $c > 0$. The rate of damping is now determined by $(k\Delta x)^3$, which means that the well-resolved waves are much less damped than the shorter, less well-resolved ones. Relative to the much less discriminating first-order scheme, for which the damping is determined by $k\Delta x$, the scheme is said to have scale-selective damping.

5.2.2 Finite volume

Assuming that c is constant the finite-volume scheme needs to estimate the flux, $H \equiv cF$, at the edges of the cells (the finite volumes), that is, H_i and H_{i-1}, in terms of the values of the cell-averaged values $\hat{F}_{i-1/2}$ which are approximated by the cell-centered values $F_{i-1/2}$. The two simplest approaches are either: (a) to average the cell centered values from the two neighboring cells to the common edge of those two cells, that is,

$$H_i = \frac{1}{2}\left(cF_{i+1/2} + cF_{i-1/2}\right),$$

and similarly for H_{i-1}; or (b) to use the cell-centered value from the cell upwind of the target edge, that is,

$$H_i = cF_{i-1/2}, \quad \text{for } c \geq 0,$$

and

$$H_i = cF_{i+1/2}, \quad \text{for } c < 0,$$

and similarly for H_{i-1}.

Applying each of these schemes to Eq. (104) (with $\hat{F}_{i-1/2} = F_{i-1/2}$) gives, respectively,

$$\Delta x \frac{\partial F_{i-1/2}}{\partial t} = \frac{1}{2}\left(cF_{i+1/2} - cF_{i-3/2}\right), \tag{121}$$

and (assuming $c \geq 0$)

$$\Delta x \frac{\partial F_{i-1/2}}{\partial t} = \left(cF_{i-1/2} - cF_{i-3/2}\right). \tag{122}$$

It is seen that these schemes are equivalent to the centered and upwind finite-difference schemes, Eqs. (94), (97), respectively. The analysis of these finite-volume schemes, therefore, follows directly that for the finite-difference schemes.

The reason for such a direct relation here between the finite-volume and finite-difference schemes lies in: the simple geometry (there would be differences in multidimensions when it would be natural for the finite-volume scheme to use a more accurate representation of the cell-edge integral); and the low-order assumption that $\hat{F}_{i-1/2} = F_{i-1/2}$ (in some schemes, e.g., the piecewise-parabolic method of Colella and Woodward (1984) and the quadratic upstream interpolation for convective kinematics (QUICK) scheme of Leonard (1979), a distinction between these two quantities is maintained, with $F_{i-1/2}$ being reconstructed from $\hat{F}_{i-1/2}$ subject to the constraint that $\hat{F}_{i-1/2} \equiv \int_{x_{i-1}}^{x_i} F dx$). Additional differences arise when positivity limiters and monotonicity schemes are imposed to ensure that the numerical schemes preserve, where appropriate, properties such as the positivity and shape of the predicted quantity. A finite-difference scheme can directly enforce a constraint, such as $F_{i-1/2} \geq 0$ on the predicted values. In contrast, if a finite-volume scheme wants to retain its conservation principle, then it has to enforce any such constraint through manipulating the fluxes, $H_{i-1/2}$. This is in general harder than in the finite-difference case and will lead to different behaviors of the schemes.

5.2.3 *Finite element*

For the finite-element scheme, consider first the continuous, piecewise-linear approach. This is an example of the CG method. Therefore, expand F as

$$F^{CG}(x,t) = \sum w_i(x) \tilde{F}_i^{CG}(t), \tag{123}$$

with w_i given by Eq. (114). Substituting this into Eq. (110) with $\tilde{H} = -c\tilde{F}$, together with $f_j = h_j = w_j$, the scheme requires that

$$\sum_i \left[\frac{d\tilde{F}_i^{CG}}{dt} \left(\int_x w_j w_i dx \right) + c\tilde{F}_i^{CG} \left(\int_x w_j \frac{dw_i}{dx} dx \right) \right] dx = 0, \quad \text{for all } j. \tag{124}$$

Since the basis functions are continuous across the domain, then the range of the integrals is taken to be the whole domain. (Although continuity of w_i does not guarantee that the derivative dw_i/dx is defined pointwise everywhere, continuity of w_i is sufficient to ensure that the integral defining the stiffness matrix is itself well defined and exists. Note that the basis functions are always chosen to be differentiable within each cell but the value of that derivative will, in general, be discontinuous across the cell edges.) Substituting for w_j from Eq. (114) and performing the integrations analytically (noting the compactness of w_j) results in

$$\frac{1}{6}\frac{d\tilde{F}_{j+1}^{CG}}{dt} + \frac{4}{6}\frac{d\tilde{F}_j^{CG}}{dt} + \frac{1}{6}\frac{d\tilde{F}_{j-1}^{CG}}{dt} + \frac{c}{2\Delta x}\left(\tilde{F}_{j+1}^{CG} - \tilde{F}_{j-1}^{CG} \right) = 0, \quad \text{for all } j. \tag{125}$$

In this case, the mass matrix is the tri-diagonal matrix with nonzero row entries $(\Delta x/6, 4\Delta x/6, \Delta x/6)$ and the stiffness matrix is also tri-diagonal with (using an appropriate ordering of $\tilde{F}_j$) nonzero row entries $(c/2, 0, -c/2)$. (It is interesting to note that if the mass matrix is replaced by a diagonal matrix with the diagonal entries obtained by summing all the column entries in each row, a procedure known as "mass lumping," then a finite-difference scheme is recovered.)

Seeking a solution to Eq. (125) of the form

$$\tilde{F}_j^{CG}(t) = \exp\left(i\omega_{CG} t - ikx_j\right), \tag{126}$$

leads to

$$i\omega_{CG}\left[\frac{1}{6}\exp(-ik\Delta x) + \frac{4}{6} + \frac{1}{6}\exp(ik\Delta x)\right] + \frac{c}{2\Delta x}\left[\exp(-ik\Delta x) - \exp(ik\Delta x)\right] = 0, \tag{127}$$

with solution

$$\begin{aligned}\omega_{CG} &= \frac{c}{\Delta x}\frac{\sin(k\Delta x)}{[2+\cos(k\Delta x)]/3}\\ &= \omega_{ana}\frac{1-(k\Delta x)^2/6+(k\Delta x)^4/120+O(k\Delta x)^6}{1-(k\Delta x)^2/6+(k\Delta x)^4/72+O(k\Delta x)^6}\\ &= \omega_{ana}\left[1+O(k\Delta x)^4\right].\end{aligned} \tag{128}$$

The scheme, therefore, has a fourth-order error. This is appealing but does comes at the cost of having to invert the mass matrix.

Now consider the discontinuous, piecewise-constant basis functions. This is an example of the DG method. Therefore, expand F as

$$F^{DG}(x,t) = \sum w_{i-1/2}(x)\tilde{F}^{DG}_{i-1/2}(t), \tag{129}$$

with $w_{i-1/2}$ given by Eq. (115). Since $w_{i-1/2}$ is now entirely confined to one cell, it is convenient to restrict the domain of integration to the relevant cell, that is, $x\in[x_{j-1}, x_j]$. Therefore, Eq. (110) becomes

$$\sum_i\left[\frac{d\tilde{F}^{DG}_{i-1/2}}{dt}\left(\int_{x_{j-1}}^{x_j} w_{j-1/2}w_{i-1/2}dx\right) + c\tilde{F}^{DG}_{i-1/2}\left(\int_{x_{j-1}}^{x_j} w_{j-1/2}\frac{dw_{i-1/2}}{dx}dx\right)\right]dx = 0, \quad \text{for all } j. \tag{130}$$

Substituting for $w_{i-1/2}$ from Eq. (115), the mass matrix is found to be Δx times the identity matrix. However, it is also seen that the term involving $dw_{i-1/2}/dx$ vanishes. The solution then is seemingly $d\tilde{F}^{DG}_{j-1/2}/dt = 0$, which would not make the scheme very accurate! This is where use of the weak form is critical to the success of the method. Returning to Eq. (106), when this equation is written for this problem it is required that

$$\int_{x_{j-1}}^{x_j} w_{j-1/2}\frac{\partial F^{DG}}{\partial t}dx + \int_{x_{j-1}}^{x_j} w_{j-1/2}c\frac{\partial F^{DG}}{\partial x}dx = 0, \quad \text{for all } j. \tag{131}$$

The integral involving $\partial F^{DG}/\partial x$ can be integrated by parts to give

$$\begin{aligned}\int_{x_{j-1}}^{x_j} w_{j-1/2}c\frac{\partial F^{DG}}{\partial x}dx &= \left[w_{j-1/2}cF^{DG}\right]_{x_{j-1}}^{x_j} - \int_{x_{j\ 1}}^{x_j}\frac{dw_{j-1/2}}{dx}cF^{DG}dx\\ &= \left[w_{j-1/2}cF^{DG}\right]_{x_{j-1}}^{x_j}\\ &= \left(cF^{DG}\right)\big|_{x_j} - \left(cF^{DG}\right)\big|_{x_{j-1}},\end{aligned} \tag{132}$$

where the integral involving $dw_{j-1/2}/dx$ vanishes because in this case $w_{j-1/2}$ is a constant within each cell. Since c has the dimensions of velocity, the term $\left(cF^{DG}\right)\big|_{x_j}$ represents the flux of F out of cell $j-1/2$ and $\left(cF^{DG}\right)\big|_{x_{j-1}}$ represents the flux into that cell.

Using the form given by Eq. (132), and with the choice (115) for $w_{j-1/2}$, Eq. (130) becomes

$$\Delta x \frac{d\tilde{F}_{j-1/2}^{DG}}{dt} + \left(c\tilde{F}^{DG}\right)\Big|_{x_j} - \left(c\tilde{F}^{DG}\right)\Big|_{x_{j-1}} = 0. \tag{133}$$

The scheme, therefore, reduces, in this piecewise-constant case, to estimating the fluxes. A simple approach is to estimate the flux at an edge as the product of c with the average of the values of $\tilde{F}^{DG}$ from each adjoining cell, that is,

$$\left(c\tilde{F}^{DG}\right)\Big|_{x_{j-1}} = \frac{c}{2}\left(\tilde{F}_{j-3/2}^{DG} + \tilde{F}_{j-1/2}^{DG}\right), \tag{134}$$

for all j. Substituting this into Eq. (133) gives

$$\frac{d\tilde{F}_{j-1/2}^{DG}}{dt} + c\left(\frac{\tilde{F}_{j+1/2}^{DG} - \tilde{F}_{j-3/2}^{DG}}{2\Delta x}\right) = 0. \tag{135}$$

With H replaced by $-cF$, this scheme is seen to result in the same discretization as the centered, unstaggered, finite-difference scheme (94) and the finite-volume scheme (104). An alternative is to estimate the flux using the value of $\tilde{F}^{DG}$ from the nearest upwind cell. For positive c,

$$\left(c\tilde{F}^{DG}\right)\Big|_{x_{j-1}} = c\tilde{F}_{j-3/2}^{DG}, \tag{136}$$

for all j. Using this in Eq. (133) gives

$$\frac{d\tilde{F}_{j-1/2}^{DG}}{dt} + c\left(\frac{\tilde{F}_{j-1/2}^{DG} - \tilde{F}_{j-3/2}^{DG}}{\Delta x}\right) = 0, \tag{137}$$

and the scheme is the same as the finite-difference scheme (97).

The key advantage of the finite-element scheme over the finite-volume one is that it gives a route to deriving higher-order representations of F within the cell by choosing higher-order basis functions w.

As noted previously, the spectral method is like a finite-element scheme but with global basis functions (i.e., they are nonzero over all the domain, other than where they change sign). In Cartesian geometry, such methods are equivalent to working in Fourier space. In spherical geometry, the transformation is a Fourier one in the latitudinal direction and a Legendre transformation in the longitudinal direction. The advantages of spectral schemes are that once the cost of the transformations has been paid then evaluation of spatial derivatives is: (a) trivial (it simply requires multiplication by the appropriate wave number) and (b) that evaluation of the derivatives is exact for a given wave number, the only approximation is in how many waves are retained in the expansion. The complexity comes in the evaluation of nonlinear terms for which careful allowance has to be made for the interaction between different wave numbers. Some early references on application of the spectral method to NWP are Bourke (1972) and Hoskins and Simmons (1975). The spatial analyses presented here have all been linear and so application of the spectral method to these problems simply reproduces the analytic spatial result.

6 (Semi-)Lagrangian approach

Having discussed both spatial and temporal discretizations, we are now in a position to introduce a third type of discretization, one which handles both spatial and temporal aspects together. This approach returns to what is arguably a physically more fundamental form of Eq. (116). Namely, it recognizes that Eq. (116) can be rewritten as

$$\frac{DF}{Dt}=0, \tag{138}$$

where $F=F(x,t)$ and D/Dt denote what is variously termed the Lagrangian, material, or full, derivative, defined as

$$\frac{D}{Dt}=\frac{\partial}{\partial t}+c\frac{\partial}{\partial x}. \tag{139}$$

Eq. (138), together with Eq. (139), says that the quantity F is unchanged along a trajectory $x=x(t)$ such that

$$\frac{dx}{dt}=c. \tag{140}$$

The approaches to discretizing Eq. (33) in time were to integrate the equation over a time period of one or two time steps. Implicit in that procedure was that the integration was performed at a fixed point in space. In contrast, Lagrangian discretizations are instead based on integrating Eq. (138) along the trajectory (in both space and time) defined by integrating Eq. (140) over the same time period. Integrating Eq. (138) along the trajectory $x(t)$ for a time-step results in

$$F[x(t+\Delta t),t+\Delta t]=F[x(t),t], \tag{141}$$

where, integrating the trajectory equation (140),

$$x(t+\Delta t)=x(t)+c\Delta t. \tag{142}$$

Therefore, if at time t the quantities $F[x(t),t]$ and $x(t)$ are known, then the value of F at time $t+\Delta t$ is known at the point $x(t+\Delta t)$ (given by Eq. 142). Note that in general c is also a function of space and time. Therefore, Eq. (142) is replaced by an integral of c along the trajectory and this integral needs to be approximated in some way, in general in both space and time.

The Lagrangian approach is sometimes used in pollution modeling. In this case, a large number of fluid parcels are released at known points and with known values. Those parcels are then tracked by solving the trajectory equation for each one and their initial properties are carried along those trajectories. However, the value of the approach for more general problems is somewhat limited. This is because if we release particles at a mesh of points then as time progresses those particles will cluster, trajectories can tangle, and gaps in coverage will form meaning that it will be hard to evolve a forecast in a reliable way. Therefore, one has to introduce some means of smoothing the particle properties so that estimates of those properties are available everywhere.

An alternative approach that has proved very successful in NWP is the semi-Lagrangian method (see Staniforth and Côté, 1991 for a review). The key aspect of this approach is that the Lagrangian method is still applied but it is applied only over one time step. At the end of that time step, quantities that have been transported so that their positions no longer coincide with a mesh point are remapped to or from the surrounding mesh points. This ensures that values are always known at all mesh points so that there is uniform coverage. It also avoids (provided a suitable time step is used) the issue of tangling of trajectories. The remapping might be to or from mesh points. This is because there are two flavors of semi-Lagrangian schemes. One is the forward trajectory approach, which mimics the particle release case. In this case, $x(t+\Delta t)$ will in general not coincide with a mesh point and, therefore, values of $F(t+\Delta t)$ at mesh points need to be reconstructed from an irregular set of trajectory end points. The alternative is the backward-trajectory approach for which $x(t+\Delta t)$ is constrained to coincide with a mesh point (the arrival point). The trajectory is then integrated backwards in time, using the velocity $-c$, to evaluate where that trajectory originated from at the previous time step (the departure point). This departure point will not in general be a mesh point and so the value of F there is obtained by interpolating from the surrounding mesh points. This is the more commonly applied approach and will be described in a little more detail.

The full equation set can, in principle, be written as the Lagrangian conservation of various properties along a number of characteristics (the trajectories determined by the phase velocity of each type of wave), see, for example, Oliger and Sundström (1978). This is known as the method of characteristics. This is a challenging and numerically expensive approach. The semi-Lagrangian method has been applied successfully by limiting its application to transport by the mean wind, that is, the case where $c=u$. The backward-trajectory form of the semi-Lagrangian discretization of Eq. (138) is then given by

$$\begin{aligned} F_A^{n+1} &\equiv F[x(t+\Delta t),t+\Delta t] \\ &\equiv F(x_A,t+\Delta t) \\ &= F[x(t),t] \\ &= F(x_D,t) \\ &\equiv F_D^n, \end{aligned} \tag{143}$$

where the subscript A indicates an arrival point, chosen in this method to be a mesh point, and subscript D denotes evaluation at the departure point defined as the solution to the equation:

$$x_D = x_A - \int_t^{t+\Delta t} u\left[x\left(t'\right),t'\right]dt'. \tag{144}$$

The integral in this equation is approximated in many different ways. An appealing approach is to use the trapezoidal rule to obtain

$$\begin{aligned} x_D &= x_A - \frac{\Delta t}{2}\{u[x(t+\Delta t),t+\Delta t] + u[x(t),t]\} \\ &= x_A - \frac{\Delta t}{2}\left(u_A^{n+1} + u_D^n\right). \end{aligned} \tag{145}$$

However, this equation is implicit in x_D for two reasons (it is doubly implicit). The first reason is that the evaluation of u_D^n requires x_D to be known. Second, in a full model (i.e., when the time evolution of u is not prescribed) u_A^{n+1} also depends on x_D. This is because the velocity at the next time step, u_A^{n+1}, will only be known at the end of the time step and that depends on knowing the value of x_D. Both of these aspects can be handled using an iterative approximation along the lines of Eq. (72).

There are no approximations made in deriving Eq. (143) and in that sense it is exact. However, to apply the equation, two approximations need to be made. The first is in the evaluation of the departure point. Unless the wind is constant or known analytically, various approximations are needed to evaluate x_D from an equation like Eq. (145). The second is that the departure point will in general not be a mesh point and, therefore, some form of interpolation is needed to estimate F_D^n from the values F_A^n at surrounding mesh points.

Some insight into the different effects that these two approximations have can be obtained by, again, considering the case when F has the form (117), that is,

$$F = \exp(i\omega t - ikx), \tag{146}$$

and is governed by Eqs. (138), (139) with c a constant phase velocity. Given an arrival point x_A, the departure point x_D is given by Eq. (142) (this is the analytic solution but also the discrete solution for any consistent discretization of the trajectory equation). From Eq. (143), it follows that the analytic solution is given by

$$\begin{aligned} F_{ana}(x_A, t+\Delta t) &= F_A^{n+1} = F_D^n = F(x_A - c\Delta t, t) \\ &= \exp(i\omega t - ikx_A + ikc\Delta t) \\ &= F_{ana}(x_A, t)\exp(ikc\Delta t) \\ &= F_{ana}(x_A, t)\exp(i\omega_{ana}\Delta t), \end{aligned} \tag{147}$$

that is, a phase shift by $\omega_{ana}\Delta t \equiv kc\Delta t$ without any change in amplitude.

Now suppose that the trajectory displacement, that is, the distance between the arrival point and the departure point, is less than one mesh space, that is, $c\Delta t < \Delta x$ and that linear interpolation is used to estimate F_D^n from the two arrival point values $F(x_A, t)$ and $F(x_A - \Delta x, t)$. This semi-Lagrangian approximation is, therefore, given by

$$\begin{aligned} F_{SL}(x_A, t+\Delta t) &= \left(\frac{c\Delta t}{\Delta x}\right)F(x_A - \Delta x, t) + \left(1 - \frac{c\Delta t}{\Delta x}\right)F(x_A, t) \\ &= \left[\left(\frac{c\Delta t}{\Delta x}\right)\exp(ik\Delta x) + \left(1 - \frac{c\Delta t}{\Delta x}\right)\right]\exp(-i\omega_{ana}\Delta t)F_{ana}(x_A, t+\Delta t) \\ &= \left\{1 - \left(\frac{c\Delta t}{\Delta x}\right)[1 - \cos(k\Delta x)] + i\left(\frac{c\Delta t}{\Delta x}\right)\sin(k\Delta x)\right\}\exp(-i\omega_{ana}\Delta t)F_{ana}(x_A, t+\Delta t) \\ &= [C\cos(K - \omega_{ana}) + (1-C)\cos\omega_{ana}]F_{ana}(x_A, t+\Delta t) \\ &\quad + i[C\sin(K - \omega_{ana}) - (1-C)\sin\omega_{ana}]F_{ana}(x_A, t+\Delta t) \\ &= A_{SL}\exp[i(\omega_{SL} - \omega_{ana})\Delta t]F_{ana}(x_A, t+\Delta t), \end{aligned} \tag{148}$$

where $C \equiv c\Delta t/\Delta x$, $K \equiv k\Delta x$,

$$A_{SL}^2 = C^2 + (1-C)^2 + 2C(1-C)\cos K, \tag{149}$$

and

$$\tan[(\omega_{SL} - \omega_{ana})\Delta t] = \frac{C\sin(K - \omega_{ana}) - (1-C)\sin\omega_{ana}}{C\cos(K - \omega_{ana}) + (1-C)\cos\omega_{ana}}. \tag{150}$$

Note the use of $F_{ana}(x_A, t+\Delta t)$ on the right-hand sides of Eq. (148) and hence the explicit appearance of ω_{ana} in Eq. (150) (the reasons for this will become clear below). Also, note that A_{SL} can be negative and stability requires that $-1 \le A_{SL} \le 1$.

Because, by assumption, $0 \le C \le 1$, the maximum value of A_{SL} is seen to be 1 and the scheme is, therefore, stable. Minimizing A_{SL} over the permissible values of $0 \le C \le 1$ shows that, in this case for linear interpolation, the greatest damping occurs when the departure is half way between mesh points, $C = 1/2$, for which A_{SL} takes the value $\cos(K/2)$. The minimum amplitude is, therefore, unity for the longest wavelengths, $K = 0$, and reduces as K increases, reaching a minimum of zero at the shortest resolvable wavelength for which $K = \pi$.

Although the previous analysis assumed that $0 \le C \le 1$, this restriction is in fact only one of convenience. Consider what happens when $C > 1$. In this case, define C' such that $C = [C] + C'$, where $[C]$ denotes the integer part of C. Then, by construction, $0 \le C' < 1$. Define $[c]$ and c' similarly. The departure point will now lie in the interval $x_A - [C+1]\Delta x < x_D \le x_A - [C]\Delta x$ and the previous analysis still holds but with the following changes:

1. $F(x_A - \Delta x, t)$ and $F(x_A, t)$ in the first line of Eq. (148) are replaced, respectively, by $F(x_A - [C+1]\Delta x, t)$ and $F(x_A - [C]\Delta x, t)$.
2. In Eqs. (148)–(150), both c (where it appears) and C are replaced by their noninteger counterparts c' and C'.
3. ω_{ana} retains its definition, KC (and not KC').

In particular, because $0 \le C' < 1$, then $A_{SL} \le 1$ for all K. The scheme is, therefore, unconditionally stable. Although this analysis has been for a constant value of c and in one dimension it turns out that the scheme is indeed very stable more generally, despite being explicit. The key issue that limits the time step is that it is important for the trajectories from neighboring arrival points not to cross. The limitation then arises from the discretization of the trajectory equation. This requires that the time step is less than the time scale implied by the shear or divergence of the flow. This can symbolically be represented by

$$\Delta t \left\| \frac{\partial u_i}{\partial x_j} \right\| \le O(1), \tag{151}$$

where here u_i indicates the ith component of the wind vector and x_j the jth component of the position vector, and $\|\cdot\|$ indicates the maximum absolute value over all possible combinations of i and j. This arises from requiring a Lipschitz condition to be satisfied for convergence of the iterative scheme needed to solve Eq. (140) (see, e.g., Pudykiewicz and Staniforth, 1984 and the appendix of Smolarkiewicz and Pudykiewicz, 1992). This can be compared with the Courant number limitation of explicit, non-Lagrangian (i.e., Eulerian) schemes $\Delta t|u|/\Delta x < O(1)$. In the explicit, Eulerian case the time step is limited by a purely numerical time scale, $\Delta x/|u|$, that depends on the absolute value of the wind speed. In contrast, the time step of the semi-Lagrangian scheme is constrained by the physical time scale $\|\partial u_i/\partial x_j\|^{-1}$ that itself does not depend on the absolute value of the wind speed, only on changes in the wind speed. This presents a significant advantage in two cases of particular relevance to NWP. The first is for transporting properties of air in the jet stream. Here, the wind will be very large and explicit Eulerian schemes will need a very small time step. Yet, the flow is very uniform so in practice little is going on. The semi-Lagrangian scheme in contrast can take very long-time steps. The second is in the vicinity of the poles where any mesh based on latitude-longitude coordinates will have mesh spacings that are much smaller than they are away from the poles. This increased resolution is generally not used, or needed, to resolve finer-scale features; it is purely there as an artifact of the geometry of the mesh. However, in an explicit, Eulerian model the time step of the whole model would be determined by this region of very small mesh lengths. In contrast, in a semi-Lagrangian scheme, the time step can be chosen to match the physical scales that are of interest. These will be much greater than the mesh spacing near the poles (provided any cascade of energy to those small scales is controlled).

A final point about semi-Lagrangian schemes is that they inherently mix spatial and temporal aspects. The errors in A_{SL} and ω_{SL} are complex functions of both C and K. K is a function only of Δx, whereas C is a function of both Δx and Δt. In analyzing numerical properties of the scheme, it is, therefore, important to be very clear about what is being held constant and what is not. For example, if one is interested in the spatial truncation error, then it would be very natural to consider what happens in the limit that $\Delta x \to 0$ while keeping Δt fixed. In an Eulerian scheme, this makes perfect sense and the analysis will generally quantify the benefits of better resolving the wavelength of interest ($K \to 0$). However, in a semi-Lagrangian scheme in this limit, the Courant number will increase. If initially the departure point is relatively close to the arrival point, interpolation to the departure point will be relatively accurate. As Δx reduces though, unless Δt also reduces, then the relative position of the departure point within a mesh interval will move further away from its arrival point and the interpolation error will increase. To isolate interest to what happens when $K \to 0$, it would be necessary to additionally reduce Δt in order to keep C constant. But then there might be an additional, purely temporal, influence from an improved departure point calculation. Each type of analysis is valid on its own terms but it is important that the analyst recognizes the different aspects, assumptions, and strengths and weaknesses, of each.

7 Multidimensional aspects

7.1 Extending the discretization to two dimensions

So far discussion of the spatial aspects of the numerical schemes has been one-dimensional (1D). The atmosphere of course is three-dimensional (3D) and so the dimensionality of these schemes needs to be extended. As a start, consider the two-dimensional (2D) advection equation

$$\frac{\partial F}{\partial t} = -u\frac{\partial F}{\partial x} - v\frac{\partial F}{\partial y}. \tag{152}$$

If the winds u and v are constant, then this equation admits solutions with wave number k in the x-direction and l in the y-direction that have the form

$$F(x,y,t) \propto \exp(i\omega t - ikx - ily), \tag{153}$$

with $\omega = ku + lv$. This means that

$$\begin{aligned} F(x,y,t+\Delta t) &= F(x,y,t)\exp(iku\Delta t + ilv\Delta t) \\ &\approx F(x,y,t)\left[1 + i(ku\Delta t + lv\Delta t) - \frac{(ku\Delta t + lv\Delta t)^2}{2} + O(\Delta t^3)\right]. \end{aligned} \tag{154}$$

Consider the forward-Euler temporal discretization (36) coupled with the first-order upwind spatial discretization (96) applied in two dimensions, that is,

$$F_{i,j}^{n+1} = F_{i,j}^n - C_x\left(F_{i,j}^n - F_{i-1,j}^n\right) - C_y\left(F_{i,j}^n - F_{i,j-1}^n\right), \tag{155}$$

where $F_{i,j}$ indicates $F(i\Delta x, j\Delta y)$, $C_x \equiv u\Delta t/\Delta x$ is the Courant number in the x-direction, and $C_y \equiv v\Delta t/\Delta y$ is that in the y-direction (both assumed positive so that the scheme is upwinded). Seeking a solution of the form (153) leads to

$$F_{i,j}^{n+1} = \left[1 - C_x(1 - \cos K - i\sin K) - C_y(1 - \cos L - i\sin L)\right]F_{i,j}^n,$$

where $L \equiv l\Delta y$. It follows that the amplification factor of the scheme, A, is given by

$$A^2 = (1 - C_x - C_y)^2 + 2(1 - C_x - C_y)(C_x\cos K + C_y\cos L) + (C_x + C_y)^2 - 2C_xC_y[1 - \cos(K - L)]. \tag{156}$$

Stability requires $-1 \le A \le 1$ which, noting that both C_x and C_y are positive, is obtained provided $0 \le C_x + C_y \le 1$ (see Durran, 2010 for proof of this). (In the more general case where u and v can take either sign, stability requires that $|C_x| + |C_y| \le 1$.) For general flows, this is a significantly tighter restriction on the time step than for the 1D case. An appealing alternative then is to update F first in one direction, x say, and then in the other direction. The scheme can be written in predictor-corrector form as

$$F_{i,j}^* = F_{i,j}^n - C_x\left(F_{i,j}^n - F_{i-1,j}^n\right), \tag{157}$$

followed by

$$F_{i,j}^{n+1} = F_{i,j}^* - C_y\left(F_{i,j}^* - F_{i,j-1}^*\right), \tag{158}$$

or as the equivalent single-stage scheme:

$$\begin{aligned} F_{i,j}^{n+1} &= (1 - C_y)\left[(1 - C_x)F_{i,j}^n + C_xF_{i-1,j}^n\right] + C_y\left[(1 - C_x)F_{i,j-1}^n + C_xF_{i-1,j-1}^n\right] \\ &= F_{i,j}^n - C_x\left(F_{i,j}^n - F_{i-1,j}^n\right) - C_y\left(F_{i,j}^n - F_{i,j-1}^n\right) + C_xC_y\left(F_{i,j}^n - F_{i-1,j}^n - F_{i,j-1}^n + F_{i-1,j-1}^n\right). \end{aligned} \tag{159}$$

By construction, stability of the scheme is assured provided both the 1D updates in each of the x- and y-directions are stable, that is, provided

$$|C_x| \leq 1 \tag{160}$$

and

$$|C_y| \leq 1. \tag{161}$$

This scheme raises two interesting aspects.

The first aspect is that Eq. (155) is symmetric in x and y, that is, if the i and j indices are swapped and C_x and C_y are swapped then the same discretization is obtained. The same is also true for Eq. (159). In other words, the result does not change if, instead of updating in the x-direction first and then updating the y-direction, the y-direction update is done first followed by the x-direction update. However, this is purely an artifact of C_x and C_y being constant (because it has been assumed that u, Δx, v, and Δy are all constant); when this is not the case then differences of C_x in the y-direction appear in Eq. (159) but differences of C_y in the x-direction do not; the scheme is not, in the general case, symmetric in x and y. This will in general lead to biases appearing in the results. Such artifacts can be avoided by either alternating on each time step which direction is applied first or, each time step, taking the average of the two schemes "x first followed by y" and "y first followed by x."

The second aspect is that the form (159) introduces a cross-term that involves the product C_xC_y. This is entirely absent from the discretization (155). Further, such terms would remain absent from that form even if higher-order estimates of the spatial derivatives were used. And further still, Leonard et al. (1996) state that, building on earlier work, schemes such as Eq. (155) are unstable for all schemes higher than first order; they suggest that the cause of this is the lack of the cross-terms. The whole issue of how to manage such issues is referred to as "operator splitting": the spatial operator, which physically has no preferred direction, is split into two separate operators, determined purely by the choice of numerical mesh, in this case operators in each of the x- and y-directions, and those are combined in some manner (e.g., simply added or multiplied together in some sense).

For the semi-Lagrangian scheme, extension from one dimension to three dimensions is relatively straightforward: once the three components of the departure point equation have been evaluated then a 3D interpolation scheme is applied to obtain the required value at the departure point. Multidimensional Lagrange interpolation or spline interpolation are both good methods. Cascade interpolation provides an alternative approach in which three 1D interpolations are performed but the directions of those interpolations are determined by the flow. This approach might be called a flow-dependent splitting strategy.

7.2 Grids

Once we leave the comfort of one dimension and venture into more dimensions, an additional complexity arises: how should the multidimensional space be split up into discrete parts? In a rectangular domain then it is usually most natural to separate the domain into rectangular cells that fit exactly into, and completely span, the domain. This approach lends itself to using Cartesian coordinates. Another very natural representation of a rectangular domain is to use a discrete Fourier representation, a spectral method. Models based on such methods though always need to evaluate fields in physical space at some point and there is a very natural relation between the pairing of wave numbers in each direction and an appropriately defined pairing of Cartesian coordinates.

When one of the boundaries of the domain is irregular, such as is the case in the presence of hills and mountains, then there are broadly two choices. The first is to retain the same rectangular cells and allow in some way for the fact that the physical surface intersects the computational cells such as the cut-cell technique (see, e.g., Lock et al., 2012). The second is to introduce a coordinate mapping that transforms the complex physical domain into a logically rectangular one and then proceed as for that case. For example, consider a 2D domain defined by the Cartesian coordinate pair (x,z) with vertical lateral boundaries at $x = \pm L$, a horizontal top boundary at $z = z_T$, and a varying surface boundary at $z = z_S(x)$. A common approach is to use the so-called "terrain-following" coordinates of Gal-Chen and Somerville (1975). For example, let the computational, transformed, coordinate pair be (ξ,η) where

$$\xi \equiv \frac{x}{L};$$
$$\eta \equiv \frac{z - z_S(x)}{z_T - z_S(x)}.$$

Then, the vertically oriented lateral boundaries are at $\xi = \pm 1$, the top boundary is at $\eta = 1$ and the surface boundary is uniformly at $\eta = 0$. Once the equations are transformed into these coordinates the computational domain can be split into a number of rectangular cells in $\xi - \eta$ space. While there are a number of advantages to this approach, one disadvantage is that the coordinate lines are not orthogonal and there is a choice as to how the wind components are defined. An alternative is to use what is referred to as conformal coordinate transformation, which retains the orthogonality of the original Cartesian coordinates. However, the vertical coordinate then is no longer aligned with gravity, which can cause problems with maintaining a good representation of hydrostatic balance, and it has been difficult to obtain good energy conservation (e.g., Clark, 1977).

Weather models though need to work on a representation of the Earth. This is usually separated into a mean component, the shape the Earth would have in the absence of hills and mountains, that is, if it were an aqua-planet, and a perturbed component, representing the hills and mountains. In the field of weather prediction, the mean component is always approximated by a sphere which is a very accurate representation (though there has been some consideration given to allowing for the ellipticity of the Earth (see, e.g., Staniforth, 2014b and references therein). The vertical aspects in the presence of the hills and mountains are usually handled using some form of transformed, terrain-following coordinate, as outlined earlier for the rectangular domain case. But that still leaves the horizontal, spherical surface that needs to be gridded in some manner. It is a fundamental geometric property (the "hairy ball theorem") that there is no way of creating a smooth grid over the entire sphere without there being a singularity somewhere. For a latitude-longitude grid, this manifests itself as the presence of two poles in the grid (usually the North and South poles), where all the meridians meet at a point. So although most of the cells of the grid are regular quadrilaterals (regular in the latitudinal and longitudinal angles), those immediately adjacent to the two poles become triangles. As well as the technical issues of how to discretize the equations in the presence of such a singularity, the convergence of the meridians means that the horizontal spacing between two adjacent meridians reduces quadratically with horizontal resolution. Specifically, consider a latitude-longitude grid with N points equally distributed in latitude along each meridian, and $2N$ points equally distributed in longitude along each latitude circle. Then, the physical grid spacing in the North-South direction is $\pi a/N$ (where a is the radius of the sphere used to model the Earth). In the East-West direction, the spacing at latitude θ is $(\pi a/N)\cos\theta$. The latitude closest to the North pole has $\theta = \pi/2 - \pi/N$. Therefore, the East-West spacing along this latitude is $(\pi a/N)\sin(\pi/N) \approx \pi^2 a/N^2$ for large N. The highest resolution global NWP models currently have North-South resolutions of around 10 km. This equates to $N \approx \pi a/10$, if a is measured in kilometers, that is, $N = O(10^3)$. Therefore, the East-West spacing next to the pole is approximately $100/a \approx 1/64$ km or 15 m. If the global resolution were to increase to 1 km then because of the quadratic nature of the spacing, the spacing next to the pole would reduce to 15 cm! Williamson (2007) gives an interesting history of the so-called "pole problem"—how models have managed the issue of the polar singularity.

The convergence of the meridians impacts on the efficiency of models employing such grids. One is simply the waste of solving the equations at so many grid points in the region of the poles with resolution much greater than over the regions of greater interest. This has been addressed by some groups by employing a reduced grid, which systematically reduces the number of latitude circles as the pole is approached. This is particularly effective for spectral models (Hortal and Simmons, 1991). Another is that if a model has some form of CFL limitation on its time step (e.g., due to the explicit handling of either wave propagation or advection) then the presence of the poles will lead to a dramatic reduction in the size of time step. Equally though, if such a limitation is removed by using an implicit scheme the presence of the poles leads to an increase in the complexity of the problem to be solved (measured as the ratio of the largest to the smallest scales in the problem, the condition number) which leads to an increase in the number of iterations needed to solve the implicit system of equations. And if a semi-Lagrangian scheme is used to circumvent the advective CFL, then updating a point near the pole will need access to data that is many grid points away from that point. On a supercomputer, this requires a lot of communication between different parts of the supercomputer, which has a significant cost. It is probably fair to say that, from a numerical efficiency perspective, there is nowhere to hide from the pole problem.

Because of the quadratic nature of the pole problem, the issue is getting much worse as resolutions continue to increase. Therefore, there has been considerable renewed interest in how to remove the singularities by adopting more uniform grids. Examples are grids built around subdividing the icosahedron, giving a grid that comprises a mixture of spherical hexagons and pentagons (resembling a football); the dual grid to this which consists entirely of spherical triangles; the cubed-sphere grid which can be thought of as putting an inflatable Rubik's cube within a sphere and pumping it up to fill the sphere (this has a similar geometry as a volleyball); or the Yin-Yang grid which is obtained by taking two equatorial regions of a latitude-longitude grid that avoid the poles and then rotating them with respect to each other before joining them together in a shape resembling a tennis ball (the two grids slightly overlap each other). There are many variants of each of these as well as many other possibilities. Staniforth and Thuburn (2012) give an extensive review of these possibilities as well as their advantages and disadvantages.

8 An outlook

The development of dynamical cores for NWP is at an interesting point in its evolution. In the early days of NWP, through the 1950s to the 1970s, numerical methods were in their infancy and computers were relatively limited in both their speed and memory. Progress principally revolved around manipulation of the underpinning governing equations through the application of deep physical understanding of the dynamics of the atmosphere. This permitted simplification of the governing equations that allowed the limited computing capacity available to focus on the leading order contributions to the weather.

Through the 1980s and 1990s, there were a number of exciting developments in the field of numerical methods for NWP. From an operational perspective, probably the most notable was the judicious combination of the implicit method for wave propagation (along the lines of the Crank-Nicolson scheme discussed earlier) with a Lagrangian approach for the transport, the semi-implicit semi-Lagrangian approach. At the same time, there was a revolution in the amount of computing power available. This is best captured by the combination of two empirical observations: Moore's Law, that the number of transistors in a circuit approximately doubles every 2 years; and Dennard scaling, that the power density remains constant as the transistors get smaller. Taken together these observations reflect that between the mid-1970s and the mid-2000s the speed of computation that each watt of power delivered, doubled every couple of years or so. This has been referred to as the "free lunch": wait a couple of years; buy a new computer; and with the same power, the same code will deliver results at twice the speed. This staggering technological development was used to routinely increase the resolution of the NWP models and to increase their complexity: more physical processes were included in the models; and slowly the approximations to the governing equation sets have been relaxed so that most of the operational models today are solving almost unapproximated equation sets. These developments are all behind what Bauer et al. (2015) call the "quiet revolution" of weather forecasting in which over the last 40 years or so, as they say "forecast skill in the range from 3 to 10 days ahead has been increasing by about one day per decade: today's 6-day forecast is as accurate as the 5-day forecast ten years ago...."

Unfortunately, the technological improvements that led to the free lunch have come to an end; Dennard scaling ceased to be realized in the mid-2000s and Moore's Law is beginning to slow. This is largely because we are reaching fundamental limitations imposed by the physics of the engineering problem. The response of the supercomputer industry has been to try to maintain Moore's Law (and hence the increase in computing speed) as much as it can by providing more and more processors. But without Dennard scaling this means that the power requirement increases as the number of processors increases. We are now at the point where the operating costs over the lifetime of a supercomputer are greater than the purchase cost. The other problem is that prior to the current era existing codes would largely work on new supercomputers more or less as efficiently as they did on previous ones, but that is no longer necessarily the case. The codes face two problems if they are to be able to access the increased computing power of new supercomputers.

The first problem is that the codes need to be able to efficiently spread their work over vast numbers of processors. Current global NWP models with resolutions of the order of 10 km typically use 500 nodes of a supercomputer, each of which will have around 30 processors.

Over the coming years that number of processors is expected to continue to increase and codes will have to run efficiently on several hundreds of thousands of processors. Any calculation that requires access to the data on all those processors (such as a global sum, or a synchronization point that makes all processors wait for each other, such as the end of a time step) has a significant cost associated with it. This favors numerical methods that only need local calculations (such as an explicit scheme) and that can use as long a time step as possible (such as an implicit scheme). It is clear that there is a tension between those two requirements: some centers are moving to explicit schemes dominated by local calculations but offset by having to do many time steps; others are focusing on implicit schemes that permit much longer time steps while minimizing global calculations by exploiting methods such as multigrid. It is also in this context that there is the move to much more uniform grids, to avoid any communication bottlenecks associated with unnecessary clustering of grid points.

The second problem is that in order to address the challenge that ever-increasing power costs present to the community, supercomputer manufacturers are exploring even more radical processor designs. For some time, this has included greater use of commodity processors, such as those used in mobile phones or game consoles, that is, graphical processing units (GPUs). These are much more light weight than traditional central processing units (CPUs), which have previously dominated supercomputer architectures and they are typically used alongside CPUs acting as accelerators. These accelerators are now being absorbed into the processing unit creating a complex, hybrid system with increasingly complex layers of different memory structures. Other options include user-configurable processors, such as field programmable gate arrays (FPGAs). This wide range of possible architectures, including no doubt as-yet-unknown developments, presents a challenge to a model developer.

It is in the context of such an uncertain future that the computational science concept of a "separation of concerns" is being explored by some groups (e.g., Adams et al., 2019). By employing a domain-specific language (DSL) coupled with automatic code generation (a form of DSL compiler), it is possible for the natural science aspects of a model (i.e., the specific choice of discretization) to be separated from the specifics of the computational implementation. This should allow the natural science aspects of a model to be able to be developed without having to target a specific architecture; the implementation on a given supercomputer and optimization of the model for that architecture is implemented through the DSL compiler. The success of this approach requires close cooperation between the natural scientists and the computational scientists—the principle of codesign. Implementation and exploitation of this approach to an operational NWP model is in its infancy but it is an exciting prospect that will hopefully see the "quiet revolution" in weather forecast accuracy continue into the coming decades.

Acknowledgments

The author would like to thank Christine Johnson of the Met Office for her careful reading of an early version of this chapter, which led to many improvements. It is also his pleasure to acknowledge the insights patiently and generously shared with him by Andrew Staniforth, Andy White (both formerly of the Met Office), and John Thuburn (University of Exeter).

References

Adams, S.V., Ford, R.W., Hambley, M., Hobson, J.M., Kavčič, I., Maynard, C.M., Melvin, T., Müller, E.H., Mullerworth, S., Porter, A.R., Rezny, M., Shipway, B.J., Wong, R., 2019. LFRic: meeting the challenges of scalability and performance portability in weather and climate models. J. Parallel Distrib. Comput. 132, 383–396. https://doi.org/10.1016/j.jpdc.2019.02.007.

Baldauf, M., 2008. Stability analysis for linear discretisations of the advection equation with Runge-Kutta time integration. J. Comput. Phys. 227, 6638–6659.

Bauer, P., Thorpe, A., Gilbert, B., 2015. The quiet revolution of numerical weather prediction. Nature 525, 47–55. https://doi.org/10.1038/nature14956.

Bourke, W., 1972. An efficient, one-level, primitive-equation spectral model. Mon. Weather Rev. 100, 683–689.

Brown, N., Weiland, M., Hill, A., Shipway, B., Maynard, C., Allen, T., Rezny, M., 2015. A highly scalable met office NERC cloud model. In: Proceedings of the 3rd International Conference on Exascale Applications and Software—EASC 2015, Edinburgh, UK, https://dl.acm.org/citation.cfm?id=2820083.2820108.

Charron, M.A., Zadra, A., 2014. On the dynamical consistency of geometrically approximated equations for geophysical fluids in arbitrary coordinates. Q. J. R. Meteorol. Soc. https://doi.org/10.1002/qj.2303.

Choi, S.-J., Hong, S.-Y., 2016. A global non-hydrostatic dynamical core using the spectral element method on a cubed-sphere grid. Asia Pac. J. Atmos. Sci. 52, 291–307. https://doi.org/10.1007/s13143-016-0005-0.

Clark, T.L., 1977. A small-scale dynamic model using a terrain following coordinate transformation. J. Comput. Phys. 24, 188–215.

Colella, P., Woodward, P.R., 1984. The piecewise parabolic method (PPM) for gas-dynamical simulations. J. Comput. Phys. 54, 174–201.

Cotter, C.J., Shipton, J., 2012. Mixed finite elements for numerical weather prediction. J. Comput. Phys. 231, 7076–7091.

Davies, T., Staniforth, A., Wood, N., Thuburn, J., 2003. Validity of anelastic and other equation sets as inferred from normal-mode analysis. Q. J. R. Meteorol. Soc. 129, 2761–2775.

Davies, T., Cullen, M., Malcolm, A., Mawson, M., Staniforth, A., White, A.A., Wood, N., 2005. A new dynamical core for the Met Office's global and regional modelling of the atmosphere. Q. J. R. Meteorol. Soc. 131, 1759–1782.

Durran, D.R., 2010. Numerical Methods for Fluid Dynamics With Applications to Geophysics, second ed. Springer-Verlag, New York, NY, p. 516.

Gal-Chen, T., Somerville, R.C.J., 1975. On the use of a coordinate transformation for the solution of the Navier-Stokes equations. J. Comput. Phys. 17, 209–228.

Gill, A.E., 1982. Atmosphere-Ocean Dynamics. Academic Press, London, p. 662+xv.

Holdaway, D., Thuburn, J., Wood, N., 2008. On the relation between order of accuracy, convergence rate and spectral slope for linear numerical methods applied to multiscale problems. Int. J. Numer. Methods Fluids 56, 1297–1303.

Holton, J.R., 1992. An Introduction to Dynamic Meteorology, third ed. Academic Press, New York, NY.

Hortal, M., Simmons, A.J., 1991. Use of reduced Gaussian grids in spectral models. Mon. Weather Rev. 119 (4), 1057–1074. https://doi.org/10.1175/1520-0493(1991)119<1057:UORGGI>2.0.CO;2.

Hoskins, B.J., Simmons, A.J., 1975. A multi-layer spectral model and the semi-implicit method. Q. J. R. Meteorol. Soc. 101, 637–655.

Lauritzen, P., Jablonowski, C., Taylor, M., Nair, R. (Eds.), 2011. In: Numerical Techniques for Global Atmospheric Models, Lecture Notes in Computational Science and Engineering, vol. 80. Springer, p. 565.

Leonard, B.P., 1979. A stable and accurate convective modelling procedure based on quadratic upstream interpolation. Comput. Methods Appl. Mech. Eng. 19 (1), 59–98. https://doi.org/10.1016/0045-7825(79)90034-3.

Leonard, B.P., Lock, A.P., MacVean, M.K., 1996. Conservative explicit unrestricted-time-step multidimensional constancy-preserving advection schemes. Mon. Weather Rev. 124, 2588–2606.

Lock, S.J., Bitzer, H.W., Coals, A., Gadian, A., Mobbs, S., 2012. Demonstration of a cut-cell representation of 3D orography for studies of atmospheric flows over very steep hills. Mon. Weather Rev. 140 (2), 411–424. https://doi.org/10.1175/MWR-D-11-00069.1.

Lock, S.J., Wood, N., Weller, H., 2014. Numerical analyses of Runge-Kutta implicit-explicit schemes for horizontally-explicit vertically-implicit solutions of atmospheric models. Q. J. R. Meteorol. Soc. 140, 1654–1669. https://doi.org/10.1002/qj.2246.

Marras, S., Kelly, J.F., Moragues, M., Mueller, A., Kopera, M.A., Vazquez, M., Giraldo, F.X., Houzeaux, G., Jorba, O., 2016. A review of element-based Galerkin methods for numerical weather prediction. Finite elements, spectral elements, and discontinuous Galerkin. Arch. Comput. Methods Eng. 23, 673–722.

Melvin, T., Staniforth, A., Thuburn, J., 2012. Dispersion analysis of the spectral element method. Q. J. R. Meteorol. Soc. 138, 1934–1947.

Nastrom, G.D., Gage, K.S., 1985. A climatology of atmospheric wavenumber spectra of wind and temperature observed by commercial aircraft. J. Atmos. Sci. 42, 950–960.

Oliger, J., Sundström, A., 1978. Theoretical and practical aspects of some initial boundary value problems in fluid dynamics. SIAM J. Appl. Math. 35, 419–446.

Persson, A., 2005. Early operational numerical weather prediction outside the USA: an historical introduction. Part 1: internationalism and engineering NWP in Sweden, 1952-69. Meteorol. Appl. 12 (2), 135–159. https://doi.org/10.1017/S1350482705001593.

Priestley, A., 1993. A quasi-conservative version of the semi-Lagrangian advection scheme. Mon. Weather Rev. 121, 621–629.

Pudykiewicz, J., Staniforth, A., 1984. Some properties and comparative performance of the semi-Lagrangian method of Robert in the solution of the advection-diffusion equation. Atmos. Ocean 22, 283–308.

Salmon, R., 1988. Hamiltonian fluid mechanics. Ann. Rev. Fluid Mech. 20, 225–256.

Shepherd, T.G., 1990. Symmetries, conservation laws and Hamiltonian structure in geophysical fluid dynamics. Adv. Geophys. 32, 287–338.

Skamarock, W.C., Klemp, J.B., Dudhia, J., Gill, D.O., Liu, Z., Berner, J., Wang, W., Powers, J.G., Duda, M.G., Barker, D.M., Huang, X.Y., 2019. A description of the Advanced Research WRF Version 4. NCAR/TN-556+STR, 145 pp. NCAR. https://doi.org/10.5065/1dfh-6p97.

Smolarkiewicz, P.K., Pudykiewicz, J.A., 1992. A class of semi-Lagrangian approximations for fluids. J. Atmos. Sci. 49, 2082–2096.

Staniforth, A., 2014a. Deriving consistent approximate models of the global atmosphere using Hamilton's principle. Q. J. R. Meteorol. Soc. 140, 2383–2387. https://doi.org/10.1002/qj.2273.

Staniforth, A., 2014b. Spheroidal and spherical geopotential approximation. Q. J. R. Meteorol. Soc. https://doi.org/10.1002/qj.2324.

Staniforth, A., Côté, J., 1991. Semi-Lagrangian integration schemes for atmospheric models—a review. Mon. Weather Rev. 119, 2206–2223.

Staniforth, A., Thuburn, J., 2012. Horizontal grids for global weather prediction and climate models: a review. Q. J. R. Meteorol. Soc. 138, 1–26.

Thuburn, J., 2006. Vertical discretizations giving optimal representation of normal modes: sensitivity to the form of the pressure gradient term. Q. J. R. Meteorol. Soc. 132, 2809–2825.

Thuburn, J., Woollings, T.J., 2005. Vertical discretizations for compressible Euler equation atmospheric models giving optimal representation of normal modes. J. Comput. Phys. 203, 386–404.

Tort, M., Dubos, T., 2014. Dynamically consistent shallow-atmosphere equations with a complete Coriolis force. Q. J. R. Meteorol. Soc. https://doi.org/10.1002/qj.2274.

Vallis, G.K., 2006. Atmospheric and Oceanic Fluid Dynamics, first ed. Cambridge University Press, Cambridge, p. 745+xxv.

Walters, D.N., Brooks, M., Boutle, I., Melvin, T., Stratton, R., Vosper, S., Wells, H., Williams, K., Wood, N., Allen, T., Bushell, A., Copsey, D., Earnshaw, P., Edwards, J., Gross, M., Hardiman, S., Harris, C., Heming, J., Klingaman, N., Levine, R., Manners, J., Martin, G., Milton, S., Mittermaier, M., Morcrette, C., Riddick, T., Roberts, M., Sanchez, C., Selwood, P., Stirling, A., Smith, C., Suri, D., Tennant, W., Vidale, P.L., Wilkinson, J., Woolnough, S., Xavier, P., 2017. The met office unified model global atmosphere 6.0/6.1 and JULES Global Land 6.0/6.1 configurations. Geosci. Model Dev. 10, 1487–1520.

Weller, H., Lock, S.J., Wood, N., 2013. Runge-Kutta IMEX schemes for the Horizontally Explicit/Vertically Implicit (HEVI) solution of wave equations. J. Comput. Phys. 252, 365–381.

White, A.A., Bromley, R.A., 1995. Dynamically consistent, quasi-hydrostatic equations for global models with a complete representation of the Coriolis force. Q. J. R. Meteorol. Soc. 121, 399–418.

White, A.A., Hoskins, B.J., Roulstone, I., Staniforth, A., 2005. Consistent approximate models of the global atmosphere: shallow, deep, hydrostatic, quasi-hydrostatic and non-hydrostatic. Q. J. R. Meteorol. Soc. 131, 2081–2107.

Williamson, D.L., 2007. The evolution of dynamical cores for global atmospheric models. J. Met. Soc. Jpn 85B, 241–269.
Wood, N., Staniforth, A., White, A., Allen, T., Diamantakis, M., Gross., M., Melvin, T., Smith, C., Vosper, S., Zerroukat, M., Thuburn, J., 2014. An inherently mass-conserving semi-implicit semi-Lagrangian discretization of the deep-atmosphere global nonhydrostatic equations. Q. J. R. Meteorol. Soc. 140, 1505–1520. https://doi.org/10.1002/qj.2235.
Zerroukat, M., Allen, T., 2015. On the monotonic and conservative transport on overset/Yin-Yang grids. J. Comput. Phys. 302, 285–299. https://doi.org/10.1016/j.jcp.2015.09.006.

CHAPTER

2

Numerical uncertainties in discretization of the shallow-water equations for weather predication models

Ning Wang

Cooperative Institute for Research in the Atmosphere, Colorado State University, and Earth System Research Laboratory, Boulder, CO, United States

1 Introduction

Numerical simulation of the fluid dynamics on the earth is a key component of weather predication models. The computational component is commonly referred to as the dynamical core, a numerical package that simulates atmosphere and ocean dynamics that are governed by a set of partial differential equations (PDEs). The governing equation set for numerical weather prediction (NWP) models derives from some form of the Euler equations expressing conservation of mass, momentum, and energy. In this chapter, we focus on the shallow-water equations (Vallis, 2006), which is a simplified conserved system derived from Navier-Stokes equations. These equations admit much of the nonlinear behavior of the Euler and Navier-Stokes equations, and are well suited to describe the horizontal motion of atmosphere or oceans, which has a high horizontal-vertical aspect ratio.

The design and implementation of dynamical core of a numerical model include creation of initial state (condition), selection of a form of the governing equations that will accurately and efficiently describe and simulate the state of atmosphere and ocean, and a spherical grid and the numerical schemes that solve the equations on the grid.

Uncertainties exist in many parts of the numerical simulation of the fluid dynamics. In this chapter, we will not discuss the uncertainties occurred in the creation of initial condition, which is the subject of the observation quality control and data assimilation technology.

https://doi.org/10.1016/B978-0-12-815491-5.00002-1

As for the model governing equations, we assume the shallow-water primitive equations can yield a good approximation of observed phenomenological waves of interest and their interaction with fluid mechanical structures. We will not dive into details of those uncertainties caused by inappropriate numerical implementations; these uncertainties, we assume, may be eliminated by choosing the correct numerical formulation and procedures that avoids significant round-off error caused by misuse of computer discrete floating point representation for continuous real solutions. We focus our discussion on the uncertainties occurred in discretization of the shallow-water equations. Furthermore, we focus our attention to the global grid model based on finite-difference or finite-volume discretization. The discretization grids we discuss in this chapter, in addition to the Cartesian grid, are the two most commonly used unstructured spherical grids in the global NWP: the icosahedral-hexagonal grid and the cubed-sphere grid.

Depending on the support of the basis functions adopted in the discretization, modern atmospheric models are classified into three different numerical types: spectral model (globally supported spherical harmonics), finite (spectral) element model (locally supported element basis), and grid-point model (Dirac delta sampling basis). There are weather forecast models that are based on locally supported basis functions such as wavelet basis functions. These multiresolution global models could be considered as intermediate between global spectral and finite (spectral) element models.

Each model type is based on a set of numerical schemes that approximate the governing equations, their spatial and temporary differential operators. Numerical schemes for these models all have their pros and cons, in terms of the numerical accuracy, numerical stability, and computational efficiency.

Spectral models are in general more accurate at the same resolution and less prone for the shortest waves to alias to the longer waves due to the truncation of the shorter wave coefficients. However, the truncation of the shorter waves can cause Gibbs effect in the regions of steep changes. The computational complexity of the spectral model is high because it requires frequent spherical spectral transforms between spectral domain and physical domain, and this $O(N^3)$ transform does not scale well on parallel computer systems due to its globally supported basis functions. Several operational NWP models are based on spectral implementations of dynamical cores (ECMWF, 2016; Sela, 2009, 2010).

As an emerging numerical technique for atmospheric modeling, the finite (spectral) element method has advantages of local high-order accuracy and flexibility in adapting to irregular domains and nonuniform resolutions. With locally supported basis functions, finite (spectral) element model can scale well on modern parallel computers. The main drawbacks of the discontinuous Galerkin (DG) algorithms, the backbone of the finite (spectral) element discretization, are their high complexity in implementation and more restricted CFL instability condition with the currently used explicit time stepping. In addition, unevenness of quadrature points within each element, especially for high-order schemes, are undesirable for the weather forecast models. There are several ongoing efforts to construct experimental or operational NWP models using spectral element discretization (Girado et al., 2013; Kopera and Giraldo, 2014; Choi et al., 2014; Choi and Hong, 2016).

The grid-point model based on finite-difference or finite-volume methods is most popular in recent years due to its simplicity in implementation and its scalability for modern parallel computer systems. The numerical disadvantages of the grid-point models are numerical accuracy, nonlinear instability and aliasing, and heavy dependence on the grid geometry.

All types of NWP models and their dynamical cores have numerical uncertainties. They are usually caused by the limited accuracy and fidelity of the numerical discretizations, which are heavily impacted by the highly nonlinear nature of the fluid to be simulated, and constrained by the complexity of the numerical schemes and available computational power. There is no best set of numerical schemes for NWP models that can satisfy all applications of various time and spatial scales. Even the question of what is a suitable set of numerical schemes for a given application is often heavily debated. From the perspective of numerical simulation, it is important for discretization schemes to not only possess the essential mimetic properties, such as conservations and dispersion relations, but also maintain the numerical consistency with desired orders of accuracy. Both properties are fundamental to reduce and control the numerical uncertainties, and one does not necessary lead to or guarantee the other.

The chapter is organized as follows. Sections 2 and 3 present some general discretization methods for the shallow-water equations; the sections describe temporal and spatial discretizations of the PDEs, wave dispersion analysis of the discretized linearized equations on various staggered grids, and numerical schemes for horizontal advection. Section 4 discusses some important numerical techniques that are used in the shallow-water equations to maintain the computational stability. In Section 5, we present global NWP models on two popular unstructured grids—the icosahedral grid and the cubed-sphere grid. Two representative models—MPAS (Model for Prediction Across Scales) and FV3 (Finite-Volume cubed-sphere model) are highlighted, and the basic numerical schemes used for their shallow-water equations are briefly described. Throughout the five sections, we focus our discussions on how numerical errors are introduced by different numerical methods, and how to increase the accuracy of the solution to the shallow-water equations and thereby reduce the numerical uncertainties of the NWP models. We summarize the chapter with some general remarks in Section 6.

2 Discretization of the governing equation I

Discretization of the shallow-water equations involves the numerical implementation of spatial and temporal differential operators. It also involves creation of numerical operators, such as averaging (mapping, or interpolating) operators that are often part of the discretization of differential operators, and diffusion and filtering operators that remove computational modes to increase numerical stability and prevent numerical solutions from violating some critical physical properties of the continuous equations. In this section, we discuss the discretization of temporal and spatial differential operators, and its impact on simulation accuracy and dispersion behavior. In the section that follows, we present numerical schemes that combine the spatial and temporal difference operators to simulate the horizontal advection. We leave the discussion of other operators in Section 4.

The shallow-water equations used in NWP consist of two equations: the momentum equation which describes the evolution of the velocity field and the continuity equation which governs the transport or advection of the fluid mass. The momentum equation can be written in a few different forms, and among them the most commonly used are advective form and vector-invariant form, where the prognostic variable is velocity. The nonlinear shallow-water equations in vector-invariant form are

$$\frac{\partial \mathbf{u}}{\partial t} = -\eta \mathbf{k} \times \mathbf{u} - \nabla(K + gh), \tag{1}$$

$$\frac{\partial h}{\partial t} = -\nabla \cdot (h\mathbf{u}). \tag{2}$$

Here $\mathbf{u}$ is the velocity vector, $\eta = \zeta + f$ is the absolute vorticity, $\zeta = \mathbf{k} \cdot (\nabla \times \mathbf{u})$ is the relative vorticity, f is the Coriolis parameter, $\mathbf{k}$ is the vertical unit vector, $K = \frac{1}{2}|\mathbf{u}|^2$ is the kinetic energy, g is the gravity constant, and h is the fluid depth (height).

Arakawa (1970) stated that there are two key problems in finite-difference discretization for atmospheric models. First one is to correctly simulate the geostrophic adjustment process, by which the atmosphere establishes a geostrophic balanced flow state, through propagation and dispersion of the gravity-inertial waves. The second one is to correctly simulate the established large-scale quasi-nondivergent flow, which is mainly achieved through a carefully designed horizontal advection scheme.

In this and the following section, we will discuss the issue of discretization uncertainty under the context of the two challenges.

2.1 Temporal differential operator

Discretization of the temporal differential operator is an important part of the numerical simulation. Assuming the independence of time and spatial dimensions, the discretization often adopts a numerical solver for coupled ordinary differential equation (ODE). The main focus for the time discretization is its numerical stability and computational efficiency.

Numerical accuracy is also a consideration in the design of the time integration for NWP models. However, order of accuracy is not necessary the first criterion to choose or devise a time discretization scheme. The accuracy of a numerical solution to a PDE often dominated by the accuracy of the discretization of spatial derivatives; time resolution is usually constrained by the stability requirement rather than the accuracy requirement, that is, time resolution is often more than adequate for accuracy purposes. Furthermore, high-order schemes are more expensive to compute, and timely numerical simulation is critical for NWP models.

2.1.1 *Euler and Crank-Nicholson methods*

Let us consider the ODE,

$$\frac{d\phi(t)}{dt} = F(t, \phi(t)), \quad \phi(t_0) = \phi_0, \tag{3}$$

where F is a function of t and $\phi(t)$.

Let the time step size be Δt, $t_n = t_0 + n\Delta t$, and $\phi_n = \phi(t_n)$. The generalized Euler method discretization has the following form:

$$\frac{\phi_{n+1}-\phi_n}{\Delta t}=(1-\theta)F(t_n,\phi_n)+\theta F(t_{n+1},\phi_{n+1}), \tag{4}$$

where $0 \leq \theta \leq 1$.

When $\theta = 0$, Eq. (4) yields the simplest straightforward method, the (standard) forward Euler method. This is a purely *explicit* scheme, in that update requires only function F be evaluated at current time level n. The method has low-accuracy ($O(\Delta t)$ truncation error) and becomes numerically unstable when the time step is too large. When $\theta = 1$, Eq. (4) becomes the backward Euler method, which is an implicit method, in which the update requires that F be evaluated at time level $n + 1$. The backward Euler method is stable and permits larger time step, at the expense of iterative computation of the ϕ_{n+1} on the right-hand side (for nonlinear F) usually with a variant of Newton's method.

Both Euler methods have low accuracy, and themselves alone are rarely used in discretization of the shallow-water equations.

The discrete equation (4) becomes the Crank-Nicholson method, when $\theta = 0.5$. Since it requires ϕ at both current and new time steps, the method is referred to as semi-implicit method. The Crank-Nicholson method is more accurate ($O(\Delta t^2)$ truncation error) than both Euler methods, and more stable than the forward Euler method.

Semi-implicit time integration has been used widely in modern NWP models (Adcroft et al., 1999; Cullen, 2001; Smolarkiewicz et al., 2014; Sandbach et al., 2015). The scheme is applied in such a way that the advective and other nonlinear terms are treated explicitly, and the linear terms, including those governing the propagation of gravity waves, are handled implicitly.

Euler methods use only the function value at current time level, and they are only first-order accurate. To achieve higher order of accuracy, we can use multi-time level schemes. There are two major classes of multi-time level numerical integration methods that are widely used in NWP models: multistep methods and multistage methods. These are discussed in turn below.

2.1.2 Multistep methods

Leapfrog method is a second-order time integration scheme, using two time levels. The scheme is easy to implement and computationally efficient; it only requires one function evaluation per time step and one saved state at last time step. In its application in NWP models, it is often used together with a high-frequency filter because of the need to damp its computational modes.

Linear multistep methods are computationally efficient. This group of methods use information from previous time steps and only evaluate function once per time step. There are both explicit and implicit forms for these methods, and they may be constructed from a general sth-order multistep discretization of Eq. (3):

$$\phi^{n+1}-\sum_{j=1}^{s}\alpha_j\phi^{n+1-j}=\Delta t\sum_{j=0}^{s}\beta_j F^{n+1-j}, \tag{5}$$

where the coefficients, α_j and β_j depend on the specific scheme. The scheme is implicit if $\beta_0 \neq 0$; otherwise it is explicit. One commonly used explicit scheme is the Adams-Bashforth method, which specifies coefficients β_j for $1 \leq j \leq s$.

In search of a better time differencing scheme, Durran (1991) compared the leapfrog time stepping method with the Adams-Bashforth multistep method and found that the third-order Adams-Bashforth method (AB3) has most attractive numerical properties among methods with comparable computational expenses. As an explicit multistep method, AB3 is computationally efficient; it is more accurate compared with leapfrog in terms of amplitude and phase-speed errors. AB3's maximum stable step size is close to that of the leapfrog scheme with Robert-Asselin filtering (Robert, 1966; Asselin, 1972). AB3 method has been used in several global models (Ringler et al., 2000; Lee and MacDonald, 2009).

Implicit multistep methods, such as Adams-Moulton schemes, may be derived from Eq. (5), by specifying β_j for $0 \leq j \leq s$. These methods are more accurate and have a larger stability domain than explicit Adams-Bashforth schemes. For a nonlinear equation, Adams-Moulton schemes require an iterative method to solve the implicit equation, and are thus more expensive in computation.

2.1.3 *Multistage methods*

Multistage methods are numerically more accurate compared to the multistep methods. Its stability domain increases as the order increases (Schneider et al., 2013). The main disadvantage for this family of methods is their high computational costs. An s-order method will usually require s function evaluations.

Among various multistage methods, the fourth-order Runge-Kutta (RK4) is the most commonly used, classic multistage scheme. In weather prediction models, the second- and third-order Runge-Kutta (RK2, RK3) schemes are used frequently as well (Wicker and Skamarock, 1998, 2002; Satoh et al., 2008; Skamarock et al., 2008, 2012).

2.2 Spatial differential operators

The spatial differential operators in the shallow-water governing equations are commonly discretized into finite-difference or finite-volume operators. These operators are defined with their specific stencils, which determine the accuracy of the discrete operators and link to their dispersion behaviors and computational modes.

One of the main tasks for dynamical core is to propagate the various modes of waves to simulate the real atmosphere and oceans fluid correctly. In the shallow-water model implementation, numerical schemes together with the discretization grids should approximate the continuous equations and best preserve the dispersion relation of waves of different wave lengths. However, to obtain a good mimetic discretization of dynamical cores is no trivial task. Here, we discuss the issue starting with the simplified one-dimensional (1D) shallow-water equations on unstaggered and staggered grids.

2.2.1 *Grid staggering and operator discretization*

Fig. 1 shows an unstaggered and a staggered grid in 1D Cartesian space. They, when generalized to two-dimensional (2D) space, are referred to as the A-grid and the C-grid (Arakawa and Lamb, 1977).

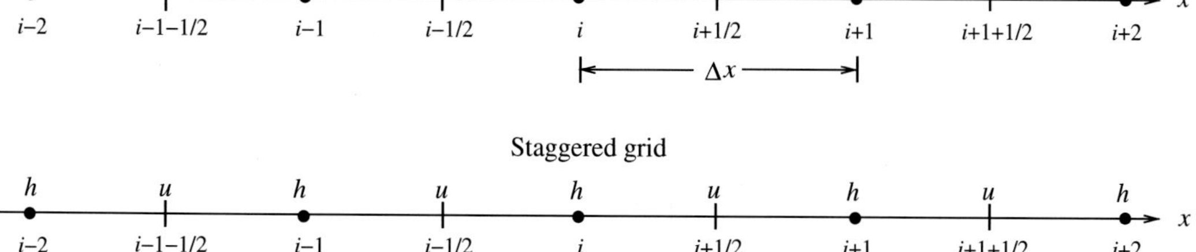

FIG. 1 Unstaggered and staggered grid arrangement. *Solid dots* are cell centers and *vertical bars* are cell boundaries.

Consider the following 1D linearized shallow-water equations:

$$\begin{aligned} \frac{\partial h}{\partial t} + H\frac{\partial u}{\partial x} &= 0, \\ \frac{\partial u}{\partial t} + g\frac{\partial h}{\partial x} &= 0, \end{aligned} \tag{6}$$

where g is the gravity constant, H is the nominal depth of the "water," u is the flow velocity, and h is the displacement of the free surface.

Applying $\frac{\partial}{\partial t}$ operator to the first equation of Eq. (6), and assuming the symmetry of second derivatives and substituting the expression of $\frac{\partial u}{\partial t}$ from the second equation, we have

$$\frac{\partial^2 h}{\partial t^2} = gH\frac{\partial^2 h}{\partial x^2}. \tag{7}$$

Substituting in a wavelike solution $h = h_0 e^{i(kx-\omega t)}$ to the equation, we get the relation between the angular frequency (phase speed) and wave number for the continuous equations (Eq. 6),

$$\omega = \pm\sqrt{gH}k. \tag{8}$$

Let us discretize the right-hand side of the continuous equation (7) using second-order centered finite difference, on the unstaggered grid,

$$\frac{\partial^2 h}{\partial t^2} = \frac{gH}{(2\Delta x)^2}(h_{i-2} - 2h_i + h_{i+2}), \tag{9}$$

and on the staggered grid,

$$\frac{\partial^2 h}{\partial t^2} = \frac{gH}{(\Delta x)^2}(h_{i-1} - 2h_i + h_{i+1}). \tag{10}$$

Following the similar procedure that obtains Eq. (8), for the semi-discrete equations, substituting in a discrete wavelike solution $h_j = h_0 e^{i(kx_j - \omega t)}$, where $x_j = j\Delta x$, we obtain dispersion relations for the semi-discrete equation on the unstaggered grid,

$$\omega = \pm\sqrt{gH}\sin(k\Delta x)/\Delta x, \tag{11}$$

and on the staggered grid,

$$\omega = \pm\sqrt{gH}2\sin(k\Delta x/2)/\Delta x. \tag{12}$$

Fig. 2 shows the dispersion relations and group velocities of wave solutions for the continuous equations and the discrete equations on unstaggered and staggered grids (for the positive ω). While the dispersion relation of the staggered grid is closer to that of the continuous equations, both discretizations have slower phase speed (Fig. 2A). Both have a nonconstant group velocity $\frac{d\omega}{dk}$, unlike the continuous case; the unstaggered grid is worse in that its group velocity is zero at $k\Delta x = \pi/2$ (Fig. 2B). The gravity waves near this wave number can be excited locally, and the wave energy that would normally propagate and disperse stays locally, causing local energy to accumulate and introduce spurious grid space noise; when $k\Delta x$ passes $\pi/2$, the group velocity of the unstaggered grid has a wrong sign, meaning gravity waves of the group will propagate and disperse in the wrong direction. These computational modes lead to an incorrect simulation and potentially greater uncertainties. Furthermore, for the unstaggered grid, the phase speed of Nyquist wave is zero, that is, the minimally resolvable wave becomes stationary.

2.2.2 Nonzero null space of discrete operators

A null space of a linear operator $L: V \to W$, where V and W are two vector spaces, is defined as

$$\mathbf{null}(L) = \{\mathbf{v} \in V | L(\mathbf{v}) = \mathbf{0}\}. \tag{13}$$

As a discrete linear operator, L is represented by a localized stencil matrix, and v is a discrete function vector.

Write Eq. (6) in a semi-discrete centered finite-difference form on the unstaggered grid:

$$\frac{\partial h_i}{\partial t} + H\frac{u_{i+1} - u_{i-1}}{2\Delta x} = 0,$$
$$\frac{\partial u_i}{\partial t} + g\frac{h_{i+1} - h_{i-1}}{2\Delta x} = 0.$$

The stencil for the finite difference operator is $\{0,\ldots,1/(2\Delta x),0,-1/(2\Delta x),\ldots,0\}$. Thus, there is a nonzero null space for the centered finite difference operator.

$$\mathbf{null}(\delta_x) = \{v | v_{i+1} = v_{i-1}\}.$$

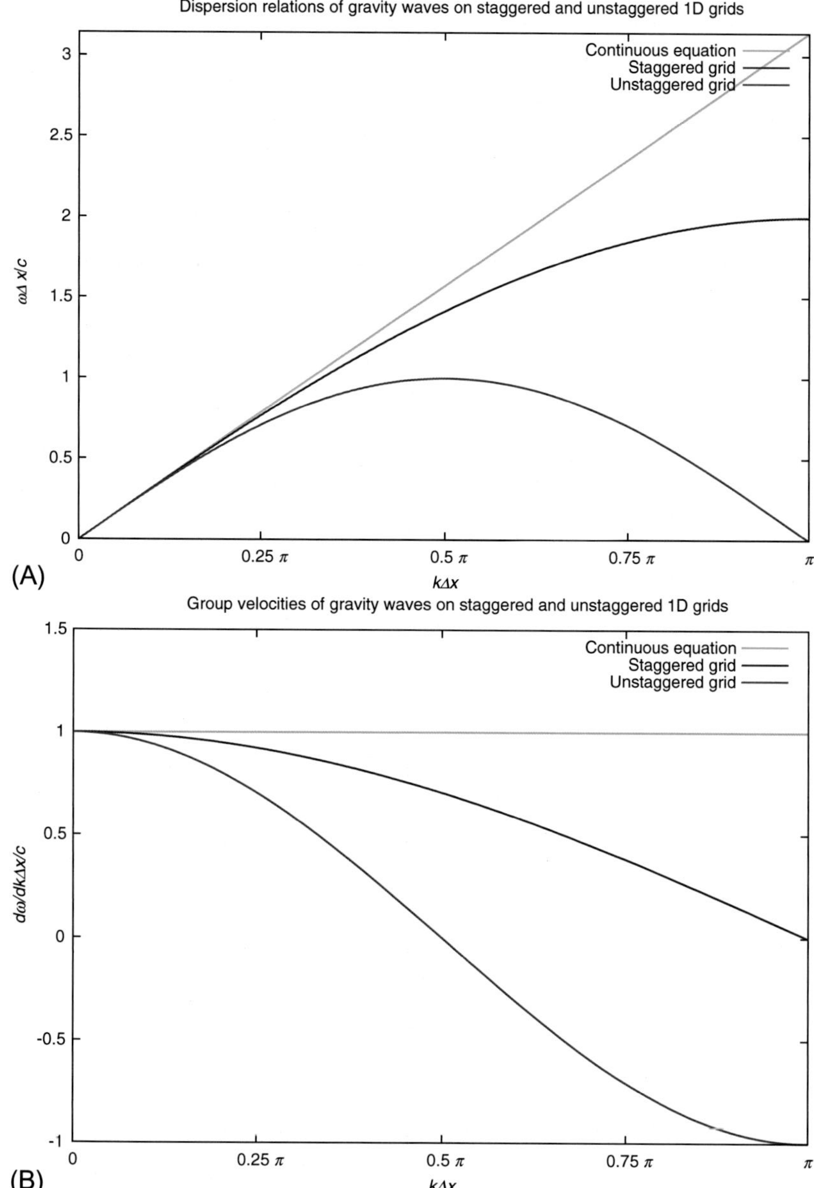

FIG. 2 Dispersion relations (A) and group velocities (B) for wave solutions of 1D gravity wave equation (6), and its discretizations on unstaggered A-grid and staggered C-grid; $c = \sqrt{gH}$.

The null space above includes wave functions of h and u with the shortest resolvable wavelength of $2\Delta x$. It implies that for a uniform velocity field, a checkerboard patterned h wave will fail to propagate and disperse.

In the following subsection, we will see in 2D C-grid, there is a null space for the averaging operator for Coriolis terms, which could lead to a checkerboard pattern for height and velocity amplitude.

In general, the nonzero null spaces for discrete linear operators cause some high wave number waves to stall and grid-scale noise to increase, which lead to an incorrect simulation of geostrophic adjustment process. Thus, from this perspective, when we design a numerical scheme with various discrete operators, we should try to minimize the associated null spaces.

2.3 Extension to two dimensions

The linearized shallow-water equations in 2D space include additional Coriolis terms from (Eq. (1)); the terms are associated with the inertial wave modes. It is a challenge to discretize all the terms accurately and mimetically on the unstaggered and staggered grids.

Winninghoff (1968) realized that for grid model the manner in which prognostic variables are distributed affects numerical simulation of the geostrophic adjustment process. Arakawa and Lamb (1977) systematically analyzed dispersion relations for the linearized shallow-water equations on various unstaggered and staggered grids on Cartesian plane (grid A–D in Fig. 3). In development of the "Z" grid model (Z-grid in Fig. 3), Randall (1994) provided detailed linear analyses for different 2D grids, and compared them with the analysis of the Z-grid. Please refer to these classic works for details about the dispersion relations for different staggering schemes.

Here, we briefly highlight the dispersion behaviors of the linearized shallow-water equations discretized on two of the most common grids, the A-grid and the C-grid.

On the square lattice, where $\Delta x = \Delta y = d$, following the similar procedure to the 1D analysis, and using the simple centered space differencing operator and 1D mid-point averaging operator, we have the following dispersion relations (Randall, 1994) for:

the continuous equations,

$$(\omega^*)^2 = 1 + \left(\frac{\lambda}{d}\right)^2 \left[(kd)^2 + (ld)^2\right]; \tag{14}$$

the A-grid,

$$(\omega^*)^2 = 1 + \left(\frac{\lambda}{d}\right)^2 \left[\sin^2(kd) + \sin^2(ld)\right]; \tag{15}$$

and the C-grid,

$$(\omega^*)^2 = \frac{1}{4}[1 + \cos(kd) + \cos(ld) + \cos(kd)\cos(ld)] + 4\left(\frac{\lambda}{d}\right)^2 \left[sin^2(kd/2) + sin^2(ld/2)\right]. \tag{16}$$

Here, $\lambda \equiv \sqrt{gH}/f$ is the Rossby radius of deformation, f is the latitude dependent Coriolis parameter, and $\omega^* = \omega/f$ is the normalized angular frequency. The typical value of λ is about 1000 km for atmosphere and 10 km for ocean. From these dispersion relations, we obtain the group velocity vector $\mathbf{v}_g/f = \nabla\omega^* = (\frac{\partial\omega^*}{\partial k}, \frac{\partial\omega^*}{\partial l})$:

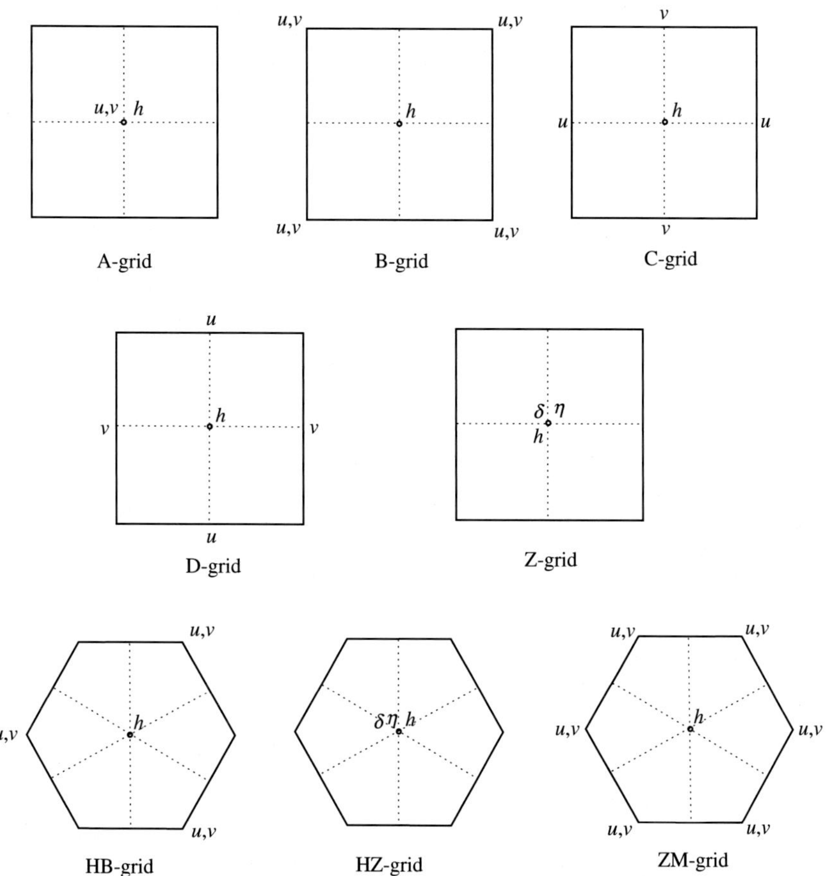

FIG. 3 Commonly used grid staggering schemes for atmosphere and ocean models.

the continuous equations,

$$\nabla\omega^* = \left[1+\lambda^2(k^2+l^2)\right]^{-1/2}\left(\lambda^2 k, \lambda^2 l\right); \tag{17}$$

the A-grid,

$$\nabla\omega^* = \frac{\lambda^2}{2d}\left[1+\frac{\lambda^2}{d^2}\left(\sin^2(kd)+\sin^2(ld)\right)\right]^{-1/2}\left(\sin(2kd), \sin(2ld)\right); \tag{18}$$

and the C-grid,

$$\begin{aligned}\nabla\omega^* &= \frac{d}{8}\beta^{-1/2}\left[\sin(kd)\left(8\left(\frac{\lambda}{d}\right)^2-1-\cos(ld)\right), \sin(ld)\left(8\left(\frac{\lambda}{d}\right)^2-1-\cos(kd)\right)\right], \text{and}\\ \beta &= \beta(kd,ld) = \frac{1}{4}[1+\cos(kd)+\cos(ld)+\cos(kd)\cos(ld)]+4\left(\frac{\lambda}{d}\right)^2\left[\sin^2(kd/2)+\sin^2(ld/2)\right].\end{aligned} \tag{19}$$

Analogous to the unstaggered 1D grid, the unstaggered 2D A-grid has all prognostic variables colocated at the center of the grid cell (Fig. 3). For the 2D shallow-water equations, with both components of the velocity vector colocated, no averaging is needed for the Coriolis terms in the momentum equation, resulting in a good accuracy for inertial oscillations and a straightforward energy conservation for the Coriolis terms, kinetic energy gradient and pressure gradient terms. It is relatively easier to interpolate and reconstruct velocity vectors at locations different from the cell centers for finite volume schemes, for example. The biggest problem with the A-grid is its poor dispersion behavior for short-wavelength gravity-inertial waves. Fig. 4 shows a comparison of dispersion relations for wave solutions of the A-grid and the C-grid discretization of the linearized shallow-water equations, and their continuum counterpart. Similar to the 1D case, the A-grid surface shows a much slower phase speed than the C-grid and the continuous case, at a relative high-resolution (Fig. 4-A). More importantly, there is a set of false maximum points along the center lines of the 2D wave number plane, where kd or $ld = \pi/2$. At these maximum frequency points, one of the group velocity components equals zero (Eq. 18). It is known that nonzero group velocity is critical for the geostrophic adjustment process. In addition, passing the maximum points into the higher wave number region, the group velocity points to a wrong direction compared with the continuum case (Eq. 17). Moving further toward higher wave numbers, at the minimally resolvable wavelength $2\Delta d$ in each dimension, a component of group velocity turns to zero again. As a result of these poor dispersion behaviors, solution h on the A-grid could be very noisy and it could damage the simulation of proper geostrophic adjustment.

C-grid has been most popular for high-resolution atmosphere models, due to its superior dispersion behavior for inertial-gravity waves of the fluid with a relative large Rossby radius of deformation λ, such as high-resolution NWP models (Fig. 4-A). The staggered grid allows more accurate discretization of the divergence and height gradient terms. The C-grid's dispersion relation does not have local or global maxima (except for the highest wave numbers, when kd or ld equals π). However, for a small λ/d ratio, group velocities have wrong (negative) signs (Eq. 19) and phase speeds are much slower (Fig. 4-B) over the entire wave number plane. The biggest problem of the C-grid is its inferior dispersion behavior for inertial waves, which has a nonneglected impact on the simulation of high inertia modes when the fluid has a small λ/d ratio, as in the case of ocean models. On a C-grid, averaging operators are used to obtain the Coriolis terms and kinetic energy gradient term, since the u, v components of the velocity are not colocated. If λ is not significantly larger than the grid space d, the classic four-point averaging operator could have a nonzero null space, which can excite a checkerboard noise in horizontal solution fields. In a different way, use of simple two points average operator for kinetic energy, in discretization of vector invariant form of the momentum equations, leads to non-conservation of momentum, which can cause Hollingsworth instability (Hollingsworth et al., 1983).

Due to its popularity in atmosphere models, the C-grid has received much attention from the NWP modeling community, and researchers have been working on the numerical schemes to fix or alleviate their weakness. Arakawa and Lamb (1981) introduced an average enstrophy and energy conserving scheme for the C-grid, and it became widely used in regional grid models. Hollingsworth et al. (1983) proposed a modification to the scheme's average operator for kinetic energy term to make sure that in a discrete form, spatial derivatives $\frac{\partial u}{\partial y}$ and $\frac{\partial v}{\partial x}$ in different terms cancel out to avoid the instability while conserving its average

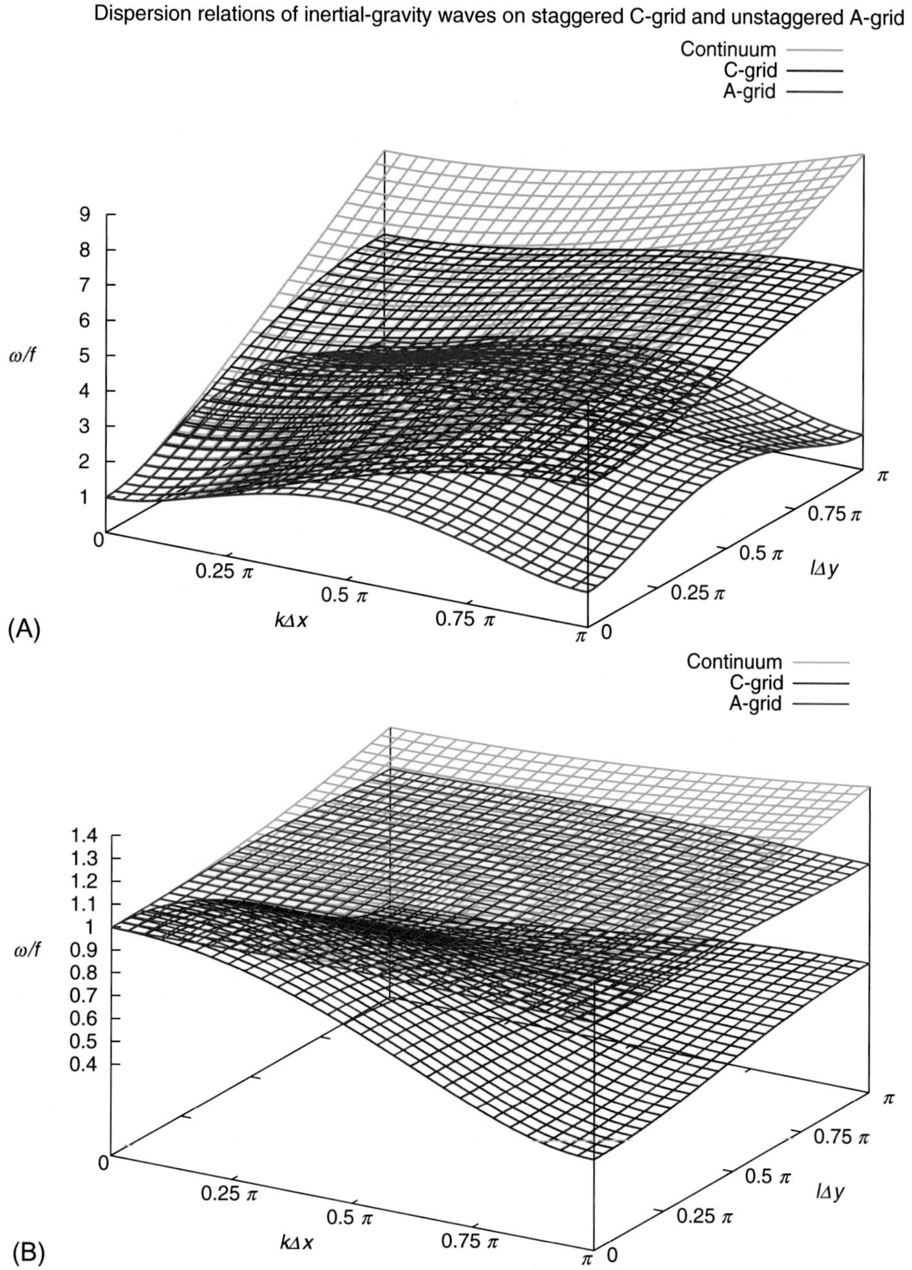

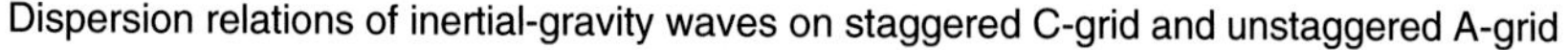

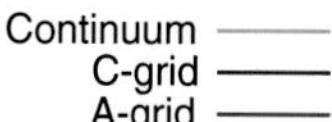

FIG. 4 Dispersion relation for wave solutions of 2D gravity-inertial wave equations. (A): $\lambda/d = 2$, (B): $\lambda/d = 0.2$, $d = \Delta x = \Delta y$.

enstrophy and energy. Adcroft et al. (1999) added tangential component of the velocity vector as an additional prognostic variable to obviate the need for the averaging operation for the Coriolis terms computation. Nechaev and Yaremchuk (2004) used 3 more independent prognosic equations for each velocity component to compute Coriolis terms to avoid averaging. Dobricic (2006) proposed to use an averaging operator of larger stencil so that it cancels the second-order term to form a fourth-order accurate approximation. Nechaev and Yaremchuk indicated that to improve the Coriolis terms in the C-grid, it is important to not only increase the accuracy but also eliminate or reduce the null-spaces associated with the averaging operators. McGregor (2005), on the other hand, suggested using the unstaggered A-grid, and only "transforming" the A-grid velocity vectors to the C-grid for calculation of the gravity wave terms.

The numerical techniques to alleviate the problems with Coriolis and kinetic energy gradient terms on the C-grid are more complex and difficult to design and implement on unstructured grids. While the discretization scheme with the most desired mimetic properties can be created, the scheme may not maintain the crucial numerical consistency on deformed grid cells, as we will see in an example in Section 5.

In proposing the Z-grid, Randall (1994) showed its superiority in simulating the geostrophic adjustment process for fluid with both large and small Rossby deformation radii. The Z-grid has an accuracy in simulating inertial oscillations that is as good as an A-grid, and gravity wave motions that is as good as a C-grid. The Z-grid model uses vorticity-divergence form of the shallow-water equations, and it requires to solve the elliptic equations every time step. That presents a computational challenge for massive parallel computation, and for ocean models which have irregular and nonperiodic domains. To overcome this difficulty, Ringler and Randall (2002) proposed an alternative, the ZM-grid for hexagonal grid (see Fig. 3). Similar to the CD-grid proposed by Adcroft et al. (1999), while it adds an extra velocity component to improve the dispersion behavior for the geostrophic adjustment process, the ZM-grid also introduces new computational modes that need to be damped in order for the model to run stably.

Ničković et al. (2002) and Thuburn (2008) studied dispersion relations on regular hexagonal grid. Nickovic analyzed various staggering schemes (Fig. 3), and found that hexagonal grid has, in general, gravity-inertial wave dispersion relations similar to or better than their rectangular counterparts due to the more isotropic geometry. Thuburn analyzed the regular hexagonal C-grid in detail, including various average operators, their impact on reducing nonzero geostrophic modes, and their energy and enstrophy conservation. Both authors found that the hexagonal C-grid schemes have an extra branch of Rossby modes, small scale modes of vorticity structures, which have much lower frequencies than the continuous counterpart. This incorrect dispersion relation was a major objection to the use of the icosahedral-hexagonal grid, which consists mostly of hexagonal Voronoi cells (Staniforth and Thuburn, 2012).

Most of the dispersion analyses are done for the linearized shallow-water equations. Konor and Randall (2018a, b) in their recent papers systematically studied three-dimensional (3D) linearized anelastic equations that are discretized on the A, C, D, CD, E, and Z horizontal grids, and the Lorenz (L) and Charney-Phillips (CP) vertical grids.

The modeling community has studied various grid staggerings for NWP models. At this point, no particular staggering and the associated discretization has been identified as perfect or optimal. Uncertainties in wave simulation, including errors in phase speed and group velocity, violations of conservation properties, and linear and nonlinear instabilities, exist in all commonly used grid staggerings and their discretizations, to various extents.

3 Discretization of the governing equation II

The advection equation (Eq. 2), or continuity equation of the shallow-water equations governs the large-scale quasi-nondivergent flow. Constructing an good advection scheme is one of the two key problems to accurately simulate the atmosphere and ocean circulations in the NWP modeling. A good advection (transport) scheme should have several numerical properties: mass conservation, high accuracy (physical veracity), monotonicity preservation (minimally sign preservation), and computational efficiency.

Connected to the horizontal mass advection is the advection or transport of various tracers. Strictly speaking, tracer transport scheme is not an integral part of the shallow-water equations discretization. However, since it is necessary to maintain the consistency between mass and tracer advection, and since active tracer transport is so closely coupled to the shallow-water equations, through the thermodynamic equation, we present some of the important advection schemes in NWP, that are used for both mass and tracer continuity equations.

In the previous section, we discussed temporal and spatial discretization schemes independently. For an advection scheme, in order to achieve these desired numerical properties stated earlier, it will be more appropriate to construct and analyze them together as a PDE solver in space-time plane. There are two common approaches that couple the spatial and temporal discretization to form advection schemes. One is based Taylor series expansion in time, using the advection equation to substitute the time derivatives with space derivatives and to form a one-step time-space coupled linear scheme. The other is the *method of lines* approach, which largely separates the spatial and temporal discretization, using multistage time stepping methods, such as the Runge-Kutta method, to couple the spatial discretization to form an advection scheme. The second approach is more suitable for the high-order monotonicity preserving advection schemes, because it is relatively easy to construct and analyze the numerical properties of the schemes.

The simplest advection scheme uses the Forward in Time Central in Space (FTCS) finite-difference method. The von Neumann stability analysis indicates the scheme is numerically unstable and numerical experiment shows it produces strong oscillations for flows with shock waves, the Gibbs effect mentioned in Section 1. A stable and popular finite-difference transport scheme, the Lax-Wendroff method, is constructed with a Taylor expansion approach, and it is second-order accurate in both time and space (Lax and Wendroff, 1960). The method, however, is not monotonicity preserving, and it includes a nonexplicitly controllable diffusion that may not be always desirable.

A concise description of the numerical task for the transport scheme is to compute the updated advection quantity for the cell i, q_i^{n+1}, from the quantity q_i^n, in the way that maintains the required accuracy, shape (sign) preservation, and conservation. We wish to create a transport scheme that meets these numerical requirements.

Enormous amounts of research and literature have been devoted to this subject—how to accurately and stably simulate the horizontal advection. Here, we highlight some of the frequently adopted, actively researched advection schemes in NWP models. These flux form shape preserving, or essentially oscillation free transport schemes are well suited for both structured and unstructured grids, and for both Eulerian and Lagrangian schemes.

3.1 Flux-corrected transport (FCT) schemes

The theory of the FCT algorithm was first developed by Boris and Book (1973). The basic idea is to compute the transportive flux in two stages, first, the advection-diffusion stage, and second, the antidiffusion stage. The solution is calculated using the fluxes from the two stages.

In the first stage, at each cell interface a low-order flux, which gives a monotonic (oscillation free), but more diffusive and low-accuracy solution, is computed; at this stage, using low-order flux at the interfaces, an updated low-order solution q_l could be computed.

In the second stage, at each cell interface a high-order flux, which gives a less diffusive, more accurate but potentially nonmonotonic solution, is computed, and then an antidiffusive flux, defined as the difference between the high-order flux and low-order flux, is obtained. A "correction" ("limit") is calculated for the antidiffusive flux, so that the high-order flux is used to the greatest extent, and the solution q_c, created with q_l and the corrected antidiffusive fluxes at interfaces, has a high-order of approximation and is free of new extrema.

Zalesak (1979) successfully improved and extended the FCT to multiple dimensions. Multidimensional FCT has become a popular numerical scheme to achieve high-order monotonic transport in numerical modeling. In application, multistage time difference schemes are often coupled with a flux-corrected divergence operator to form an FCT advection scheme. Zalesak (1979) used a two-stage leapfrog-trapezoidal time differencing for the proposed FCT scheme; the trapezoidal stepping is used every few time steps to damp the computational mode generated by leapfrog stepping. Skamarock and Gassmann (2011) constructed an advection scheme using FCT for unstructured grids. The scheme uses the third-order Runge-Kutta method (RK3) for time differencing, and the correction is applied in the last RK step.

3.2 Flux limiter transport schemes

Godunov (1959) proposed a conservative first-order transport scheme, donor-cell advection, which assumes piecewise constant over each cell and at each time step. Godunov's scheme is conservative and shape preserving, and computationally efficient. However, the scheme is considered too diffusive for NWP models.

To approximate the q at current time step more accurately, we can assume q varies linearly or even parabolically over the cell, which leads to second-order or higher-order schemes. To prevent any unphysical oscillation from high-order schemes, we require the scheme preserve monotonicity, which is equivalent to having total variation diminishing (TVD) (Harten, 1983). A numerical scheme is to be TVD if

$$TV(q^{n+1}) \leq TV(q^n), \tag{20}$$

where $TV(q) = \sum_{i=1}^{N} |q_i - q_{i-1}|$ is the total variation.

Godunov's theorem states any linear numerical schemes for PDEs that do not generate new extrema can be at most first-order accurate. The theorem precludes any possibility to find a high-order (second or higher) linear numerical scheme that is TVD.

From this perspective, introduction of a limiter function is necessary, since the nonlinear limiter that works with linear high-order numerical schemes makes it possible to be TVD.

The idea to construct a flux limiter for piecewise linear scheme is to limit the spatial derivatives (slopes) of each cell to a value so that the advection preserves monotonicity near the regions of shock wave while achieving second order of accuracy in the regions of smoothly varying q.

There are several ways to construct a piecewise linear scheme and different ways to define a limiter function. Among them, the Monotonic Upstream-centered Scheme for Conservation Laws (MUSCL) scheme proposed by van Leer (1977, 1979) is widely used as a high-order extension to the Godunov scheme. The MUSCL scheme defines a limiter $\phi(r_i)$, which is a function of ratio, r, of the two consecutive slopes, to limit the transported quantity at the cell boundary so that the solution of the transport scheme is TVD. The similar limiter can be applied to a piecewise parabolic scheme that reconstructs a third-order accurate solution.

The most broadly cited and adopted piecewise parabolic method (PPM) was by Colella and Woodward (1984). It essentially also grew out of the earlier work of van Leer on the MUSCL scheme. It is introduced to the atmospheric modeling in early 1990s (Carpenter et al., 1990), and has been used in weather and climate models. The method involves several steps: first, a linear distribution of each cell is determined by choosing the slope of the least magnitude among the left, middle, and right slopes; second, a first guess parabolas using two-cell average values and two slopes is created; the third step "steepens" the parabolas for the cells containing a discontinuity; the final step removes under- and overshoots by a monotonization procedure.

Because of its high accuracy and robustness in dealing with strong shock waves, the algorithmic complexity of PPM in implementation as well as the time expense to run PPM are relatively high. The third step of the method is expensive and may be unnecessary for atmospheric modeling. The final step tends to flatten smoothly varying physical extrema, which may be simplified and improved if minor over- and undershoots can be tolerated.

Limiters of higher-orders are available, such as Weighted Essentially Non-Oscillatory (WENO)-based limiters. As the name suggested, WENO schemes are not exactly oscillation free, i.e., they are not TVD or bound preserving. WENO schemes use a dynamic set of stencils, which is created with a nonlinear convex combination of approximations consisting of different stencils (Liu et al., 1994; Jiang and Shu, 1996). The major advantage of the WENO method is its high order of accuracy in smooth regions and its "essentially" nonoscillatory and sharp transition near discontinuity. The method has been extended to multidimensions, with a general framework for the smoothness indicators and nonlinear weights described in Jiang and Shu (1996). For a more recent review, see Shu (2009). The disadvantage of the WENO-method is its high computational expense due to the large stencil size, which also leads to deep halo depth that potentially impacts parallel efficiency. Because of its capability of achieving high-order accuracy, WENO limiters can be suitable for high-order finite (spectral) element methods.

Most 1D flux limiter transport schemes can be extended to multidimensions without significant changes of numerical properties. For piecewise linear second-order schemes, the 1D scheme can be directly extended to two dimensions by calculating each dimension separately. For piecewise parabolic (PPM) schemes, the extension to two dimensions can be obtained using piecewise biparabolic polynomial that are continuous at selected points along the cell interfaces. At the same time, since multidimensional Riemann problem is intractable, so it is preferred to formulate the scheme in 1D form and extend it to a multidimension scheme using a technique such as *Strang splitting*. The challenge for operator splitting is to appropriately symmetrize the cross-derivative in order to minimize the flow-to-grid line dependence and anisotropic distortion. Rančić (1992) extended Colella and Woodward's PPM scheme to two dimensions for use in a semi-Lagrangian scheme. Lin et al. (1996) introduced a general 2D scheme that uses MUSCL-type 1D schemes as building blocks. Hubbard (1999) extended MUSCL-type limiter to unstructured grids. Thuburn extended the Uniformly Third-Order Polynomial Interpolation Algorithm (UTOPIA) scheme, another flux limiter scheme (Leonard et al., 1993), to 2D unstructured grids (Thuburn, 1995, 1996).

The spatial discretization using flux limiters of the MUSCL method and the WENO limiters are often coupled with RK-type multi-stage time stepping methods. To ensure that TVD property for the spatial discretization is maintained in the high order temporal discretization, a TVD RK (aka Strong-Stability-Preserving (SSP) RK) scheme can be used (Gottlieb and Shu, 1998). In NWP modeling, Miura (2007, 2013) developed upwind-biased third-order schemes for the icosahedral grid with the multidimensional flux limiter by Thuburn (1996), and coupled spatial schemes with a forward Euler time stepping method and a RK3 time stepping method. Lunet et al. (2017) combined WENO limiters with explicit RK methods, including TVD RK, to form a transport scheme for flux form of wind. Zhang and Nair (2012) used the discontinuous Galerkin (DG) spatial discretization and RK3 time stepping to form a high-order advection scheme for a DG cubed-sphere shallow water model. To remove spurious oscillations, a variant of WENO limiter with a more compact stencil was implemented for the RKDG transport scheme.

3.3 Multidimensional positive definite advection transport algorithm (MPDATA)

While the methods discussed earlier were developed in the field of general computational fluid dynamics, MPDATA was developed for meteorological modeling (Smolarkiewicz, 1983, 1984). The method is more suitable for high Reynolds number and low Mach number atmospheric flows. The algorithm consists of multiple of donor-cell passes; the first pass obtains a first-order approximation; the second and subsequent passes increase accuracy by successively estimating and compensating the high-order truncation error of the previous pass. The standard MPDATA method is second-order accurate, sign-preserving, low in phase error, and computationally efficient. In addition, since MPDATA's second (and subsequent) donor-cell pass uses a diffusion velocity of opposite sign to reduce the diffusion and increase the accuracy, the scheme effectively limits the magnitude of the velocity vector as opposed to scalar flux component; thus, the scheme is dimension unsplit.

Since its introduction, MPDATA has evolved and enhanced to include many options for various applications in atmospheric model development. These options include a third-order accuracy option (Margolin and Smolarkiewicz, 1998), and full monotonicity option (Smolarkiewicz and Grabowski, 1990). The enhanced MPDATA can be applied to transport schemes on unstructured grids (Smolarkiewicz and Szmelter, 2005) and can be applied in curvilinear coordinate systems (Smolarkiewicz and Clark, 1986).

In general and in principle, any conservative, monotonicity preserving (or positive definite, sign preserving) advection schemes must contain some diffusive components. More accurate schemes control and contain the diffusion in the narrower regions of the strong shock waves, and maintain higher-order of accuracy transport for regions of smooth flows. Almost all the limits or corrections to the flux are applied locally, and thus the accuracy affected and error introduced should be locally as well. Numerical experiments confirmed that while the limiter has minimum influence on the accuracy in L_2 norm, it has significant impact on the accuracy in L_∞ norm (Miura, 2013). Numerical diffusion in a conservative shape preserving transport scheme is physically necessary, and this diffusion will impact the accuracy of transport scheme, and increase the uncertainties in the discretization of the shallow-water equations.

The time stepping for the mass and active tracer advection is synchronized with the momentum equation, thus the size of the time step is restricted by the CFL condition for all the characteristic speeds admitted by the underlying PDEs. To overcome the restriction, semi-Lagrangian transport schemes have been developed and applied together with semi-implicit schemes to the primitive equations (Staniforth and Côté, 1991; Robert, 1982). The semi-Lagrangian scheme calculates (a) the trajectories of the upstream air parcel's vertices that correspond to the vertices of the Eulerian grid cell at the next time step and (b) the average value of the upstream air parcel using the subgrid functions associated with Eulerian grid cells that overlap with the air parcel; this average value will be the average value of the Eulerian cell at the next time step. The subgrid function of each Eulerian cell is reconstructed with the aforementioned conservative, shape-preserving (or positive definite) schemes. Semi-Lagrangian transport could significantly increase the size of time step, since it is no longer CFL limited. However, semi-Lagrangian transport introduces discretization errors that are associated with the computation of trajectories and areas (volumes) of the upstream air parcels. These errors contribute to the uncertainties of advection simulation and put a limit on the size of time step.

4 Filtering, damping, and limiting techniques

Many numerical techniques that have been devised in the discretization of the shallow-water equations to reduce numerical noise and oscillation at different scales, improve the model's stability, and increase the computational efficiency. For systematical presentations and discussions, please see Durran (1999), Holton (1992), and Lauritzen et al. (2010). We list a few of the most commonly adopted techniques in modern modeling practice to illustrate the benefits and potential uncertainties they bring to the discretization of the shallow-water equations.

There are two types of diffusion, explicit and implicit. Explicit diffusion applies the diffusion operator directly to the prognostic variable to remove high frequency noise. Implicit diffusion is embedded in the numerical schemes that are designed to suppress nonphysical signals in the prognostic variables.

4.1 Horizontal diffusion

The explicit horizontal diffusion operator $\mathcal{D}_i$ takes the form of $(-1)^{i+1}K_{2i}\nabla^{2i}$, where K_{2i} is the diffusion coefficient, ∇^2 is a scalar or vector Laplacian operator, and $i=1,2,3,\ldots$.

The operator is appended to the right-hand side of the evolution equation and the process is referred to as $2i$th-order explicit diffusion.

$$\frac{\partial \psi}{\partial t} = \cdots \mathcal{D}_i \psi. \tag{21}$$

When $i > 1$, it is often called hyperdiffusion. The higher the order, the better the scale selectivity, i.e., the larger the wave number modes selected for dissipation, and the less the diffusion. The explicit horizontal diffusion effectively reduces high wave number noise and helps maintain the numerical stability. At the same time, it could also impact the accuracy of the solutions due to too much diffusion, causing energy spectrum of both rotational and divergent flow to decay too fast.

4.2 Divergence damping (2D)

The divergence damping term takes a form of $\mathbf{D}_i = (-1)^{i+1} v_{2i} \nabla (\nabla^{2i-1} \cdot \mathbf{u})$, where $2i$ is the order of the damping and v_{2i} is the spatial-time discretization scale-dependent diffusion coefficient. As in horizontal scalar diffusion, the vectorial damping term is appended to the right-hand side of the momentum equation. The divergence damping is used to damp gravity waves and prevent noise in the divergence field from contaminating the simulation. This damping is equivalent to adding the horizontal diffusion of the same order to the divergence equation, with a diffusion coefficient of v_{2i} (Eq. 22). Using the above definition, divergence damping can be conveniently added to the momentum equations in (u, v) form.

$$\frac{\partial D}{\partial t} = \cdots \mathcal{D}_i D. \tag{22}$$

Divergence damping targets specific dynamical modes, the divergent modes; it controls the spurious accumulation of grid scale energy for the purpose of maintaining numerical stability. For the high resolution NWP models, which are designed to retain important physical divergent modes, divergence damping should be used cautiously. A higher-order damping scheme should be used if the damping is necessary (Skamarock, 2004).

4.3 Smagorinsky horizontal diffusion

Smagorinsky (1963) proposed a nonlinear horizontal diffusion to prevent model from shear instabilities. The diffusion operator is defined as $\nabla \cdot (K_H \nabla \psi)$, with ψ being the component of flow field, and

$$\begin{aligned} K_H &= l_s^2 \left(T^2 + S^2\right)^{1/2}, \text{and} \\ T &= \frac{\partial u}{\partial x} - \frac{\partial v}{\partial y}, S = \frac{\partial u}{\partial y} + \frac{\partial v}{\partial x} \end{aligned} \tag{1}$$

$$l_s^2 = c^2 \cdot (\Delta x \Delta y). \tag{2}$$

Here, c, the 'Smagorinsky-constant', has been suggested to set in the range of [0.1, 0.3] (Smagorinsky, 1963; Skamarock et al., 2008); Δx, Δy are the grid spaces in the respective dimensions.

4.4 Shapiro filters

These are classic 1D symmetric low-pass filters (Shapiro, 1975, 2004) to remove $2\Delta x$ waves. They may be used to eliminate $2\Delta x$ noise caused by the A-grid discretization, and to damp high latitudinal (but not polar region) CFL unstable waves. Classic Shapiro filter is a five-tap symmetric digital filter. It is simple to implement and computationally inexpensive to apply. In a longitude-latitude grid, when application of the filtering is needed in lower latitudinal regions, it is necessary to use high-order filters to avoid over smoothing.

4.5 Polar spectral filtering

For global latitude-longitude models, a polar spectral filter is necessary. An fast Fourier transform (FFT) polar filter is commonly used to remove unstable short waves in the polar regions. The zonal data of an entire latitude circle are first transformed to Fourier coefficients using FFT; the coefficients for the wave numbers exceeding a prescribed threshold are changed to reflect the damping effect; finally, an inverse transform is applied to the modified coefficients to complete the procedure. The procedure is more expensive, especially on modern distributed parallel computer systems since nonlocal transforms are required.

4.6 Robert-Asselin time filtering

A Robert-Asselin filtering is often used together with the leapfrog time integration method to damp the computational mode occurred due to the different numerical effects at even and odd time steps (Robert, 1966; Asselin, 1972). The Robert-Asselin filter effectively damps the computational mode of the leapfrog time stepping method, but it also affects the physical mode, the phase, and amplitude of the resolved waves. The use of filter reduces the formal order of accuracy of the leapfrog method from second-order to first-order. Williams proposed a modification to the Robert-Asselin filter to reduce its negative impact on the physical mode and recovered accuracy to second-order, at the price of slight instability (Williams, 2009).

4.7 Lax-Wendroff advection method

Lax and Wendroff (1960) proposed a stable, second-order accurate scheme for horizontal advection. The scheme has an implicit damping of $2\Delta x$ waves embedded in it. Richtmeyer (1963) suggested use of the scheme for the purpose of suppression of the shortest waves. It was found that an intermittent application of Lax-Wendroff step at a rather long interval is sufficient to suppress the nonlinear instability (Kasahara, 1969). The problem with using Lax-Wendroff method to prevent instability is its lack of control of the implicit dissipation amount specifically. As mentioned in the previous section, Lax-Wendroff method is not a monotonicity preserving advection scheme. To achieve oscillation free advection, Lax-Wendroff method can be used together with a flux limiter, which potentially introduces further diffusion to the solution.

4.8 Shape preserving advection methods

The shape preserving and oscillation suppressing mechanisms for transport schemes described in Section 3 are forms of implicit diffusion. They are built into the advection schemes to avoid instabilities caused by oscillations that arise from numerically created new extrema. In other words, each of these transport schemes contains a diffusive component that is activated when it is numerically necessary. As discussed in the Section 3, the order of accuracy is reduced when the diffusion is applied, and how large the spatial and time domains that are impacted and by how much are determined by the specific scheme.

Numerical diffusion techniques are a necessary part of NWP models. They are required to remove unphysical signals, the computational artifacts, in order to better represent some unresolved physical process and to obtain a more physical and realistic solution. They are also required to damp high wave number computational modes that could cause local energy accumulation and lead to the instability of NWP models. However, application of these techniques inevitably removes some important physical signals and reduces the accuracies of the resolvable physical modes of the model, and therefore introduces additional uncertainties to the simulation.

5 Global models on unstructured grids

Global models have been constructed historically on latitude-longitude grid or Gaussian grid (for spectral models). Latitude-longitude grid is periodic in zonal dimension and evenly spaced in meridional dimension. It has a convenient Cartesian topology for most numerical schemes. Gaussian grid is similar to the latitude-longitude grid, except that in the meridional dimension, grid points are taken to be the numerically efficient Gaussian quadrature points. However, with the increase of horizontal and vertical resolutions, the Cartesian grids show their major weakness, the polar singularity and, for spectral models, the unscalable expensive spectral transform. To overcome these problems, the modeling community, in the recent years, start to develop global models on unstructured grids, quasi-uniform spherical grids, that are typically constructed based on regular polyhedrons.

Descartes' theorem states that if a polyhedron is homeomorphic to a sphere the total angular defect of the polyhedron is $2\pi\chi$, where χ is the Euler characteristic for the polyhedron. The Euler characteristic for convex polyhedron and sphere is 2. Thus, there is always a 4π total angular defect for polyhedron vertices.

The theorem is a special case of Gauss-Bonnet theorem, where all Gaussian curvature and geodesic curvature are concentrated discretely at the vertices. In case of cubed-sphere grid, each vertex of the cube is associated with three faces of $\pi/2$ angle; when all three faces deform into the sphere, they need to stretch $3 \times \pi/6 = \pi/2$ angle, which is called angular defect for the vertex. There are eight vertices for a cube, thus the total angular defect is 4π. For an icosahedron, there are 12 vertices, according to the theorem, each vertex will have $4\pi/12$ angular defect. Indeed, each vertex of an icosahedron is associated with five faces of $\pi/3$ angle, and a total of $5\pi/3$, resulting in an angular defect of $2\pi - 5\pi/3 = \pi/3$.

We could consider the classic latitude-longitude grid as a special degenerate case of convex polyhedron grid, where the polyhedron is degenerated into a line segment whose two end points (vertices) are north and south poles. In this sense, we can imagine that the 4π total angular defect are distributed to only these two vertices, and thus each vertex (the pole point) has a 2π angular defect.

It is conceivable and intuitively reasonable that we want to distribute the 4π total angular defect to more vertices so that at each vertex the deformation of the manifold is smaller, and so are the numerical uncertainties near each vertex. From this perspective, spherical grid based on an icosahedron, which has the most vertices among all recursively subdivisible regular convex polyhedrons, should be the first choice. On the other hand, spherical grid based on subdivisions of a cube has a convenient topology to adopt existing modeling technology, since each of the six panels has a Cartesian grid structure. In the following, we will discuss these two commonly used unstructured grids and the models implemented on them in some detail.

5.1 Icosahedral-hexagonal grid

Icosahedral grid generation starts from an icosahedron, which inscribes the unit sphere. Connecting the neighboring vertices with geodesic curves, we create the initial triangular tessellation of the sphere. Recursively bisecting the spherical triangle edges and connecting bisecting points, we obtain the icosahedral grid. Using the vertices of the icosahedral grid as generators, we create the hexagonal/pentagonal Voronoi regions (cells). Together, the combination of triangular mesh and the hexagonal/pentagonal mesh are referred to as the icosahedral-hexagonal/pentagonal grid (IHP grid) (Fig. 5A). The resolution of the IHP grid is often denoted by the number of recursive levels, like *Gn*. The left grid in Fig. 5 has a resolution of *G*4.

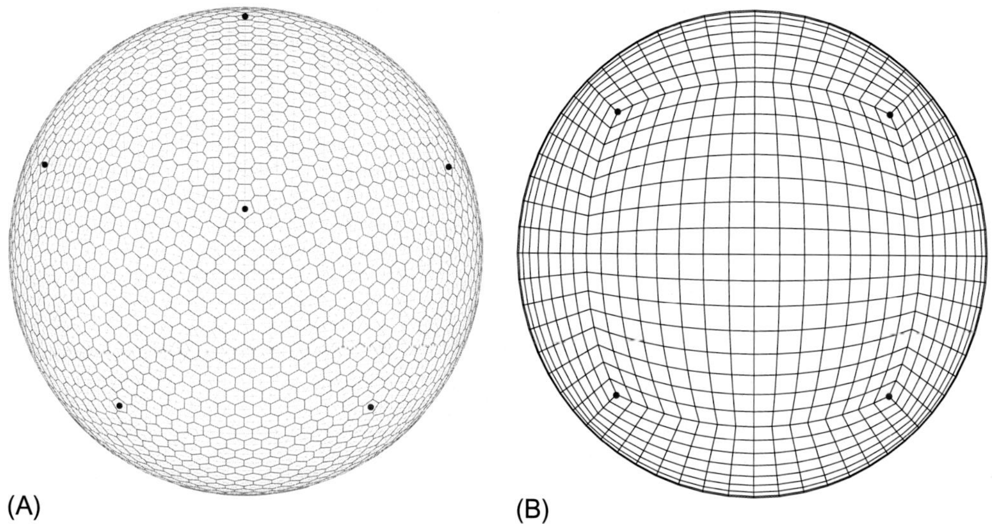

FIG. 5 The icosahedral-hexagonal grid and the cubed-sphere grid. *Left*: an icosahedral-hexagonal grid at G4 resolution (2562 cells). *Right*: a cubed-sphere grid at C16 resolution (1536 cells). *Blue dots* indicate the vertices of icosahedron and cube.

The IHP grid is largely uniform, and its Voronoi cells, when not in the vicinity of 12 pentagon cells, are largely isotropic. However, the icosahedral grid created using above classic construction has a few undesirable geometric properties, which lead to several undesirable numerical properties. Detailed geometric properties of the classic IHP grid and a few alternative constructions of the grid are analyzed and discussed in Wang and Lee (2011).

The IHP grids are adopted by several research institutes as well as operational centers for their community research models and operational forecast models, including Model for Prediction Across Scales (MPAS) at National Center for Atmospheric Research (NCAR) (Skamarock et al., 2012), Nonhydrostatic Icosahedral Atmospheric Model (NICAM) at Japan Agency for Marine-earth Science and TECHnology (JAMSTEC) (Tomita and Satoh, 2004; Satoh et al., 2008), the CSU model at the Colorado State University (Heikes and Randall, 1995a; Heikes et al., 2013), and Flow-following Icosahedral grid Model (FIM) at Earth System Research Laboratory (ESRL) (Lee and MacDonald, 2009; Bleck et al., 2014).

5.1.1 MPAS: A community global model

To improve the geometric properties of the IHP grid, numerous optimizations are proposed. Tomita et al. (2001, 2002) proposed spring dynamics algorithms, which iteratively minimize the overall tensions of the imaginary springs between adjacent icosahedral grid points. The optimization generally reduces the abrupt change of grid geometric properties near 12 pentagons and make the grid alignment smoother. Heikes and Randall (1995b) (HR95) proposed an algorithm to reduce r_e (Fig. 6) the distance between the midpoint of each Voronoi cell edge and the intersection of the edge and the line segment connecting the generators of the two corresponding Voronoi cells. HR95 algorithm reduces the overall error of the Poisson solver, since the derivative along the normal direction of each edge is defined at

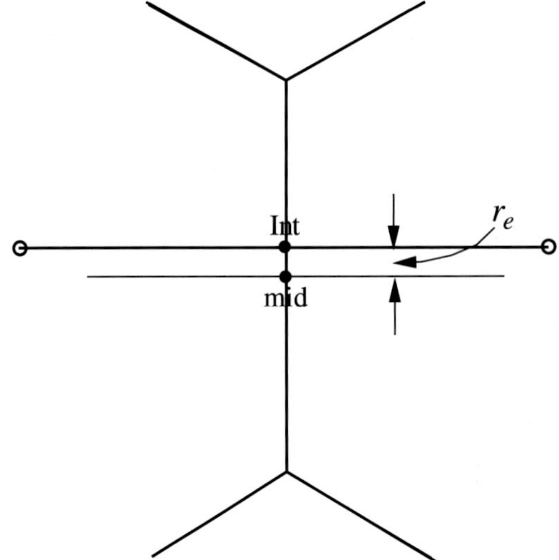

FIG. 6 Illustration of r_e, the distance between midpoint of the edge and intersection of the edge and the line segment connecting the two cells' generators.

the middle point. For the C-grid models on the IHP grid, r_e is also critical to its accuracy, because the original C-grid design is based on the assumption that dual grid cells intersect with primal grid cells symmetrically, and this assumption is not valid for the IHP grid. To improve the accuracy for finite-volume operators on the IHP grid, Du et al. (2003) proposed an iterative method to optmize the icosahedral grid such that each Voronoi cell's generator colocates with its centroid; this method is termed the IHP *centroidal Voronoi tessellation* (CVT). A comparison of different grid optimization methods and their impact on the geometric properties of the IHP grid has been documented in Miura and Kimoto (2005).

Based on the work for arbitrary orthogonal C-grid by Thuburn et al. (2009) and Ringler et al. (2010) (TRiSK method), MPAS is implemented on a CVT optimized IHP C-grid. The TRiSK method offers many desirable properties, including accurate representation of fast waves and geostrophic adjustment process, *f*-plane stationary geostrophic mode, curl-free pressure gradient, and energy conservation of pressure gradient and Coriolis terms. The main problem with TRiSK is its low order of accuracy, especially its failure to converge with some classic test cases (Peixoto, 2016; Yu et al., 2020).

Based on the geometric analysis of Voronoi tessellation and HR95 optimization, and accuracy analysis of the shallow-water model discretization on the optimized grids (Miura and Kimoto, 2005; Peixoto and Barros, 2011), Peixoto (2016) proposed some modifications to the TRiSK method. The main ideas proposed are to place the normal component of the prognostic velocity at the middle of the Voronoi cell edge, to linearly interpolate the height to the same position using barycentric coordinate, to reconstruct the kinetic energy term using second-order method, and to have an alternative way to compute the tangential component of the velocity. Peixoto's scheme achieves first-order accuracy in L_∞ norm at the expense of some mimetic properties of TRiSK, such as total energy conservation and the steady geostrophic modes on the *f*-plane.

Dispersion analysis for hexagonal grid indicates that IHP C-grid supports an extra branch of spurious Rossby modes (cf. Section 2.3). This nonphysical dispersion relation leads to erroneous grid-scale potential vorticity oscillations. In TRiSK, an Anticipated Potential Vorticity Method (APVM) (Sadourny and Basdevant, 1985) was used to remove the grid-scale noise. Weller (2012) proposed to use a continuous, linear-upwind stabilized transport (CLUST) scheme, together with a new diagnostic that better detects the grid-scale oscillations, in place of APVM to more discriminatingly remove the grid-scale oscillations.

The TRiSK method suffers from Hollingsworth instability due to reasons similar to those in the discretization of the Cartesian C-grid, the energy-enstrophy conserving scheme in the TRiSK method is momentum nonconserving. Gassmann (2013, 2018) discussed and proposed changes to the TRiSK method to reduce and remove this instability.

5.2 Cubed-sphere grid

The cubed-sphere grid was first introduced by Sadourny in the 1970s (Sadourny, 1972), and it is created by projecting an "equidistant" grid on each face of the inscribed cube to the sphere (Fig. 5B). In his pioneering work, Sadourny formulated the shallow-water equations and a corresponding conservative finite-difference scheme on the cubed-sphere grid, using covariant and contravariant components of the velocity vector, which is represented in a local nonorthogonal coordinate system defined at each cell of the cubed-sphere grid. This formulation is still used in modern cubed-sphere NWP models, including FV3 (Putman and Lin, 2007). Ronchi et al. (1996) advocated a different type of the cubed-sphere grid, an "equiangular" grid, which consists of a family of grid lines equally spaced in angle. Both grids are gnomonic grids, named after an old map projection, gnomonic projection, which projects a geodesic on the sphere to a straight line on the projection plane. Apparently, the equiangular grid is more uniform. In addition, the equiangular grid has an extra computational bonus that grid point extended to the neighboring panels can be interpolated along one grid line (Purser, 2018).

Rančić et al. (1996) compared both cubed-sphere grids and concluded that, due to the loss of accuracy at the edges of the cubed-sphere panels, equiangular grid, which has a more uniform grid spacing within each panel, might not be an improvement. Putman and Lin (2007) proposed and compared various cubed-sphere grids that have been optimized for transport schemes.

The cubed-sphere grid has been adopted by several global models, including FV3, the choice of the Next Generation Global Predication System (NGGPS) at National Centers for Environment Predication (NCEP), the Navy Environmental Prediction sysTem Utilizing the NUMA corE (NEPTUNE) model at Naval Research Laboratory (NRL), and Korean Integrated Model (KIM) at Korea Institute of Atmospheric Prediction Systems (KIAPS) (Hong et al., 2018). The latter two are spectral element models.

5.2.1 FV3: NGGPS model at NCEP

FV3 is evolved from the finite-volume dynamical core of a NASA/Goddard Space Flight Center (GSFC) transport model, and a subsequent shallow-water model by Lin and Rood (1997). This shallow-water model is a full finite-volume implementation using high-order monotonic advection. The full model is based on Lin and Rood transport algorithm using the PPM method of Lin et al. (1996) and Colella and Woodward (1984), Lin and Rood (1997) shallow-water scheme, Lin's pressure gradient force finite-volume method (Lin, 1997), and Lin's vertically Lagrangian scheme for vertical discretization (Lin, 2004).

For the shallow-water model, FV3 emphasizes the mimetic property of vorticity generation, so a prognostic D-grid is used. To compute the time mean fluxes, the C-grid wind components are first interpolated from the D-grid wind, as with most staggered grid methods, and then advanced a half time step; this time advancement for the C-grid velocity component uses the gradient of height which is updated with the divergence interpolated from cell corners. Diagnostically computing the C-grid wind avoids computational mode caused by additional momentum equation for the extra prognostic variable. FV3 developers believe that this CD-grid formulation achieves similar results as the Z-grid or true CD-grid (with both the C-grid and the D-grid wind components being prognostic variables), but without the computational complexity of solving a global elliptic equation, or integrating an extra momentum equation and damping its computational mode. However, modeling community has some different opinions about this assessment.

Skamarock (2008) conducted a linear analysis of a similar CD-grid algorithm in the NCAR's CCSM's dynamical core for its dispersion behavior, and found that it does not behave as a Z-grid or a true CD-grid, rather it behaves basically as a D-grid, which is not a recommended staggering scheme by Arakawa and Lamb (1977) due to its poor dispersion behavior. In their analysis for the linearized anelastic equations on Cartesian grids, Konor and Randall (2018b) also came to the same conclusion for the CD-grid with the D-grid prognostic variables. The linear analysis of the FV3 shallow-water model shows that it has a D-grid like dispersion behavior, which is caused by considerable averaging used in computing height gradient, divergence, and Coriolis terms. In addition, to compute the nonlinear kinetic energy term, FV3 CD-grid method also needs more averaging operations, compared with the C-grid. The use of PPM transport scheme helps damp the $2\Delta x$ modes; when it is not adequate, a divergence damping is applied to suppress the grid scale noise. As discussed in Section 4.2, for modern high resolution NWP models, divergence damping could also suppress the meaningful divergent modes, and thus it should be only used when necessary.

The convenient topological structure within each panel of the cubed-sphere grid makes it easy to adopt Cartesian grid algorithms and schemes, such as high-order, monotonicity constrained transport schemes, using a curvilinear formulation. At the same time, however, every edge of the six panels of the cubed-sphere grid becomes a discontinuity zone for these algorithms. The order of the accuracy drops at these edges and special treatments are required to maintain the validity of the algorithms and schemes. Special treatments at the edges and corners of the six panels make parallelization of the model more difficult, and the resultant parallel code less efficient.

5.3 Some remarks on unstructured grid

We believe that fundamentally there are two key geometric properties for quasi-uniform and unstructured grids on the sphere: grid space uniformity and grid cell shape regularity. More uniform grid can have a larger time step for explicit Eulerian schemes, since for these schemes the maximum time step is constrained by the minimum grid space in the entire grid. More regular cell shape implies almost all the geometric properties we have been seeking. For example, a more regular IHP grid will minimize the maximum and average r_e (Fig. 6), which is critical to the numerical accuracies of the C- and Z-grid discretization on the IHP grid; a more regular cubed-sphere grid increases the accuracy of transport scheme and reduces grid imprint at the vertices (and edges) of the six panels.

For parallel computing, unstructured spherical grids can be naturally divided into desired numbers of computational domains, which are compact in shape (thus a small halo region) and even in grid cell numbers (thus a balanced workload). Therefore unstructured grids are better suited for the future ultra-high resolution weather simulation. On the other hand, the unstructured grids have several weaknesses. First, they no longer have the simple grid structure and grid lines orthogonality of the Cartesian grid. This implies that some efficient Cartesian grid algorithms cannot be adopted. Second, grid line alignment and topology have a discontinuity near the vertices of the starting polyhedrons. This implies that treatments for the special cases are required; these treatments will affect the parallel efficiency. Third, due to the inevitable geometric irregularity to achieve the similar mimetic discretization of Cartesian grid, more sophisticated algorithms have to be used. These algorithms are more expensive in computation because of their higher algorithmic complexity and increased communication cost due to the enlarged stencils.

In terms of discrete uncertainty, for a given grid staggering, the unstructured grids have more difficulties to achieve the same order of accuracy while maintain a mimetic discretization. Fundamentally, interpolations are required to reduce the truncation error for the unstructured grids, because some prognostic variables are no longer colocated due to grid cell irregularity; on the other hand, to maintain a mimetic discretization one needs to avoid any additional interpolations for the discretization scheme.

6 Summary

The most important properties of discretization of an initial value problem are consistency, stability, and convergence. The following is a simple definition for these properties:

- *Consistency*: The discretized PDE becomes the exact continuous PDE as the grid size approaches zero, i.e., local truncation error vanishes.
- *Stability*: Numerical errors generated during the process of solving the discretized equations are not amplified.
- *Convergence*: Numerical solution becomes arbitrarily close to the exact solution of the PDE as the grid size approaches zero.

The Lax-Richtmyer Equivalence Theorem states that linear numerical methods for well-posed, linear initial value problems have the following relationship:

$$\textit{Consistency} + \textit{Stability} \Leftrightarrow \textit{Convergence}.$$

The end goal of a numerical simulation is its convergence, which assures that the uncertainties of the simulation are reduced with the increase of resolution. Therefore, we must first make sure that the adopted numerical methods are consistent and stable. Under these two fundamental conditions, we seek more efficient and accurate methods, that is, the methods that can converge faster at the same resolution.

Lax-Richtmyer theorem only applies to linear schemes for linear PDEs. On the other hand, the theorem is quite general, making no assumption about the type of differential equations; it applies to both linear advection and diffusion equations. NWP is not a linear initial value problem, and the shallow-water equations is not linear either. Nonetheless, the linear part of the equations are most significant, and thus the theorem still provides us with a basic guideline to design and implement numerical schemes for the equations.

Consistency is relatively straightforward to achieve and verify. The numerical methods for a given equation have to be designed and implemented with this property, and otherwise we solve the wrong equation. Stability thus is fundamental for time integration, and critial for weather simulation in NWP. The motions of atmosphere air and ocean current are highly nonlinear and even chaotic. The PDEs that govern these motions are nonlinear and chaotic in nature as well, and their solutions are superpositions of many interacting waves and eddies of different wavelengths and velocities. To reduce the uncertainties of simulation, one must use either a low-order scheme at high spatial resolution, or a high-order scheme, for the most nonlinear processes. Both approaches face the challenge of instability.

For explicit time stepping schemes, linear instability is commonly caused by the violation of Courant-Friedrichs-Levy (CFL) condition, that is, a too large time step for the spatial resolution. To alleviate the constraints due to the CFL conditions for fast-moving waves, implicit or semi-implicit formulations of time stepping can be used. For horizontal advection, a semi-Lagrangian scheme can be used to overcome the CFL constraints that arise from the scheme based on Eulerian reference frame. The semi-Lagrangian scheme is especially effective for high-latitude regions in a global Cartesian grid, where Δx (grid space in longitudinal dimension) are too small for most Eulerian schemes to operate efficiently, or for any high-aspect ratio grid whose CFL condition is determined by the "smaller" dimension.

Within the stability region, significantly increased time step could decrease the accuracy of the solution. For example, as the dispersion relation indicates, using semi-implicit method with large time step could result in significantly reduced phase speeds for large wave numbers (Durran, 1999). For semi-Lagrangian schemes, too large time step will cause the departing region to deform so much that the computational errors in approximating the area and average value of the departing region are too large, making the transport scheme unusable.

Certainly, we can reduce the size of time step if necessary; however, a smaller time step implies not only higher computational expense, but also larger accumulated truncation errors, assuming that spatial discretization error dominates the overall error of the simulation.

The nonlinear instability, as the name suggests, comes from the nonlinear terms of the primitive equations for the NWP models. It is usually caused by the erroneous computational energy cascade. The nonlinear instability is not alleviated by the time step reduction. There are several ways to remove or reduce nonlinear instability. The simple and straightforward way is to remove high wave number components from the numerical solution explicitly with filters or implicitly using a differencing scheme that has a built-in damping mechanism for the shortest waves. A more elegant approach could be to control systematic energy cascade into the higher wave numbers by means of conserving average potential enstrophy and energy (Arakawa and Lamb, 1981). However, conservation of energy and potential enstrophy will not stop cascading of either to higher wave numbers entirely. Some dissipation of high wave number energy are still required in order for the simulation to remain stable.

There are debates about whether high order methods should play a larger role in atmospheric models, and in NWP models in particular. High-order numerics helps to reduce the uncertainties in simulating intrinsically nonlinear advection process most relevant to the local weather forecast. High accuracies of the large magnitude acceleration terms, such as Coriolis force and pressure gradient terms, help to achieve an accurate simulation of the geostrophic adjustment process. On the other hand, high-order methods can also introduce unphysical computational modes, especially high frequency oscillations, which needs to be controlled or damped in order to maintain the stability of simulation. The methods that remove the instability of simulation modify the numerical solution. While the undesirable computational modes are suppressed, meaningful physical modes are impacted as well. In addition, higher order methods are usually associated with a high computational complexity and a large computational stencil. The former is always undesirable for timely NWP requirement and the latter can be a challenge for the modern parallel computing, due to the increased communication cost.

To construct an accurate, stable dynamical core for NWP model that scales well on modern parallel computer systems is still a great challenge for the modeling community. It requires the combined efforts of researchers from atmospheric and oceanic dynamics, computational fluid dynamics, applied mathematics, computer science, and software engineering to tackle this challenge successfully.

Acknowledgments

Thanks to Dr. D. Rosenberg (Cooperative Institute for Research in the Atmosphere, Colorado State University) for his detailed review, insightful discussion, and constructive comments.

References

Adcroft, A.C., Hill, C., Marshall, J., 1999. A new treatment of the Coriolis terms in C-grid model at both high and low resolutions. Mon. Weather Rev. 127, 1928–1936.

Arakawa, A., 1970. Numerical Simulation of Large-Scale Atmospheric Motions. Department of Meteorology, Los Angeles, pp. 24–40 Originally published Symposium in Applied Mathematics (1968: Durham, NC) Numerical solution of field problems in continuum physics. Providence, RI: American Mathematical Society, 1970.

Arakawa, A., Lamb, V.R., 1977. Computational design of the basic dynamical processes of the UCLA general circulation model. Methods Comput. Phys. 17, 172–256.

Arakawa, A., Lamb, V.R., 1981. A potential enstrophy and energy conserving scheme for the shallow water equations. Mon. Weather Rev. 109, 18–36.

Asselin, R., 1972. Frequency filter for time integrations. Mon. Weather Rev. 100, 487–490.

Bleck, R., et al., 2014. A vertically flow-following, icosahedral-grid model for medium-range and seasonal prediction. Part 1: model description. Mon. Weather Rev. 143, 2386–2403 (in review).

Boris, J.P., Book, D.L., 1973. Flux-corrected transport. I. SHASTA, a fluid transport algorithm that works. J. Comput. Phys. 11, 38–69.

Carpenter, R., Droegemeier, K.K., Woodward, P., Hane, C.E., 1990. Application of the piecewise parabolic method (PPM) to meteorological modeling. Mon. Weather Rev. 118, 586–612.

Choi, S.J., Hong, S.Y., 2016. A global non-hydrostatic dynamical core using the spectral element method on a cubed-sphere grid. Asia-Pac. J. Atmos. Sci. 52, 291–307.

Choi, S.J., Giraldo, F.X., Shin, S., 2014. Verification of a nonhydrostatic dynamical core using horizontally spectral element vertically finite difference method: 2-D aspects. Geosci. Model Dev. 7, 2717–2731.

Colella, P., Woodward, P., 1984. The piecewise parabolic method (PPM) for gas-dynamical simulations. J. Comput. Phys. 54, 174–201.

Cullen, M.J.P., 2001. Alternative implementations of the semi-Largrangian semi-implicit schemes in the ECMWF model. Q. J. R. Meteorol. Soc. 127, 2787–2802.

Dobricic, S., 2006. An improved calculation of Coriolis terms on the C-grid. Mon. Weather Rev. 134, 3764–3773.

Du, Q., Gunzburger, M., Ju, L., 2003. Constrained centroidal Voronoi tessellations for surfaces. SIAM J. Sci. Comput. 24, 1488–1506.

Durran, D.R., 1991. The third-order Adams-Bashforth method: an attractive alternative to leapfrog time differencing. Mon. Weather Rev. 119, 702–720.

Durran, D.R., 1999. Numerical Methods for Fluid Dynamics: With Applications to Geophysics. Springer, New York.

ECMWF, 2016. IFS Documentation CY41R2. ECMWF, Reading, UK NA-91-07.

Gassmann, A., 2013. A global hexagonal C-grid non-hydrostatic dynamical core (ICON-IAP) designed for energetic consistency. Q. J. R. Meteorol. Soc. 139, 152–175.

Gassmann, A., 2018. Discretization of generalized Coriolis and friction terms on the deformed hexagonal C-grid. Q. J. R. Meteorol. Soc. 144, 2038–2053.

Girado, F.X., Kelly, J.F., Constantinescu, E., 2013. Implicit-explicit formulations of a three-dimensional non-hydrostatic unified model of the atmosphere (NUMA). SIAM J. Sci. Comput. 35, B1162–B1194.

Godunov, S.K., 1959. A difference scheme for numerical calculation of discontinuous solutions of the hydrodynamic equations. Matematichaskiy Sbornik 47 (3), 271–306.

Gottlieb, S., Shu, C.-W., 1998. Total variation diminishing Runge-Kutta schemes. Math. Comput. 67 (221), 73–85.

Harten, A., 1983. High resolution schemes for hyperbolic conservation laws. J. Comput. Phys. 49, 357–393.

Heikes, R.P., Randall, D.A., 1995a. Numerical integration of the shallow-water equations on a twisted icosahedral grid. I. Basic design and results of tests. Mon. Weather Rev. 123, 1862–1880.

Heikes, R.P., Randall, D.A., 1995b. Numerical integration of the shallow-water equations on a twisted icosahedral grid. II. A detailed description of the grid and analysis of numerical accuracy. Mon. Weather Rev. 123, 1881.

Heikes, R.H., Randall, D.A., Konor, C.S., 2013. Optimized icosahedral grids: performance of finite-difference operators and multi-grid solver. Mon. Weather Rev. 141, 4450–4469.

Hollingsworth, A., Kållberg, P., Renner, V., Burridge, M., 1983. An internal symmetric computational instability. Q. J. R. Meteorol. Soc. 109, 417–428.

Holton, J.R., 1992. An Introduction to Dynamic Meteorology, 3rd Academic Press, New York.

Hong, S.Y., Kwon, Y.C., Kim, T., et al., 2018. The Korean Integrated Model (KIM) system for global weather forecasting. Asia-Pac. J. Atmos. Sci. 54, 267–292.

Hubbard, M.E., 1999. Multidimensional slope limiters for MUSCL-type finite volume schemes on unstructured grids. J. Comput. Phys. 155, 54–74.

Jiang, G., Shu, C.W., 1996. Efficient implementation of weighted ENO schemes. J. Comput. Phys. 126, 202–228.

Kasahara, A., 1969. Computer simulation of the earth's atmosphere. NCAR Manuscript 70-22, 16 pp National Center for Atmospheric Research, Boulder, CO.

Konor, C.S., Randall, D., 2018a. Impacts of the horizontal and vertical grids on the numerical solutions of the dynamical equations—part 1: nonhydrostatic inertia-gravity modes. Geosci. Model Dev. 11, 1753–1784.

Konor, C.S., Randall, D., 2018b. Impacts of the horizontal and vertical grids on the numerical solutions of the dynamical equations—part 2: quasi-geostrophic Rossby modes. Geosci. Model Dev. 11, 1785–1797.

Kopera, M.A., Giraldo, F.X., 2014. Analysis of adaptive mesh refinement for IMEX discontinuous Galerkin solutions of the compressible Euler equations with application to atmospheric simulations. J. Comput. Phys. 275, 92–117.

Lauritzen, P.H., Jablonowski, C., Tayor, M.A., Nair, R.D., 2010. Numerical Techniques for Global Atmospheric Models. Springer, New York, NY.

Lax, P.D., Wendroff, B., 1960. Systems of conservation laws. Commun. Pure Appl. Math. 13, 217–237.

Lee, J.L., MacDonald, A.E., 2009. A finite-volume icosahedral shallow water model on local coordinate. Mon. Weather Rev. 137, 1422–1437.

Leonard, B., MacVean, M., Lock, A., 1993. Positivity-preserving numerical schemes for multidimensional advection. NASA STI/Recon Tech. Rep. N 93, 27091.

Lin, S.J., 1997. A finite-volume integration method for computing pressure gradient force in general vertical coordinates. Q. J. R. Meteorol. Soc. 123, 1749–1762.

Lin, S.J., 2004. A vertically Lagrangian finite-volume dynamical core for global models. Mon. Weather Rev. 132, 2293–2307.

Lin, S.J., Rood, R.B., 1996. Multidimensional flux-form semi-Lagrangian transport schemes. Mon. Weather Rev. 124, 1749–1762.

Lin, S.J., Rood, R.B., 1997. An explicit flux-form semi-Lagrangian shallow-water model on the sphere. Q. J. R. Meteorol. Soc. 123, 2477–2498.

Liu, X.D., Osher, S., Chan, T., 1994. Weighted essentially non-oscillatory schemes. J. Comput. Phys. 115, 200–212.

Lunet, T., Lac, C., Auguste, F., Visentin, F., Masson, V., Escobar, J., 2017. Combination of WENO and explicit Runge–Kutta methods for wind transport in the Meso-NH model. Mon. Weather Rev. 145, 3817–3838.

Margolin, L., Smolarkiewicz, P.K., 1998. Antidiffusive velocities for multipass donor cell advection. SIAM J. Sci. Comput. 20, 907–929.

McGregor, J.L., 2005. Geostrophic adjustment for reversibly staggered grids. Mon. Weather Rev. 133, 1119–1128.

Miura, H., 2007. An upwind biased conservative scheme for spherical hexagonal-pentagonal grids. Mon. Weather Rev. 135, 4038–4044.

Miura, H., 2013. An upwind-biased conservative transport scheme for multistage temporal integrations on spherical icosahedral grids. Mon. Weather Rev. 141, 4049–4068.

Miura, H., Kimoto, M., 2005. A comparison of grid quality of optimized spherical hexagonal-pentagonal geodesic grids. Mon. Weather Rev. 123, 2817–2833.

Nechaev, D., Yaremchuk, M., 2004. On the approximation of the coriolis terms in C-grid models. Mon. Weather Rev.. 132.

Ničković, S., Gavrilov, M.B., Tosic', I.A., 2002. Geostrophic adjustment on hexagonal grids. Mon. Weather Rev. 130, 668–683.

Peixoto, P.S., 2016. Accuracy analysis of mimetic finite volume operators on geodesic grids and a consistent alternative. J. Comput. Phys. 310, 127–160.

Peixoto, P.S., Barros, S.R.M., 2011. Analysis of grid imprinting on geodesic spherical icosahedral grids. J. Comput. Phys. 237, 61–78.

Purser, R.J., 2018. Möbius Net Cubed-Sphere Gnomonic Grids. ON496. NCEP, Camp Springs, MD.

Putman, W.M., Lin, S.J., 2007. Finite-volume transport on various cubed-sphere grids. J. Comput. Phys. 227, 55–78.

Randall, D., 1994. Geostrophic adjustment and the finite-difference shallow-water equations. Mon. Weather Rev. 122, 1371–1377.

Rančić, M., 1992. Semi-Lagrangian piecewise biparabolic scheme for two-dimensional horizontal advection of a passive scalar. Mon. Weather Rev. 120, 1394–1406.

Rančić, C., Purser, J., Messinger, M., 1996. A global shallow water model using an expanded spherical cube: gnomonic versus conformal coordinates. Q. J. R. Meteorol. Soc. 122, 959–982.

Richtmeyer, R.D., 1963. A survey of difference methods for non-steady fluid dynamics. NCAR Tech. Rep. 63-2, 25 pp National Center for Atmospheric Research.

Ringler, T.D., Randall, D., 2002. The ZM grid: an alternative to the Z grid. Mon. Weather Rev. 130, 1411–1422.

Ringler, T.D., Heikes, R.P., Randall, D.A., 2000. Modeling the atmospheric general circulation using a spherical geodesic grid: a new class of dynamical cores. Mon. Weather Rev. 128, 2471–2490.

Ringler, T.D., Thuburn, J., Klemp, J.B., Skamarock, W.C., 2010. A unified approach to energy conservation and potential vorticity dynamics for arbitrarily structured C-grids. J. Comput. Phys. 229, 3065–3090.

Robert, A., 1966. The integration of a low order spectral form of the primitive meteorological equations. J. Meteorol. Soc. Jpn 44, 237–245.

Robert, A., 1982. A semi-Lagrangian and semi-implicit numerical integration scheme for the primitive meteorological equation. J. Meteorol. Soc. Jpn 60, 319–325.

Ronchi, C., Iaconoi, R., Paolucci, P.S., 1996. The "Cubed-Sphere:" a new method for the solution of partial differential equations in spherical geometry. J. Comput. Phys. 124, 93–114.

Sadourny, R., 1972. Conservative finite-difference approximations of the primitive equations on quasi-uniform spherical grids. Mon. Weather Rev. 100, 136–144.

Sadourny, R., Basdevant, C., 1985. Parameterization of subgrid scale barotropic and baroclinic eddies in quasi-geostrophic models: anticipated potential vorticity method. J. Atmos. Sci. 42 (13), 1353–1363.

Sandbach, S., Thuburn, J., Vassilev, D., 2015. A semi-implicit version of the MPAS-atmosphere dynamical core. Mon. Weather Rev. 143, 3838–3855.

Satoh, M., Matsuno, T., Tomita, H., Miura, H., Nasuno, T., Iga, S., 2008. Nonhydrostatic Icosahedral Atmospheric Model (NICAM) for global cloud resolving simulations. J. Comput. Phys. 227, 3486–3514.

Schneider, K., Kolomenskiy, D., Deriaz, E., 2013. Is the CFL condition sufficient? Some remarks. In: de Moura, C.A., Kubrusly, C.S. (Eds.), The Courant-Friedrichs-Lewy Condition.pp. 139–146.

Sela, J.G., 2009. The Implementation of the Sigma Pressure Hybrid Coordinate into the GFS. NCEP, Camp Springs, MD ON461.

Sela, J.G., 2010. The Derivation of Sigma Pressure Hybrid Coordinate Semi-Lagrangian Model Equations for GFS. NCEP, Camp Springs, MD ON462.

Shapiro, R., 1975. Linear filtering. Math. Comput. 29 (132), 1094–1097.

Shapiro, R., 2004. The use of linear filtering as a parameterization of atmospheric diffusion. J. Atmos. Sci. 28, 523–531.

Shu, C.W., 2009. High order weighted essentially non-oscillatory schemes for convection dominated problems. SIAM Rev. 51, 82–126.

Skamarock, W.C., 2004. Evaluate mesoscale NWP models using kinetic energy spectra. Mon. Weather Rev. 132, 3019–3032.

Skamarock, W.C., 2008. A linear analysis of the NCAR CCSM finite-volume dynamical core. Mon. Weather Rev. 136, 2112–2119.

Skamarock, W.C., et al., 2008. A description of the advanced research WRF version 3. NCAR/TN.

Skamarock, W.C., Gassmann, A., 2011. Conservative transport schemes for spherical geodesic grids: high-order flux operators for ODE-based time integration. Mon. Weather Rev. 139, 2962–2975.

Skamarock, W.C., Klemp, J.B., Duda, M.G., Fowler, L.D., Park, S.H., Ringler, T.D., 2012. A multiscale nonhydrostatic atmospheric model using centroidal voromoi tesselations and C-grid staggering. Mon. Weather Rev. 140, 3090–3105.

Smagorinsky, J., 1963. General circulation experiments with the primitive equations. I. The basic experiment. Mon. Weather Rev. 91, 99–164.

Smolarkiewicz, P.K., 1983. A simple positive definite advection scheme with small implicit diffusion. Mon. Weather Rev. 111, 479–486.

Smolarkiewicz, P.K., 1984. A fully multidimensional positive definite transport algorithm with small implicit diffusion. J. Comput. Phys. 54, 325–362.

Smolarkiewicz, P.K., Clark, T.L., 1986. The multidimensional positive definite transport algorithm: further development and applications. J. Comput. Phys. 67, 396–438.

Smolarkiewicz, P.K., Grabowski, W.W., 1990. The multidimensional positive definite advection transport algorithm: nonoscillatory option. J. Comput. Phys. 86, 355–375.

Smolarkiewicz, P.K., Szmelter, J., 2005. MPDATA: an edge-based unstructured-grid formulation. J. Comput. Phys. 206, 624–649.

Smolarkiewicz, P.K., Kuhnlein, C., Wedi, N.P., 2014. A consistent framework for discrete integrations of soundproof and compressible PDEs of atmospheric dynamics. J. Comput. Phys. 263, 185–205.

Staniforth, A., Côté, J., 1991. Semi-Lagrangian integration schemes for atmospheric models—a review. Mon. Weather Rev. 119, 2206–2223.

Staniforth, A., Thuburn, J., 2012. Horizontal grids for global weather and climate prediction models: a review. Q. J. R. Meteorol. Soc. 138 (662), 1–26.

Thuburn, J., 1995. Dissipation and cascades to small scales in numerical models using a shape-preserving advection scheme. Mon. Weather Rev. 123, 1888–1903.

Thuburn, J., 1996. Multidimensional flux-limited advection schemes. J. Comput. Phys. 123, 74–83.

Thuburn, J., 2008. Numerical wave propagation on the hexagonal C-grid. J. Comput. Phys. 227, 5836–5858.

Thuburn, J., Ringler, T.D., Skamarock, W.C., Klemp, J.B., 2009. Numerical representation of geostrophic modes on arbitrarily structured C-grids. J. Comput. Phys. 228, 8321–8335.

Tomita, H., Satoh, M., 2004. A new dynamical framework of nonhydrostatic global model using the icosahedral grid. Fluid Dyn. Res. 6, 357–400.

Tomita, H., Tsugawa, M., Satoh, M., Goto, K., 2001. Shallow water model on a modified icosahedral geodesic grid by using spring dynamics. J. Comput. Phys. 174, 579–613.

Tomita, H., Satoh, M., Goto, K., 2002. An optimization of the icosahedral grid modified by spring dynamics. J. Comput. Phys. 183, 307–331.

Vallis, G.K., 2006. Atmospheric and Oceanic Fluid Dynamics: Fundamentals and Large-Scale Circulation. Cambridge University Press, Cambridge, UK.

van Leer, B., 1977. Towards the ultimate conservative difference scheme. III. Upstream-centered finite-difference schemes for ideal compressible flow. J. Comput. Phys. 23, 263–275.

van Leer, B., 1979. Towards the ultimate conservative difference scheme. V. A second-order sequel to Godunov's method. J. Comput. Phys. 32, 101–136.

Wang, N., Lee, J., 2011. Geometric properties of the icosahedral-hexagonal grid on the two-sphere. SIAM J. Sci. Comput. 33, 2536–2559.

Weller, H., 2012. Controlling the computational modes of the arbitrarily structured C-grid. Mon. Weather Rev. 140, 3220–3234.

Wicker, L.J., Skamarock, W.C., 1998. A time-splitting scheme for elastic equations incorporating second-order Runge-Kutta time differencing. Mon. Weather Rev. 126, 1992–1999.

Wicker, L.J., Skamarock, W.C., 2002. Time-splitting method for elastic models using fast forward time schemes. Mon. Weather Rev. 130, 2088–2097.

Williams, P., 2009. A proposed modification to the Robert-Asselin time filters. Mon. Weather Rev. 137, 2538–2546.

Winninghoff, F.J., 1968. On the Adjustment Toward a Geostrophic Balance in a Simple Primitive-Equation Model With Application to the Problems of Initialization and Objective Analysis (Ph.D. thesis). University of California, Los Angeles.

Yu, Y.G., Wang, N., Middlecoff, J., Peixoto, P., Govett, M., 2020. Comparing numerical accuracy of icosahedral A-grid and C-grid schemes in solving the shallow-water model. Mon. Weather Rev. 148, 4009–4033.

Zalesak, S., 1979. Fully multidimensional flux-corrected transport algorithms for fluids. J. Comput. Phys. 31, 335–362.

Zhang, Y, Nair, R.D., 2012. A nonoscillatory discontinuous Galerkin transport scheme on the cubed sphere. Mon. Weather Rev. 140, 3106–3126.

CHAPTER

3

Probabilistic view of numerical weather prediction and ensemble prediction

Roberto Buizza

Scuola Superiore Sant'Anna and Center for CLimate Change Studies and Sustainable Actions (3CSA), Pisa, Italy

Abbreviations

AMSU-A	Advanced Microwave Sounding Unit-A
AIR, AIREP	aircraft data
ASR	satellite all-sky radiance observations
BV	bred vector
DA	data assimilation
DRI, DRIBU	drifting buoys
ECMWF	European Centre for Medium-Range Weather Forecast
EDA	ensemble of data assimilations
EFI	the ECMWF Extreme Forecast Index product
ENIAC	electronic numerical integrator and compute
ENS	the ECMWF medium-range/monthly ensemble
ERA5	the most recent version of the ECMWF reanalysis, version 5
EUMETSAT	European Organization for the Exploitation of Meteorological Satellites
GPC	ground-based precipitation composites
hPa	hectoPascal
HRES	the ECMWF single high-resolution version of the model
ICs	initial conditions
LIMB	satellite limb observations
METOP	METeorological OPerational satellite
NAO	the North Atlantic Oscillation, a low-frequency weather regime that characterizes the circulation over the Euro-Atlantic sector
NOAA	National Oceanic and Atmospheric Administration
NWP	numerical weather prediction

Uncertainties in Numerical Weather Prediction
https://doi.org/10.1016/B978-0-12-815491-5.00003-3

ORAS5	the most recent version of the ECMWF ocean reanalysis version 5
PIL, PILOT	balloons and profile observation
SATE, SATEM	satellite sounding observations
SATO, SATOB	satellite atmospheric motion vector observations
SCA, SCATT	satellite scatterometer observations
SEAS	the ECMWF seasonal ensemble
SKEB	stochastic kinetic energy backscatter scheme
SYN, SYNOP	land station and ship observations
SPPT	stochastically perturbed parameterized tendency model error scheme
STTP	stochastic total tendency perturbation
SV	singular vector
TEM, TEMP	radiosonde observations
TOA	top of the atmosphere
UTC	coordinated universal time

1 The numerical weather prediction problem

The atmosphere is an intricate dynamical system with many degrees of freedom, the state of which can be described by the spatial distribution of the system's state variables (normally, wind, temperature, surface pressure, and specific humidity). The time evolution of these variables is described by mathematical differential equations deduced by the physical laws that describe the behavior of the atmosphere, fluid on a rotating sphere.

Richardson (1922) is considered the first one to have shown that the weather could be predicted numerically. At that time, when computers were not available, he approximated the differential equations governing the atmospheric motions with a set of algebraic difference equations for the tendencies of various field variables at a finite number of grid points in space. By extrapolating the computed tendencies ahead in time, he could predict the field variables in the future. Unfortunately, his results were very poor, both because of deficient initial data and because of serious problems in his approach.

After the World War II, interest in numerical weather prediction (NWP) revived, partly because of an expansion of the meteorological observation network, and also because of the development and availability of digital computers. Charney (1948) developed a model applying a filtering approximation to Richardson's equations, based on the so-called geostrophic and hydrostatic equations. In 1950, an electronic computer (ENIAC) was installed at Princeton University, and Charney, Fjørtoft, and Von Neumann and Richtmeyer (1950) made the first numerical prediction using the equivalent barotropic version of his model. This model was extremely simple, compared to the ones used today, and was deduced by simplifying the equation of motion of a fluid on a rotating sphere (the Earth) to a manageable set, that would still capture some of the key aspects of the atmosphere dynamics. This model was used to provide forecasts of the geopotential height near the mid-troposphere, at 500 h-Pascals (hPa), and could be used as an aid to provide explicit predictions of other variables as surface pressure and temperature distributions.

Charney's results led to the development of more complex models of the atmospheric circulations, the so-called global circulation models. An example of these models was the one developed and implemented at the end of the 1970s at the European Centre for Medium-Range Weather Forecast (ECMWF). Today, most of the major national meteorological centers of the world have developed and use in operation a global, or a limited area model of the atmosphere, which includes also key land-surface and air-sea interaction processes.

1.1 What are the key processes of numerical weather prediction?

A weather forecast valid for time t, is generated by integrating in time, numerically, a set of equations that describes its evolution, starting from an initial state. In other words, NWP is an initial value problem, that requires a model (of the real system), observations (that describe the current state of the system), a computer (to integrate numerically the equations), and scientists (who are capable to develop a realistic model and procedures to compute the initial state of the system and integrate the model equations in time; ...).

Let's consider the time t of a typical day d. Generally speaking, the NWP process to generate a forecast that starts at time t includes three main steps:

1. *Collect observations*—As many observations as possible are collected in a time window prior to time t, usually during the past 6-to-12h for the atmosphere (this time window depends on data availability, computer power resources, and operational constraints); these data are accessed using a global telecommunication network, used by the data providers (e.g., national meteorological services, satellite agencies, ...) to share data among themselves.
2. *Generate the initial conditions (ICs)*—Once the observations within the time window have been collected, a data assimilation (DA) procedure is performed to estimate the state of the system at time t, i.e., to compute the ICs required to integrate numerically the equation of motions, and thus generate a numerical weather forecast; this DA procedure merges in an optimum way a first guess state of the atmosphere, given by a short-range forecast, and all the collected observations.
3. *Generate the forecasts*—From the ICs, a forecast valid for the next hours/days/months is launched; the short-range part of this forecast (up to the time that covers the next DA cycle) is used as an input to the next DA procedure while the whole forecast is used to generate all forecast products.

These three steps are repeated few times a day, usually every 6-to-12h, at times coinciding with what are called the synoptic times: 00, 06, 12, and 18 UTC, where UTC stands for Coordinated Universal Time.

Consider, for example, the following four successive times/dates:

- t_0, defined as 00 UTC on the 14th of May;
- t_1, defined as 12 UTC on the 14th of May;
- t_2 and t_3 defined as 00 and 12 UTC on the 15th of May, 2019.

Fig. 1 is a schematic of the NWP process used at the ECMWF to produce analyses and forecasts at t_1 and t_2:

- t_1 (12 UTC on the 14th of May)—At around 14 UTC on the 14th of May, once observations taken between t_0 and t_1 (00 and 12 UTC on the 14th of May) have been received (the system usually waits about 2h before starting the DA procedure, to allow observations to be received and quality controlled), the DA procedure starts; this procedure also uses the

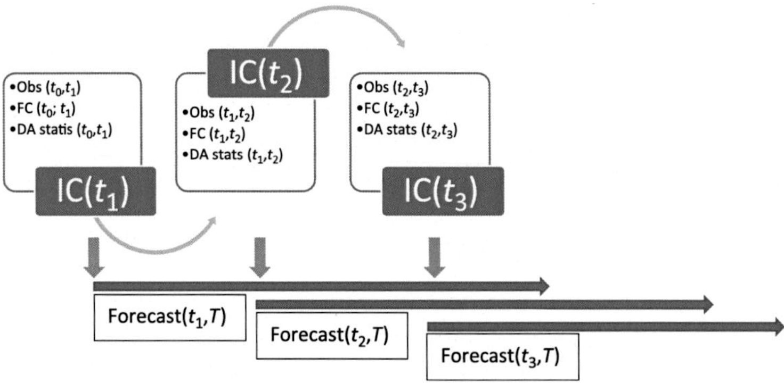

FIG. 1 Schematic of the numerical weather prediction (NWP) process.

short-range forecast (for a 12-h window DA, up to +12h forecast) which started at t_0 as input; in about 2h the ICs, valid for t_1, are generated; then, the numerical integrations, to generate the forecasts, are started, and after another 2h they are completed; thus, at about 18 UTC, the forecasts started at t_1 are completed, and after product generation and quality control (which takes about another hour) they are made available; thus after about 7h from the synoptic time, the forecasts are available.

- t_2 (00 UTC on the 15th of May)—At t_2, the same procedure followed at t_1 is repeated; at about 02 UTC on the 15th of May, when the short-range forecast started at t_1 and the observations covering the time window between t_1 and t_2 (12 UTC of the 14th and 00 UTC on the 15th) are available, the DA procedure starts; at about 04 UTC, the ICs are available; and at about 07 UTC, all forecasts are available;
- t_3 (12 UTC on the 15th of May) and subsequent times—At t_3, and any subsequent time, the same procedure is followed.

1.2 An integrated suite of analysis and forecasts

At most national meteorological centers, not only one but many different forecasts are generated and made available during a day. At the ECMWF, for example, analyses and forecasts are produced with four initial times a day: 00, 06, 12, and 18 UTC; all with different characteristics (this is necessary to take into account both the computer cost of generating the forecasts and the users' demands).

Thus, rather than talking about a *"forecast system,"* it is better to think about an *"integrated analysis and forecast system"* (at ECMWF, it is indeed called the IFS) that includes a range of analyses and forecasts with different characteristics. As an example, Table 1 lists the main characteristics (at the time of writing, January 2020) of the six key components of the ECMWF IFS, which includes:

- *DA*: the single, higher-resolution DA procedure, which has a horizontal resolution of 9km, and 137 vertical levels;
- *Ensemble Data Assimilation (EDA)*: the EDA procedure, which includes 51 members and has a horizontal resolution of 18km, and 137 vertical levels;

TABLE 1 List of the analyses and the forecasts generated operationally on a daily basis at the four synoptic times (00, 6, 12, and 18 UTC) by the European Centre for Medium-Range Weather Forecasts (ECMWF) (see text for more details).

Time (UTC)	DA	EDA	ORAS5	HRES	ENS			SEAS	
	(an)	(an)	(an)	+10d	+6.5d	+15d	+46d	+7m	+13m
00	Y	Y	Y	Y	Y	Y	Mon+Thu	First of the month	Each quarter
06	Y			Y	Y				
12	Y	Y	Y	Y	Y	Y			
18	Y			Y	Y				

- *Ocean ReAnalysis System 5 (ORAS5)*: the ensemble ocean DA procedure, which includes five members and has an ocean resolution of 0.25 degrees and 75 vertical levels;
- *Hybrid Renewable Energy Systems (HRES)*: the single, high-resolution forecast, which has a horizontal resolution of 9 km, and 137 vertical layers;
- *Ensemble forecasts (ENS)*: the 51-member medium-range/monthly ensemble, which has a horizontal resolution of 18 km up to forecast day 15, and of 36 km from day 15 to day 46, and 91 vertical layers.
- *ECMWF seasonal ensemble (SEAS)*: the 51-member seasonal ensemble, which at the time of writing (May 2019) has a horizontal resolution of 35 km, and 91 vertical layers.

HRES forecasts, valid for up to 10 days, are issued four times a day, at 00, 06, 12, and 18 UTC. Medium-range/monthly ensemble forecasts (ENS) valid up to +132 h are issued four times a day so that users interested to issue short- and medium-range forecasts, and to use the global forecasts to drive limited area ensembles, can do so whenever it is convenient for their needs. ENS forecasts are extended to 15 days at 00 and 12 UTC, and 46 days twice a week (on Mondays and Thursdays) but with a lower resolution (36 instead of 18 km). Ensemble seasonal forecasts (SEAS; Stockdale et al., 2018; Mogensen et al., 2012a, b) are issued once a month for up to 7 months; these forecasts are extended to 13 months once every quarter (on the 1st of February, May, August, and November).

The most advanced national meteorological centers have a similar suite of analysis and forecasts and generate analysis and forecasts a few times a day.

1.3 Observations

As it should have been clear from the previous section, observations are key to estimate the state of the atmosphere and compute the ICs, from which forecasts are computed.

The last 50 years have seen a huge increase in the quality and quantity of the Earth-system observations, especially of the atmosphere. Since the end of the 1930s, surface and subsurface (for the ocean) data have been complemented by an increasing number of upper-air data, e.g., from soundings. Then, the 1970s have seen the start of satellite soundings, and today observations taken by instruments on satellite platforms constitute 95% of all the available data.

Before the satellite era, every day the number of atmosphere observations was of the order of a few hundreds of thousands. Today, meteorological centers receive daily few hundreds of millions of atmospheric observations. Fig. 2 shows an example of the relative number of observations from a range of platforms (source: ECMWF; data received on March 1st, 2017):

- SATE: SATEM **satellite** sounding data: ~1882 M (81%)
- ASR: **satellite** all-sky radiances ~243 M (10%)
- AIR (AIREP): aircraft data ~69 M (3%)
- SATO (SATOB): **satellite** atmospheric motion vectors ~27 M (1.2%)
- LIMB: **satellite** limb observations ~22 M (1%)
- TEM (TEMP): radiosondes ~21 M (0.95%)
- SYN (SYNOP): land station and ships ~14 M (0.6%)
- PIL (PILOT): balloons and profilers ~10 M (0.4%)
- SCA: SCATT **satellite** scatterometer data ~7.3 M (0.3%)
- DRI (DRIBU): drifting buoys ~1.1 M (0.05%)
- GPC: ground-based precipitation composites ~0.4 M (0.02%)

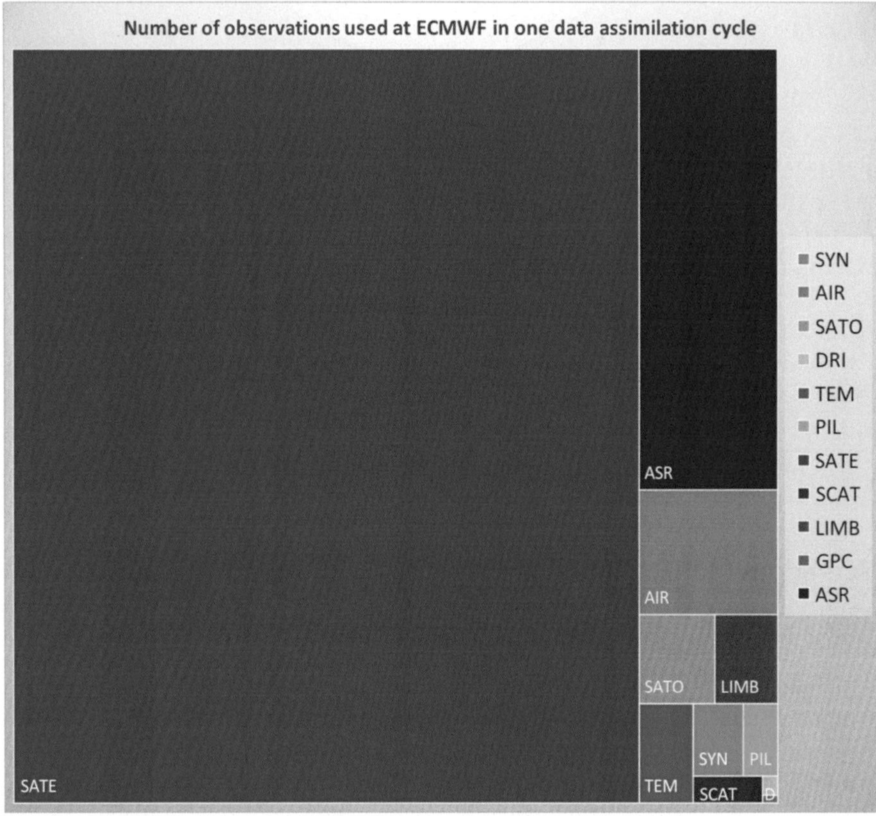

FIG. 2 Relative number (percentage) of observation received daily at ECMWF (see text for a detailed explanation of the acronyms). *Source: ECMWF; data received on March 1st, 2017.*

Only about 10% of these data are used to estimate the state of the atmosphere, and generate the ICs. Partly, this is due to quality control procedures, which reject the observations of a poor quality (e.g., because they have been contaminated by the presence of water vapor and/or water droplets in the clouds); and partly is due to the coarse model resolution and the available computer power.

Fig. 3 shows an example of the number of AMSU-A satellite observations received (~659,000) and used (~84,000) at ECMWF during the DA procedure of April 23, 2019. The observations were taken by the AMSU-A instruments onboard seven satellites: NOAA-15, NOAA-18, NOAA-19, METOP-A, AQUA, METOP-B, and METOP-C. AMSU-A is the Advanced Microwave Sounding Unit-A, designed to measure global atmospheric temperature profiles; NOAA is the US National Oceanic and Atmospheric Administration, and NOAA-15, NOAA-18, and NOAA-19 are three US satellites; METOP is a satellite of the European Organization for the Exploitation of Meteorological Satellites, EUMETSAT; AQUA is a satellite of the US National Aeronautics and Space Administration, NASA. The difference between the two maps, and the reduction of the number of used data to about 12%, is partly due to data thinning (i.e., data are used at a lower spatial resolution), and partly due to data rejection by the quality control procedure.

Although we can say that today the atmosphere is well observed, this cannot be said for the other components of the Earth-system that we need to initialize, especially if we want to produce subseasonal and seasonal forecasts. In particular, the ocean is much less observed: considering the subsurface ocean, for example, today we only receive a few thousand observations a day, compared to few millions a day for the atmosphere. An example is given in Fig. 4, which shows that there is a factor of about 2000 the atmospheric surface observations of type synoptic observations (SYNOP) (land observations), SHIP (data from ships) and METeorological Aerodrome Report (METAR, surface observations from airport stations), received between 0 and 12 UTC of April 23, 2019 (*top panel*), and the number of oceanic salinity observations received during April 22, 2019.

Figs. 3 and 4 have given us some evidence of the fact that observations coverage varies, and that only a small percentage of the received observations are used. Fig. 5 gives us some evidence of the fact that observations are affected by errors, and that errors depend on the observation type and the pressure level of the measurement. For example, Fig. 5 shows that for the horizontal wind components, the root-mean-square error of TEMP/PILOT and SYNOP types of observations is smaller than the error of satellite (SATOB) observations. TEMP reports are given by radiosondes, which take a standard set of observations; PILOT reports of wind only involve tracking an ascending balloon.

1.4 The model equations

The Earth-system is a dynamical system governed by Newton's laws of physics. In particular, the ocean and the atmosphere are governed by the laws of physics written for fluid on a rotating sphere (the Earth). The starting point from which the equations are deduced is given by Newton's second law of physics:

$$\boldsymbol{F} = m\boldsymbol{a} \tag{1}$$

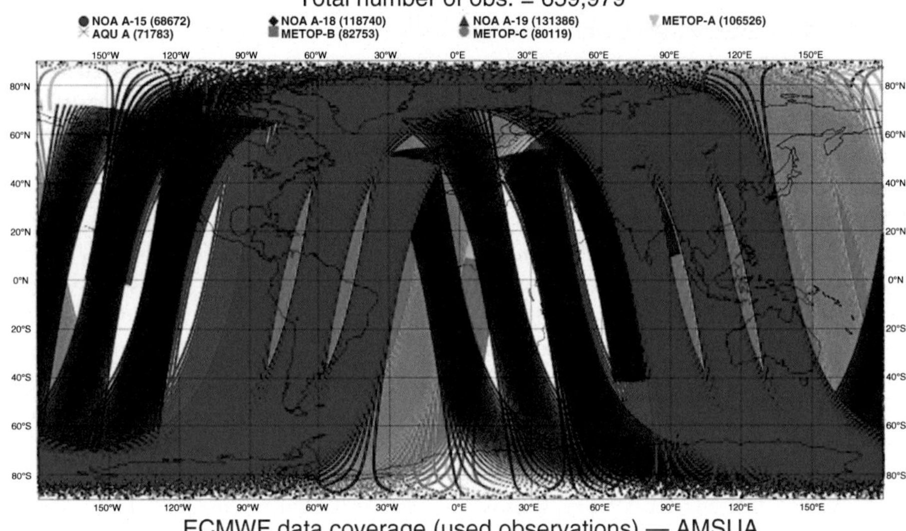

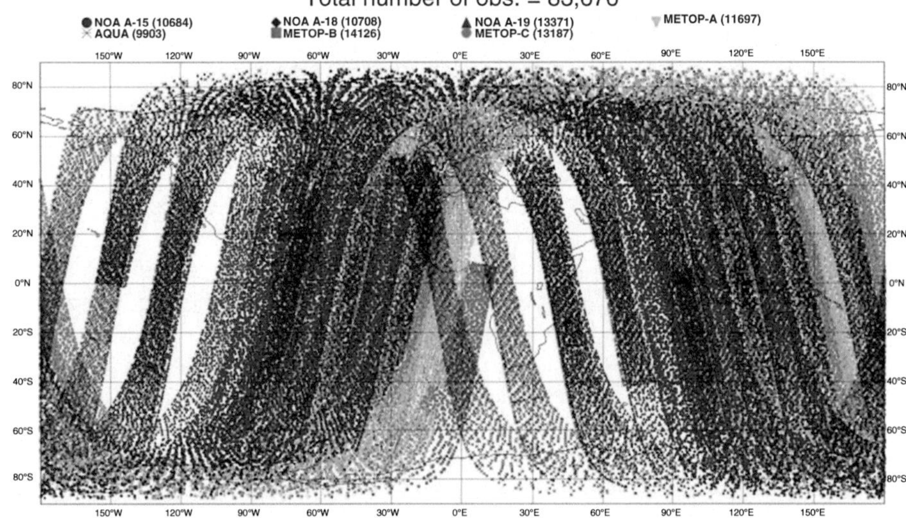

FIG. 3 AMSU-A (Advanced Microwave Sounding Unit-A) data received on April 23, 2019 between 0 and 12 UTC (~659,000; *top*), and used (~84,000; *bottom*) by the ECMWF data assimilation procedure to generate the analysis valid for at 12 UTC.

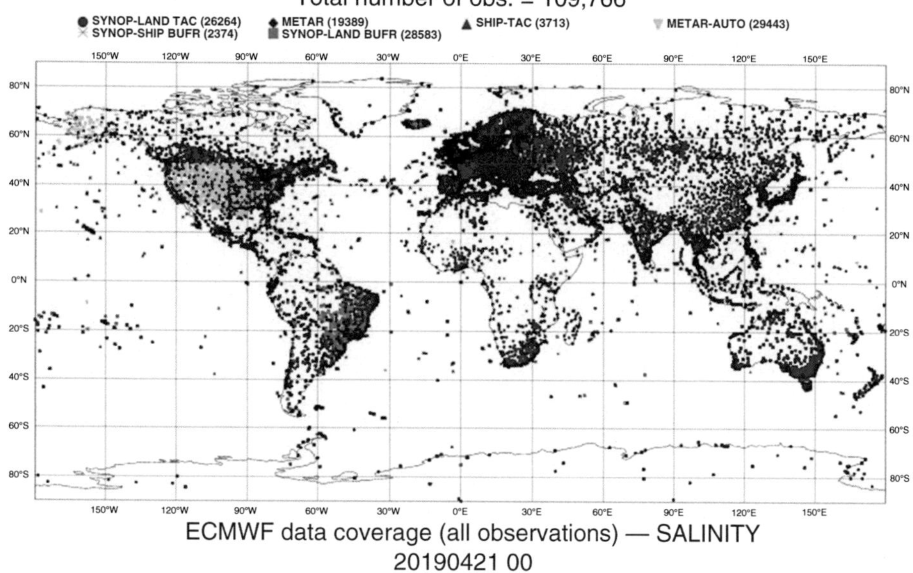

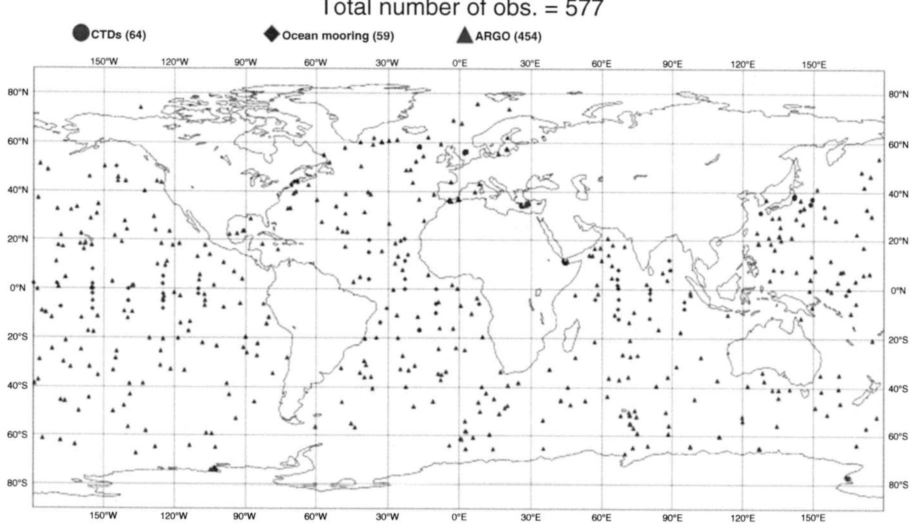

FIG. 4 SYNOP, SHIP, and METAR atmospheric data received on April 23, 2019 between 0 and 12 UTC (~109,000; *top*), and oceanic salinity data received on April 22, 2019 (~577; *bottom*) at ECMWF.

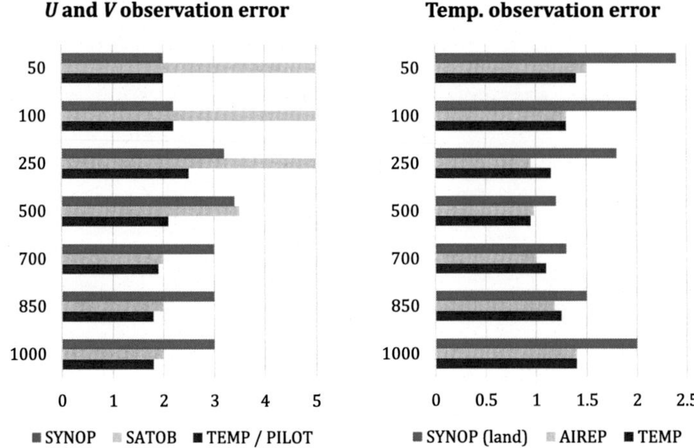

FIG. 5 Prescribed root-mean-square errors for observations of the horizontal wind components (in m/s, *left*) and temperature (in K, *right*), at seven different pressure levels (from the surface at 1000–50 hPa), used in the ECMWF DA procedure.

where ***F*** stands for the three-dimensional force (a vector), *m* for mass, and ***a*** for the three-dimensional acceleration (a vector), and the energy and mass conservation equations for a fluid (see Holton and Hakim, 2012; Hoskins and James, 2014).

Following Holton and Hakim (2012), the dynamic equations for a unitary fluid volume of the atmosphere on a rotating sphere (the Earth), take the following form:

$$\frac{dv}{dt} = -2\Omega \times v - \frac{1}{\rho}\nabla p + g + P_v(\text{momentum equation}) \tag{2}$$

$$c_v\frac{dT}{dt} + p\frac{d\alpha}{dt} = P_T(\text{thermodynamic energy equation}) \tag{3}$$

$$\frac{dq}{dt} = P_q(\text{water vapor conservation}) \tag{4}$$

$$\frac{1}{\rho}\frac{d\rho}{dt} + \nabla \cdot v = 0(\text{continuity equation}) \tag{5}$$

$$\frac{dp}{dz} = -\rho g(\text{hydrostatic balance}) \tag{6}$$

where

- v is the two-dimensional horizontal wind vector;
- $\boldsymbol{\Omega}$ is the Earth angular velocity vector, directed along the Earth rotation axis;
- ρ is the atmospheric density;
- p is the pressure;
- g is the gravity vector, with magnitude g and directed toward the Earth center;
- T is the temperature;
- R is the gas constant for dry air ($=287\,\text{J}\,\text{kg}^{-1}\,\text{K}^{-1}$);

- c_v is the specific heat at constant volume ($=717\,\mathrm{J\,kg^{-1}\,K^{-1}}$);
- $\alpha=\frac{1}{\rho}$ is the specific volume;
- q is the specific humidity;
- $\boldsymbol{P}_v$, P_T, and P_q are the tendencies of, respectively, the horizontal wind components, temperature, and specific humidity, due to physical processes (such as convection, the interaction of clouds with radiation, ...).

These equations are expressed in terms of the atmospheric state variables, the horizontal wind components, temperature, surface pressure, and specific humidity q.

The terms ($\boldsymbol{P}_v$, P_T, P_q) on the right-hand-side of Eqs. (2–4) represent the effect of physical processes on the (tendencies of) the state variables. For example, they include the impact of convection on energy and moisture fluxes, the effect of clouds on the absorption, refraction, and reflection of short-wave solar radiation and long-wave radiation emitted by the Earth surface, the effect of mountains on atmospheric flows, the effect of turbulence on the energy and momentum transport. They are the most difficult terms to be defined and are some of the sources of forecast errors.

These equations are solved on a three-dimensional grid that covers the whole globe, with a horizontal spacing that varies, for global models, from about 9 km (for the ECMWF high-resolution model version) to about 200 km (for models used for seasonal prediction and climate projections). Vertically, these models use between few tens and about 200 levels, spanning the first 40-to-80 km of the atmosphere. In the ECMWF global model, for example, the number of vertical levels ranges from 91, in the model version used by all the ensemble systems and to generate the ECMWF reanalysis, to 137 for the operational single, high-resolution analysis and 10-day forecast.

The equations are integrated in time using finite-difference methods, on very powerful supercomputers. The finer the resolution, the higher the number of floating-point operations have to be completed to solve the equations, since a higher-resolution model include more grid points.

Thus, computer power availability is one of the key elements that determine the choice of a model resolution: given that a forecast is valuable if it is produced in a reasonable amount of time, it puts a cap on the resolution. Generally speaking, let's say that one usually aims to produce a 72-h forecast in a few hours, and a few weeks forecast in less than a day.

The second element that determines the choice of a model resolution is the scales and forecast range that are the main focus of the forecasting system. If one aims to predict phenomena characterized by very fine resolution (e.g., wind gusts linked to the passage of a frontal system across a mountain range characterized by complex orography), generally speaking the model needs to be able to "resolve" the relevant scales involved in the phenomena (although it is true to say that some of the scales can be parameterized and/or simulated using stochastic methods). By contrast, if one aims to predict slowly varying, large-scale patterns, a coarser resolution might be sufficient.

The third element is whether the forecasting system aims to provide not only a forecast of the most likely scenario but also an estimate of the forecast confidence, for example expressed in terms of a range of possible forecast scenarios. In this case, one needs to devote a computer.

As an example, let's look at the resolution of the ECMWF IFS (Table 2) used to produce the single, high-resolution, and the ensemble analyses and forecasts at the time of writing (January 2020).

TABLE 2 Key configuration of the seven components of the ECMWF operational suite run at the time of writing (January 2020; see main text for more details): description (column 2), number of members (column 3), horizontal and vertical resolution (column 4; all models have a 0.01 hPa top of the atmosphere), forecast length (column 5), name of the coupled ocean/sea-ice model and resolution of the ocean model (column 6) and method used to simulate initial and/or model uncertainties.

IFS component	Description	#	Atm/Land resolution	FC length	Ocean/Sea-ice model	Uncertainty simulation
4DVar analysis	Atm/land/wave High-resolution analysis	1	Tco1279 (9 km) L137	-	no	no
EDA[51] analyses	Atm/land/wave Ensemble of Data Assimilation	51	Tco639 (18 km) L137	-	No	Yes: - Observations - Model: SPPT(1)
ORAS5[5] analyses	Ocean Ensemble of analyses	5	-	-	NEMO/LIM2 0.25 degree 75 layers	Yes: - Observation
ERA5[11] analysis	Atm/land/wave Reanalysis	1	Tco319 (36 km) L91	-	no	Yes: - Observations - Model: SPPT(1)
HRES forecast	Atm/land/wave/ocean High-resolution	1	Tco1279 (9 km) L137	10 d	NEMO/LIM2 0.25 degree 75 layers	No
ENS[51] forecast	Atm/land/wave/ocean Medium-range and monthly ensemble	51	Tco639 (18 km) L91	15 d	NEMO/LIM2 0.25 degree 75 layers	Yes: - ICs: EDA, SVs, ORAS5 - Model: SPPT(3)
			Tco319 (36 km) L91	15-46 d		
SEAS5[51] forecast	Atm/land/wave/ocean Seasonal ensemble	51	Tco319 (36 km) L91	7 m 13 m	NEMO/LIM2 0.25 degree 75 layers	Yes: - ICs: EDA, SVs, ORAS4 - Model: SPPT(3), SKEB

Let's remind the reader that ECMWF aims (a) to provide accurate and reliable global forecasts, including confidence intervals, valid for up to 1 year and (b) to update its forecasts as fast as possible, so that its users can have frequent, fresh updates. These two aims imply that ECMWF has to generate not only single forecasts but also ensembles. Furthermore, the fact that ECMWF aims to provide forecasts up to 1 year, implies that it cannot use a very high-resolution model for all forecast ranges since this would make forecast generation too long (maybe impossible) and too computing demanding.

Compromises have to be made: this is achieved by using not only a single resolution model, but a suite of models with different resolutions that, combined, allow the generation of products covering the 1-day-to-1-year forecast range. The ECMWF suite of models includes the six key components already mentioned above plus the reanalysis suite ERA5 (see Table 2), which are all required to provide users with estimates of the probability density function (PDF) of analyses and forecast states:

1. **EDA[51]**: the 51-member, 18-km, L137 (137 vertical levels) Ensemble of Data Assimilation, which provides flow-dependent statistics and estimates of the analysis PDF (it uses the same horizontal resolution of the first 15-days of ENS);
2. **4DVar**: the single, 9-km single L137 analysis;
3. **ORAS5[5]**: the 5-member ensemble of ocean analyses, version "5:" with the Nucleus for European Modeling of the Ocean (NEMO) ocean model run with a 0.25-degree resolution and 75 vertical layers, and the Louvain-la-Neuve Sea Ice Model (LIM2) sea-ice model (it uses the same resolution of the NEMO model used in all the forecasts);
4. **ERA5[11]**: the 11-member, 36-km, L91 reanalysis, which is used to generate the ICs for the ENS and SEAS5 reforecast suites (it uses the same resolution as the monthly extension of ENS and as SEAS5);
5. **HRES**: the single, 9-km resolution, L137, 10-day forecast; it is coupled to the NEMO ocean model run with a 0.25-degree resolution and 75 vertical layers, and the LIM2 sea-ice model;
6. **ENS[51]**: the 51-member, L91 coupled ensemble, which provides forecasts at an 18-km resolution up to day 15, and at a 36-km resolution from day 15 to 46 (only at 00 UTC, on Mondays and Thursdays); it is coupled to the NEMO ocean model run with a25-degree resolution and 75 vertical layers, and the LIM2 sea-ice model;
7. **SEAS5[51]**: the 51-member, L91, 36-km coupled seasonal ensemble System-5, which provides forecasts and estimates of the forecast PDF for the seasonal time scale; it is coupled to the NEMO ocean model run with a 0.25-degree resolution and 75 vertical layers, and the LIM2 sea-ice model (this is the same configuration as the one of the monthly extension of ENS).

In particular, it should be noted that ECMWF uses a spectral model, with a corresponding cubic-octahedral grid in geographical space (Wedi et al., 2015). If we consider, for example, the two daily ensembles, the Ensemble of Data Assimilation (EDA) and the first leg (i.e., up to forecast day 15) of the medium-range/monthly ensemble (ENS), they use a Tco639 horizontal resolution (corresponding to about 18 km in grid-point space). In the vertical, the EDA uses 137 levels, while ENS uses 91, both with the top of the atmosphere (TOA) at 0.01 hPa. The ENS monthly extension and the seasonal ensemble SEAS5 use a Tco319 resolution (corresponding to about 36 km in grid-point space). The high-resolution analysis (4DVAR) and forecast (HRES) use a Tco1279 horizontal resolution (corresponding to about 9 km in grid-point space) and 137 vertical levels (with the TOA at 0.01 hPa).

1.5 The definition of the initial conditions

Generating weather forecasts requires integrating the equation of motions of the Earth-system, starting from an initial state. In other words, weather prediction is an initial value problem, whereby accurate ICs are required to be able to generate accurate forecasts.

Determining the initial state is very complex and time-consuming, given that the system has many degrees of freedom, and that many observations have to be assimilated to compute a state as close as possible to reality, if not all of them, need to be initialized properly.

Considering the ECMWF IFS, for example, the atmosphere of its ENS component has about 100 million degrees of freedom. This number increases to about 450 million for a simulation performed with the ECMWF high-resolution version, which has a 2-times finer horizontal resolution than ENS and 137 instead of 91 vertical levels.

The ICs (the so-called analysis) are computed by comparing a short-term forecast, started from the last available ICs, with all the available observations within a time window (order 10 million), using a procedure called "data assimilation."

For example, at ECMWF, a four-dimensional variational assimilation procedure (4d-Var) is used to compute the analysis (Rabier et al., 1999), by comparing a 12-h forecast started from the previously available analysis with all the observations collected inside this window. 4d-Var computes the analysis by finding the minimum of a cost function that measures the distance between the short-range forecast trajectory and the observations.

Let us denote by $x=(u,v,T,q,p)$ the state vector that characterizes the state of the atmosphere, where (u,v) are the horizontal wind components, T is temperature, q is the specific humidity, and p is pressure, defined on the three-dimensional model grid.

Let's denote by $x_b(t)$ the short-range forecast started from the previously available analysis, also called the first guess, and by $x_a(t)$ the forecast started from the analysis $x_a(0)$ (i.e., the initial state of the system) we want to compute.

Define:

$$\delta x = x_a(t) - x_b(t) \tag{7}$$

the correction that needs to be added to the short-range forecast to compute the analysis.

Let us also define the cost function of 4d-Var that measures the distance between the first guess and the observations, from the forecast started from the analysis:

$$J(\delta x) = \frac{1}{2}\left[\delta x \cdot B^{-1} \cdot \delta x\right] + \frac{1}{2}\left[(H \cdot \delta x - d)^T \cdot R^{-1} \cdot (H \cdot \delta x - d)\right] \tag{8}$$

where

- B is a matrix defined by the forecast error statistics (and B^{-1} is its inverse);
- R is a matrix defined by the observation error statistics;
- H is the observation operator, that map the state vector x from the model phase space onto the observation space (e.g., it is the operator that would compute the temperature at a station location from the model temperature values at the closest grid points);
- $d=o-H \bullet x_b$ is the distance between the observation and the first guess, also called the innovation vector.

The first term of the cost function measures the distance of the solution from the first guess, while the second term measures the distance of the solution from the observations.

Note that the observation operator B depends on the model quality: a high-quality, high-resolution model will allow the use of many observations, and will guarantee that they are "translated" into accurate information in terms of model variables. By contrast, a poor model will make it difficult to use observations and could lead to inaccurate ICs.

The inverse of the two matrices B and R define the relative weight given to the two terms: it depends on the estimated relative accuracy (i.e., confidence) of the first guess and the observations. Since, as we discussed above, both the first guess (which is a short-range forecast) and the observations are affected by errors, and these errors depend, e.g., on the state of the system and the location, these two matrices are not constant. They do indeed depend on the state of the system, and the daily availability and accuracy of the observations.

4d-Var aims to compute the correction δx that minimizes the cost function J. This minimum is computed by applying complex minimization procedures that involve the definition of a tangent forward and its adjoint operator. The reader is referred to Daley (1991) and Kalnay (2002) for an introduction to DA. Other DA procedure would solve a different minimization problem, using different methods, but the essence of the problem remains the same: merge all the available information (observations and first guess estimates of the state of the system) to compute the analysis.

From Eq. (8) it is clear that the quality of the analysis depends on the quality, quantity, and coverage of the observations, on the model, which is used to define the first guess and to define the observation operators, and on the DA assumptions. This latter dependency is linked to the choice of the assimilation methods, and, for 4d-Var, on the cost function definition and on the way the weight matrices are defined and computed. Uncertainties in any of these three areas will affect the knowledge of the ICs and will propagate in time during the time integration of the forecast equations.

2 Sources of forecast errors and the chaotic nature of the atmospheric flow

The fact that, at any time t_0, only a limited number (limited with respect to the degrees of freedom of the system: let's say at least a factor of 100-to-1000) of observations are available, that all observations are characterized by an error, and that they do not similarly cover the whole globe, is one of the main sources of the IC, and thus a forecast error. Observational errors, usually larger at the smallest scales, amplify and through nonlinear interactions spread to the large scales and eventually affect the skill of these latter ones. The presence of uncertainties in the ICs is one of the sources of forecast errors.

The second source of forecast error is related to the approximations made in the numerical models of the atmospheric system. These models do not include all the physical processes that affect the time evolution of weather systems and simulate the ones that they include only in an approximate way. Furthermore, models have a finite resolution: thus, they do not include scales below the model resolution, and they are capable to represent realistically only scales up to a multiple of the model resolution, estimated to be about 5 times the grid spacing. Thus, if we take for example the ECMWF, HRES, and ENS model versions, which have a horizontal resolution of about 9 and 18 km, diagnostics indicate that they can resolve realistically (in the sense that they are capable to simulate the right variability) only for scales courser than about 45 and 90 km.

These two sources of forecast errors cause weather forecasts to deteriorate with forecast time. A further complication is due to the fact that the atmosphere is a chaotic system (Lorenz, 1965, 1969a, b, 1993), and exhibits error growth rates that are sensitive dependence to ICs.

Small uncertainties related to the characteristics of the atmospheric observing system will always affect the ICs, and model approximations will always affect the quality of model integration. This means that forecast skill will always be finite! We can aim to extend it, and indeed in the past three decades, since we started the operational production of ensemble-based probabilistic forecasts, we had been extending the forecast skill horizon.

For example, Buizza and Leutbecher (2015) computed the average forecast skill of the ECMWF ensemble forecasts for 1 year (July 2012 to July 2013) and concluded that forecasts of instantaneous, grid-point fields are skillful up to between 16 and 23 days, and that forecasts of large-scale, low-frequency filtered fields are skillful between 23 and 32 days (at that time, the ECMWF ensemble forecast length was 32 days). Looking at the MJO, Vitart et al. (2014) showed that the 2013 version of the ECMWF monthly ensemble predicted it skillfully up to about 27 days, indicating a 1-day gain in skill per year since 2004. Considering the North Atlantic Oscillation (NAO), they reported that the ECMWF ensemble showed skill in 2013 up to about forecast day 13, compared to about day 9 ten years earlier.

3 Ensemble-based probabilistic prediction

A complete description of the weather prediction problem can be stated in terms of the time evolution of an appropriate PDF in the atmosphere's phase space. The predicted PDF of possible future atmospheric states is a function greater than zero in the phase space regions where the atmospheric state can be, with maximum values identifying the most probable future states.

Fig. 6 is a schematic of a probabilistic forecast, expressed in terms of a PDF, from which one could compute the most likely value (e.g., the mean of the PDF) and a confidence interval (e.g., expressed as the interval, which has an 80% probability to include the observed value), or as a probability that the observed value will be above or below a specific value. Fig. 6 shows the example of a PDF forecast for a single location, London, of surface temperature, at a specific time. For simplicity, it has been assumed that the forecast distributions are Gaussian, or very close to Gaussian, so that they can be characterized by their mean value and standard deviation. The left panel shows the PDF $f(x)$, and the right panel its related cumulative distribution function (CDF):

$$g(x) = \int_{-\infty}^{x} f(y)dy \tag{9}$$

wherein this case the variables x and y denote temperature.

The first estimate of the forecast temperature could be given by the statistics of the observed field during the past years: using all these observations, one could construct a PDF, characterized a mean value and a certain spread (the *dotted black curve*). In this case, it has a mean value of 18°C and a standard deviation of 3°C. From the climatological CDF, we can compute that there is a 30% probability that the temperature could be lower than 16°C, and only a 20% probability that it could be higher than 20°C. The red curves represent a forecast issued 12 days before the event. Compared to the climatological PDF, the +12d PDF is shifted to the right and has a lower dispersion: it has a mean of 19°C and a standard

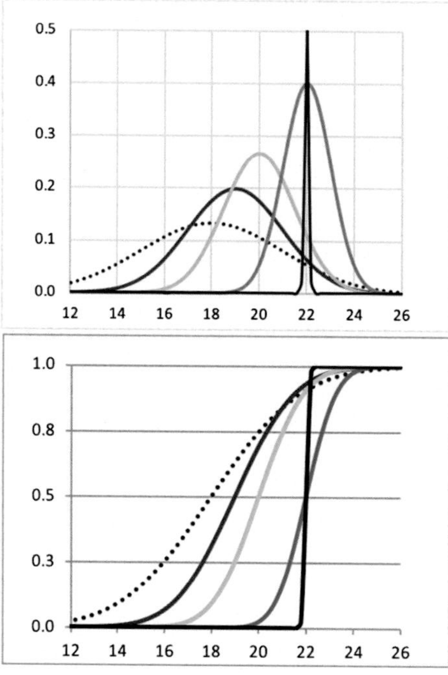

FIG. 6 Schematic of probabilistic forecasting. In both panels, the x-axis refers to a typical meteorological variable: surface temperature and the y-axis refers to probabilities (from 0 to 1). The *left panel* shows probability density functions (PDFs) and the *right panel* the corresponding cumulative distribution functions (CDFs). The *black dotted curves* represent the climatological PDF/CDF. The *red*, *green*, and *blue curves* represent subsequent PDF/CDF forecasts with decreasing forecast length, all valid for the same verification time (e.g., a 12 day, an 8 day, and a 4-day forecast). The *black curves* represent the observed PDF/CDF, at verification time: the PDF is centered on the observed value (22°C) and has a width that depends on the observation error (0.1°C); the CDF is very close to a step function, with a zero value before the observed value and 1 afterward.

deviation of 2°C. The *green* and *blue curves* represent forecasts issued 6 and 3 days before the event. As we are getting closer to the event, the PDFs become narrower, indicating that there is less uncertainty (and more confidence) in the forecast. This is reflected in the CDFs, which become steeper. The +3d forecast has a mean of 22°C and a 1°C standard deviation: from the CDF we can deduce that there is only a 20% probability that temperature will be lower than 21°C, and a 20% probability that it will be higher than 23°C. The solid black PDF represents the observed value: 22°C, with an observation error standard deviation of 0.1°C; the corresponding observed CDF is very close to a step function, equal to 0 before 22°C and equal to 1 afterward.

The climatological (reference) PDF (*dotted black line*) has only a small overlap with the observation, whereas the forecast PDFs match the observation PDF increasingly well as the forecast time shortens. Consistently, the forecast CDFs (defined by integrating the PDFs; not shown) also approach the observation CDF.

3.1 How can a forecast PDF be generated?

The problem of the prediction of the PDF can be formulated exactly through the continuity equation for probability (Liouville equation). Unfortunately, with the current computer power availability, the Liouville equation that describes the PDF time evolution can only be solved for simple systems characterized by a very limited number of degrees of freedom (say order few tens, compared to the tens of millions of the degrees of freedom of the atmosphere as simulated by the current generation of models).

The alternative method investigated first in the 1960s (Leith, 1965; Epstein, 1969), and developed in the 1980s, and implemented operationally in the 1990s is based on an ensemble of integrations, which can be used to estimate the PDF. The idea is very simple: instead of generating one single forecast, generate as many as it is feasible, and use them to compute the future statistics. The number of ensemble members is a function of the model resolution and the available computer power: typically, operational ensembles include between 10 and 50 members initialized at the same time. Some centers manage to increase this number of mixing forecasts issued at different times/days.

Although the idea behind is simple, the practical development and implementation of "reliable" ensembles is not trivial, precisely because of the fact that a very small number of forecasts are used to estimate the future PDF of a system that has tens of millions of degrees of freedom. This will be discussed in some details in Section 3. Up to date, ensembles are the only feasible methods to predict the PDF beyond the range of linear error growth.

Fig. 8 illustrates the difference between having a full PDF forecast (the *continuous red curve*) or an ensemble with a small number (51 in this case) of forecasts. In this case, for each temperature value T, one could build a histogram by counting the number of forecasts with a value within an interval ($T-0.1$; $T+0.1$°C). The histogram could then be used to find the closest continuous distribution (e.g., a Gaussian curve), to compute the step-wise CDF and thus to compute the probability of any event of interest.

3.2 Ensemble methods: How should one design an ensemble?

Ensemble prediction systems should be designed to simulate all sources of forecast errors, in particular, errors due to initial and model uncertainties. The relative importance of these two sources of forecast errors depends on the characteristics (e.g., spatial and temporal scales) of the phenomena under investigation. For large-scale atmospheric patterns, research studies performed with state-of-the-art numerical models have indicated that for short and medium time ranges (say for forecasts up to 3–5 days) errors are mainly due to initial uncertainties. Model errors due to the parameterized physical processes start having a nonnegligible effect after forecast day 3. By contrast, for the prediction of small scale low-pressure systems and associated precipitation fields, for example, model errors can be as important as initial uncertainties at forecast day 2 or even earlier. For forecast times longer than 10 days (monthly and seasonal prediction) other error sources should be simulated. For example, the possible influence of uncertainties in the boundary conditions (e.g., in the soil moisture content, in the ice and vegetation coverage) should be taken into account.

3.3 Simulation of initial condition uncertainties

IC uncertainties are due to the fact that observations are affected by observation errors, and do not cover the whole globe with the same quality and frequency. Furthermore, any DA procedure used to estimate the initial state of the system, from which a forecast is computed, is based on some statistical assumptions and approximations.

There is not a unique perturbation strategy to simulate initial uncertainties, and different centers have developed alternative methods or variants of existing methods. In general terms, in my view we can cluster these methods into three groups:

- The first group of methods followed a pure "Monte Carlo approach," and tried to generate initial perturbations that would include all possible sources of analysis errors: they perturbed randomly the observations, the model and the statistics used in the DA, trying to be as general as possible (i.e., trying to perturb randomly along with all the directions of the phase space of the system).
- The second group decided that, given the small number of the ensemble members compared to the degrees of freedom of the problem (i.e., of the phase space of the system), it would have been more efficient to try to perturbed along "selected directions," for example, identified by the directions characterized by the fastest growth over a finite time interval.
- Today, few centers have moved toward a combination of both approaches, to try to benefit from both.

To provide few examples, we can look at the main characteristics of the initial perturbation' strategy used in the first three operational ensembles, implemented at ECMWF, in the United States and Canada (see Buizza et al., 2005 for a comparison of the performance of these three global ensembles, and a discussion of the impact of their assumptions on the ensemble's performance).

In the first version of the ECMWF ensemble, initial uncertainties were simulated using singular vectors (SVs), which are perturbations with the fastest growth over a finite time interval (Buizza et al., 1993; Buizza and Palmer, 1995; Molteni et al., 1996). SVs remained the only type of initial perturbations in the ECMWF ensemble until 2008 when the ensemble of data assimilations (EDA) was added to improve the way observations' quality affects ICs (Buizza et al., 2008, Isaksen et al., 2010).

By contrast, the first version of the NCEP ensemble used bred vectors (BVs) to simulate initial uncertainties. The BV cycle (Toth and Kalnay, 1997) was designed to emulate the DA cycle, and it is based on the notion that analyses generated by DA will accumulate growing errors by the virtue of perturbation dynamics.

In 1995, Canada implemented a new type of ensemble, designed to simulate a wider range of error sources, linked to initial uncertainties due to observation errors and DA assumptions, and also linked to model uncertainties (Houtekamer et al., 1996). Because of this, the Canadian initial perturbation strategy could take into account uncertainties linked to observations' quality and coverage, and IC errors linked to model uncertainties.

In rather general terms, the initial perturbation strategies used by all the other ensembles operational today (see Table 3) have been inspired by these three approaches, although their detailed implementation differ from them, and include upgrades and changes.

TABLE 3 Key characteristics of the seven global, medium-range ensembles operational at the time of writing (January 2020), and of the UK Met Office ensemble that had been operational up to 2014, listed in alphabetic order: initial uncertainty method, model uncertainty simulation, and key references.

Center	Simulation of initial uncertainties (in brackets the area where the perturbations are computed)	Simulation of model uncertainties	Few key references
CMA (China Meteorological Administration)	Global Bred Vectors [BV (globe)]	No	Su et al. (2014)
CPTEC (Centro de Previsao de Tempo e Estudos Climatico; Brazil)	Localized Empirical Orthogonal Function method [EOF(40S:30N)]	No	Coutinho (1999) and Zhang and Krishnamurti (1999)
ECMWF (European Centre for Medium-Range Weather Forecasts; Europe)	Localized singular vectors [SV(NH, SH, TC)] and global Ensemble of Data Assimilation [EDA(globe)]	Stochastically Perturbed Parameterized Tendencies (SPPT)	Buizza and Palmer (1995), Molteni et al. (1996), Buizza et al. (1999, 2008), Leutbecher and Palmer (2008), Palmer et al. (2009), and Leutbecher et al. (2017)
JMA (Japan Meteorological Administration; Japan)	Localized Ensemble Transformed Kalman Filter [LETKF] and localized singular vectors [SV(NH, TR, SH)]	SPPT	Miyoshi and Sato (2007) and Yamaguchi and Majumdar (2010)
KMA (Korea Meteorological Administration; Korea)	Global Ensemble Transformed Kalman Filter [ETKF(globe)]	SPPT	Goo et al. (2003), Bowler et al. (2008), and Kai and Kim (2014)
MSC (Meteorological Service of Canada; Canada)	Global Ensemble Kalman Filter [EnKF(globe)]	SPPT	Houtekamer et al. (1996, 2009, 2014)
NCEP (National Centers for Environmental Prediction; USA)	Global Ensemble Transformed Kalman Filter with Rotation [ETR(globe)]	Stochastic Total Perturbation Scheme (STTP)	Toth and Kalnay (1993, 1997), Boffetta et al. (1998), Wei et al. (2006, 2008), and Hou et al. (2008)
UKMO (United Kingdom Meteorological Office—Note that the UKMO global, medium-range ensemble was operational up to 2014)	Ensemble Tranformed Kalman Filter (ETKF)	Random Parameter Scheme (RP) and Stochastic Convective Vorticity Scheme (SCV)	Wei et al. (2006), Bishop et al. (2001), Bowler et al. (2007, 2008), Lin and Neelin (2000), Bright and Mullen (2002), and Gray and Shutts (2002)

3.4 Simulation of model uncertainties

Model uncertainties arise because the models that we use to generate weather forecasts are imperfect, simulate only certain physical processes on a finite mesh, and do not resolve all the scales and phenomena that occur in the real world.

The Canadian ensemble implemented in 1995 was the first one to include also a simulation of model uncertainties (Houtekamer et al., 1996). Following the Canadian example, the simulation of model uncertainties was introduced in the ECMWF ensemble in 1999, using a stochastic approach to simulate the effect of model errors linked to the physical parameterization schemes (Buizza et al., 1999). This was the first time that a stochastic term was introduced in NWP.

At present, four main approaches are followed in ensemble prediction to represent model uncertainties (for a review, see, e.g., Palmer et al., 2009; Buizza, 2014):

- A multimodel approach, where different models are used in each of the ensemble members; models can differ entirely or only in some components (e.g., in the convection and turbulence schemes).
- A perturbed parameter approach, where all ensemble integrations are made with the same model but with different parameters defining the settings of the model components; one example is the Canadian ensemble (Houtekamer et al., 1996).
- A perturbed-tendency approach, where stochastic schemes designed to simulate the random model error component are used to simulate the fact that tendencies are known only approximately: one example is the ECMWF Stochastically Perturbed Parametrization Tendency scheme (SPPT, Buizza et al., 1999).
- A stochastic backscatter approach, where a Stochastic Kinetic Energy Backscatter scheme (SKEB; Shutts, 2005; Berner et al., 2008) is used to simulate processes that the model cannot resolve, e.g., the upscale energy transfer from the scales below the model resolution to the resolved scales.

3.5 Ensemble of analyses to estimate initial condition uncertainties

The Canadian ensemble implemented in 1995 (Houtekamer et al., 1996), included an ensemble of analyses, generated using an ensemble Kalman filter (EnKF). The ICs of the ensemble forecasts were defined by one of the members of the EnKF. The EnKF has been providing MSC Canada with information about uncertainties in the analysis.

The Canadian example was followed at the ECMWF (Buizza et al., 2008; Isaksen et al., 2010) and Météo-France (Berre et al., 2007), who started producing an Ensemble of Data Assimilation in 2008.

At the ECMWF, the EDA is based on an ensemble of N separate DA procedures (where N is 51 at the time of writing), each using perturbed observations and the SPPT model uncertainty scheme. Observations are perturbed to simulate observation errors, linked to the instruments' characteristics and their representativeness. Model uncertainties are also simulated to take into account the fact that the models used to define the analysis (i.e., the forecast ICs) are not perfect.

The addition of EDA-based perturbations has had a major impact on the ensemble reliability and accuracy in the short forecast range over the extratropics, and for the whole forecast range over the tropics (Buizza et al., 2008).

3.6 Ensemble of forecasts (and reforecasts) to estimate uncertainty

In Section 3.1 we discussed the idea behind ensemble prediction, and Figs. 7 and 8 illustrated how ensembles can be used to estimate the PDF of forecast states and the forecast uncertainty. This is computed assuming that the ensemble is reliable, and that, if this is the case, one can use the dispersion of the ensemble members as a predictor of forecast skill.

An ensemble is reliable if, on average over many cases, when it predicts that an event will occur with a probability p, it is observed p-times. This can be verified by comparing the forecast probabilities to the observed frequencies. One can visualize the average couples of (forecast probability; observed frequency) in a reliability diagram, like the one shown in Fig. 9: the closest to the diagonal the points are, the better is the probabilistic forecast system. In this case, the blue forecasts are closer to the diagonal. This means that there is a more accurate relationship between the forecast probability and the frequency of occurrence: when the forecast probability is very high, in other words when the forecasts lie close to each other and the ensemble spread is very small, the forecast is more accurate.

To measure how close the points are to the diagonal, one could compute the Brier score (Wilks, 2005), which is the mean squared distance between the points and the diagonal:

$$BS = \frac{1}{N}\sum_{j=1}^{N}(f_j - o_j)^2 \tag{10}$$

where N is the number of intervals (10 in our case), f_j is the forecast probability and o_j is the observed frequency.

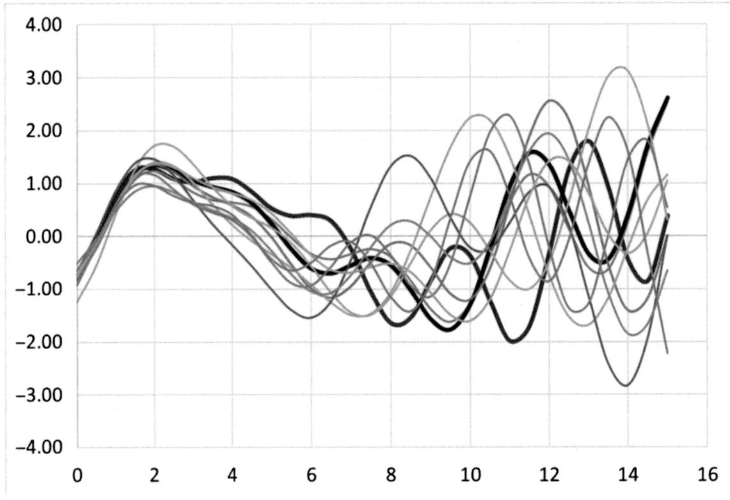

FIG. 7 Schematic of ensemble prediction. The x-axis shows forecast time (in days) and the vertical axis surface temperature anomaly with respect to climatology, at a specific location. The *solid black curve* is the single unperturbed "control" forecast and the *solid red curve* is the observed anomaly. The *thin gray curves* are 10 perturbed forecasts.

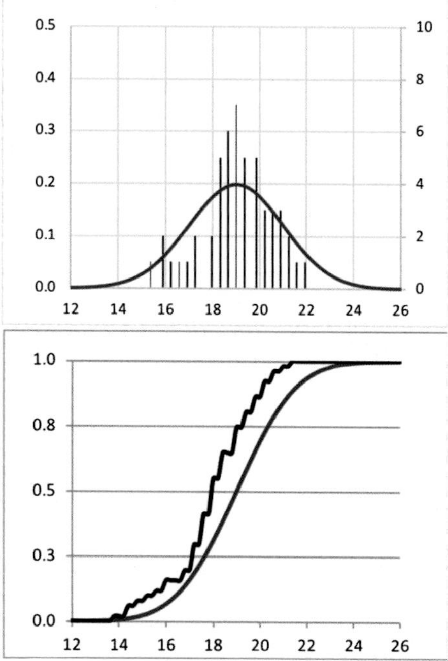

FIG. 8 Schematic of ensemble prediction. In both panels, the *x*-axis refers to a typical meteorological variable: surface temperature and the *y*-axis refers to probabilities (from 0 to 1). The *left panel* shows an ensemble of 51 forecasts (*black bars*; their values are to be read from the secondary axis, ranging from 0 to 10) and a probability density function (PDF; *red curve*). The *right panel* the corresponding cumulative distribution functions (CDFs). In the *left panel*, the *black bars* denote how many ensemble members predicted a certain temperature value. The *red* PDF.

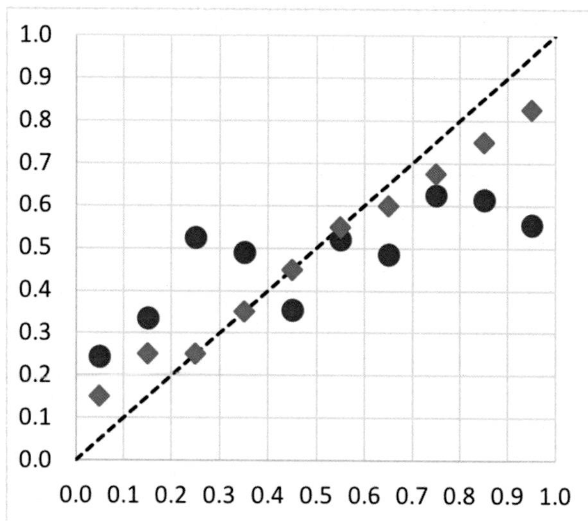

FIG. 9 Reliability diagram for two probabilistic forecasting systems, which issue forecasts in a 10% category. The *x*-axis shows the observed frequencies and the *y*-axis the forecast probabilities, of two different forecasts. For example, the diagram tells us that when the blue forecast gives a probability of 35% of an event to occur, the event happens in 35% of the cases; by contrast, when the red forecast gives a probability of 35%, the event frequency can be either 15% or 45%. If we look at the distribution of the blue forecasts, they lie closer to the diagonal than the red forecasts. In other words, the red forecast is less capable to discriminate between the occurrence and nonoccurrence of events. A perfectly reliable forecast would have all points aligned along the diagonal (*black dashed line*).

In a reliable probabilistic forecasting system, when the probability of an event is high, all the ensemble members are close to each other, and the event has a high probability to occur. In other words, in this case the spread among the members is low, and the forecast turns out to be frequently accurate. If the probability is very high, it means that the spread is very low, and the forecast turns out to occur very frequently. By contrast, when the probability is low, the frequency of occurrence is low: in other words, the forecast could be right, but could also be wrong. In this case, one can say that the forecast is less predictable.

As we discussed in Section 1, one of the problems of the NWP models is their finite resolution, and thus the fact they do not resolve the local phenomena. This has an impact on the phenomena that one can predict with accuracy and realism: while the larger scales can skillfully predict days and weeks ahead, and for the very large-scale patterns even months ahead, predicting local phenomena, extreme rainfall events, which are sensitive to the orographic forcing, is very difficult. In other words, let's say that the model climate and the "real climate" are very close to the large-scale, low-frequency phenomena, but they are not exactly the same for the small scale, fast ones.

The result is that, if one were to try to predict, for example, extreme precipitation events, a model with an ~18 km resolution (as the ECMWF ENS) would maybe give maximum values of about 50 mm of rain over a 6-h period, whereas in reality one could experience maybe few hundreds of mm. Still, 50 mm/6 h is a very large value if compared to the model climatology, to all the precipitation events that the ENS model can predict. A way to use this information, and translate it into a warning that extreme precipitation values could occur in reality, is to design products that compare the latest forecast PDF with the model climate PDF.

One way to estimate the model climate PDF is to generate ensembles with ICs of many years, clearly using the same model, and an ensemble configuration, which is as close as possible to the one used in operations. At the ECMWF, for example, reforecasts are generated for both the medium-range/monthly ensemble (ENS) and the seasonal ensemble (SEAS5) using smaller-size ensembles, using smaller-size ensembles spanning the past decades. For ENS, given that the ENS model is updated quite frequently (say about every 12 months), the reforecasts are generated "on the fly," just 1 week before they are needed to generate the operational forecasts. For SEAS5, since the model is frozen for a few years, they are all generated when the model is upgraded (i.e., once about every 5 years).

Table 4 gives the details of these "reforecast" ensembles (they are called reforecasts since they are forecast rerun for past cases, using the same configuration of the operational forecasts; see, e.g., Hamill and Whitaker, 2007; Hagedorn et al., 2012).

Generating reforecasts requires a substantial amount of computer power, although this is not required "in real time," since reforecasts can be generated before they are required (i.e., when the supercomputers are available). To give you an idea of the cost of generating ensemble forecast and reforecast:

- *ENS reforecasts*—Let's suppose that ENS is upgraded once a year; then in each year the ECMWF generates 37,230 forecasts up to 15 days ($365 \times 2 \times 51$), and 5304 forecasts up to 46 days ($52 \times 2 \times 51$); in parallel, ECMWF generates 22,880 reforecasts up to 46 days ($52 \times 2 \times 11 \times 20$); if we consider the first 15days, the reforecasts add ~62% (22,880/37,230) to the cost of the operational forecasts;

TABLE 4 Key characteristics of the ECMWF ENS and SEAS5 ensemble of forecasts(fc) and reforecasts (refc), at the time of writing (January 2020; see main text for more details): number of members included in each integration (column 3), start dates and times (columns 3 and 4), number of years spanned by the reforecasts (column 5), number of members generated every week for ENS, and every month for SEAS5 (column 6), and number of reforecast members used to generate the operational products that require them (column 7).

IFS component	Type	Members	Start dates	Start times	# Years	# members generated	# of refc members used for product generation
ENS 15d	fc	51	Daily	00-12	--	714 per week (51 × 2 × 7)	Operational products that need them, use refc run for past 20 years, for 3 weeks centered on the day, thus 1100 members (5 × 11 × 20)
ENS 46d	fc	51	Mon-Thu	00	--	102 per week (51 × 2)	Operational products that need them, use refc run for past 20 years, for 3 weeks centered on day, thus 1100 members (5 × 11 × 20)
ENS 15d/46d	refc	11	Mon-Thu	00	20	420 per week (11 × 2 × 20)	--
SEAS5 7m	fc	51	First of the month	00	--	51 per month, 612 per year, 3060 in 5 years	Operational products that need them, use refc run for past 23 years, thus 575 members (25 × 23); some products use the full range of available refc spanning 35 years, 612 members
SEAS5 7m	refc	25	First of the month	00	35	875 per month, 10,500 to cover a whole year	--

- *SEAS5 reforecasts*—Let's suppose that SEAS is upgraded once every 5 years; then in the 5 years ECMWF generates 3060 forecasts (12 × 51 × 5); in parallel, ECMWF generates 10,500 reforecasts (12 × 25 × 35); thus, the reforecasts add ~340% (10,500/3060) to the cost of the operational forecasts.

4 Examples of ensemble-based probabilistic products

In this section, we give some examples of what it means to issue a probabilistic forecast and show some examples of ECMWF probabilistic forecasts. Generally speaking, we can categorize them in six main types:

- *Ensemble-grams*: they show at a selected location and for a specific variable/index, the full range of the ensemble forecast distribution, highlighting the minimum and maximum values, the mean and/or the median, and come percentiles;
- *Probability maps*: they show a map of the probability of occurrence of a specific event or the probability that a certain variable could be either above a threshold $THRES_1$, or below a threshold $THRES_2$;
- *Maps of indices of anomalous weather*: they are computed by comparing the PDF of the latest forecast and the model climatology, and identify areas where anomalous weather conditions (in terms of few surface weather parameters) could occur;
- *Ensemble-mean maps*: they show the average variable predicted by the ensemble;
- *Ensemble spread maps*: they show the average spread, defined usually by the ensemble standard deviation; usually, ensemble-mean and ensemble-spread maps are

superimposed, so that forecasters can identify areas where the forecast uncertainty is higher, and relate it to the atmospheric flow;
- *Stamp-maps*: they visualize maps of all ensemble forecasts, so that a user can immediately spot whether there is agreement or disagreement among them;
- *Cluster centroids maps*: they show the centroid (e.g., defined by the mean) of groups of ensemble members, which have been grouped into a cluster because their fields were very similar.

We show hereafter few of these maps from ECMWF: without ensembles, these maps that contain flow-dependent estimates of the forecast confidence could not have been generated.

Let's start with Fig. 10, which shows two maps of the ECMWF Extreme Forecast Index (Lalaurette, 2003). The EFI indicates how different the forecast CDF is from the climatological CDF, and it is an integral measure of the distance between the two curves (see discussion later on refer Fig. 12). The two maps in Fig. 10 show the +24 h and the +120 h forecasts generated at 00 UTC of February 25, 2020, for three surface variables: 24-h accumulated precipitation, 10-m wind gusts, and 2-m temperature. These maps highlight the areas where extreme events, can occur.

Users can identify one region of interest, and then generate probabilistic forecasts for a single location. Fig. 11 shows an example: the PDF forecast for Copenhagen, valid for the next 10 days, generated at 00 UTC on February 25, 2020. For each of the four shown variables, at 6-hourly intervals, the plot shows some of the forecast PDF percentiles in the form of a box-and-whiskers' diagram: the minimum and maximum values, and the 10%, 25%, 75%, and 90% percentiles. The forecast uncertainties/confidence can be gaged by looking at the size of the box-and-whiskers: the smaller it is, the smaller the ensemble spread is, and the more predictable the situation is.

Similarly, although slightly different information can be deduced by looking at the forecast CDFs for Copenhagen, shown in Fig. 12 for three of the four variables shown in Fig. 11. For each variable, the left panel shows the CDF forecasts for the next 6 days, and the climatological CDF computed from the ENS reforecasts; the right panel shows the Extreme Forecast Index product (EFIs) for the different forecast times (note that the EFI maps of Fig. 11 show the +24 h and + 120 h values of all grid points inside the Euro-Atlantic map). From the CDFs, for each variable one could very easily look at the value predicted with a probability of 20%, or 80%. Alternatively, one can select a value, and look for the probability at which it is forecast. Furthermore, by looking at the steepness of the CDF curves, one can gage the ensemble spread (and thus the forecast uncertainty/confidence): the steeper the CDF is, the smaller the spread and the more predictable the event is (if the ensemble is reliable, as discussed above).

Fig. 13 shows longer-range probabilistic forecasts of weekly average anomalies of 2-m temperature and precipitation. The maps show the ECMWF ENS ensemble-mean forecast anomaly, computed with respect to the climatological mean: areas are colored only where the forecast and the climatological PDFs are statistically significantly different. These maps give again not only information on the most likely outcome (the forecast anomaly), but it does it taking into account the distribution of forecast states, compared to climatology. Note that the map has "more colored" areas for temperature than for total precipitation, confirmation of the fact that predicting precipitation is more difficult. Such a difficulty translates into having a very large ensemble spread, very similar to climatology in many different areas.

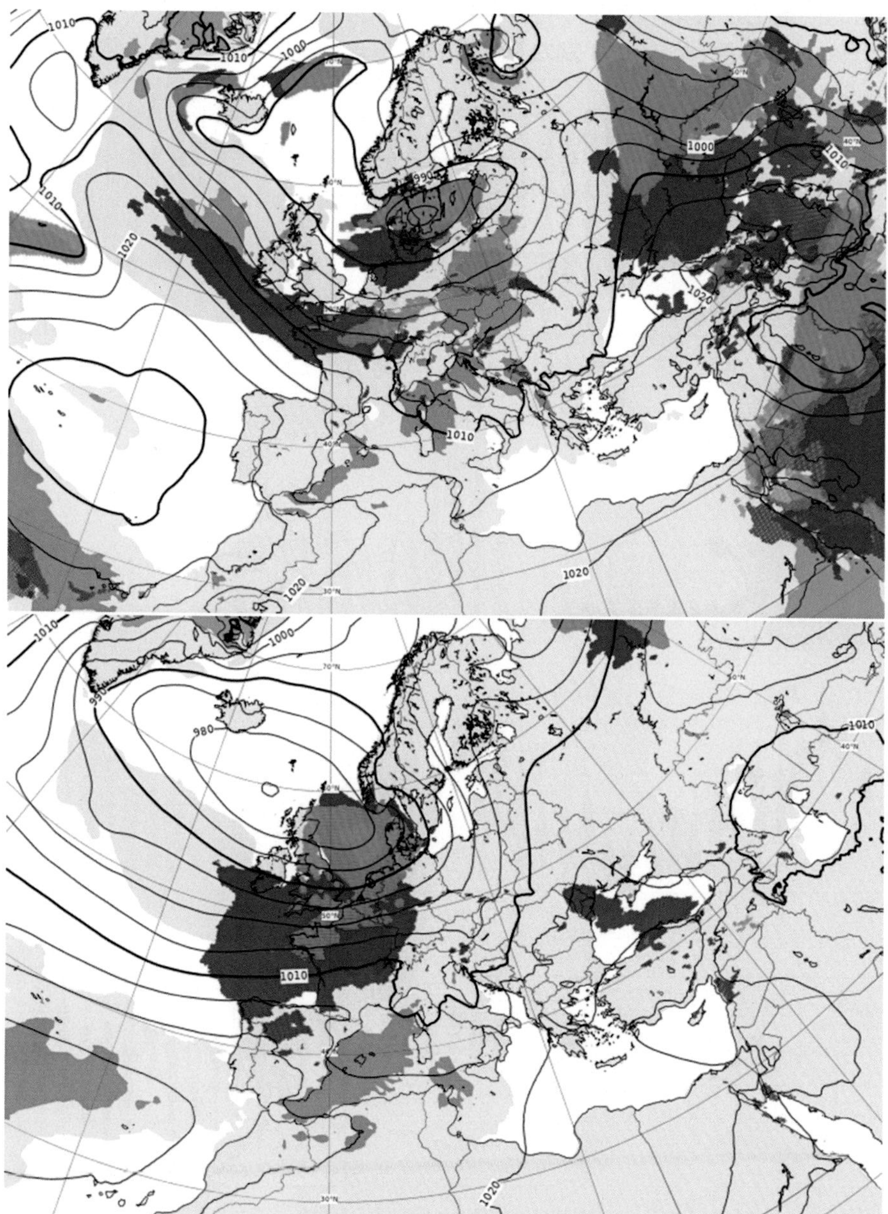

FIG. 10 Maps of the ECMWF Extreme Forecast Index (EFI) generated at 00 UTC of February 25, 2020: the *top panel* shows the +24 h forecast and the *bottom panel* the +120 h forecast. 2-m temperature: *yellow/orange shadings* refer to the 2-m temperature: *yellow* for EFI between 0.5 and 0.75; *orange* for EFI between 0.75 and 1.00. 24-h accumulated total precipitation: *Green* refers to precipitation: light green for EFI between 0.6 and 0.8; *dark green* for EFI between 0.8 and 1. 10-m wind gusts: *purple* refers to wind gusts: *light purple* for EFI between 0.6 and 0.8; *dark purple* for EFI between 0.8 and 1. *Source: ECMWF.*

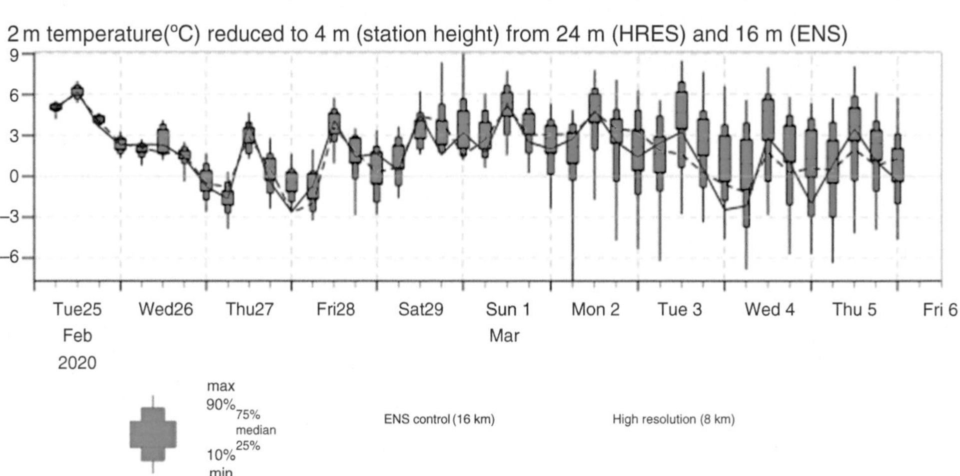

FIG. 11 ECMWF ensemble-gram for one specific location, Copenhagen, generated at 00 UTC of February 25, 2020. The ENS-gram shows 6-hourly forecast probability density function (PDFs), expressed in terms of few percentiles, for four variables: total cloud cover (*top*), 6-h accumulated precipitation (*second panel*), 10-m wind speed (*third panel*), and 2-m temperature (*bottom panel*). For each forecast time, a *bar-and-whiskers symbol* shows the median, the minimum, and maximum values, and the 10%, 25%, 75%, and 90% percentiles. *Source: ECMWF.*

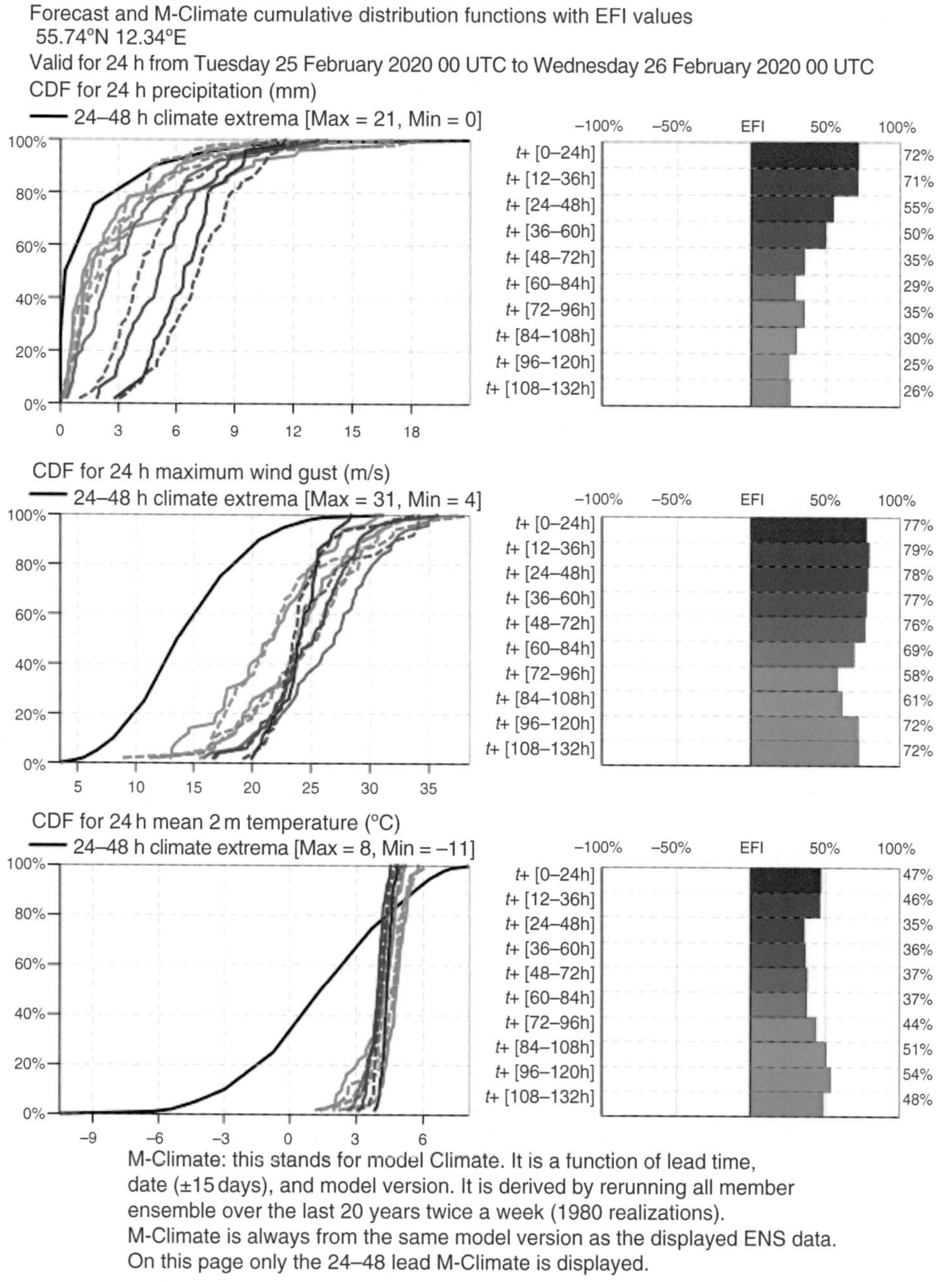

Forecast and M-Climate cumulative distribution functions with EFI values
55.74°N 12.34°E
Valid for 24 h from Tuesday 25 February 2020 00 UTC to Wednesday 26 February 2020 00 UTC

CDF for 24 h precipitation (mm)
— 24–48 h climate extrema [Max = 21, Min = 0]

EFI	
t+ [0–24h]	72%
t+ [12–36h]	71%
t+ [24–48h]	55%
t+ [36–60h]	50%
t+ [48–72h]	35%
t+ [60–84h]	29%
t+ [72–96h]	35%
t+ [84–108h]	30%
t+ [96–120h]	25%
t+ [108–132h]	26%

CDF for 24 h maximum wind gust (m/s)
— 24–48 h climate extrema [Max = 31, Min = 4]

EFI	
t+ [0–24h]	77%
t+ [12–36h]	79%
t+ [24–48h]	78%
t+ [36–60h]	77%
t+ [48–72h]	76%
t+ [60–84h]	69%
t+ [72–96h]	58%
t+ [84–108h]	61%
t+ [96–120h]	72%
t+ [108–132h]	72%

CDF for 24 h mean 2 m temperature (°C)
— 24–48 h climate extrema [Max = 8, Min = −11]

EFI	
t+ [0–24h]	47%
t+ [12–36h]	46%
t+ [24–48h]	35%
t+ [36–60h]	36%
t+ [48–72h]	37%
t+ [60–84h]	37%
t+ [72–96h]	44%
t+ [84–108h]	51%
t+ [96–120h]	54%
t+ [108–132h]	48%

M-Climate: this stands for model Climate. It is a function of lead time, date (±15 days), and model version. It is derived by rerunning all member ensemble over the last 20 years twice a week (1980 realizations).
M-Climate is always from the same model version as the displayed ENS data.
On this page only the 24–48 lead M-Climate is displayed.

FIG. 12 ECMWF cumulative distribution functions (CDFs) for one specific location, Copenhagen, generated at 00 UTC of February 25, 2020. The CDFs show the latest forecasts (up to +132 h) for three variables: 24-h accumulated precipitation (*top panel*), 10–24-h maximum 10-m wind gusts (*middle panel*), and 24-h mean 2-m temperature (*bottom panel*). For each variable, the *left panel* shows the 10 forecast CDFs (in *colors*) and the model climatology (*black solid line*); the *right panel* shows the EFI values for the 10 different forecast ranges (from 0 to 24 h, to 108–132 h, every 12 h). *Source: ECMWF.*

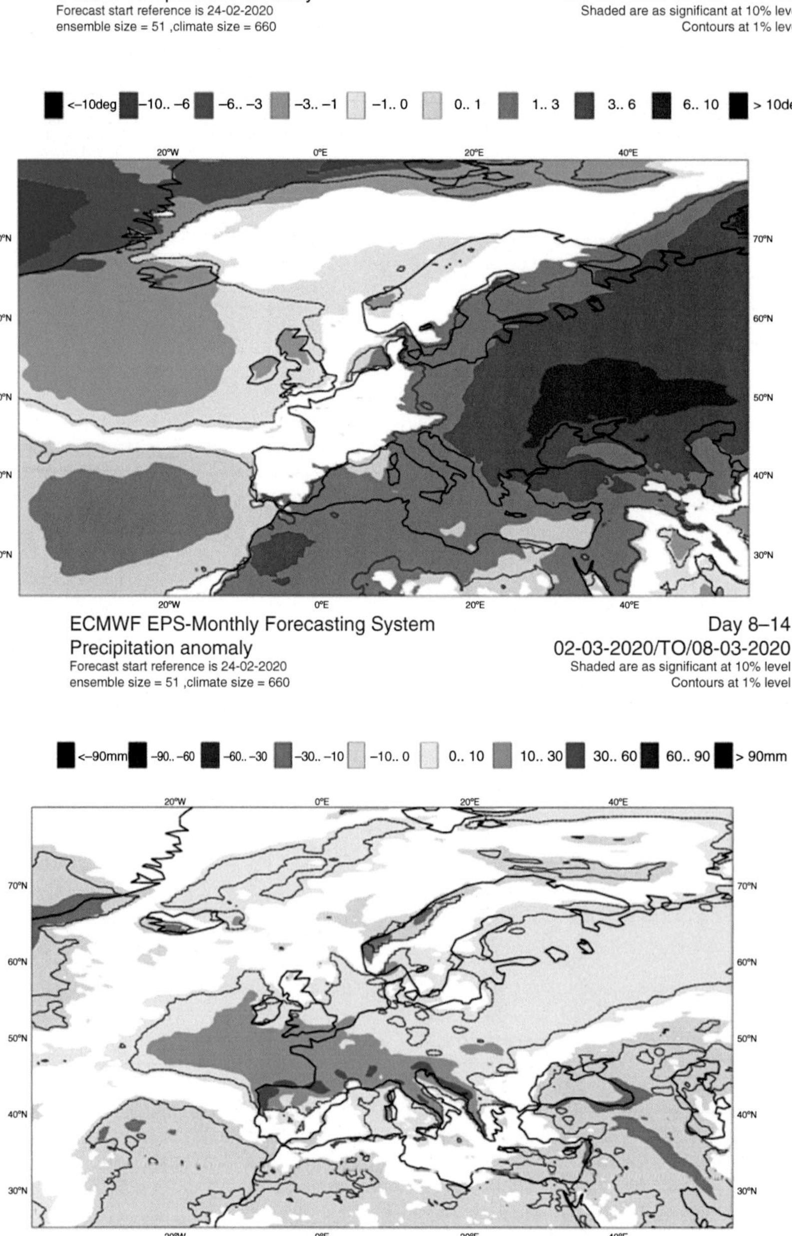

FIG. 13 ECMWF forecast map of the weekly average anomaly of 2-m temperature (*top*) and accumulated precipitation (*bottom*), generated at 00 UTC of February 24, 2020, and valid for the week 2–8 March (+8 to +14 days). *Shaded values identify the areas where the forecast PDF is statistically significantly different from the model climatology. For temperature, red shading* identifies areas of positive (hot) anomalies, while *blue* identify areas with negative (cold) anomalies. For precipitation, *blue shading* identifies areas of positive (wet) anomalies, while *blue* identify areas with negative (dry) anomalies. *Source: ECMWF.*

Figs. 14 and 15 show two examples of the ECMWF SEAS5 seasonal forecasts of large-scale, low-frequency phenomena. Fig. 14 shows the seasonal forecast of the NAO (Hurrell et al., 2003): a positive value denotes a weather pattern over the Euro-Atlantic sector with increased westerlies and, consequently, cool summers and mild and wet winters in Central Europe and its Atlantic facade. In contrast, negative NAO values denote situations with suppressed westerlies: as a consequence, northern European areas suffer cold dry winters and storms track southwards toward the Mediterranean Sea. In this case, the PDF forecast is superimposed to the climatological PDF (both represented with a box-and-whiskers diagram). Several pieces of information can be extracted from this diagram: for example, the spread of the PDF indicates the forecast confidence, the median gives the most likely forecast, the distance between the two PDFs indicates how extreme/rare the forecast is, compared to climatology. Note, for example, the large shift toward positive NAO values, of the forecast for February 2020.

Fig. 15 shows the ECMWF SEAS5 13-month SST forecast anomaly in the El Nino Index 3.4 area: this is an area in the central tropical Pacific. Positive/negative SST anomalies indicate the presence of El Nino/La Nina conditions. The plot shows the 15 members of the 51 SEAS5 members that have been extended beyond the 7-month forecast length of SEAS5: these extended forecasts are generated every quarter, on the 1st of February, May, August, and November. In this case, the single SST-anomaly forecasts are shown, rather than PDF's percentiles, since the computation of percentiles is less accurate when the ensemble membership is low. Users can gage uncertainty and confidence by looking at the dispersion between the 15 forecasts. As for most of the monthly and seasonal forecasts, they are expressed in

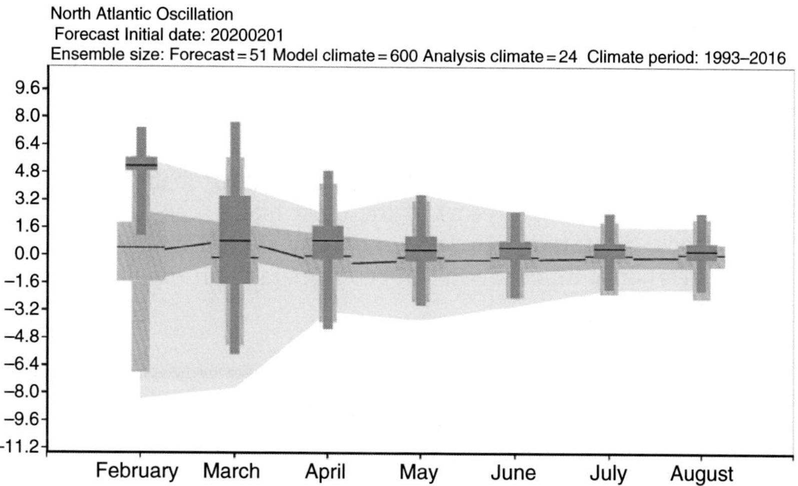

FIG. 14 ECMWF seasonal forecast of the North Atlantic Oscillation (NAO), generated at 00 UTC of February 01, 2020, and valid for the next 7 months. The forecast is expressed in terms of the NAO index. The plot shows in *dark purple color* the seasonal PDF forecast in terms of the median, 10%, 25%, 75%, and 90% percentiles; in *gray color*, the plot shows the PDF of model climatology. The *yellow shading* connects the model climatological minimum and maximum value, and the *orange shading* connects the model climatological 25% and 75% percentiles. *Source: ECMWF.*

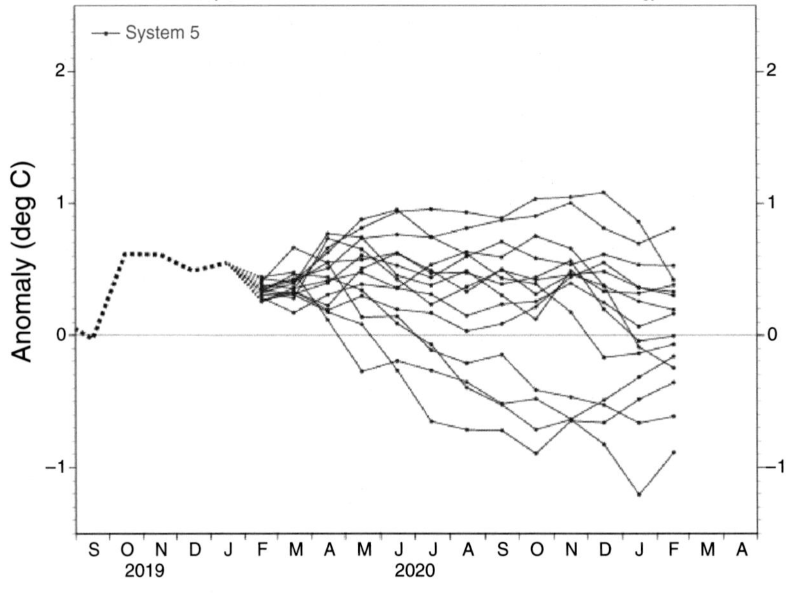

FIG. 15 ECMWF seasonal forecasts of the sea surface temperature (SST) in the Nino 3.4 area, generated at 00 UTC of February 01, 2020 and valid for the next 13 months. The forecast anomalies from the 15 members (only 15 of the 51 SEAS5 members are extended beyond 7 months) are shown, with the anomalies computed with respect to the model climate. *Source: ECMWF.*

terms of anomalies, computed with respect to the model climatology computed using the seasonal reforecasts.

These six figures illustrate the value of ensemble-based, probabilistic forecasts: the fact they allow not only to provide a forecast of the most likely scenario but also to estimate the forecast confidence, the forecast uncertainty. They should have also given the reader examples of how uncertainty information can be estimated and displayed.

5 A look into the future

The move away from single analysis and forecast systems, toward ensemble-based probabilistic systems is clear and will continue. Ensembles are, and for the foreseeable future will remain, the only feasible tools capable to provide accurate and reliable estimates of the PDF of initial and forecast states.

In terms of development, Buizza (2018) identified five areas of active research and development that could lead to further advances.

The first area of development is toward a more integrated Earth-system approach. This is justified by the results obtained in the past two decades that showed that by adding relevant processes we can improve the quality of the existing forecasts, and further extend the forecast skill horizon at which dynamical forecasts lose their value. Buizza and Leutbecher (2015), for example, looked at the evolution of the skill of the ECMWF ensemble from 1994 to date and concluded that "*Forecast skill horizons beyond 2 weeks are now achievable thanks to major advances in numerical weather prediction. More specifically, they are made possible by the synergies of better and more complete models, which include more accurate simulation of relevant physical processes (e.g. the coupling to a dynamical ocean and ocean waves), improved data-assimilation methods that allowed a more accurate estimation of the initial conditions, and advances in ensemble techniques.*"

The second area is the design of better model uncertainty schemes to be used in all model components. To date, for example, the dynamical ocean, the ocean waves, and the sea-ice components do not include any model error scheme. Work is progressing also in trying to develop model error schemes, not any more adds-on, but as an integral part of each parameterization scheme (see, e.g., Raynaud et al., 2012; Piccolo and Cullen, 2016; Leutbecher et al., 2017; Lock et al., 2019).

The third area of development is a move toward seamless approaches. Such a move comes partly from scientific reasons and partly for technical reasons. From the scientific point of view, for example, there is evidence that processes that were thought to be relevant mainly for the extended range are also relevant for the short-range. An example comes from the introduction of a dynamical ocean in the ECMWF ensembles, which was initially included in the seasonal ensemble, then in the medium-range/monthly ENS and eventually also in the HRES system. From the technical point of view, using the same model in analysis and prediction model and from day 0 to year 1, simplifies production, maintenance, and the implementation of upgrades. It also helps the diagnostics and evaluation of a model version, since tests carried out over different time scales can help to identify undesirable behaviors that could lead to forecast errors.

The fourth area is the development and testing of higher-resolution models, to be able to resolve better the smaller scales, and their interaction with the slightly-less-smaller-scales. Furthermore, high-resolution is required if one wants to generate accurate and reliable forecasts of small scale, fast phenomena, which are affected, for example, by orographic forcing and sea-land contrasts.

The fifth area where progress should lead to improvements is in the computation of the ensemble ICs. Zagar (2017) highlighted the fact that, at the start of forecasts, "*the growth of forecast uncertainties takes place in all wavenumbers, and that the growth of uncertainty at large scales appears dominant over the impact of errors cascading up from small scales.*" This suggests that more work is still required to reduce the initial uncertainty not only for the small scales but also for the large scales. If this is achieved, Zhang et al. (2019), concluded that "*reducing the current-day initial-condition uncertainty by an order of magnitude extends the deterministic forecast lead times of day-to-day weather by up to 5 days.*" Ensembles of analyses and forecasts should be linked closer together, to improve the consistency in the simulation of initial uncertainties for all scales. To date, ICs for the different model components, more specifically for the atmosphere and the ocean, are generated using uncoupled or weakly coupled DA (Laloyaux et al., 2018; Browne et al., 2019). Furthermore, the ensembles' initial perturbations are generated in an uncoupled way. A move toward fully coupled ensembles of DA should lead to better ICs, and thus more reliable and accurate forecasts.

Let me conclude by saying that in the past 25+ years, ensembles have demonstrated to be extremely valuable and essential sources of information. In weather prediction, ensemble performance has been improving by about 2 days per decade in the medium-range (Buizza and Leutbecher, 2015), and even by up to 1 week per decade for the monthly time scale for large-scale phenomena such as the Madden Julian Oscillation (Vitart et al., 2012, 2014; Vitart, 2013). Their performance will continue to improve, provided that we can advance in the areas discussed above (modeling, including model error simulation, DA, and ensemble initialization, ensemble design, and membership).

6 Key learning points

In this chapter, we have:

- Introduced the process of NWP;
- Looked at the role of observations;
- Discussed what is an NWP model;
- Analyzed what are the key sources of forecast error, and highlighted the chaotic nature of the atmosphere;
- Introduced the concept of a probabilistic forecast, and how ensembles can be used to generate probabilistic forecasts;
- Reviewed the principal methods applied to simulate IC and model uncertainties;
- Shown examples of ensemble-based, probabilistic forecasts, and looked at how uncertainty/confidence information can be estimated and displayed in different ways.

Acknowledgments

The weather forecast maps shown in this chapter have been generated by the European Centre for Medium-Range Weather Forecasts (ECMWF) ensembles, whose role is acknowledged.

References

Berner, J., Shutts, G., Leutbecher, M., Palmer, T.N., 2008. A spectral stochastic kinetic energy backscatter scheme and its impact on flow-dependent predictability in the ECMWF ensemble prediction system. J. Atmos. Sci. 66, 603–626.

Berre, L., Pannekoucke, O., Desroziers, G., Stefanescu, S.E., Chapnik, B., Raynaud, L., 2007. A variational assimilation ensemble and the spatial filtering of its error covariances: increase of sample size by local spatial averaging. In: The Proceedings of the ECMWF Workshop on Flow-Dependent Aspects of Data Assimilationpp. 151–168 available from ECMWF, Shinfield Park, Reading RG2-9AX, U.K.

Bishop, C.H., Etherton, B.J., Majumdar, S.J., 2001. Adaptive sampling with the ensemble transform Kalman filter. Part I: theoretical aspects. Mon. Weather Rev. 129, 420–436.

Boffetta, G., Guliani, P., Paladin, G., Vulpiani, A., 1998. An extension of the Lyapunov analysis for the predictability problem. J. Atmos. Sci. 55, 3409–3416.

Bowler, N.E., Arribas, A., Mylne, K.R., Robertson, K.B., 2007. Numerical Weather Prediction: The MOGREPS Short-Range Ensemble Prediction System. Part I: System Description. UK Met. Office NWP Technical Report No. 497p. 18.

Bowler, N.E., Arribas, A., Mylne, K.R., Robertson, K.B., Shutts, G.J., 2008. The MOGREPS short-range ensemble prediction system. Q. J. Roy. Meteorol. Soc. 134, 703–722.

Bright, D.R., Mullen, S.L., 2002. Short-range ensemble forecasts of precipitation during the southwest monsoon. Weather Forecast. 17, 1080–1100.

Browne, P., de Rosnay, P., Zuo, H., Bennett, A., Dawson, A., 2019. Weakly coupled ocean-atmosphere data assimilation in the ECMWF NWP system. In: ECMWF Research Department Technical Memorandum n. 836.p. 28 available from ECMWF, Shinfield Park, Reading RG2 9AX, UK.

Buizza, R., 2014. The TIGGE medium-range, global ensembles. In: ECMWF Research Department Technical Memorandum n. 739. ECMWF, Shinfield Park, Reading RG2-9AX, UK, p. 53. http://www.ecmwf.int/sites/default/files/elibrary/2014/7529-tigge-global-medium-range-ensembles.pdf.

Buizza, R., 2018. Introduction to the Special Issue on '25 years of ensemble forecasting'. Q. J. Roy. Meteorol. Soc. 1–11. https://doi.org/10.1002/qj.3370.

Buizza, R., Leutbecher, M., 2015. The forecast skill horizon. Q. J. Roy. Meteorol. Soc. 141 (693 Pt. B), 3366–3382.

Buizza, R., Palmer, T.N., 1995. The singular vector structure of the atmospheric general circulation. J. Atmos. Sci. 52, 1434–1456.

Buizza, R., Tribbia, J., Molteni, F., Palmer, T.N., 1993. Computation of optimal unstable structures for a numerical weather prediction model. Tellus 45A, 388–407.

Buizza, R., Miller, M., Palmer, T.N., 1999. Stochastic representation of model uncertainties in the ECMWF EPS. Q. J. Roy. Meteorol. Soc. 125, 2887–2908.

Buizza, R., Houtekamer, P.L., Toth, Z., Pellerin, G., Wei, M., Zhu, Y., 2005. A comparison of the ECMWF, MSC, and NCEP global ensemble prediction systems. Mon. Weather Rev. 133, 1076–1097.

Buizza, R., Leutbecher, M., Isaksen, L., 2008. Potential use of an ensemble of analyses in the ECMWF ensemble prediction system. Q. J. Roy. Meteorol. Soc. 134, 2051–2066.

Charney, J.G., 1948. On the scale of atmospheric motions. Geofys. Publ. 17 (2), 1–17.

Coutinho, M.M., 1999. Ensemble prediction using principal-component-based perturbations. Thesis in Meteorology-National Institute for Space Research (INPE), p. 136 (in Portuguese).

Daley, R., 1991. Atmospheric Data Analysis. Cambridge University Press.

Epstein, E.S., 1969. A scoring system for probability forecasts of ranked categories. J. Appl. Meteorol. 8, 985–987.

Goo, T.-Y., Moon, S.-O., Cho, J.-Y., Cheong, H.-B., Lee, W.-J., 2003. Preliminary results of medium-range ensemble prediction at KMA: implementation and performance evaluation as of 2001. Korean J. Atmos. Sci. 6, 27–36.

Gray, M.E.B., Shutts, G.J., 2002. A stochastic scheme for representing convectively generated vorticity sources in general circulation models. In: APR Turbulence and Diffusion Note No. 285. Met Office, UK.

Hagedorn, R., Buizza, R., Hamill, M.T., Leutbecher, M., Palmer, T.N., 2012. Comparing TIGGE multi-model forecasts with re-forecast calibrated ECMWF ensemble forecasts. Q. J. Roy. Meteorol. Soc. 138, 1814–1827.

Hamill, T.M., Whitaker, J.S., 2007. Ensemble calibration of 500 hPa geopotential height and 850 hPa and 2-meter temperatures using re-forecasts. Mon. Weather Rev. 135, 3273–3280.

Holton, J.R., Hakim, G.J., 2012. An Introduction to Dynamic Meteorology, fifth ed. International Geophysics Series, vol. 88. Academic Press, p. 552 (ISBN 978-0123848666).

Hoskins, B.J., James, I.N., 2014. Fluid Dynamics of the Mid Latitude Atmosphere. Wiley, p. 423 (ISBN 978-0470795194).

Hou, D., Toth, Z., Zhu, Y., Yang, W., 2008. Impact of a stochastic perturbation scheme on NCEP global ensemble forecast system. In: Proceedings of the 19th AMS Conference on Probability and Statistics, 21–24 January 2008, New Orleans, Louisiana.

Houtekamer, P.L., Derome, J., Ritchie, H., Mitchell, H.L., 1996. A system simulation approach to ensemble prediction. Mon. Weather Rev. 124, 1225–1242.

Houtekamer, P.L., Mitchell, H.L., Deng, X., 2009. Model error representation in an operational ensemble Kalman filter. Mon. Weather Rev. 137, 2126–2143.

Houtekamer, P.L., Deng, X., Mitchell, H.L., Baek, S.-J., Gagnon, N., 2014. Higher resolution in an operational ensemble Kalman filter. Mon. Weather Rev. 142, 1143–1162.

Hurrell, J., Kushnir, Y., Ottersen, G., Visneck, M., 2003. An overview of the North Atlantic oscillation. In: Climatic Significance and Environmental Impact Geophysical Monograph 134. https://doi.org/10.1029/134GM01 Copyright 2003 by the American Geophysical Union.

Isaksen, L., Bonavita, M., Buizza, R., Fisher, M., Haseler, J., Leutbecher, M., Raynaud, L., 2010. Ensemble of data assimilations at ECMWF. In: ECMWF Research Department Technical Memorandum n. 636. Available from ECMWF, Shinfield Park, Reading RG2-9AX (see also http://old.ecmwf.int/publications/).

Kai, J., Kim, H., 2014. Characteristics of initial perturbations in the ensemble prediction system of the Korea Meteorological Administration. Weather Forecast. 29, 563–581.

Kalnay, E., 2002. Atmospheric Modelling, Data Assimilation and Predictability. Cambridge University Press, p. 368 ISBN-10: 0521796296, ISBN-13: 978-0521796293.

Lalaurette, F., 2003. Early detection of abnormal weather conditions using a probabilistic extreme forecast index. Q. J. Roy. Meteorol. Soc. 129, 3037–3057. https://doi.org/10.1256/qj.02.152.

Laloyaux, P., Balmaseda, M., Brönnimann, S., Buizza, R., Dahlgren, P., de Boisseson, E., Dee, D., Kosaka, Y., Haimberger, L., Hersbach, H., Martin, M., Poli, P., Scheppers, D., 2018. CERA-20C: a coupled reanalysis of the twentieth century. J. Adv. Model. Earth Syst. 10, 1–24. https://doi.org/10.1029/2018MS001273.

Leith, C.E., 1965. Theoretical skill of Monte Carlo forecasts. Mon. Weather Rev. 102, 409–418.

Leutbecher, M., Palmer, T.N., 2008. Ensemble forecasting. J. Comput. Phys. 227, 3515–3539.

Leutbecher, M., Lock, S.J., Ollinaho, P., Lang, S.T.K., Balsamo, G., Bechtold, P., Bonavita, M., Christensen, H.M., Diamantakis, M., Dutra, E., English, S., Fisher, M., Forbes, R.M., Goddard, J., Haiden, T., Hogan, R.J., Juricke, S., Lawrence, H., MacLeod, D., Magnusson, L., Malardel, S., Massart, S., Sandu, I., Smolarkiewicz, P.K., Subramanian, A., Vitart, F., Wedi, N., Weisheimer, A., 2017. Stochastic representations of model uncertainties at ECMWF: state of the art and future vision. Q. J. Roy. Meteorol. Soc. 143, 2315–2339.

Lin, J.W.B., Neelin, J.D., 2000. Influence of a stochastic moist convective parameterization on tropical climate variability. Geophys. Res. Lett. 27, 3691–3694.

Lock, S.J., Lang, S.T.K., Hogan, R.J., Vitart, F., 2019. Treatment of model uncertainty from radiation by the Stochastically Perturbed Parametrization Tendencies (SPPT) scheme and associated revisions in the ECMWF ensembles. Q. J. Roy. Meteorol. Soc. 145 (Suppl. 1), 75–89. https://doi.org/10.1002/qj.3570).

Lorenz, E.N., 1965. A study of the predictability of a 28-variable atmospheric model. Tellus 17, 321–333.

Lorenz, E.N., 1969a. The predictability of a flow which possess many scales of motion. Tellus XXI (3), 289–307.

Lorenz, E.N., 1969b. Atmospheric predictability as revealed by naturally occurring analogues. J. Atmos. Sci. 26, 636–646.

Lorenz, E.N., 1993. The Essence of Chaos. University of Washington Press, Seattle, p. 227.

Miyoshi, T., Sato, Y., 2007. Assimilating satellite radiances with a local ensemble transform Kalman filter (LETKF) applied to the JMA global model (GSM). SOLA 3, 37–40.

Mogensen, K., Alonso Balmaseda, M., Weaver, A., 2012a. The NEMOVAR ocean data assimilation system as implemented in the ECMWF ocean analysis for System 4. In: ECMWF Research Department Technical Memorandum n. 668.p. 59 Available from ECMWF, Shinfield Park, Reading RG2-9AX (see also http://old.ecmwf.int/publications/).

Mogensen, K., Keeley, S., Towers, P., 2012b. Coupling of the NEMO and IFS models in a single executable. In: ECMWF Research Department Technical Memorandum n. 673.p. 23 Available from ECMWF, Shinfield Park, Reading RG2-9AX (see also http://old.ecmwf.int/publications/).

Molteni, F., Buizza, R., Palmer, T.N., Petroliagis, T., 1996. The ECMWF ensemble prediction system: methodology and validation. Q. J. Roy. Meteorol. Soc. 122, 73–119.

Palmer, T.N., Buizza, R., Doblas-Reyes, F., Jung, T., Leutbecher, M., Shutts, G.J., Steinheimer, M., Weisheimer, A., 2009. Stochastic parametrization and model uncertainty. In: ECMWF Research Department Technical Memorandum No. 598.p. 42 available from ECMWF, Shinfield Park, Reading RG2-9AX, UK.

Piccolo, C., Cullen, M., 2016. Ensemble data assimilation using a unified representation of model error. Mon. Weather Rev. 144, 213–224.

Rabier, F., Järvinen, H., Klinker, E., Mahfouf, J.F., Simmons, A., 1999. The ECMWF operational implementation of four dimensional variational assimilation. Part I: experimental results with simplified physics. In: ECMWF Research Department Technical Memorandum No. 271.p. 26 available from the ECMWF website: http://www.ecmwf.int/en/elibrary/technical-memoranda.

Raynaud, L., Berre, L., Desrozier, G., 2012. Accounting for model error in the Météo-France ensemble data assimilation system. Q. J. Roy. Meteorol. Soc. 138, 249–262.

Richardson, L.F., 1922. Weather Prediction by Numerical Process. Cambridge University Press, Dover, New York Report.

Shutts, G., 2005. A kinetic energy backscatter algorithm for use in ensemble prediction systems. Q. J. Roy. Meteorol. Soc. 131, 3079–3100.

Stockdale, T., Alonso-Balmaseda, M., Johnson, S., Ferranti, L., Molteni, F., Magnusson, L., Tietsche, S., Vitart, F., Decremer, D., Weisheimer, A., Roberts, C.D., Balsamo, G., Keeley, S., Mogensen, K., Zuo, H., Mayer, M., Monge-Sanz, B.M., 2018. SEAS5 and the future evolution of the long-range forecast system. In: ECMWF Research Department Technical Memorandum n. 835.p. 81 available from ECMWF, Shinfield Park, Reading RG2 9AX, UK.

Su, X., Yuan, H., Zhu, Y., Luo, Y., Wang, Y., 2014. Evaluation of TIGGE ensemble predictions of Northern hemisphere summer precipitation during 2008-2012. J. Geophys. Res. Atmos. 119 (12), 7292–7310.

Toth, Z., Kalnay, E., 1993. Ensemble forecasting at NMC: the generation of initial perturbations. Bull. Am. Meteorol. Soc. 74, 2317–2330.

Toth, Z., Kalnay, E., 1997. Ensemble orecasting at NCEP and the breeding method. Mon. Weather Rev. 125, 3297–3319.

Vitart, F., 2013. Evolution of ECMWF sub-seasonal forecast skill scores over the past 10 years. In: ECMWF Research Department Technical Memorandum No. 694.p. 28 available from ECMWF, Shinfield Park, Reading RG2-9AX, UK.

Vitart, F., Robertson, A., Anderson, D., 2012. Sub-seasonal to seasonal prediction project: bridging the gap between weather and climate. WMO Bull. 61 (2), 23–28.

Vitart, F., Balsamo, G., Buizza, R., Ferranti, L., Keeley, S., Magnusson, L., Molteni, F., Weisheimer, A., 2014. Sub-seasonal predictions. In: ECMWF Research Department Technical Memorandum No. 738.p. 45 Available from ECMWF, Shinfield Park, Reading, RG2-9AX, UK (see also: http://www.ecmwf.int/sites/default/files/elibrary/2014/12943-sub-seasonal-predictions.pdf).

Von Neumann, J., Richtmeyer, R.D., 1950. A method for the numerical calculation of hydrodynamical shocks. J. Appl. Phys. 21, 232.

Wedi, N.P., Bauer, P., Deconinck, W., Diamantakis, M., Hamrud, M., Kühnlein, C., Malardel, S., Mogensen, K., Mozdzynski, G., Smolarkiewicz, P.K., 2015. The modelling infrastructure of the integrated forecasting system: recent advances and future challenges. In: ECMWF Research Department Technical Memorandum No. 760..

Wei, M., Toth, Z., Wobus, R., Zhu, Y., Bishop, C., Wang, X., 2006. Ensemble Transform Kalman Filter-based ensemble perturbations in an operational global prediction system at NCEP. Tellus A 58, 28–44.

Wei, M., Toth, Z., Wobus, R., Zhu, Y., 2008. Initial perturbations based on the ensemble transform (ET) technique in the NCEP global operational forecast system. Tellus A 60, 62–79.

Wilks, D., 2005. Statistical Methods in the Atmospheric Sciences, second ed. vol. 100. Academic Press, p. 648 ISBN: 9780080456225.

Yamaguchi, M., Majumdar, S.J., 2010. Using TIGGE data to diagnose initial perturbations and their growth for tropical cyclone ensemble forecasts. Mon. Weather Rev. 138, 3634–3655.

Zagar, N., 2017. A global perspective of the limits of prediction skill of NWP models. Tellus A. 69, 131573.

Zhang, Z., Krishnamurti, T.N., 1999. A perturbation method for hurricane ensemble predictions. Mon. Weather Rev. 127, 447–469.

Zhang, F., Sun, Y.Q., Magnusson, L., Buizza, R., Lin, S.-J., Chen, J.-H., Emanuel, K., 2019. What is the predictability limit of midlatitude weather? J. Atmos. Sci. 76, 1077–1091. https://doi.org/10.1175/JAS-D-18-0269.1.

CHAPTER

4

Predictability

Roberto Buizza

Scuola Superiore Sant'Anna, Pisa, Italy Centre for Climate Change studies and Sustainable Actions (3CSA), Pisa, Italy

1 Predictability, error growth and uncertainty

We discuss predictability in the context of a deterministic dynamical system, for which we can write a set of equations that describes its evolution:

$$\frac{dx}{dt} = f(x, t) \tag{1}$$

where x denotes the state variables of the system, t denotes time, and $f(x,t)$ is the forcing term. The system is "deterministic", in the sense that the evolution equation does not contain any stochastic term. The system can be simple, with few degrees of freedom, or complex with many millions of degrees of freedom, like the real atmosphere. The forcing term $f(x,t)$ can include nonlinear terms.

If we know the forcing term $f(x,t)$ and we know the state of the system at a time t_0, $x_0 \equiv x(t_0)$, we can compute the states of the system at the future times $t > t_0$by integrating in time the set of dynamical equations (1):

$$x(t) = x_0 + \int_{t_0}^{t} f(x, \tau)\, d\tau \tag{2}$$

If we knew the initial state x_0 ***exactly***, and we could compute the time integral ***exactly***, we would have infinite predictability. In other words, we would be able to predict all future states with no error.

This is not the case for real systems, such as the atmosphere: (a) we do not know exactly the initial state, (b1) we do not know the function $f(x,t)$ but we can only write a set of approximate equations that describes the real physical processes relevant to the system's evolution, and (b2) we can solve the approximate set of equations only approximately using numerical methods. In other words, there are initial uncertainties [point (a) above] and model

Uncertainties in Numerical Weather Prediction
https://doi.org/10.1016/B978-0-12-815491-5.00004-5

uncertainties, due either to poor knowledge of the real dynamics [point (b1) above] or to the fact that the model equations can only be integrated using numerical methods, which all have inherent sources of approximation [point (b2) above].

Since we can only solve numerically a finite-difference approximation of Eq. (2), our numerical solution, our forecast:

$$\tilde{x}(t) = \tilde{x}_0 + \sum_{j=1}^{N} F(x, \tau_j) \triangle \tau_j \tag{3}$$

will not be perfect, and we will be able to predict the system time evolution only for a limited amount of time. In other words, predictability will be limited in time.

Let's define the initial and the forecast errors, respectively, as:

$$\delta x_0 \equiv \delta x(t_0) \equiv \tilde{x}(t_0) - x(t_0) \tag{4a}$$

$$\delta x(t) \equiv \tilde{x}(t) - x(t) \tag{4b}$$

Let's introduce the norm $\|\ldots\|_E^2$:

$$\|x\|_E^2 = \langle x|\, Ex\rangle \tag{5}$$

that will be used to measure how close two states are, and thus also how close a forecast is to the real state of the system. This norm is defined by a metric E, which could be the total energy, or the Euclidean norm of the state vector x (i.e., E is a diagonal matrix with unit values for all the diagonal terms).

Define two states "almost identical", or "close enough" if they are at a distance which is smaller than ε. In other words, if one of the two states is a forecast and the other the real state of the system, let's say that we consider the forecast "of a good enough quality", if its distance from reality is smaller than ε. We can then define the predictability limit as the forecast time $T_{lim} > t_0$ for which the error reaches, for the first time, ε:

$$\begin{cases} \|\delta(t)\|_E^2 < \varepsilon \quad for \quad t \le T_{lim} \\ \|\delta(t)\|_E^2 \ge \varepsilon \;\; for \quad t = T_{lim} \end{cases} \tag{6}$$

The forecast error $\delta x(t) \equiv \tilde{x}(t) - x(t)$ can be different from zero and become larger than the acceptable value ε, for the three reasons (a, b1 and b2) mentioned above, which can be linked to two types of uncertainties:

- ***Initial uncertainties***: we do not know exactly the initial conditions, or in other words there are initial uncertainties (linked to reason a);
- ***Model uncertainties***: we do not know the function $f(x,t)$, and or we have solved the set of equations in an approximate way (linked to reasons b1 and b2).

Fig. 1 is a schematic of the impact of initial and model uncertainties. It shows four different curves, defined to illustrate the impact of initial and model uncertainties:

- The solid black curve (in all four panels) shows the "correct" system evolution: it has been computed by starting from unperturbed initial conditions, and integrating the unperturbed model equation;

- The dotted black curve and the three red curves show three forecasts, made by integrating slightly different model equations starting from slightly different initial conditions;

The black curve could represent reality, and the red curves three forecasts, all affected by initial and/or model uncertainties. We can define the predictability time limit as the time when the error (i.e., the difference between the "correct" trajectory and an approximate one) becomes larger than 1 unit.

Fig. 2 shows that the squared-error evolution for the three forecasts: it shows that the time limit is 60 time units for the black dotted forecast, and 37, 15, and 20 time unites for the three red forecasts.

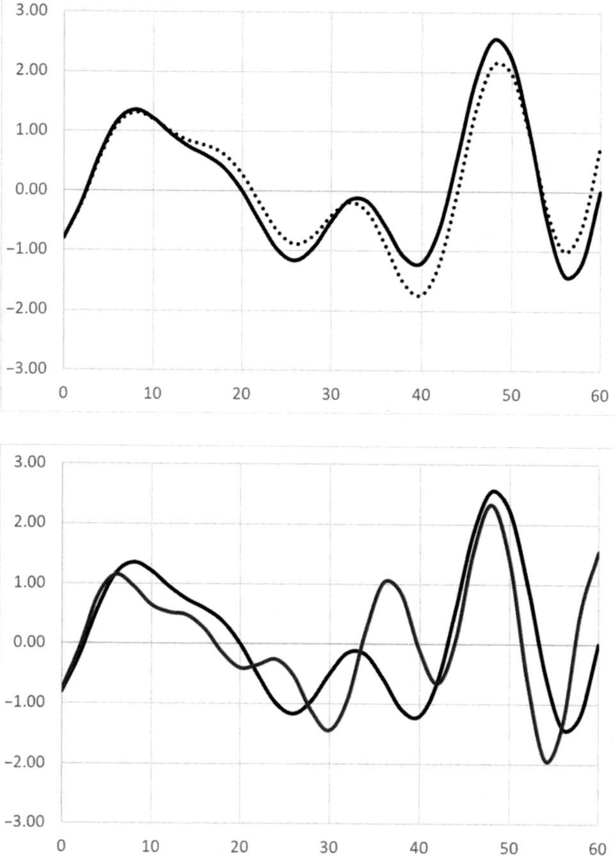

FIG. 1 Examples of sensitivity to initial and model uncertainties on the evolution of a dynamical system. The black solid lines in all panels show the "correct" system trajectory, computed by solving the "correct" equations from the "correct" initial state, while the other lines represent integrations with initial and/or model uncertainties. The dotted-black line in the top-left panel shows a trajectory started from an initial state that has a small error, computed by solving the "correct" equations. The red lines in the other three panels shows the effect of both starting from an initial state that has a small error, and using an approximate set of equations.

(continued)

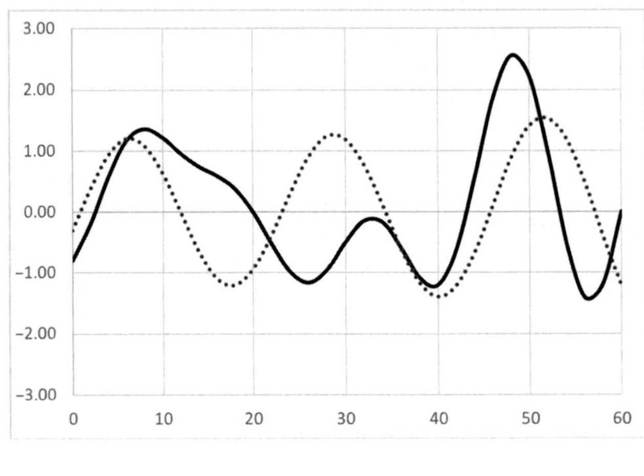

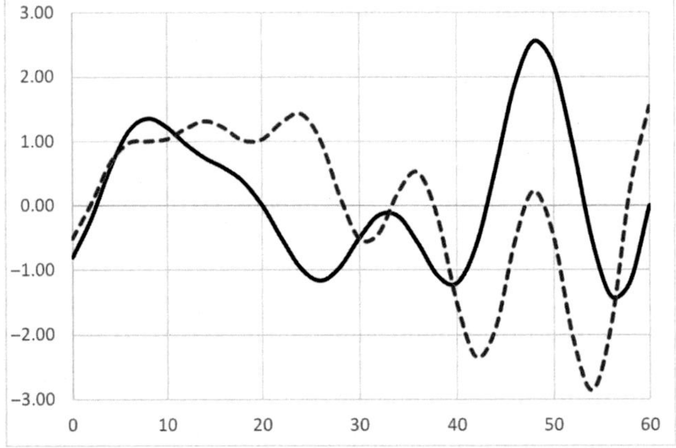

FIG. 1, cont'd If we define the predictability time limit as the time when the error (i.e., the difference between the "correct" trajectory and an approximate one) becomes larger than 1 unit, we can see that in the top-left panel 1 $T_{lim} = 60$, in the top-right panel $T_{lim} = 36$, in the bottom-left panel $T_{lim} = 14$ and in the bottom-right panel $T_{lim} = 20$.

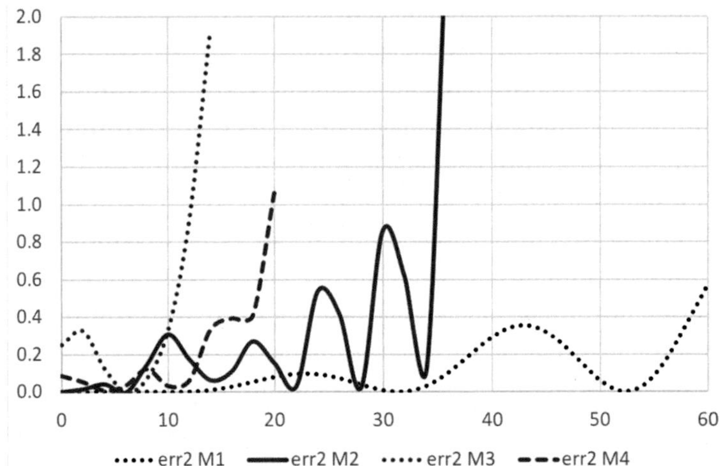

FIG. 2 Squared differences between the correct trajectory and the trajectories computed with initial and/or model uncertainties, shown in Fig. 1. Each curve has been interrupted at the time when the predictability limit, defined as when $\|\delta x(T_{lim}\|_E^2 = 1$, was reached: $T_{lim} = 60$ for the first curve shown in the top-left panel of Fig. 1 (black-dotted line), $T_{lim} = 36$ for the second curve shown in the top-right panel of Fig. 1 (solid red line), $T_{lim} = 14$ for the third curve shown in the bottom-left panel of Fig. 1 (red dotted line), and $T_{lim} = 20$ for the fourth curve shown in the bottom-right panel of Fig. 1 (red dashed line).

This simple example illustrates the link between predictability, error growth, and sources of uncertainty. These are the topics that we will be discussing, with a focus on weather prediction.

2 Error growth, and scale-dependent predictability

Charney et al. (1966) and Smagorinsky (1969) were among the first to study the error growth rate of weather forecasts, and to investigate how predictable the weather is. They estimated, respectively, error doubling times of weather forecasts of 5 days and 3 days. With the magnitude of the initial error that was characterizing weather forecasts in the 1960s, these doubling-time estimates implied a predictability limit of about two weeks.

How difficult it is to improve weather prediction was pointed out already in the 1960s. Lorenz (1969), for example, using a two-dimensional vorticity equation, showed that "*... there may be some systems where a reduction of the initial error will not increase the range of predictability at all.*" He pointed out that whether this happens or not, will depend on the steepness of the spectrum of kinetic energy in the flow.

Rotunno and Snyder (2008) also discussed this point, and concluded that for spectra shallower than K^{-3} with K being the total wave number, the predictability is limited to a finite time as long as there is some initial error regardless how small. Going back to Lorenz (1969), he confirmed Charney et al. (1966) and Smagorinsky (1969)'s results that predictability was about two weeks: more precisely, he estimated the limit being 16.8 days. In his discussion, Lorenz (1969) said that "*... if the theory of atmospheric instability were correct, one flap of a sea gull's wings would forever change the future course of the weather*". This is the first time the famous "butterfly effect" is mentioned, although originally the text talked about a sea gull's wings rather than a butterfly's ones.

An update of the error growth rates came from Lorenz (1982), who examined the error of forecasts from the European Centre for Medium-Range Weather Forecasts (ECMWF). He fitted an error growth model to the 10-day forecast error curves, and estimated error doubling times of about 2 days for the small scales. He thought that his estimates were more correct than the earlier one, since the earlier ones did not include all scales of motion, and neglected the fact that errors grow faster for the smaller the spatial scale.

Dalcher and Kalnay (1987) extended the Lorenz (1982) model and applied it also to ECMWF forecasts. They estimated the theoretical limit of dynamical predictability for different spatial scales, identified by their total wave number, and investigated its seasonal sensitivity. They concluded that the limit is longer in winter than in summer: for the low-frequency, larger waves, they estimated that in winter it is longer than 10 days, while in summer it is about 10 days. Simmons and Hollingsworth (2002) also revisited the error growth model of Lorenz (1982) and estimated error doubling times as low as 1.5 d for 500 hPa geopotential in the Northern Hemisphere extra-tropics in boreal winter. Despite the much shorter error doubling times of 1.5 days, compared to the estimates of 5 days by Charney et al. (1966), the 3 days by Smagorinsky (1969), and the 2 days of Lorenz (1982), they discussed the possibility that there could be predictive skill up to about 3 weeks, if one applied bias-correction to the forecasts.

Thus, while in the 1960s skillful forecasts beyond two weeks were thought impossible, at the end of the 1990s, more statistically robust evidence started to emerge that subseasonal forecasts could be achieved. Scientists started discussing the fact that predictability can vary substantially for different phenomena, and in the 1970s and 1980s several studies (see e.g., Schukla, 1981) showed that predictability is scale dependent, with the long-waves being more predictable than the short ones, if they could be properly initialized.

Schukla and Kinter III (2006) review how the slowly varying components of the Earth system could bring long-range predictability. It is worth quoting directly from their Chapter 12.3 which Earth-system's components could bring it:

- "The sea surface temperature, which governs the convergence of moisture flux as well as sensible and latent heat fluxes between the ocean and the atmosphere;
- The soil moisture, which alters the heat capacity of the land surface and governs the latent heat flux between continents and the atmosphere;
- Vegetation, which regulates surface temperature, as well as the latent heat flux to the atmosphere from land surfaces;
- Snow, which participates in the surface radiative balance through its effect on surface albedo and in the latent heat flux, introducing a lag due to the storage of water in solid form in winter which is melted and evaporated in the spring and changes the soil wetness;
- Sea ice, which likewise participates in the energy balance and inhibits latent heat from the ocean."

As evidence of potential sources of long-range predictability became evident, meteorological centers started developing operational capabilities for these time scales, to complement the existing operational short- (up to 72 h) and medium-range (up to 15 days) operational forecasting systems. ECMWF was one of the key centers at the forefront in subseasonal and seasonal prediction since the mid 1980's, when they started experimentation to develop a monthly prediction system (Molteni et al., 1987; Brankovic et al., 1990). The monthly prediction system was introduced in operation few years later, in 2002, using a coupled ocean–atmosphere model (Vitart, 2004) as for the seasonal system.

In parallel to the development of a subseasonal system, ECMWF explored also the possibility to develop a seasonal prediction system, and in 1997 it implemented the seasonal prediction system-1 (Anderson, 2006). System-1 was followed by upgrades every 5-7 years: in 2002, system-1 was replaced by system 2, in 2007 by system-3, in 2011 by system-4 and in 2017 by system-5 (Stockdale et al., 2019).

One reasonable question to ask would be: why use different models to predict the different time scales? Is there a "predictability" reason why one should follow this approach?

Provided that a model does not show any significant drift and bias increase with forecast time, in principal there is no reason why a forecasting system designed for the medium, subseasonal or seasonal ranges should not be seamless, possibly even the same. Actually, it could simplify the development, testing, and diagnostic.

The main reason why forecasts for these ranges are often done using disjoint/stand-alone system is linked to computer resources' availability. A simple back-of-the-envelope calculation illustrates why such an approach is required. Let's say that a typical national meteorological service running high-resolution, global ensembles devote about 50% of their supercomputing facilities for 2 h to generate a 15-day ensemble forecast. Suppose this is done twice a day, and suppose that about the same time and resources are needed to generate the ensemble's initial conditions. This means that about 50% of their super-computer is allocated for these two tasks for about 8h. If they were to generate a subseasonal (say a 6-week) ensemble forecast with the same resolution, say once a day, they would need to extend the medium-range ensemble by another 4 weeks, which would require another 4h of 50% of the super computer. This brings the total to about 12h.

If they wanted also to cover the seasonal range, they would require to extend the subseasonal forecasts for 18 more weeks, which would require another 18h of 50% of the super computer, which are not available! Please also note that these estimates have not taken into consideration other operational tasks, such as observations' processing, products' generation, verification, and diagnostics: for example, they have not taken into account the generation of the ensemble re-forecasts, required to generate some of the probabilistic forecasts, which would increase the cost by about 25%. Thus, it should be evident that using the same resolution up to the seasonal time range would require more computing resources than what is usually allocated.

The second reason why resolution is gradually reduced to produce longer range forecasts, is the fact that forecasts loose predictability on the small scales first, and thus it is questionable whether they should be included in the forecast system also for forecast lenghts after the time when the model was not caple to skilfully predict them. It would still make sense if the impact of explicitly resolving the small scales on the predictability of the larger scales is strong: results suggest that this is the case only for a very short time after the model lost predictability, say up to 1-2 days after that time (see, e.g., the discussions in Buizza et al., 2007 on the impact of using variable resolution on ensemble forecasts).

Given that computer resources are limited, the combination of these two aspects makes it cost-effective to use lower resolution to produce longer range forecasts.

Scale-dependent predictability is illustrated in Fig. 3, which visualizes the fact that the forecast skill horizon (vertical axis, from short to long times) skill depends on the spatial and temporal scales of the forecast phenomena (horizontal axis, from small/fast to large/slow scales). In the diagram, the x-axis is the horizontal spatial scale of predicted aspects of the weather (in km), and the y-axis is the forecast skill range, in days. The diagram was built using results from Buizza and Leutbecher (2015), when the ECMWF ensemble forecasts, used to produce most of the estimates, was 32 km: the gray rectangular to the left of $x = 32$ km identifies the scales that are definitely unresolved in the ECMWF medium-range ensemble at the time when this diagram was produced (at that time, it had a grid spacing of about 32 km, while now it has a grid spacing of about 18 km up to forecast day 15, and 36 km from day 15 to the seasonal time scale). See the section "Further readings" to read more about how this diagram was designed to summarize our current view of predictability.

Fig. 3 shows skill estimates for short- and extended-range forecasts of different fields, from the surface (e.g.. the 2-m temperature and precipitation) to upper levels, and large-scale patters identified by variability indices such as the North Atlantic Oscillation (NAO, Hurrell et al., 2003), the Madden Julian Oscillation (MJO, Madden and Julian, 1971) and El Nino Southern Oscillation (Trenberth, 1997). The red lines of the fast and finer-scale surface variables are closer to the x-axis, illustrating the fact that surface variables are less predictable. By contrast, the blue lines of the large-scale patterns identified by teleconnection indices (e.g., linked to the NAO and the MJO), or monthly average sea-surface temperature (SST) in the Pacific regions affected by El Niño are further away from the x-axis and closer to the top-right part of the diagram, since they can be skilfully predicted months ahead.

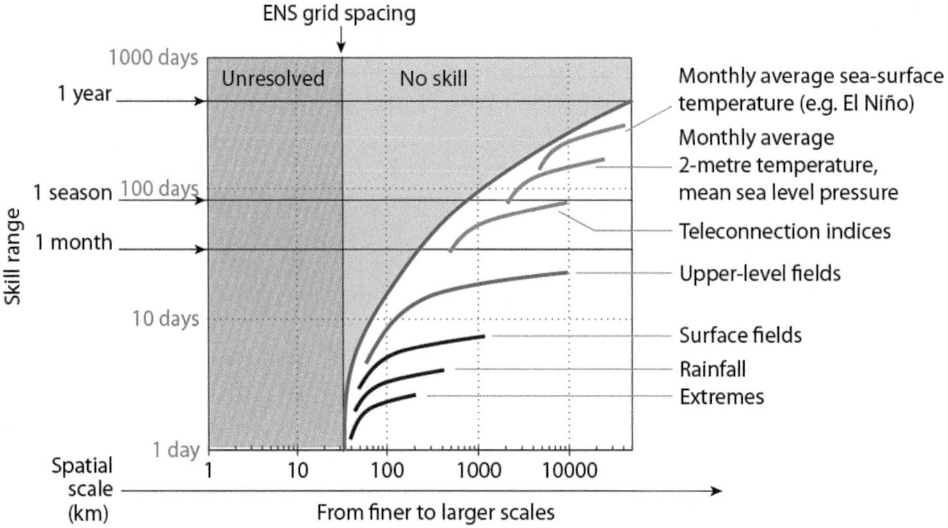

FIG. 3 The Forecast Skill Horizon diagram, and predictability estimates for different time scales and phenomena. *From Buizza, R., Leutbecher, M., Thorpe, A., 2015. Living with the butterfly effect: a seamless view of predictability. ECMWF newsletter no. 145, autumn 2015, pp. 18–23 (Available from ECMWF, Shinfield Park, Reading RG2-9AX, UK); see text for more details.*

Two further features have been drawn in the diagram: a solid blue line envelope that includes all the individual lines, and a pink "no-skill" region. The blue line represents the forecast skill horizon today: still rather short, less than 10 days, for very detailed forecasts of surface variables such as precipitation, but longer and up to a year for monthly average SST forecasts for regions in the tropical Pacific.

As already mentioned above, in the 1960s and 1970s, it was thought that the forecast skill horizon was about 2 weeks (see, e.g., Lorenz, 1965, 1969). It is only later that people started realizing that it is scale dependent, and that certain low-frequency, large scales could be predicted for longer than 2 weeks. Thus, we have moved from an image of the forecast skill horizon being a straight line parallel to the x-axis with a y-value of about 2 weeks, to the blue line shown in Fig. 3. A line that is not straight and parallel to the x-axis, but is curved, reflecting the fact that the forecast skill horizon is scale-dependent, variable-dependent, area-dependent, and season-dependent.

This has been achieved because, today, the initial conditions are more accurate, especially for the larger scales, models are better and include more processes, and we have moved from single to ensemble-based probabilistic forecasts (Buizza, 2018). The combination of these facts led to improved initial condition, especially for the planetary and synoptic scales, to slower error growth rates, and to more consistent and reliable forecasts. Furthermore, the inclusion of more relevant processes (such as the coupling of atmospheric models to a wave, sea ice, and 3-dimensional dynamical ocean, and improvements in the simulation of land-surface processes), proper (although still weakly) coupled initialization, make it possible to extract predictable signals from the initial conditions, and contrast the error growth from the smallest scales.

Let us just remind that the fact that predictability was extended beyond what was thought to be the predictability limit in the 1970s, does not negate that it is finite. This was confirmed, for example, by Rotunno and Snyder (2008) and Durran and Gingrich (2014), who linked it to the fact that in the meso-scales the kinetic energy spectrum is shallow and depends on horizontal wave number as $K^{-5/3}$.

3 Metrics to measure forecast error and forecast skill

Four of the most-commonly used metrics applied to compute forecast error are the mean-absolute error (MAE), the root-mean-square-error (RMSE), and the anomaly correlation coefficient (ACC) for single forecasts, and the continuous ranked probability score (CRPS, Brown, 1974, Hersbach, 2000) for probabilistic forecasts. The reader is referred to Wilks (2019) for a review of accuracy measures.

The CRPS is an extension of the Ranked Probability Score (Epstein, 1969; Murphy, 1971). Compared to the RPS, the CRPS has two advantages: it is sensitive to the entire range of the parameter of interest and it does not rely on predefined intervals (i.e., it is not restricted to fixed intervals as the RPS is). The CRPS is the limit of the RPS for an infinite number of intervals, and is one of the most commonly used metrics of probabilistic forecast accuracy.

Consider an area Σ that includes N grid points x_j, a single forecast $f(x_j;t_0,t)$, issued at time t_0 and valid for time + t, a corresponding probabilistic forecast $p(x_j;t_0,t)$, and a verification field $a(x_j;t_0+t)$, defined either directly by observations or by an analysis field (see, e.g., Chapter 2 for a definition of the analysis).

The mean-absolute error is defined as the mean absolute difference, computed inside the area Σ, between the forecast and the verification:

$$MAE=\frac{1}{N}\sum_{j=1}^{N}\left|f(x_j;t_0,t)-a(x_j;t_0+t)\right| \tag{7}$$

The root-mean-square-error is defined as the root of the mean-squared difference:

$$RMSE=\sqrt[2]{\frac{1}{N}\sum_{j=1}^{N}\left(f(x_j;t_0,t)-a(x_j;t_0+t)\right)^2} \tag{8}$$

Another measure used to assess forecast accuracy is the anomaly correlation coefficient:

$$ACC=\frac{\sum_{j=1}^{N}\left(f(x_j;t_0,t)-c(x_j;t_0+t)\right)\left(a(x_j;t_0+t)-c(x_j;t_0+t)\right)}{\sqrt{\sum_{j=1}^{N}\left(f(x_j;t_0,t)-c(x_j;t_0+t)\right)2}\sqrt{\sum_{j=1}^{N}\left(a(x_j;t_0+,t)-c(x_j;t_0+t)\right)2}} \tag{9}$$

Consider now a probabilistic forecast $p(x_j;t_0,t)$, defined by a probability density function $g(x_j;t_0,t;p)$. Compute the forecast and verification cumulative distribution functions at each grid point x:

$$CDF_g(x_j;t_0,t;p)=\int_{-\infty}^{p}g(x_j;t_0,t;q)dq \tag{10a}$$

$$CDF_a(x_j; t_0, t; p) = \int_{-\infty}^{p} a(x_j; t_0, t; q)dq \tag{10b}$$

The verification PDF is a very narrow distribution, with a width that depends on the analysis, or observation, error/uncertainty. If the verification is treated as perfect (i.e., without any error), then the PDF is a delta-function centered on the verification value, and the CDF_a becomes a Heaviside function.

The CRPS at each grid point is defined as:

$$CRPS(x_j; t_0, t) = \int_{-\infty}^{\infty} [CDF_g(x_j; t_0, t; p) - CDF_a(x_j; t_0 + t; p)]^2 dp \tag{11}$$

The CRPS inside the area Σ is given by:

$$CRPS = \frac{1}{N}\sum_{j=1}^{N} CRPS(x_j; t_0, t) \tag{12}$$

The CRPS is zero for a perfect forecast, when the forecast and observed CDFs are identical. For any score SC, the skill score SS_{sc} is defined as:

$$SS_{sc} = \frac{SC(ref) - SC(fc)}{SC(ref)} \tag{13}$$

an index based on the relative difference between the score of the forecast and the score of a reference, for example, climatology. A perfect forecast has skill score 1, while a forecast that has the same score that the reference has a zero skill score.

If we consider, for example, the CRPS, its equivalent skill score is given by:

$$CRPSS = \frac{CRPS(clim) - CRPS(fc)}{CRPS(clim)} = 1 - \frac{CRPS(fc)}{CRPS(clim)} \tag{14}$$

4 An error growth model

The forecast error time evolution can be studied applying the Simmons et al. (1995)'s version of the error growth logistic equation of Dalcher and Kalnay (1987). These authors included in the Lorenz (1982) model both the systematic and the random error components, and nonlinear error saturation, and they described the time evolution of the forecast error E is given by the following equation:

$$\frac{dE}{dt} = \propto E - \beta E^2 + \gamma \tag{15}$$

where E is a measure of forecast error.

Error growth saturates due to nonlinear processes, and the quadratic terms, which for example, "*... represent the advection of the temperature and velocity fields...*" (Lorenz, 1969), are of primary importance. The quadratic term is required since "*... Under the assumption that the principal nonlinear terms in the atmospheric equations are quadratic, the nonlinear terms in the equations governing the field of errors will also be quadratic.*".

The error E can be measured by the RMSE or the MAE for single forecasts, or the CRPS for probabilistic forecasts. The parameters of Eq. (15) can be written in a discretized form, and the three parameters (α,β,γ) can be estimated by a least-squares fit of the error.

Note that Eq. (15) can also be written as:

$$\frac{dE}{dt} = (\propto E + S)\left(1 - \frac{E}{E_\infty}\right) \tag{16}$$

where

$$\begin{cases} S \equiv \gamma \\ E_\infty \equiv \dfrac{\alpha}{2\beta} + \sqrt{\dfrac{\alpha^2}{4\beta^2} + \dfrac{\gamma}{\beta}} \\ a \equiv \beta E_\infty \end{cases} \tag{17}$$

The solution of Eq. (15) is given by:

$$\eta(t) \equiv \frac{E(t)}{E_\infty} = 1 - \frac{1}{1 + C_1 e^{C_2 t}}\left(1 + \frac{S}{aE_\infty}\right) \tag{18}$$

with

$$\begin{cases} C_1 \equiv \dfrac{1}{E_\infty - E_0}\left(E_0 + \dfrac{S}{a}\right) \\ C_2 \equiv a + \dfrac{S}{E_\infty} \end{cases} \tag{19}$$

where E_0 is the initial-time error.

As Dalcher and Kalnay (1987) indicated, we can interpret a as the initial rate of growth of the forecast error, S as the effect of model error on the error growth, and E_∞ as the error asymptotic value. The (α,β,γ) parameters of Eq. (15) can be determined by fitting the curve to the error data: once these parameters have been computed the coefficients a, S, *and* E_∞ can be determined, and the analytical solution can be computed (see, e.g., Buizza, 2010). Then, one can compute the predictability limit by finding the forecast time τ (95%), when forecast error equals 95% of the asymptotic value E_∞, that is, when, Alternatively, one can define the time limit as τ (71%), the time when the forecast error equals $1/\sqrt{2}$ of the asymptotic value E_∞.

The error-growth model can also be used to extrapolate the error curves beyond the forecast range covered by the forecast data, and normalized error-growth curves can be used to compare the error growth for different years, or for sensitivity studies.

As an example, the error growth model has been applied to investigate the trend in error growth of the ECMWF medium-range ensemble (ENS), between 1996 and 2016. Figs. 4 and 5 show the time evolution of the MAE of the unperturbed member (ENS-control) and the CRPS for the 500 hPa geopotential height over Northern Hemisphere extra-tropics for the summer (June–July–August, JJA) and winter (December–January–February, DJF) seasons of few years. Normalized curves are shown: they have been computed by fitting the Eq. (12) to annual summer and winter average values, and then by normalizing the curve by the asymptotic value, as in Eq. (15).

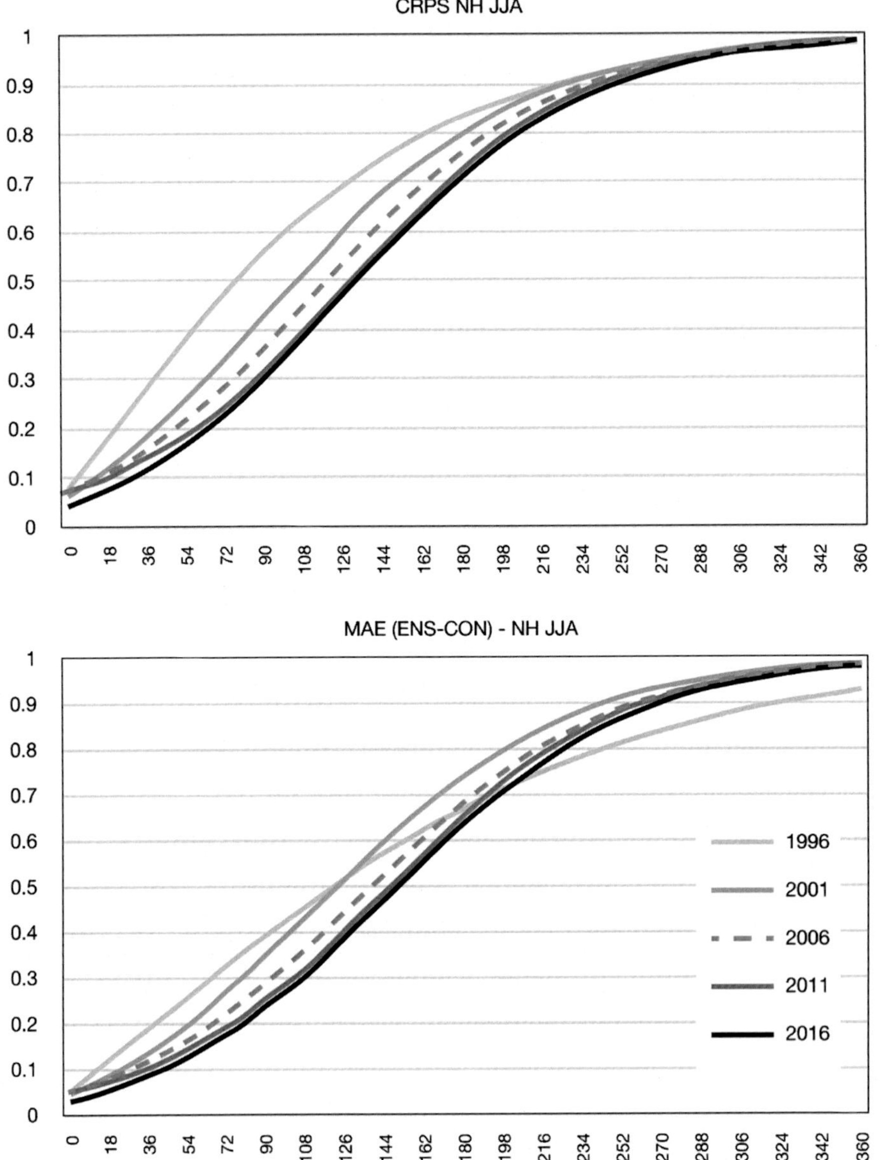

FIG. 4 Top panel: forecast-time evolution of the summer (JJA) average CRPS of the 500hPa geopotential height over the NH, every 5years from 1996 (lighter curve) to 2016 (darkest curve). For each season, the normalized error-growth curve defined in Eq. (6), with coefficients computed by fitting data between forecast t+24h and t+240h, is shown. Each curve has been extended beyond t+240h. Bottom panel: as top panel but for the MAE of the ENS-control. Forecasts have all been verified against analyses (see text for more details).

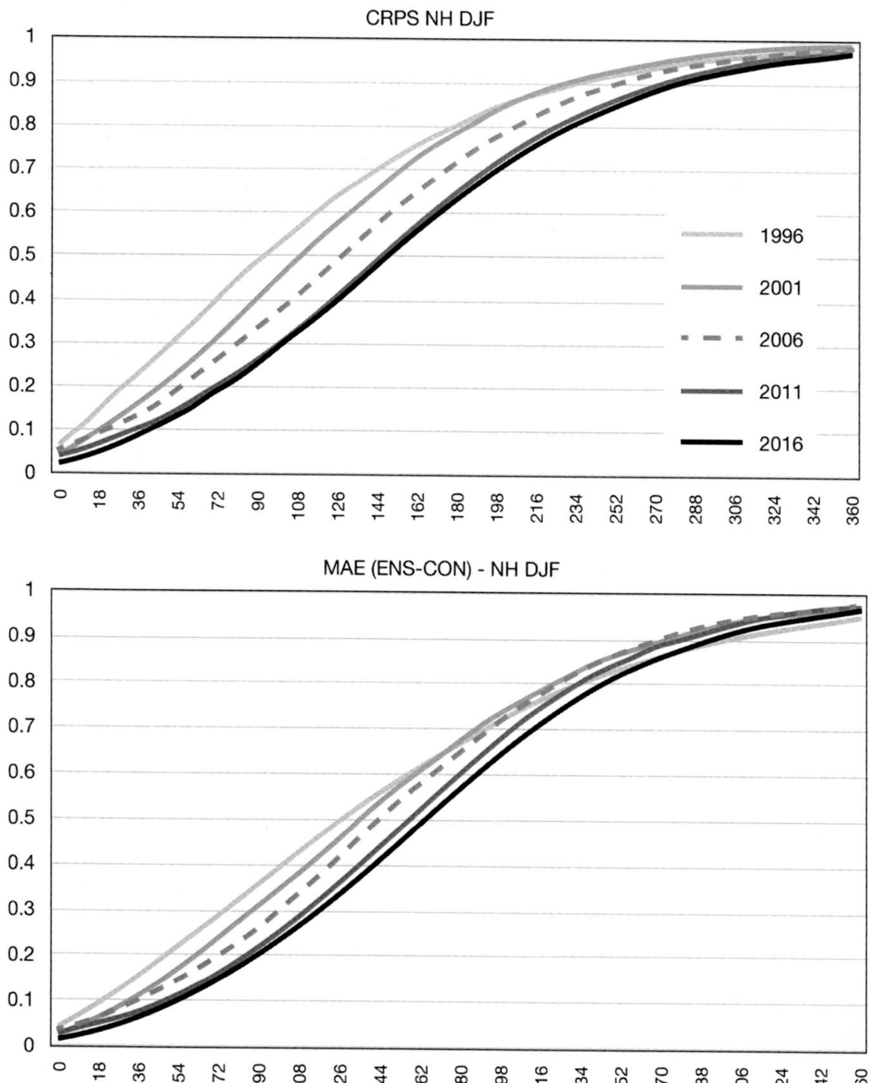

FIG. 5 As Fig. 4 but for winter (DJF).

Comparing the normalized curves allows to identify whether developments and changes implemented throughout the years have led to a lengthening of the forecast skill horizon. Figs. 4 and 5 show that there have been clear improvements, especially during the earlier periods.

Note that the MAE curve for JJA of 1996 starts as for the other years, but then has a much slower approach to saturation than all the other curves. Inspection of the fitted coefficients shows that for this case only, the error curve has a large "linear growth" term and a small error growth rate (Fig. 6). This appears to be a genuine feature of the data, and could be related

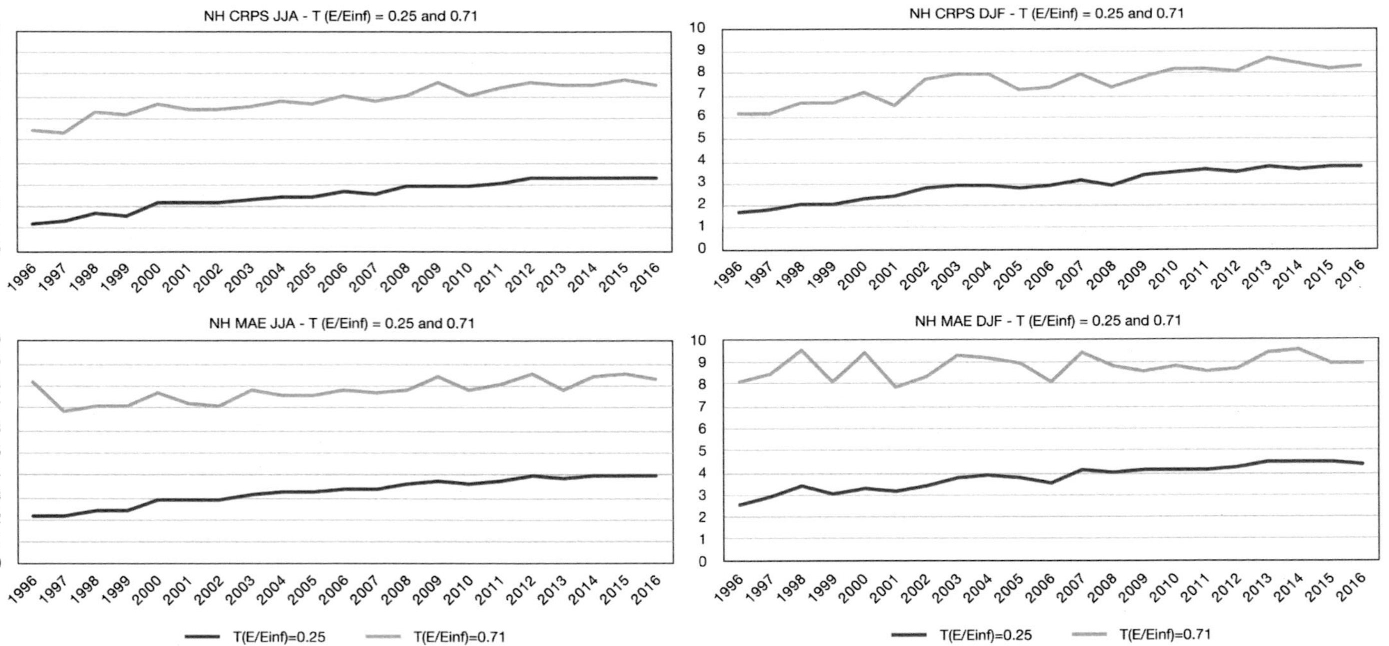

FIG. 6 Summer (JJA, left panels) and winter (DJF, right panels) time evolution of the predictability time limits τ (25%) (black line) and τ (71%) (gray line) in days for the ENS probabilistic forecasts over the NH (top), and for the ENS-control forecasts (bottom), of the 500 hPa geopotential height. Forecasts have been verified against analyses.

TABLE 1 Predictability limits τ (25%) and τ (71%) in 1996 and in 2016, computed for the ENS probabilistic forecasts over the NH and for the two seasons. The predictability limits have been computing using the error-growth curves fitted to the seasonal average CRPS values for the probabilistic prediction of the 500 hPa geopotential height.

		τ (25%) (fc days)		τ (71%) (fc days)	
		1996	**2016**	**1996**	**2016**
MAE	JJA	2.2	3.9 (+1.7)	8.2	8.3 (+0.1)
	DJF	2.5	4.3 (+1.8)	8.1	9.0 (+0.9)
CRPS	JJA	1.3	3.2 (+1.9)	5.5	7.5 (+2.0)
	DJF	1.7	3.7 (+2.0)	6.1	8.4 (+2.3)

to the characteristics of the low-resolution model used during that year only. More generally, we note that the overall improvement in scores is associated with a change in shape in the curve, an inflection point becoming clearly visible. The curves also show a substantial reduction in the error growth (i.e., in the first derivative) during the early part of the forecast. Similar considerations could have been drawn by looking at the RMSE of the ensemble-mean or the control forecast (not shown).

Fig. 6 shows the time evolution of the predictability limits τ (25%) and τ (71%), computed from the CRPS and the MAE (ENS-control) fitted curves, and Table 1 lists the forecast times when the two thresholds were reached in 1996 and in 2016. The comparison confirms that the performance of the 500 hPa geopotential height forecasts has been improving, especially at shorter lead times.

Considering the probabilistic forecasts, for example, improvements in absolute lead time have been similar in the shorter and longer time range, with a predictability gain of about 2.0 days in all cases. In relative terms, the predictability limit τ (25%) has been more than doubled, while for τ (71%) it has been lengthened by about 40%. At longer lead times the gains are smaller, as e.g. illustrated by looking at the extension of τ (95%) from, for example, 11.7 to 12.9 days for the not shown (NH) (although care should be taken in comparing values for longer lead times, since values this close to saturation are less stable, so the estimates are less certain).

5 Predictability estimates

It is interesting to look at the evolution of forecast skill over many years, to assess the impact of model developments, increased resolution, the adoption of more accurate and reliable systems and the use of more and better observations. At ECMWF, for example, every year they publish a detailed report on the performance of all operational forecasts (Haiden et al., 2019).

Fig. 7 shows the time evolution, from 1998 to 2020, of the accuracy of ECMWF high-resolution forecasts of the 500 hPa geopotential height over the NH and the SH, and of the 850 hPa vector wind over the tropics. Accuracy is measured by the anomaly correlation coefficient: the plots show the forecast time when it crosses the 80% threshold. The trend lines indicate, for these variables, improvements of about 1.5 days over 20 years for the NH, and about 2 days over 20 years for the other two variables.

Fig. 8 shows similar plots, but for the ECMWF probabilistic prediction of the 850 hPa temperature over the three areas. The accuracy measure in this case is the CRPSS: the plot shows

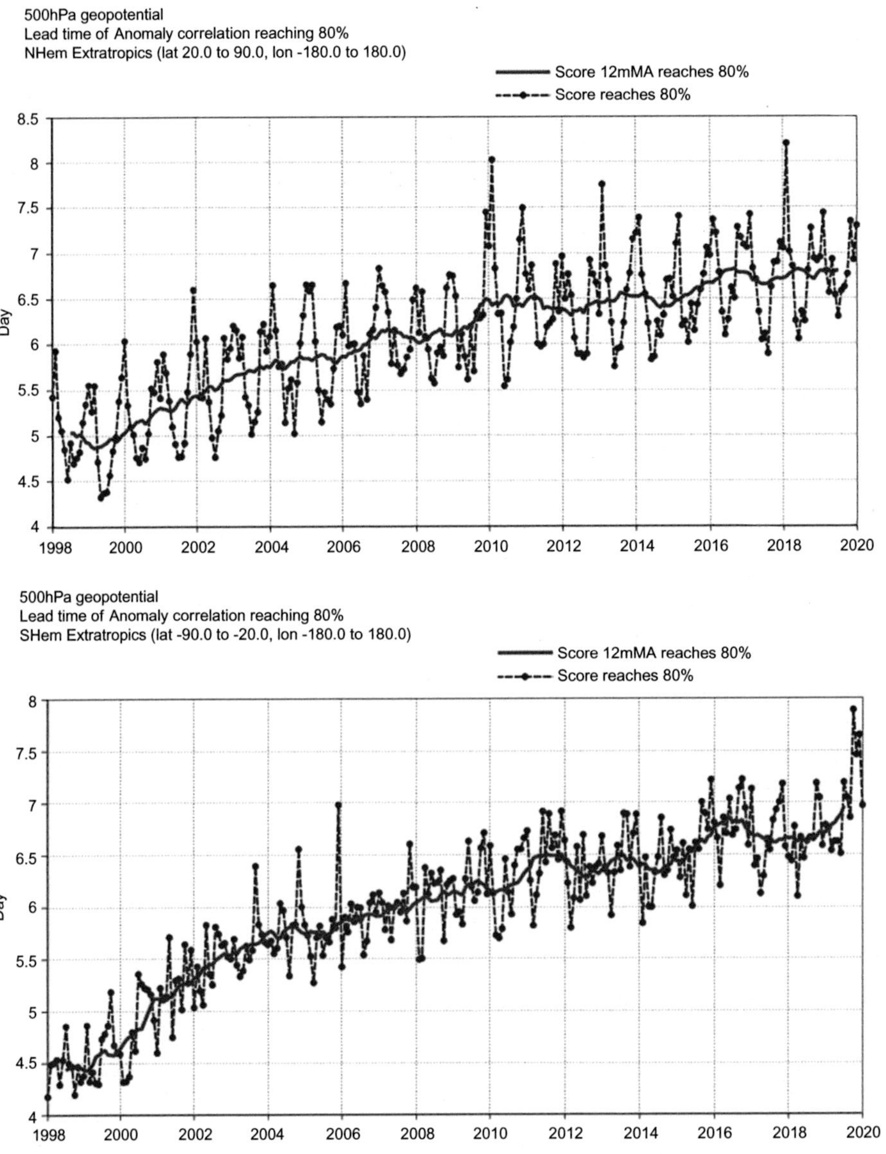

FIG. 7 Forecast time when the monthly average (blue line) anomaly correlation coefficient (ACC) of ECMWF forecasts were reaching 80%. The red lines show the running annual mean. The top panel refers to forecasts of the 500 hPa geopotential height over the Northern Hemisphere (NH).

(continued)

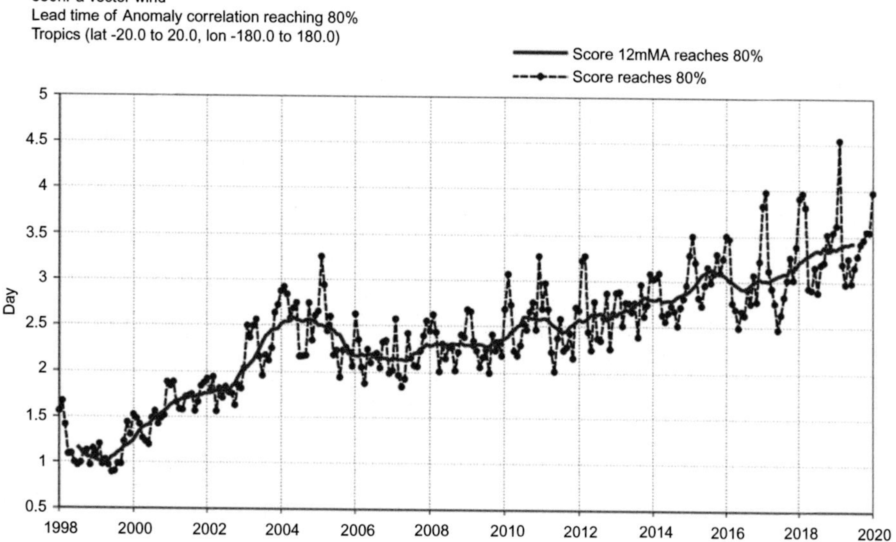

FIG. 7, cont'd The middle panel refers to forecasts of the 500 hPa geopotential height over the Southern Hemisphere (SH). The bottom panel refers to forecasts of the 850 hPa vector wind over the tropics (TR). *Source: ECMWF.*

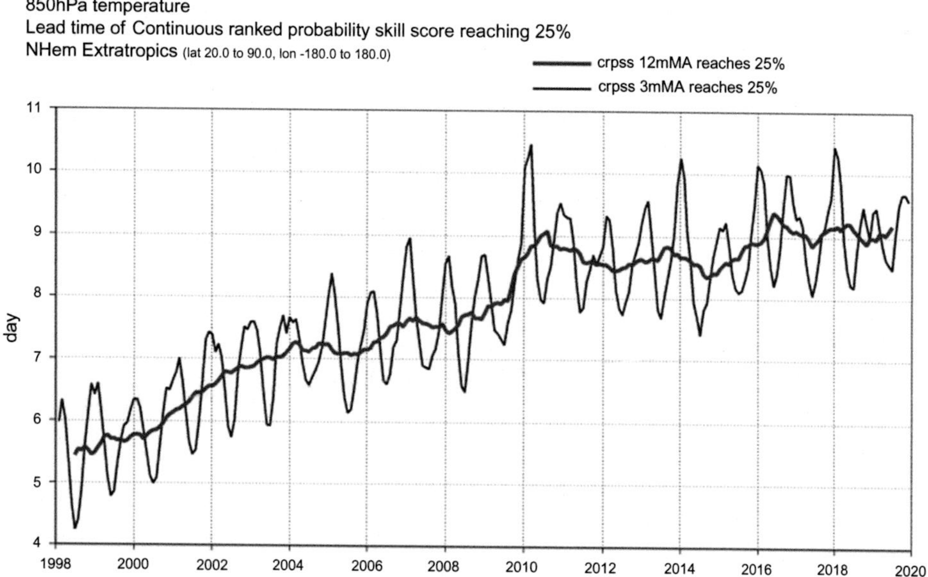

FIG. 8 Forecast time when the monthly average (blue line) continuous ranked probability skill score (CRPS) of ECMWF probabilistic forecasts were reaching 25%. The red lines show the running annual mean.

continued

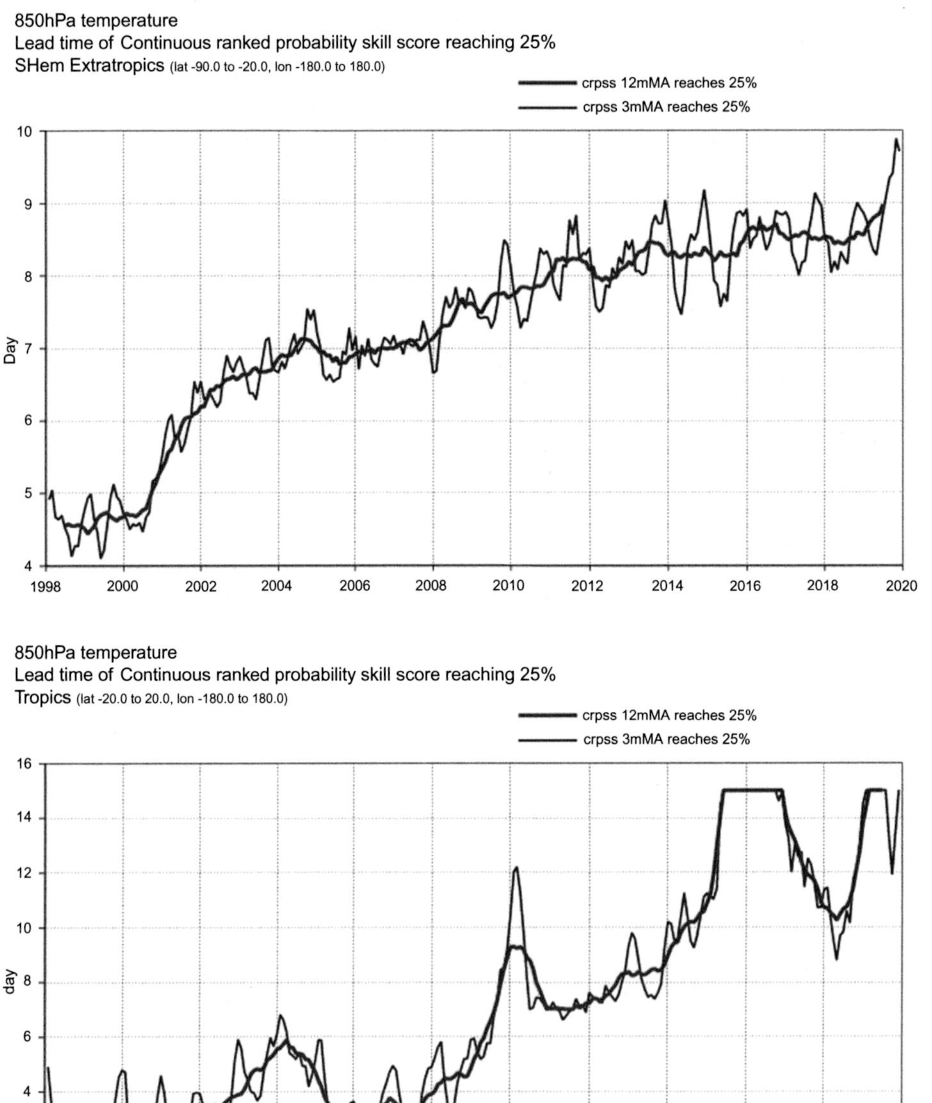

FIG. 8, cont'd The three panels refer to forecasts of the 850 hPa temperature over the Northern Hemisphere (N; top), the Southern Hemisphere (SH; middle), and the tropics (TR; bottom). *Source: ECMWF.*

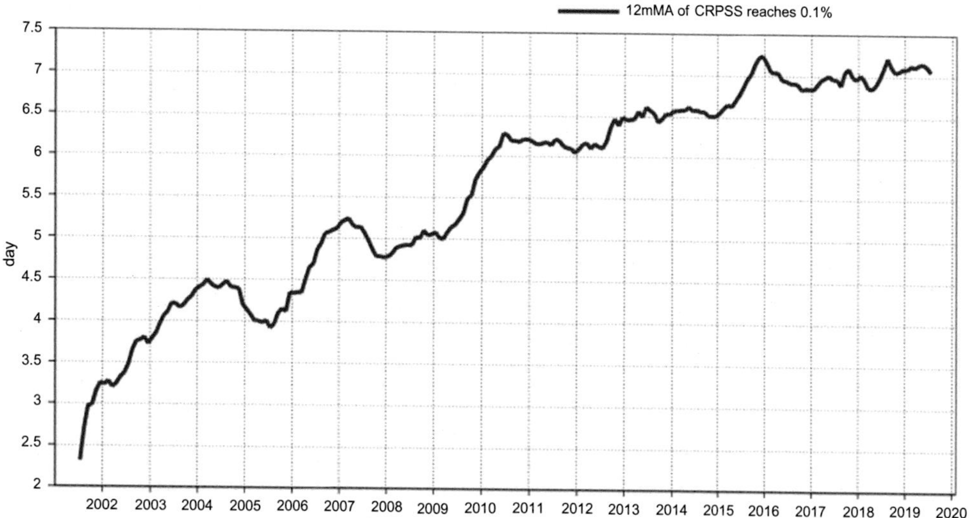

FIG. 9 Forecast time when the monthly average continuous ranked probability skill score (CRPS) of ECMWF probabilistic forecasts of total precipitation over the NH were reaching 10%. *Source: ECMWF.*

the forecast time when it crosses the 25% value. For these variables, they indicate that the improvements have been of about 3 days over 20 years for the NH, about 4 days over 20 years over the SH and about 12 days over the tropics.

The main reason why predictability of the ensemble-based probabilistic has been extended by longer times is that ensembles were implemented in operations only at the end of 1992, and for many years they were not very reliable, because they were under sampling the initial uncertainties and were not including any scheme to simulate model uncertainties (see e.g., Buizza et al., 2005). Major improvements were included in the years up to 2008 (increased resolution, increased membership, inclusion of stochastic model error schemes, inclusion also of the ensemble of data assimilation to generate the initial conditions): indeed, compared to the (single deterministic) high-resolution curves, the ensemble forecast skill curves are steeper up to about 2010, and then the two curves improve by similar percentages.

If we consider again the forecast skill horizon diagram of Fig. 3, we said that the predictability of phenomena/variables characterized by fast and small scales is shorter. To provide some more evidence of this, we show in Fig. 9 the CRPSS for the ECMWF probabilistic prediction of precipitation. The plot, in this case, shows the forecast time when the CRPSS crosses the 10% value (i.e., a lower skill level than for temperature at 850 hPa, for which the threshold was 25%). First, it is interesting to note that compared to the 850 hPa temperature, the threshold is crossed earlier, despite the fact that the threshold is lower: this confirms that predicting this variable, which is characterized by smaller/faster scales, is more difficult. Second, in terms of improvements, this plot shows that there has been a gain of about 5 days over 20 years. The reason why the improvement has been proportionally larger than for the

850 hPa temperature is because the resolution used 20 years was still very coarse, and insufficient to describe realistically the scales required to predict precipitation. Today, models have a resolution which is about a factor of 10 finer than in the 1990s, and thus can better capture these scales. In other words, we should not have expected skilful forecasts in the past, when the model did not have the resolution required to resolve the scales relevant to the phenomena we are trying to predict (small/fast vertical motions and converge/diverge patterns, linked to precipitation).

Considering the subseasonal time scale, Vitart et al. (2014) discuss the predictability of ECMWF monthly forecasts, and show that skill has improved significantly since operational production started in 2004, in particular for the prediction of large-scale, low-frequency events. Considering for example the Madden-Julian Oscillation (MJO, Madden and Julian, 1971), a very important source of predictability on the subseasonal time scale, Vitart et al. (2014) show that the 2013 version of the ECMWF monthly ensemble predicted it skillfully up to about 27 days (see their Fig. 4), compared to up to about day 17 in 2004: this is a skill gain of an impressive 1-day per year. Looking at another large-scale pattern that influences the weather over the Euro-Atlantic sector, the NAO, Vitart et al. (2014) report that the ECMWF ensemble also showed clear improvements, with skill in 2013 up to about forecast day 13, compared to about day 9 ten years earlier (see their Fig. 7). Similar, although slightly smaller, improvements were reported for the prediction of sudden-stratospheric warming events (see their Fig. 10). These results provide statistically significant evidence that some phenomena can be predicted up to a few weeks ahead.

Considering the seasonal time scale, Fig. 10 shows the current predictability of ECMWF seasonal forecasts of the sea surface temperature in the central Pacific, more specifically in an area called the El Nino 3.4 region. In this case accuracy is measured by the RMSE. One interesting point to remind is that predictability is seasonal dependent, with seasonal forecasts showing an RMSE similar to persistence earlier for the October start dates than for the other months. Fig. 11, from Stockdale et al. (2019), shows the progress in the prediction of the SST in the El Nino 3.4 region. For this particular index, which can be considered to represent the prediction of the low-frequency/large-scale oscillations in the tropical Pacific between El Nino and La Nina conditions, there has been a gain of about 2 months between 1997 (the time when seasonal system-1 was implemented) and 2017 (the time when seasonal system-5 was implemented).

6 Sources of predictability

The forecast skill diagram of Fig. 3 highlights very clearly that low-frequency, large-scale phenomena can be predicted with longer lead times than fast, small scale phenomena. The results discussed in Section 5 confirm this view, and provided further evidence. Let's close this chapter by discussing how we can explain this predictability scale dependency by looking at the potential sources of predictability.

How do we explain this?

Schukla and Kinter III (2006), for example, talking about long-range (seasonal) predictability, wrote that "... *the main determinant of seasonal atmospheric predictability is the slowly varying*

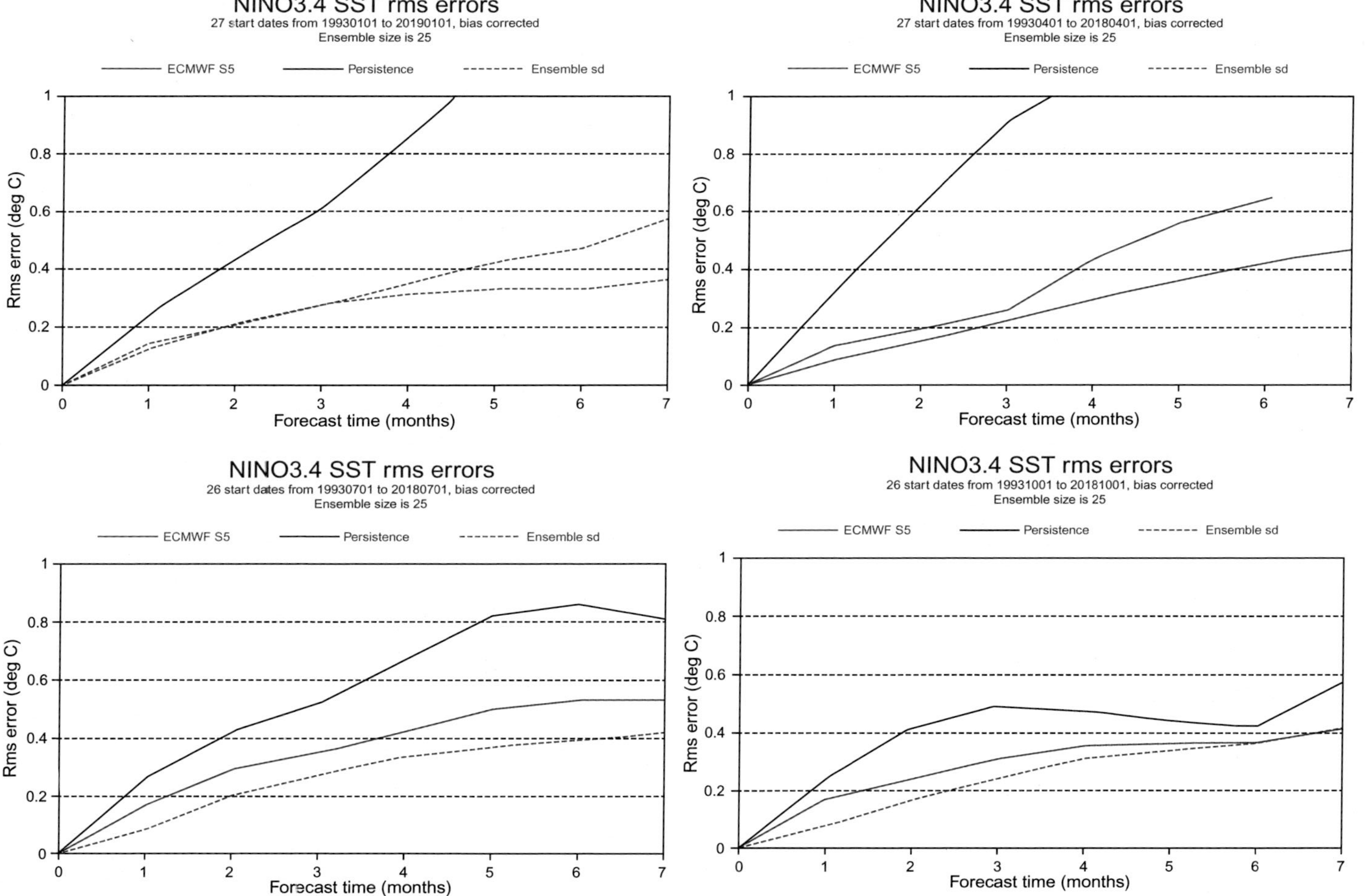

FIG. 10 27-Year average RMSE of persistence (black dashed lines) and of the ECMWF ensemble-mean seasonal forecasts of the sea surface temperature (SST) in the el Nino 3.4 region (solid red lines); the dashed red lines show the ensemble standard deviation. The top-left panel refers to forecasts initialized on the 1st January; the top-right panel refers to forecasts initialized on the 1st April; the bottom-left panel refers to forecasts initialized on the 1st of July; the bottom-right panel refers to forecasts initialized on the 1st of October. *Source: ECMWF.*

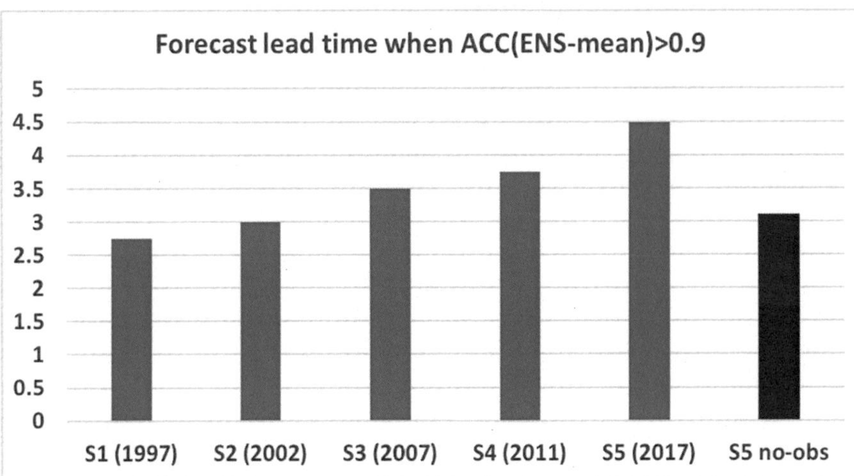

FIG. 11 Forecast lead time (in months) when the anomaly correlation coefficient of the ensemble-mean forecast of the sea surface temperature (SST) in the el Nino 3.4 region crosses the 90% threshold. The performance of the five ECMWF operational seasonal systems have been compared for the common period 1987-2002. *Source: Data from Stockdale, T., Alonso-Balmaseda, M., Johnson, S., Ferranti, L., Molteni, F., Magnusson, L., Tietsche, S., Vitart, F., Decremer, D., Weisheimer, A., Roberts, C., Balsamo, G., Keeley, S., Mogensen, K., Zuo, H., Mayer, M, and Mongez-Sanz, B., 2019. SEAS5 and the future evolution of the long-range forecast system. ECMWF Research Department Tech. Memorandum no. 835, pp. 81 (Available from ECMWF, Shienfield Park, Reading, RG2-9AX, UK).*

boundary conditions at the Earth's surface", and that "*… insofar as the boundary conditions can influence the large-scale atmospheric circulation, the time and space averages of the atmosphere are predictable at time scales beyond the upper limit of instantaneous predictability of weather*".

More generally, if we consider the Earth system as composed of the atmosphere, the ocean, the sea-ice and the land surface, we know that some of these components, for example, the ocean, evolve on slower time scales than the atmosphere. Thus, given that the ocean has an impact on atmospheric phenomena, if we can properly initialize it and we have a model that can describe accurately its evolution, we should be able to improve the atmospheric prediction. In other words, we can exploit a source of predictability (the ocean) to predict, if not the small/fast scales, at least the slow/large scales. In other words, we can extract "predictable signals", if not every day at least in some conditions.

Predictable signals can also be extracted also from the atmosphere itself. For example, if we can properly initialize the upper level, large-scale flow (e.g. changes in the stratospheric circulation), we have a model that can evolve it skillfully for long time, and changes upper levels influence the lower levels and the surface weather, we should be able to extend the prediction of surface weather. Similarly, we could extract predictable signals from part of the globe (e.g., the tropics), and exploit them to predict the weather elsewhere (e.g., Europe).

Ocean and sea-ice anomalies, if properly initialized and predicted in time, can be a source of predictability.

The influence of the sea surface temperature and sea ice anomalies on the atmosphere has been documented in many works (see, e.g., the list of references in Schukla and Kinter III,

2006). The mechanism leading to extended predictably is that, under favorable conditions, SST anomalies can change the large scale atmospheric flow, and this in turn can induce anomalies in the surface weather.

The most striking example is the influence of SST anomalies in the central Pacific, linked to El Nino/La Nina conditions, on the global weather (see, e.g., Domeisen et al., 2018). El Nino/La Nina SST anomalies have a strong impact of the tropical circulation in the Pacific, that extends also to the Atlantic and the Indian Oceans (Webster et al., 1999; Giannini et al., 2001). In the Norther hemisphere extra-tropics, for example, SST anomalies generates quasi-stationary Rossby waves (see, e.g., Hoskins and Karoly, 1981), which can be identified by high and low pressure anomaly patters (Franzke and Feldstein, 2005).

Similarly, large-scale anomalies in sea ice cover can change the sea albedo and the heat fluxes with the atmosphere, change the large scale circulation and impact distant local weather. Ruggeri et al. (2017), for example, showed that a reduction in the sea ice coverage in the Barents and Kara Seas, lead to an anomalous circulation first locally, then over the polar region and finally over the Euro-Atlantic sector. They linked this to the atmospheric response to surface diabatic heating, via troposphere-stratosphere interactions.

If the ocean and sea states can be initialized properly, and the Earth-system model is capable to simulate in a realistic way ocean/sea ice and atmosphere interactions, and does not show any significant drift with forecast time, then the prediction of weather anomalies in regions sensitivity to these anomalies can be extended.

Anomalies in land moisture, vegetation, and snow cover can also be sources of predictability, if properly initialized and predicted in time.

Anomalies in soil moisture, vegetation, and snow cover can affect the weather, and since their variations are characterized by long time scales, they can act as predictability sources, especially in regions where the coupling between the land surface and the atmosphere is stronger. For example, Koster et al. (2004) studying precipitation identified some "hot spots" regions, where the coupling is stronger: these are usually transition zones where between dry and wet climatic zones, where evaporation is large and sensitive to soil moisture, and where the boundary layer moisture can trigger convection. Proper initialization and modeling of convection can thus lead to increased long-range predictability in these regions.

Ferranti and Viterbo (2006), investigating the European heat wave of 2003, concluded that the response of large initial dry soil anomalies greatly exceeds the impact of sea surface anomalies in the Mediterranean sea, and that dry soil conditions in spring have contributed to the amplification of local temperature anomalies in summer.

Teleconnections can lead to increased predictability.

With the term "teleconnection," we mean that anomalies in one location can be connected to anomalies in others. In some cases, the identification of teleconnections can lead to the understanding of how phenomena in one area can influence the weather in another, and thus to determine a cause-effect relation. An example is given by the teleconnections linked to SST anomalies in the central Pacific, caused by El Nino/La Nina phenomena, which can influence the weather not only in the tropical band, but also in the extra-tropics. Another example is the MJO, the most important source of variability with a subseasonal time scale. Active phases of the MJO in certain tropical locations can influence the weather in the extra-tropics, for

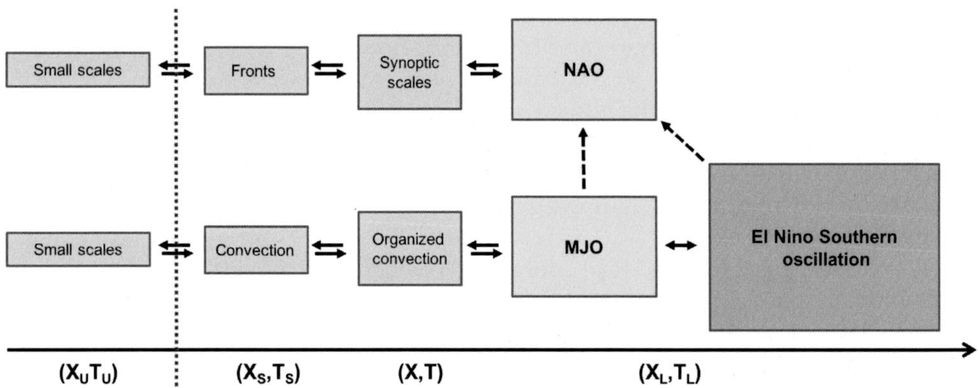

FIG. 12 Schematic of scales' interactions and teleconnections, (inspired by Fig. 2 of Hoskins, 2013). The dotted red line separates the model's unresolved scales (X_U,T_U), from the resolved scales. The bottom row shows scales' interactions in the tropics, from the small/fast scales (X_s,T_s), to the scales of organized phenomena (X,T) and the large/slow scales (X_L,T_L). The top row shows scales interactions in the extra-tropics. The dashed black arrows show the teleconnections between large scale patterns in the tropics, the MJO and El Nino Southern Oscillation, and large scale patterns in the extra-tropics, like the NAO.

example in the Euro-Atlantic sector (Vitart and Molteni, 2009), and lead to increased predictability in the subseasonal time range. If one model is capable to describe in a realistic way the teleconnections, and anomalies are properly initialized, they could lead to improved prediction in certain areas. Theoretical work by Held et al. (2002) linked teleconnections to the propagation of linear, planetary waves, triggered by anomalies in the source region.

Fig. 12, inspired by Hoskins (2013), illustrates the concepts of scales' interactions and teleconnections. The red dotted line separates a prediction model's unresolved scales (X_U, T_U) from the resolved scales: as the model's resolution increase, finer/faster scales can be simulated. The small black arrows represent the scales' interactions in the atmosphere, in the tropics (lower row) and the extra-tropics (upper row). Large/slow scale phenomena, such as the MJO and El Nino in the tropics and the NAO in the extra-tropics, affect and are affected by organized phenomena with spatial/temporal scale (X,T) and the smaller/faster scales (X_S,T_S). The dashed black lines represent the teleconnections between the tropics and the extra-tropics. EL Nino can influence directly both the MJO propagation in tropics, and the NAO over the extra-tropics. The MJO itself can also have an influence on the NAO. These two large scale phenomena propagate, are maintained or dissipated by smaller scales (synoptic scales in the extra-tropics, organized convection in the tropics), and so on down to the resolved scales.

A model capable to correctly simulate all these processes and their interactions, and to initialize in an accurate way all scales, should be able to make better predictions than a model not capable to do so.

In most cases, forecast errors would propagate upscale from the small/fast scales to the scales of the synoptic features and organized convection, and then to the large/slow scales, mainly because initial errors are usually larger, and grow faster, than for the larger/slower scales. If the large/slow scales initial conditions are small and their evolution is well represented, they can bring predictable signals downscale, to the synoptic scales and/or to organized convection. The forecast skill horizon is determined by the forecast time when the upscale error propagation overcomes the downscale propagation of the predictive signals. This time is not the same but is flow dependent, due to the chaotic nature of the atmosphere: in some cases, the error growth rate could be slower, and predictability longer, while in others it can immediately affect predictability. In other cases, the large/slow scales error could dominate, and could propagate down-scale from the large/slow scales, and could follow precisely the same teleconnection route as the predictable signals (Zagar, 2017; Zagar et al., 2017).

A continuous improvement in the models, the inclusion of all the relevant processes, increased resolution, the utilization of more and better observations, improved data assimilation, the understand of the sources of predictability and the diagnostic of how well they are represented in the models, can lead to further extensions of the predictability limits. Zhang et al. (2019), for example, looking at the small/fast scales (X_S,T_S) and the synoptic scales (X,T), estimated that a reduction of a factor of 10 of the initial conditions' errors could lead to a predictability gain of the about 5 days.

7 Conclusions

Forecast skill horizons beyond 2 weeks are now achievable thanks to major advances in numerical weather prediction. They have been made possible by the combination of better and more complete models, which include more accurate simulation of relevant physical processes (e.g., the coupling to a dynamical ocean and ocean waves), improved data-assimilation methods that allowed a more accurate estimation of the initial conditions, advances in ensemble techniques and understanding of the sources of predictability.

A number of factors need to be taken into account to explain why these recent estimates extend the range of predictive skill into several weeks while early estimates by Lorenz (1969) suggested that perhaps there is a finite time of predictability of about 2 weeks regardless how small the initial uncertainty is in the atmosphere.

Predictive skill depends on the processes represented by the numerical model. Some of the more idealized work that provided the earlier estimates included only dynamical processes and a very crude representation of physical processes. For example, they did not have a convection scheme that could simulate realistically the small-scale variability (Bechtold et al., 2008) and the diurnal cycle (Bechtold et al., 2013): a correct simulation of these effects is crucial to simulate realistically the tropical atmosphere.

A correct simulation of low-frequency, large-scale phenomena such as the MJO can lead to increased predictability in the tropics and the extra-tropics, provided that they are properly simulated. Vitart et al. (2014) showed that improvements in the simulation of convection, led to a better prediction of the propagation of organized convection in the tropics, including the MJO, and this, in turn, led to improvements in the skill over Europe. This implies that predictive skill estimated using a model that can describe the MJO propagation and its teleconnection with the extra-tropics is going to be longer than estimates based on a model that cannot describe them. Not only convection matters: other processes play also an important role, for example:

- Tietsche et al. (2014), showed that sea-ice prediction may induce predictable signals up to a lead time of three years;

– Guo et al. (2012) showed that in those regions and seasons with a sufficiently strong coupling between land surface and atmosphere, the land surface can also carry predictable signals.

While there is no generally established theoretical predictability horizon for coupled atmosphere–ocean system, there is sufficient evidence that the modulation of the atmospheric PDF by the slower ocean extends to months and seasons. As models improve and become more realistic, predictability will be further extended, and estimates will become more correct.

Predictability is finite, but we do not think that the predictability limit has yet been reached, since we believe that there are ways to further reduce the initial conditions' and model errors.

8 List of acronyms

- ACC: anomaly correlation coefficient;
- CDF: cumulative distribution function;
- CRPS: continuous ranked probability score;
- CRPS: continuous ranked probability skill score;
- DA: Data Assimilation;
- DJF: December, January, February (winter season);
- ECMWF: European Centre for Medium-Range Weather Forecast;
- hPa: hector Pascal;
- JJA: June July August (summer seasons);
- MAE: mean absolute error;
- MJO: Madden Julian Oscillation, the dominant mode of subseasonal variability 9n the tropics;
- NAO: the North Atlantic Oscillation, a low-frequency weather regime that characterizes the circulation over the Euro-Atlantic sector;
- NH: Northern Hemisphere;
- PDF: probability density function;
- RMSE: root mean square error;
- SH: Southern Hemisphere;
- SS: skill score;
- SST: Sea Surface Temperature;
- TR: tropics.

References

Anderson, D.L.T., 2006. Operational seasonal prediction. In: Predictability of Weather and Climate. Cambridge University Press pp. 702, Chapter 19. (ISBN-13: 978-0-521-84882-4).

Bechtold, P., Köhler, M., Jung, T., Doblas-Reyes, F., Leutbecher, M., Rodwell, M., Vitart, F., Balsamo, G., 2008. Advances in simulating atmospheric variability with the ECMWF model: from synoptic to decadal time-scales. Q. J. Roy. Meteorol. Soc. 134, 1337–1351.

Bechtold, P., Semane, N., Lopez, P., Chaboureau, J.-P., Beljaars, A., Bormann, N., 2013. Representing equilibrium and non-equilibrium convection in large-scale models. In: ECMWF Research Department Technical Memorandum no. 705. pp 27 (available from ECMWF, Shinfield Park, Reading RG2-9AX, U.K.; see also www.ecmwf.int.

Brankovic, C., Palmer, T.N., Molteni, F., Tibaldi, S., Cubasch, U., 1990. Extended-range predictions with ECMWF models: time-lagged ensemble forecasting. Q. J. Roy. Meteorol. Soc. 116, 867–912.

Brown, T.A., 1974. Admissible scoring systems for continuous distributions. Manuscript P-5235 The Rand Corporation, Santa Monica, CA pp. 22. Available from The Rand Corporation, 1700 Main Street, Santa Monica, CA 90470-2138.

Buizza, R., 2010. Horizontal resolution impact on short- and long-range forecast error. Q. J. Roy. Meteorol. Soc. 136, 1020–1035.

Buizza, R., 2018. Introduction to the special issue on '25 years of ensemble forecasting. Q. J. Roy. Meteorol. Soc. 1–11. https://doi.org/10.1002/qj.3370.

Buizza, R., Houtekamer, P.L., Toth, Z., Pellerin, G., Wei, M., Zhu, Y., 2005. A comparison of the ECMWF, MSC and NCEP Global Ensemble Prediction Systems. Mon. Wea. Rev 133 (5), 1076–1097. https://doi.org/10.1175/MWR2905.1.

Buizza, R., Leutbecher, M., 2015. The forecast skill horizon. Q. J. Roy. Meteorol. Soc. 141 (Issue 693, Part B), 3366–3382. https://doi.org/10.1002/qj.2619.

Buizza, R., Bidlot, J.-R., Wedi, N., Fuentes, M., Hamrud, M., Holt, G., Vitart, F., 2007. The new ECMWF VAREPS (variable resolution ensemble prediction system). Q. J. Roy. Meteorol. Soc. 133, 681–695. https://doi.org/10.1002/qj.75.

Charney, J.G., et al., 1966. The feasibility of a global observation and analysis experiment. Bull. Am. Meteorol. Soc. 47, 200–220.

Dalcher, A., Kalnay, E., 1987. Error growth and predictability in operational ECMWF forecasts. Tellus 39A, 474–491.

Domeisen, D., Garfinkel, C.I., Butler, A.H., 2018. The teleconnections of El Nino Southern Oscillation to the Stratosphere. Rev. Geophys.. https://doi.org/10.1029/2018RG000596 pp. 47.

Durran, D.R., Gingrich, M., 2014. Atmospheric predictability: why butterflies are not of practical importance. J. Atmos. Sci. 71, 2476–2488.

Epstein, E.S., 1969. A scoring system for probability forecasts of ranked categories. J. Appl. Meteorol. 8, 985–987.

Ferranti, L., Viterbo, P., 2006. The European summer of 2003: sensitivity to soil water initial conditions. ECMWF Research Department Tech. Memorandum no. 483, pp. 29 (available from ECMWF, Shinfield Park, Reading RG2 9AX, UK).

Franzke, C., Feldstein, S.B., 2005. The continuum and dynamics of northern hemisphere teleconnections. J. Atmos. Sci. 62 (9), 3250–3267.

Giannini, A., Chiang, J.C., Cane, M.A., Kushnir, Y., Seager, R., 2001. The ENSO teleconnection to the tropical Atlantic Ocean: contributions of the remote and local SSTS to rainfall variability in the tropical Americas. J. Climate 14 (24), 4530–4544.

Guo, Z., Dirmeyer, P.A., DelSole, T., Koster, R.D., 2012. Rebound in atmospheric predictability and the role of the land surface. J. Climate 25, 4744–4749.

Haiden, T., Janousek, M., Vitart, F., Ferranti, L., Prates, F., 2019. Evaluation of ECMWF forecasts, including the 2019 upgrade. ECMWF Research Department Technical Memorandum # 853, pp. 54. Available from ECMWF, Shinfield Park, Reading, UK (see also the ECMWF qweb site: https://www.ecmwf.int/en/publications/technical-memoranda).

Held, I.M., Mingfang, T., Hailan, W., 2002. Northern winter stationary waves: theory and modelling. J. Climate 15 (16), 2125–2144. https://doi.org/10.1175/1520-0442(2002)015<2125:NWSWTA>2.0.CO;2.

Hersbach, H., 2000. Decomposition of the continuous ranked probability score for ensemble prediction systems. Weather Forecast. 15, 559–570.

Hoskins, B.J., 2013. The potential for skill across the range of the seamless weather-climate prediction problem: a stimulus for our science. Q. J. Roy. Meteorol. Soc. 139, 573–584. https://doi.org/10.1002/qj.1991.

Hoskins, B.J., Karoly, D.J., 1981. The steady linear response of a spherical atmosphere to thermal and orographic forcing. J. Atmos. Sci. 38 (6), 1179–1196.

Hurrell, J.W., Kushnir, Y., Ottersen, G., Visbeck, M., 2003. An overview of the North Atlantic oscillation. In: Geophysical Monograph 134. American Geophysical Union.

Koster, R.D., Dirmeyer, P.A., Guo, Z., et al., 2004. Regions of strong coupling between soil moisture and precipitation. Science 305, 1138–1140.

Lorenz, E.N., 1965. A study of the predictability of a 28-variable model. Tellus 17, 321–333.

Lorenz, E.N., 1969. The predictability of a flow which possess many scales of motion. Tellus, XXI 3, 289–307.

Lorenz, E.N., 1982. Atmospheric predictability experiments with a large numerical model. Tellus 34, 505–513.

Madden, R.A., Julian, P.R., 1971. Detection of a 40-50 day oscillation in the zonal wind in the tropical Pacific. J. Atmos. Sci. 5, 702–708.

Molteni, F., Cubasch, U., Tibaldi, S., 1987. 30 and 60-day forecast experiments with the ECMWF spectral models. In: Proceedings of the first ECMWF Workshop on Predictability in the medium and extended range (ECMWF, Reading, U.K., 17–19 Mar. 1986), pg. 51–108. Available from ECMWF, Shinfield Park, Reading RG2-9AX.

Murphy, A.H., 1971. A note on the ranked probability score. J. Appl. Meteorol. 10, 155–156.

Rotunno, R., Snyder, C., 2008. A generalization of Lorenz's model for the predictability of flows with many scales of motion. J. Atmos. Sci. 65, 1063–1076.

Ruggeri, P., Kurchaski, F., Buizza, R., Ambaum, M., 2017. The transient atmospheric response to a reduction of sea-ice cover in the Barents and Kara seas. Q. J. Roy. Meteorol. Soc. 143, 1632–1640. https://doi.org/10.1002/qj.3034.

Schukla, J., 1981. Dynamical predictability of monthly means. J. Atmos. Sci. 38, 2547–2572.

Schukla, J., Kinter III, J.L., 2006. Predictability of seasonal climate variations: a pedagogical review. In: Palmer, Hagedorn (Eds.), Predictability of Weather and Climate. Cambridge University Press pp. 702, published as Chapter 12 (ISBN13 978-0-521-84882-4).

Simmons, A.J., Hollingsworth, A., 2002. Some aspects of the improvement in skill of numerical weather prediction. Q. J. Roy. Meteorol. Soc. 128, 647–677.

Simmons, A.J., Mureau, R., Petroliagis, T., 1995. Error growth and predictability estimates for the ECMWF forecasting system. Q. J. Roy. Meteorol. Soc. 121, 1739–1771.

Smagorinsky, J., 1969. Problems and promises of deterministic extended-range forecasting. Bull. Amer. Meteor. Soc. 50, 286–311.

Stockdale, T., Alonso-Balmaseda, M., Johnson, S., Ferranti, L., Molteni, F., Magnusson, L., Tietsche, S., Vitart, F., Decremer, D., Weisheimer, A., Roberts, C., Balsamo, G., Keeley, S., Mogensen, K., Zuo, H., Mayer, M., Mongez-Sanz, B., 2019. SEAS5 and the future evolution of the long-range forecast system. In: ECMWF Research Department Tech. Memorandum no. 835, pp 81 (available from ECMWF, Shienfield Park, Reading, RG2-9AX, UK).

Tietsche, S., Day, J.J., Guemas, V., Hurlin, W.J., Keeley, S.P.E., Matei, D., Msadek, R., Collins, M., Hawkins, E., 2014. Seasonal to interannual Arctic Sea ice predictability in current global climate models. Geophys. Res. Lett. 41, 1035–1043.

Trenberth, K.E., 1997. The definition of El Niño. Bull. Am. Meteorol. Soc. 78 (12), 2771–2777.

Vitart, F., 2004. Monthly forecasting at ECMWF. Mon. Weather Rev. 132, 2761–2779.

Vitart, F., Molteni, F., 2009. Simulation of the MJO and its teleconnections in an ensemble of 46-day EPS hindcasts. In: ECMWF Research Department Technical Memorandum no. 597, pp. 60 (Available from ECMWF, Shinfield Park, Reading RG2-9AX, UK).

Vitart, F., Balsamo, G., Buizza, R., Ferranti, L., Keeley, S., Magnusson, L., Molteni, F., Weisheimer, A., 2014. Sub-seasonal predictions. In: ECMWF Research Department Technical Memorandum no. 734, pp. 47 (available from ECMWF, Shinfield Park, Reading RG2-9AX, UK).

Webster, P.J., Moore, A.M., Loschnigg, J.P., Leben, R.R., 1999. Coupled ocean-atmosphere dynamics in the Indian Ocean during 1997-98. Nature 401 (6751), 356.

Wilks, D., 2019. Statistical Methods in the Atmospheric Sciences, fourth ed. Elsevier. ISBN 978-0-12-815823-4 pp. 818.

Zagar, N., 2017. A global perspective of the limits of prediction skill of NWP models. Tellus A 69 (1317), 573.

Zagar, N., Horvat, M., Zaplotnik, Z., Magnusson, L., 2017. Scale-dependent estimates of the growth of forecast uncertainties in a global prediction system. Tellus A 69 (1287), 492.

Zhang, F., Sun, Y.Q., Magnusson, L., Buizza, R., Lin, S.-J., Chen, J.-H., Emanuel, K., 2019. What is the predictability limit of midlatitude weather? J. Atmos. Sci. 76, 1077–1091.

Further reading

Buizza, R., Leutbecher, M., Thorpe, A., 2015. Living with the butterfly effect: a seamless view of predictability. ECMWF newsletter no. 145, autumn 2015, pp. 18–23 (available from ECMWF, Shinfield Park, Reading RG2-9AX, UK).

Vitart, F., 2013. Evolution of ECMWF sub-seasonal forecast skill scores over the past 10 years. In: ECMWF Research Department Technical Memorandum n. 694, pp. 28 (Available from ECMWF, Shinfield Park, Reading RG2-9AX, UK).

CHAPTER

5

Modeling moist dynamics on subgrid

Peter Bechtold

European Centre of Medium-Range Weather Forecast, Reading, United Kingdom

1 Introduction

Moist processes in the atmosphere strongly interact with the large-scale flow through latent heat release and the absorption and emission of radiation. Naturally, a large uncertainty in numerical weather prediction (NWP) and climate simulations stems from the representation of unresolved (subgrid) microphysical processes and moist vertical mixing through convective updraughts and downdraughts.

There exists a large variety of microphysical schemes and parametrizations for representing moist turbulent and convective transport. Experience shows that not only the physical formulation of the subgrid schemes matters, e.g., how many prognostic variables for water species are included in a microphysical scheme or what are the assumptions in a turbulence scheme that link the unknown subgrid turbulent and convective fluxes to the known resolved variables. The details in the schemes can be equally important, e.g., the complicated mixed-phase processes and the ice sedimentation rate in a microphysical scheme, e.g., Wu et al. (2013) and White et al. (2017) or the mixing length formulation in a turbulent diffusion scheme or the mass exchange of the convective updraught with the environment in a convective mass flux scheme, e.g., Gregory and Miller (1989), Shin and Dudhia (2016), and Tan et al. (2018). Furthermore, the numerical formulations of the schemes and the interactions between the various physical parametrization schemes and the model dynamics are also critically important (Beljaars et al., 2018; Williamson, 2013). Therefore, I decided not to discuss any particular subgrid parametrization scheme in this chapter.

Times are changing in NWP! With a steady increase in computer power and faster algorithms, global cloud-resolving simulations at 1–2 km horizontal grid-spacing that do not require a parametrization for deep convection are already feasible. Yet these simulations cannot be performed in real-time and are mainly used to explore climate type sensitivity studies

Uncertainties in Numerical Weather Prediction
https://doi.org/10.1016/B978-0-12-815491-5.00005-7

and severe weather events (Guichard and Couvreux, 2017; Satoh et al., 2018; The DYAMOND Initiative, n.d.). However, for real-time applications such as medium-range NWP these resolutions are unlikely to be affordable in the coming years, as e.g., European Centre of Medium-range Weather Forecast (ECMWF) is planning for a 5 km global ensemble system by 2025. Therefore, there is currently still considerable work dedicated at the main global prediction centers to the physical parameterization of deep convection in the 2–10 km resolution range. Important research efforts also concern the modeling of convective boundary-layer clouds, which typically require horizontal and vertical resolutions of about 50 m to be resolved (Guichard and Couvreux, 2017). Finally, representing forecast errors and uncertainty through ensembles becomes an even more important subject as random and systematic forecast errors are likely to be different at higher resolutions.

So where does all this lead us in terms of forecast uncertainty from moist physical processes? The answer is certainly somewhat subjective, but I have chosen to focus on processes where the physics-dynamics interaction is strong, which are crucially important in terms of potential forecast improvements and understanding and which remain challenging at future even higher O (1–5 km) model resolutions: (1) the coupling of deep convection with the larger-scale circulation, (2) vertically propagating gravity waves and their impact on the stratosphere, (3) the diurnal cycle of convection and the propagation of mesoscale convective systems, and (4) the cloudy convective boundary-layer and its impact on the global shortwave radiation budget. Apart from stated separately, all of the following numerical examples have been produced with the ECMWF Integrated Forecasting System (IFS) and if possible with the latest operational model version from 2019. We think that the results can be also considered as representative for other global atmospheric prediction systems.

2 Large-scale vs convective precipitation

The motivation for focusing and convective processes and the tropics stems from Fig. 1, which shows the skill of the IFS precipitation high-resolution forecasts (9 km horizontal resolution) as a function of forecast lead time during summer 2018 and winter 2018/19 over the northern hemisphere extratropical region and the tropics. The verification is against synoptic observations from rain gages over land. Clearly, the extratropical precipitation forecasts are more skillful by about 2 days during winter than during summer. In winter the precipitation is predominantly of the large-scale or stratiform type, while during summer convective precipitation dominates. Furthermore, a day-6 precipitation forecast during the northern hemisphere summer is as skillful as a day-2 precipitation forecast in the tropics. Interestingly, the precipitation skill for the extratropics drops more rapidly with lead time than over the tropics. As a result, beyond the medium-range a tropical precipitation forecast becomes more skillful than the extratropical forecasts, although the skill over land is rather small (the skill over water would be higher as it correlates with the sea surface temperatures). A relatively higher extended range forecast skill for the tropical regions compared to the middle latitudes reflects the fact that the tropical circulations are dominated by large-scale waves which provide an important source of predictability up to the subseasonal time scale (Vitart and Molteni, 2010a), while the middle latitude weather becomes unpredictable.

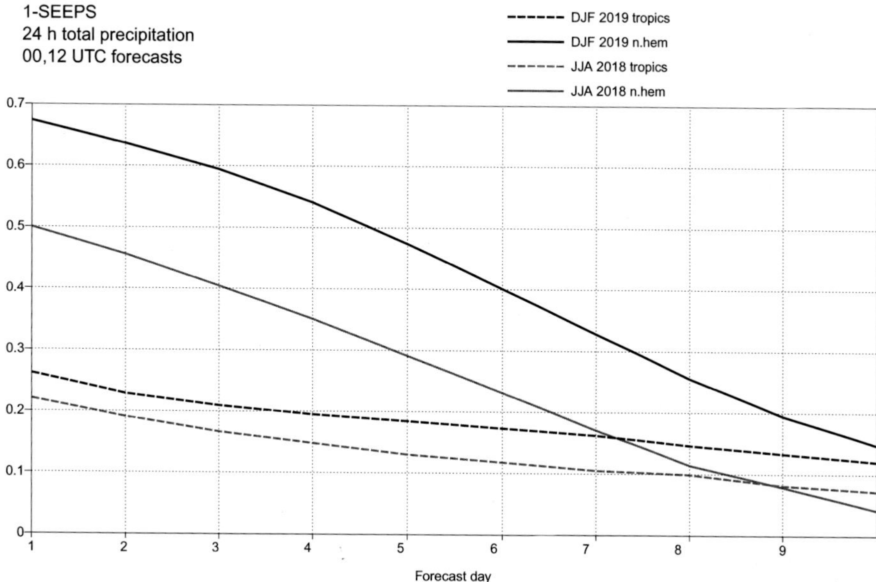

FIG. 1 The skill of the IFS high-resolution (9km) 24-h accumulated precipitation forecasts against synoptic observations for the northern hemisphere and tropical region for summer (JJA) 2018 and winter DJF 2018/19. The skill is given by the Stable Equitable Error in Probability Space (SEEPS) (Rodwell et al., 2010) as a function of forecast lead time. *Figure courtesy Martin Janousek, ECMWF. Figure courtesy Martin Janousek, ECMWF.*

3 Convection, waves, and the large-scale circulation

On synoptic space and time scales, the convection can be considered in equilibrium with the larger circulation as postulated by Arakawa and Schubert (1974). Although there is a strong interaction between the convective heating and the large-scale circulation the convection can, to a sufficient approximation, be considered as to respond rapidly to changes in the large-scale forcing. This assumption, however, breaks down in the case of summer-time diurnal convection over land that is forced by surface heating, and where the convective heating generates the large-scale response (Bechtold et al., 2014; Vizy and Cook, 2017). The quasiequilibrium assumption also breaks down for self-propagating mesoscale convective systems such as squall lines. We, therefore, expect the forecasts of convection to be better over oceans and when synoptically forced.

This is illustrated in Fig. 2 where we compare the synthetic satellite images (infrared brightness temperatures) as produced by the operational IFS t+9h forecast at 9km resolution from April 26, 2019 00 UTC over the tropical Indian Ocean and Eastern Africa to the corresponding observed infrared satellite images from the SEVIRI instrument onboard Meteosat-8 and Metesosat-11. We have also added streamlines at 850hPa on the synthetic satellite image to visualize the large-scale flow. Clearly, the short-range forecast reasonably reproduces the cloud structures over the Indian Ocean, including the developing twin tropical cyclones. However, there are differences in the details of the cloud structures and

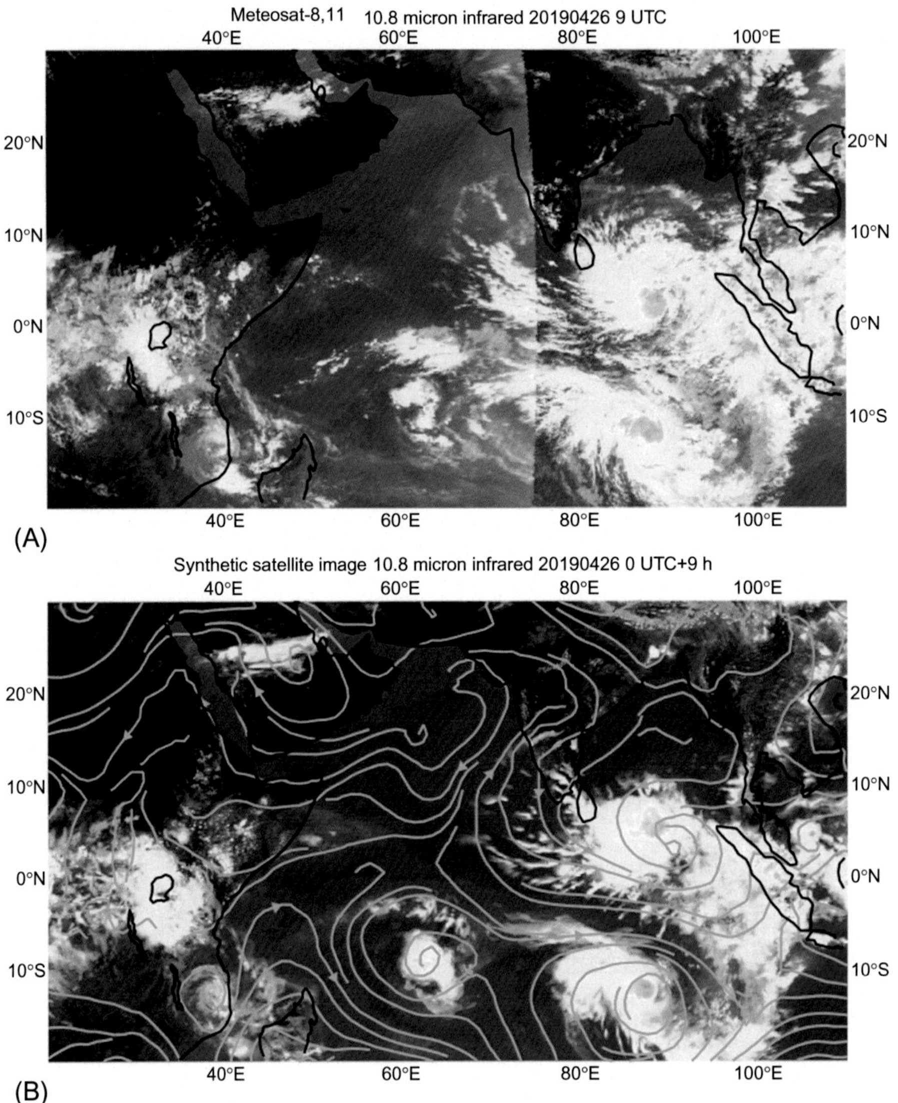

FIG. 2 The observed infrared satellite image at 10.9 μm from the SEVIRI instrument onboard Meteosat-8 and Meteosat-11 for the 26th of May 2019 9 UTC and the corresponding synthetic satellite image as obtained from the IFS forecast from 26th of May 2019 0 UTC. The large-scale flow is rendered by the 850 hPa streamlines.

particularly in the convective cloud structures over continental areas. Actually, the twin tropical cyclones that are symmetric to the equator are part of a westward propagating Rossby wave. Together with the maximum westerly winds at the equator which are a Kelvin wave signal and the cloud/circulation structures over the Maritime Continent, these compose the Madden Julian oscillation. All these tropical modes will be discussed in Sections 3.1 and 3.2.

3.1 Assimilation of convective features, microphysics, and heating

We can easily visualize the zonally propagating wave modes in the tropics with the aid of so-called Hovmöller diagrams as in Fig. 3, where we have averaged over the tropical belt of ±10° latitude the brightness temperatures from the observations from the short-range high-resolution (9 km) IFS forecasts. Data and forecast values are those obtained in every 12-h window during the data assimilation cycle. The resulting longitude vs time plot shows microwave brightness temperatures from the Advanced Microwave Scanning Radiometer-2 (AMSR-2) instrument. We have chosen this instrument as it is sensitive to both the frozen and liquid condensate of the cloud. The data is a good indicator of deep convection and can be used reliably over both water and land. In Fig. 3, four eastward propagating systems can be identified between 50 and 150°E, i.e., over the Indian Ocean and Maritime Continent region. These are envelopes of high microwave brightness temperatures that have a period of roughly 30–40 days. Inside these envelopes that can be qualified as the Madden and Julian oscillation (Madden and Julian, 1971) can be identified waves with westward and eastward phase speeds (Matsuno, 1966). Overall, the short-range IFS forecasts reasonably reproduce the observations, but there is uncertainty concerning the signal amplitude and therefore the ice content of the systems. An underestimation of brightness temperature anomalies can be due to not deep/active enough convection and/or errors in the microphysics. The microwave observation operator does not only require the in-cloud ice and snow concentrations from the prognostic cloud scheme, but also from the convection scheme. For the latter, this poses a real challenge in current NWP as the subgrid convective snow and rain content is generally unknown and must be deduced empirically from the diagnostic convective precipitation fluxes and by inferring the fractional cover of rain and snow at each height (Geer et al., 2009). In summary, the correct representation of the dynamics and (subgrid) mixed-phase microphysics of convection is of crucial importance not only to represent the convective heating and drying but also to successfully assimilate satellite data from microwave humidity sounders, which are becoming the most valuable data source in global operational forecast systems (Geer et al., 2018). It is expected that having available prognostic convective snow and rain fluxes will further improve the analyses.

The heating structure as produced by the forecasts has been recently evaluated over a roughly 1000 × 1000 km domain in the central Indian Ocean with the aid of radiosonde data collected during the DYNAMO campaign (Kim et al., 2018). The data period from October 2011 to January 2012 included three major MJO events. The total heating tendencies from the model, also known as apparent heat and moisture sinks Q1 and Q2 (Yanai et al., 1973), are simply computed as the sum of all physical parametrization tendencies (convection, cloud, radiation, boundary-layer turbulence), while the observed heating rates computed from the radiosonde data correspond to the area-averaged total tendencies including the local temperature change plus the change due to horizontal and vertical advection. The comparison in Fig. 4 for the profiles of the temperature and humidity tendencies reveals a reasonable correspondence between the model and the radiosonde observations in terms of mean tendencies and variability (standard deviations). Noticeable differences exist around the melting level at 4 km, where the variability in heating/moistening is particularly large. This again shows the importance and uncertainty related to mixed-phase microphysical processes. Differences in physical heating rates are compensated, through dynamical adjustments in the

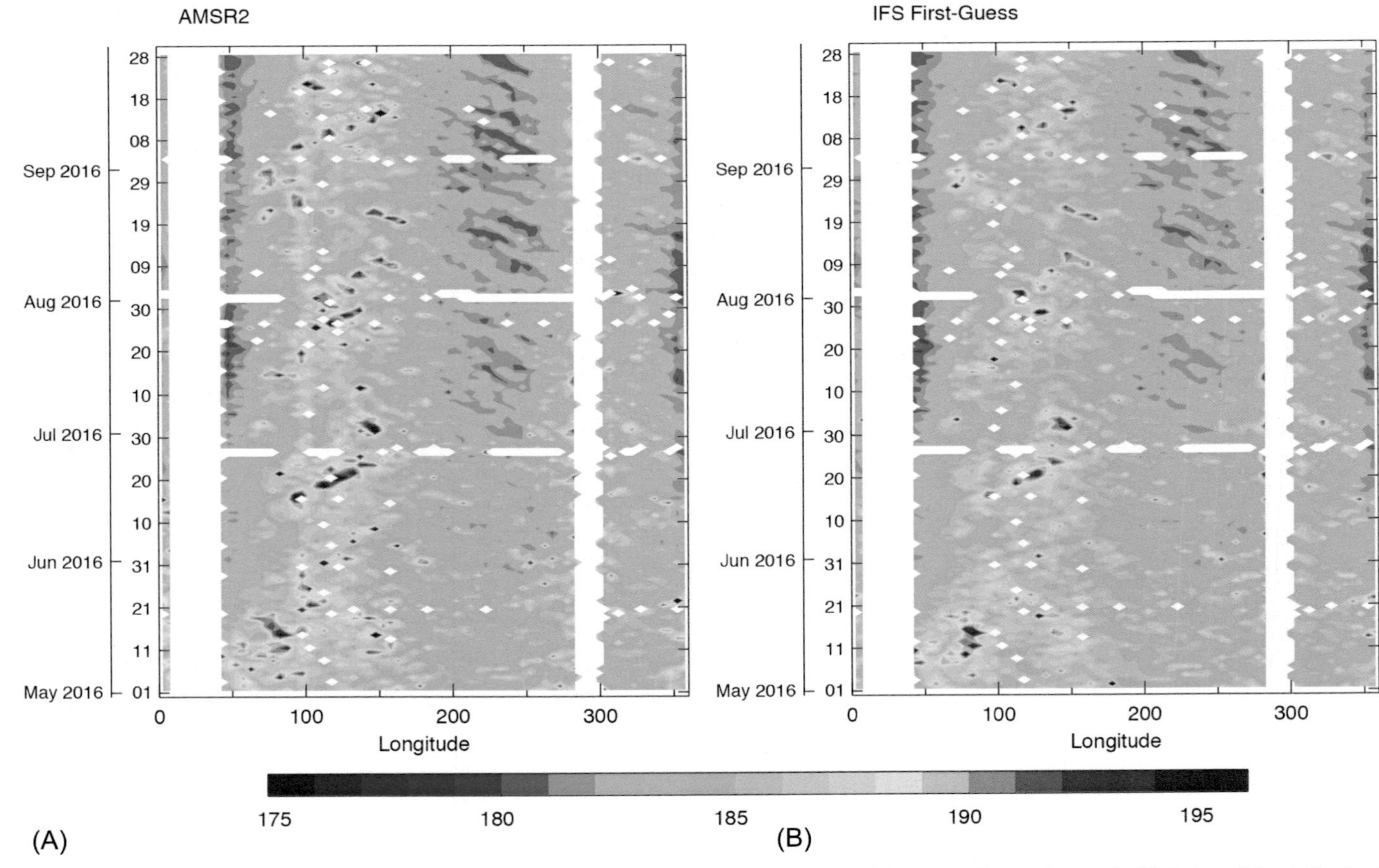

FIG. 3 Hovmöller diagrams (longitude vs time) of the microwave brightness temperatures (K) averaged over the tropical belt from May to September 2016 as obtained from the passive microwave sounder AMSR2 instrument and from the first-guess (background) forecasts of the IFS. *Figure courtesy Alan Geer, ECMWF.*

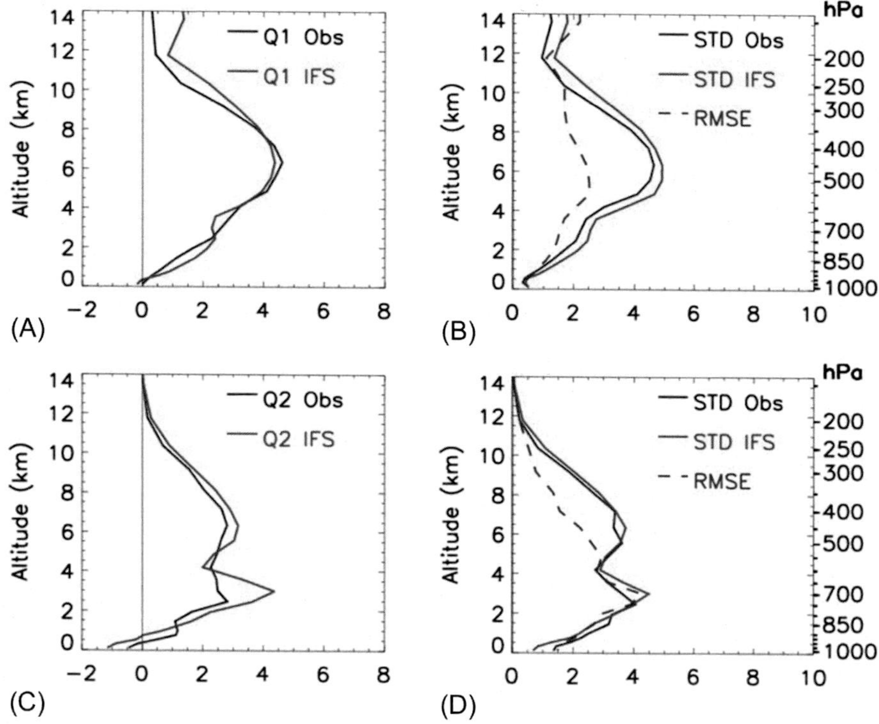

FIG. 4 Profiles of total heating and moistening tendencies (K day^{-1}) as derived from radiosonde observations during the dynamics of the MJO (DYNAMO) campaign during October-December 2011 over the Central Indian Ocean, and from the corresponding IFS short-range forecasts. These tendencies are generally denoted by Q1 for heat (temperature) and by Q2 for humidity and correspond to the sum of all respective physical tendencies in the model and equivalently to the total advective tendencies in the observations.

model, by differences in the mean vertical motion. Model errors in vertical motion become particularly critical near inversions such as the boundary-layer top and the melting level, as they strongly affect the vertical stability and the mixing.

3.2 Convectively coupled waves and the MJO

We now expand more on the different types of convectively coupled waves in the tropics as they are an important source of tropical predictability. A convenient way to display the different wave types is with the aid of frequency vs wavenumber diagrams of the outgoing longwave radiation as obtained by a two-dimensional Fourier transform (Wheeler and Kiladis, 1999). A comparison between the spectra as obtained from National Oceanic and Atmospheric Administration (NOAA) satellite data and as obtained from multiannual coupled integrations with the IFS is given in Fig. 5. The theoretical wave dispersion lines derive from the shallow water equations for a dry atmosphere and have also been added in Fig. 5.

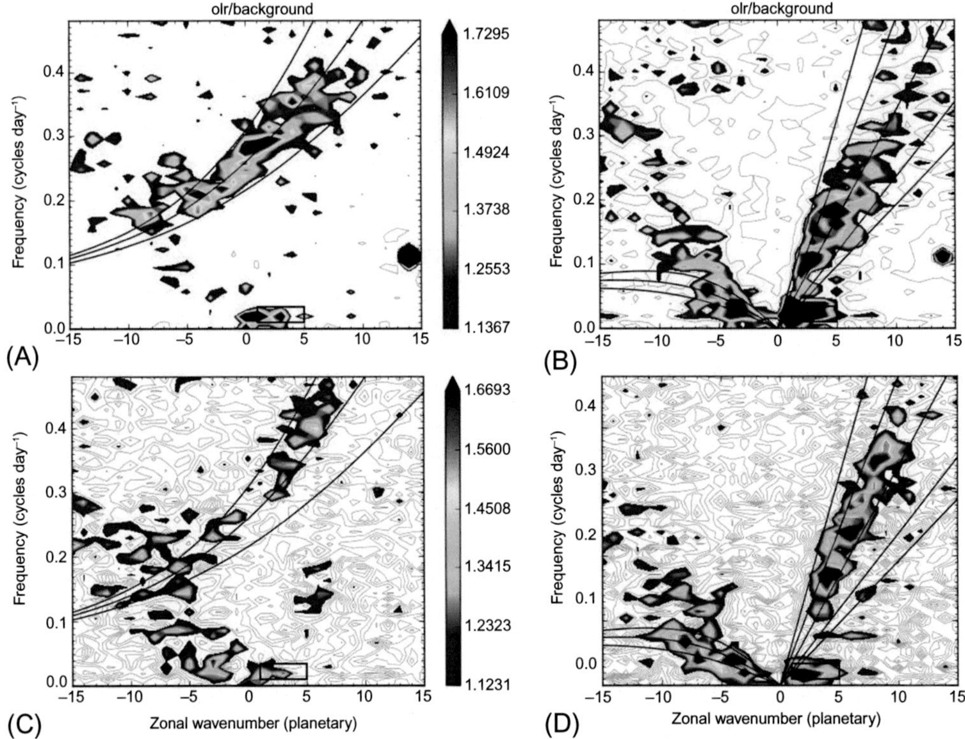

FIG. 5 Wavenumber-frequency diagrams of anomalies of daily data of outgoing longwave radiation from NOAA and from multiannual coupled integrations with the IFS. The spectra have been divided by their background values. The symmetric spectra (spectra meridionally averaged over the $\pm 15°$ tropical band) are shown in (B) and (D) and the antisymmetric spectra (raw data minus meridionally averaged data) in (A) and (C). The theoretical dispersion curves for the symmetric spectra contain the westward propagating Rossby wave, eastward propagating Kelvins waves (straight lines as for gravity waves), and the MJO, while the antisymmetric spectra contain the mixed Rossby-gravity wave modes.

The model reasonably reproduces the main wave types that are the eastward propagating Kelvin waves, the westward propagating Rossby waves, and the MJO which is a wavenumber 1–3 and frequency 30–60 days mode denoted by the black boxes in Fig. 5B and D. However, the model underestimates the MJO signal, has too much power in the high-frequency Kelvin/gravity waveband, and has difficulties in representing the mixed Rossby-gravity waves (Fig. 5A and C). The results in Fig. 5 actually reflect important improvements in 2006–08 (Fig. 6) in the IFS models representation of tropical modes that have been obtained by a thorough revision of the convection parametrization, including the convective closure and the entrainment rates (Bechtold et al., 2008). Before, the model was not able to correctly represent the Kelvin wave mode and the MJO. These deficiencies have been or are still present many in other global models (Williamson et al., 2013). The reasons for the improvements in tropical variability are subtle and have been discussed for the Kelvin wave and the MJO in Hirons et al. (2013), Shutts (2008), and Herman et al. (2016) and can be summarized as follows: over the oceans, the model (convection parametrization) must represent

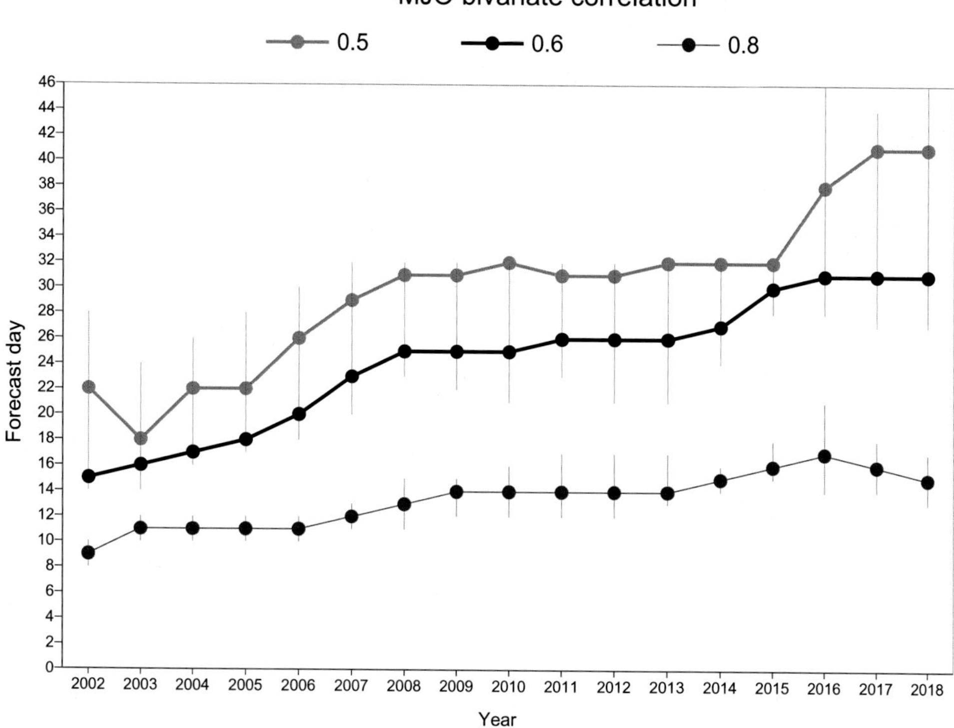

FIG. 6 Evolution of the skill of the IFS 11-member 30-year reforecasts of the MJO for different thresholds of the correlation coefficient between the forecast MJO index and the observed index. The MJO index as defined by (Malardel and Wedi, 2016) consists of the large-scale longitudinal empirical orthogonal functions for the upper- and lower-tropospheric winds and the outgoing longwave radiation. *Figure courtesy Frédéric Vitart, ECMWF.*

the observed sensitivity of rainfall to the environmental relative humidity, in particular, it must allow a gradual moistening of the lower to mid-troposphere with an evolution from a shallow convective to a cumulus congestus stage and eventually a deep convective precipitating stage. In the deep convective stage, the heating in the upper troposphere must be in phase with the temperature anomalies of the large-scale wave in order to amplify the wave. Generated mainly in the upper troposphere, the wave energy flux and also the kinetic energy flux is directed both downward into the lower troposphere and upward into the stratosphere (Sun et al., 2017).

The physical parametrizations directly affect multiple resolved scales and therefore the energy cascade through dissipation and subgrid heating (Malardel and Wedi, 2016). It appears that global models have been more successful in representing the rotational and barotropic Rossby mode than the Kelvin wave mode and the MJO (Dias et al., 2018), likely due to the baroclinic tilted temperature/moisture structure of the latter with significant vertical wind shear.

As documented in Fig. 6, the predictability of the MJO has improved a lot in the IFS. We use the standard definition of MJO predictability, i.e., the forecast day when the correlation between the ensemble mean forecasts of the large-scale MJO structure and the analysis drops below a value of 0.6. The MJO phase and amplitude are defined by empirical orthogonal functions including wind anomalies in the upper and lower troposphere (Wheeler and Hendon, 2004) and anomalies of the outgoing longwave radiation. The MJO skill of the IFS is monitored using reforecasts over a 30-year period with a 11-member ensemble. The predictability range of the MJO has increased from around 15 days in 2002 to 28 days in 2018. In particular, the 6-day predictability increase between 2006 and 2008 can be explained by more accurate model heating rates in space and time through a new shortwave radiation scheme and revisions to the convective entrainment and closure (Bechtold et al., 2014). As discussed before, the improvements in the predictability of the MJO come along with improved predictions of the Kelvin waves. There have been further improvements since in the heating rates from the cloud and convection parametrizations as well as improvements in the ocean coupling, which together with an increase in horizontal resolution from 36 to 18 km in 2016 explain some of the further improvements. However, the strong increase in forecast days with skill above the 0.5 correlation mark in 2016 is not model related but is mainly due to the extension of the monthly forecast system from 32 to 46 days.

Finally, in Fig. 7 we compare the MJO predictability of the IFS to other global operational model systems using the database from the seasonal to subseasonal international model project S2S (Vitart et al., 2017). There is a large variety in predictive skill between the models

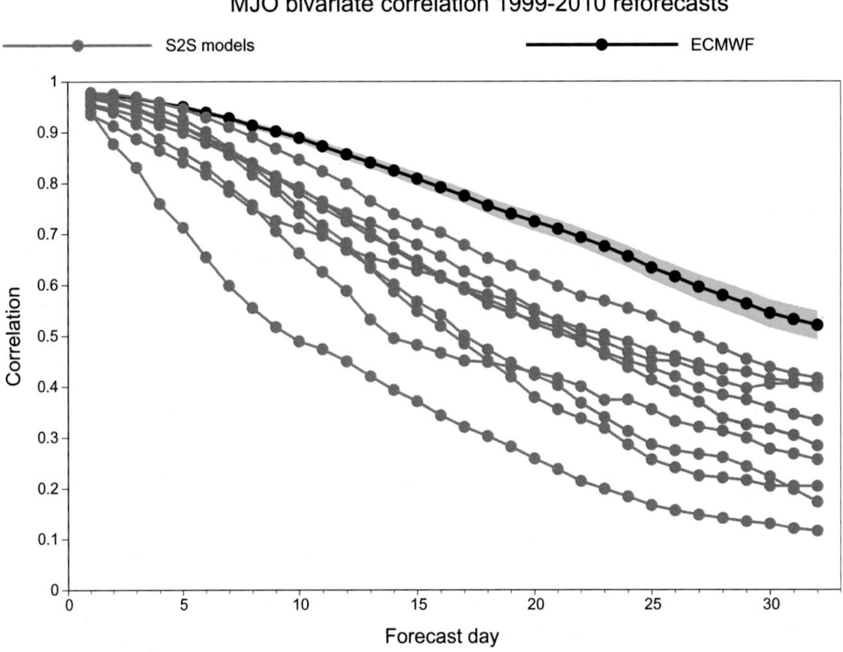

FIG. 7 As in Fig. 6 but comparing the correlation between the observed MJO anomalies and those produced by the IFS reforecasts for 1999–2010 (*black line*) to those of the other global prediction systems in the S2S database (*gray lines*).

of up to 10 days, with the IFS leading the other models. However, the important point is that there is still a large potential in improving the MJO through improvements in the physical parametrizations (heating and momentum transport) as its predictability limit can be estimated to be about one period, so around 45 days. As the MJO as a global tropical anomaly has strong teleconnections and therefore is a major source of predictability of the middle latitude weather anomalies beyond the medium-range, improving physical heating rates in the tropics is of the fundamental importance of reducing uncertainties in middle latitude weather predictions.

4 Gravity waves and the stratosphere

The characterization of atmospheric waves and their impact on the forecasts would not be complete without a short discussion on gravity waves. This is to illustrate the particular challenges gravity waves pose in the short-range forecasts and the data assimilation. There is a large amount of literature on orographic gravity waves and nonorographic gravity and inertia-gravity waves and their impact on the stratospheric flow, see e.g., Alexander et al. (2010). Here, we cannot get into a detailed discussion of the characteristics of gravity waves and their forcing of the stratospheric circulations, but content ourselves with a short summary based on the results presented in Alexander et al. (2010), Bacmeister et al. (1996), Preusse et al. (2014), and Žagar et al. (2017): (i) gravity waves have typical horizontal scales from 1 to 1000 km and several hundred m to several kilometers in the vertical, (ii) these waves can be generated by flow over orography, by convective heating and resonant forcing, by quasigeostrophic adjustments in the jet regions and frontal discontinuities among others, (iii) the distinction between orographic and nonorographic gravity waves is generally made on the basis of the ground-based phase speed which is zero for orographic gravity waves and typically in the range of 10–30 m/s for nonorographic gravity waves, and (iv) in the stratosphere the energy of inertia-gravity waves exceeds above a zonal wavenumber of 35 or for scales <500 km the energy of Rossby waves.

Preusse et al. (2014) have examined the characteristics of gravity waves in the IFS using fitted three-dimensional wave vectors and applying a ray-tracing method. They identified several sources of gravity waves such as orography, frontal systems, and convection but pointed to regions with large gravity wave momentum fluxes such as the southern tip of South America, the higher latitudes of the southern hemisphere and the tropics, where the gravity wave characteristics can show large errors.

Indeed, as illustrated in Fig. 8A the satellite stratospheric infrared sounding channels show recurrently for the May season in the lower stratosphere large first-guess temperature departures (forecast errors) between 30 and 70°S in the IFS analysis system that peak East of the Andes. As examined in Preusse et al. (2014), there are large-amplitude gravity waves over the southern tip of South America and the Antarctic peninsula are orographically generated. In contrast, the source of the gravity waves over the southern oceans around 60°S is not well understood. These waves can be of either tropospheric or stratospheric origin and can also include oblique and horizontal propagation and convergence of gravity waves from the north and south (Peter Preusse, personal communication).

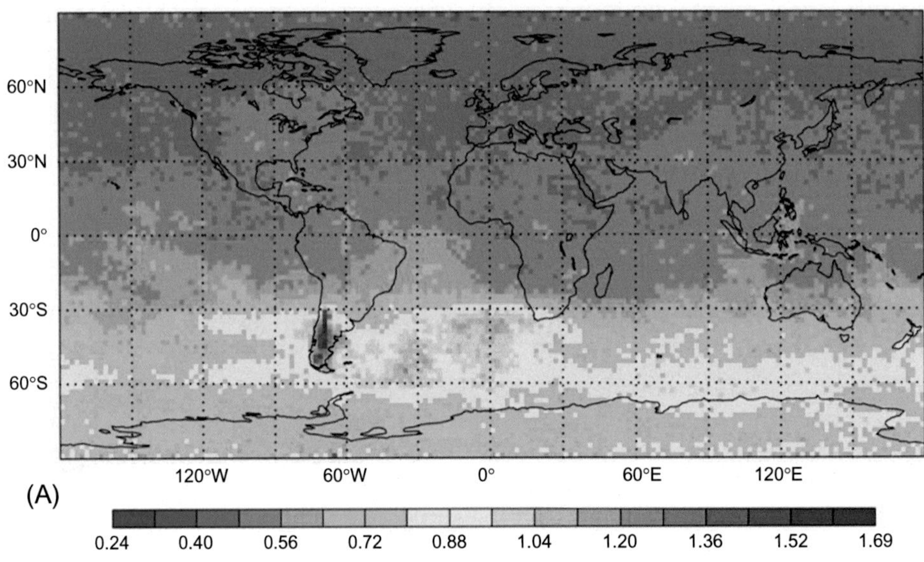

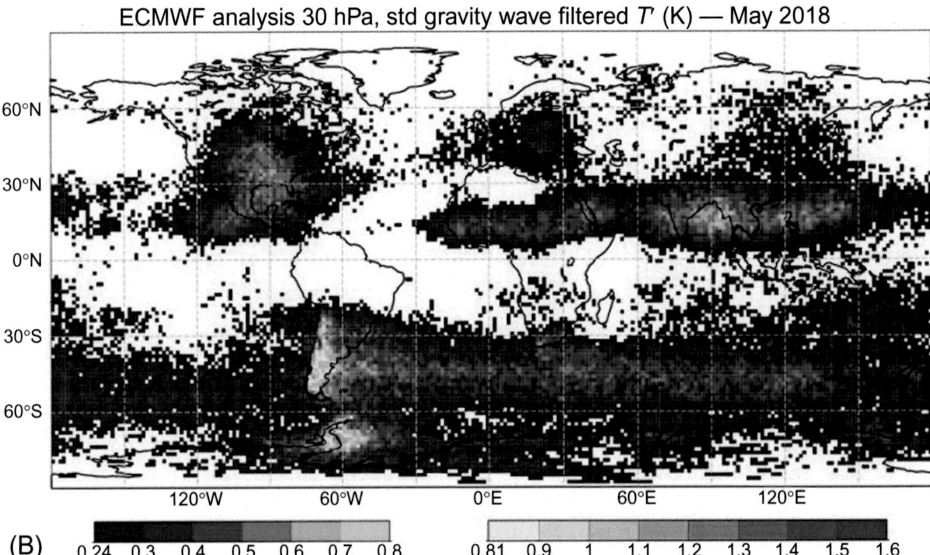

FIG. 8 Standard deviations of lower stratospheric temperature departures during May 2018: (A) first-guess temperature departures of the IFS short-range forecast against the Atmospheric Infrared Sounder (AIRS) channel 75 onboard NASA's AQUA satellite during May 2018 and (B) standard deviation of the gravity wave (total wavenumber > 100) filtered IFS temperature analysis T' (K) at 30 hPa. *Figure courtesy Mohamed Dahoui and Tony McNally, ECMWF.*

Interestingly, the IFS is able to generate the characteristic distribution of orographic and convective gravity waves in the lower stratosphere. This is documented in Fig. 8B, where we have applied on the daily analyzed temperature fields during May 2018 a high-pass spectral filter retaining total wavenumbers exceeding 100. The correspondence between the first-guess temperature departures in Fig. 8A and the gravity wave T' anomalies in Fig. 8B is clearly remarkable; the main differences concern the convective regions in the tropics and the middle latitudes, where the satellite instruments cannot resolve the small-scale (<200–300 km wavelength) convective gravity waves.

But where do these large first-guess departures in the southern hemisphere then come from if the model apparently represents these waves. We think that phase and amplitude errors play an important role.

This is illustrated in Fig. 9 which shows temperature data above 250 hPa from high-resolution radiosonde ascents at Punta Arena at the southern tip of Chile as well as the corresponding temperature profiles from the short-range IFS forecasts. While the IFS successfully forecasted the first sounding (Fig. 9A), it was not able to reproduce the very large amplitude of 20 K and the phase of the vertical waves in the second example (Fig. 9B).

Large forecast errors due to gravity waves result not only in a degraded analysis as less satellite data is assimilated in the stratosphere, but likely also in degraded longer-range predictions due to an incorrect representation of the resolved gravity wave drag on the mean flow. What can be done to improve the representation of gravity waves? The obvious answer is a further increase in both the horizontal and vertical resolution of the model in order to better resolve both the physical sources of gravity waves and the propagation of the waves. Experimentation with the IFS has shown that its current vertical resolution of 137 levels with a model top at 1 Pa that is 300–500 m vertical resolution in the stratosphere is still insufficient to resolve gravity waves with a vertical wavelength of O (1 km). Preliminary experimentation suggests reduced model biases when the number of vertical model levels is increased to around 150–175. In order to investigate the influence of horizontal resolution and convection on the gravity wave spectrum, we have performed in addition to the operational 9 km forecasts that use a deep convection parametrization, model simulations at 4 km horizontal resolution with and without the deep convection scheme. The resulting wave spectra for the horizontal kinetic energy at 50 hPa are split into their rotational and divergent components as depicted in Fig. 10. It is clear from Fig. 10 that the increase in resolution has a major impact on the spectra beyond wavenumber 400, i.e., waves with horizontal scales <50–80 km. The increase in horizontal resolution leads to a flattening of the spectrum with a spectral slope that becomes closer to the theoretical $-5/3$ spectral slope. The impact of convection on gravity waves is documented by switching off the deep convection parametrization at 4 km that is running in convection-permitting mode. In this experiment, the wave energy in the stratosphere is significantly increased, while the spectral slope remains essentially unaffected. An equivalent but more extensive study has been conducted by Stephan et al. (2019) using the ICON model of the Deutsche Wetterdienst. The authors documented similar wave spectra to Fig. 10 but note a 30%–60% increase in the gravity wave momentum flux in the stratosphere in the convection-permitting run compared to their run with parametrized convection. Although the activity of the convection-permitting run is likely too high as deduced from rainfall statistics.

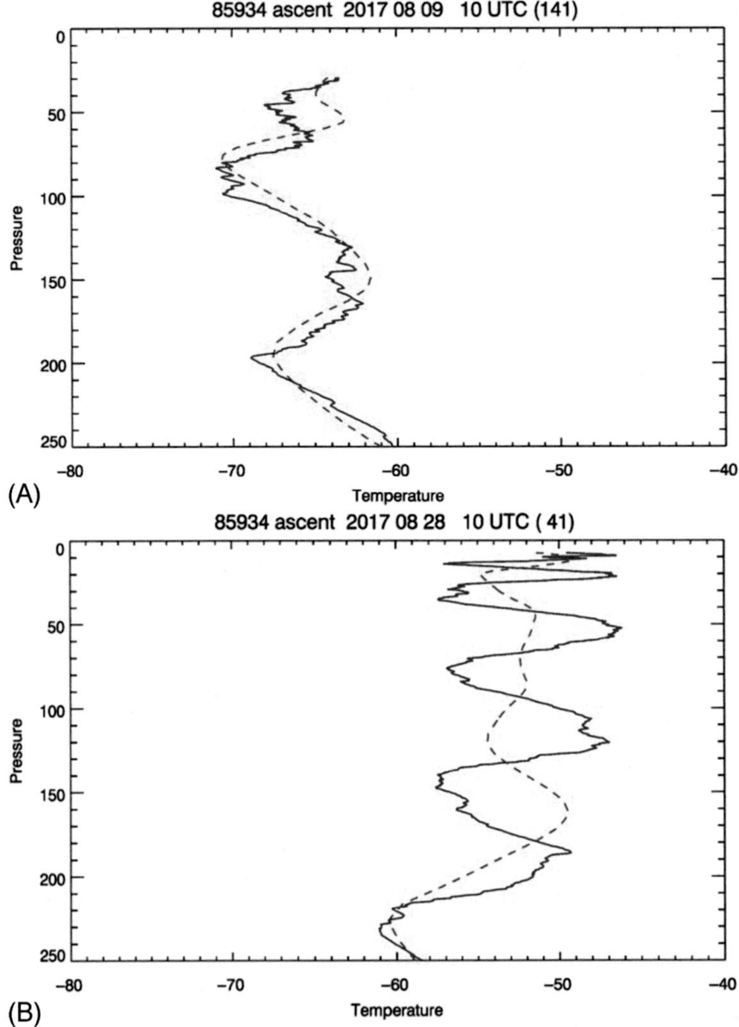

FIG. 9 High-resolution temperature profiles above 250 hPa from radiosonde ascents at Punta Arenas in Chile (53°S, 70.84°W) for the August 9, 2017 and August 28, 2017 (*solid lines*) and corresponding temperature profiles from the IFS short-range forecasts (*dashed lines*). *Figure courtesy Bruce Ingleby, ECMWF.*

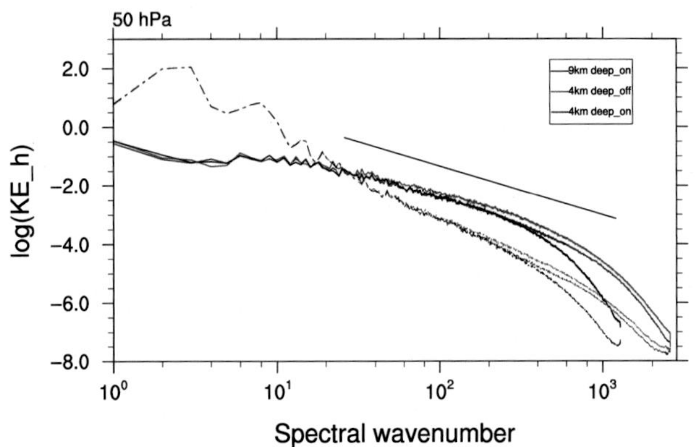

FIG. 10 Global spectra of the rotational (*dashed*) and divergent part (*solid*) of the horizontal kinetic energy at 50 hPa as a function of the global wavenumber for different model configurations: 9 and 4 km horizontal resolutions with the deep convection scheme and 4 km without the deep convection scheme. The *straight black line* denotes the −5/3 spectral slope. *Figure courtesy Nils Wedi and Inna Polichtchouk, ECMWF.*

The influence of the different physical processes on the kinetic energy spectra and the spectral energy transfer has been analyzed in Malardel and Wedi (2016). Concerning the tropospheric spectra, the important conclusions of this study are that the spectra of kinetic energy and wind variance and the predictability and realism of the forecasts are strongly impacted by (i) dissipative processes such as surface drag, convective momentum transport in the upper troposphere and numerical diffusion and (ii) by the potential energy input and conversion at the mesoscales through convective heating, the latter is particularly higher with explicit convection.

5 Mesoscale convective systems and the diurnal cycle

The representation of continental mesoscale convective systems and nocturnal convection represents major physical and conceptual challenges in models. To date, global general circulation models and NWP systems have similar difficulties in accurately predicting the timing of convective initiation and in predicting the evolution and duration of mesoscale convective systems throughout the night (Stelten and Gallus Jr, 2017). Even at convection-permitting resolutions, it remains challenging to reproduce the observed diurnal cycle of convective cells (Brousseau et al., 2016).

Accurately forecasting continental convection is immediately relevant to warn of severe weather, but the impact of mesoscale convective systems can also be far-reaching through their impact on global circulations. In the tropics, for example, convective activity impacts the monsoon propagation via its impact on the meridional heating gradients and the dynamics of the intertropical front (Vizy and Cook, 2017; Birch et al., 2014).

Mesoscale convective systems are particularly large and intense in regions with a persistent low-level nocturnal jet like the Great Planes of North America and the Sahel region. Field experiments like the African Monsoon Multidisciplinary Analysis (Redelsperger et al., 2006) and the 2015 Great Plains Elevated Convection At Night (PECAN) (Geerts et al., 2017) have been dedicated to this topic. These authors point to many physical mechanisms that trigger and maintain convective systems during the night such as propagating cold pools or gravity waves, wave or frontal features like undular bores, dynamical elevated mesoscale lifting. In addition, processes such as moisture heterogeneity, advection by low-level jets, convective organization, and convective momentum transport influence the life cycle of these systems. Notably, cold pools have been brought forward as an important process in the transition from shallow to deep convection and the propagation and triggering of mesoscale convective systems (Kurowski, 2018). As reviewed by Zuidema et al. (2017) the main effect of cold pools on deep convection is through organized moist pockets in the boundary layer, leading to larger and less entraining convective plumes. The three-dimensional and nonstationary effects of cold pools are notoriously difficult to parametrize and although some global circulation models have some form of cold pool parametrization none is currently in use in actual weather prediction models.

5.1 Diurnal cycle

We cannot assess all these processes here, but instead we will assess the diurnal cycle of convection in the IFS globally against observations and discuss its link with the formation and propagation of mesoscale convective systems. In Fig. 11 is displayed the diurnal cycle of precipitation (mm) for the summer seasons of 2011 and 2012 as a function of local solar time as obtained from radar observations (Central Europe, North America East of the Rockies) and Tropical Rainfall Measurement Mission (TRMM) climatology (Sahel zone Africa) against a model version of the IFS that was operational before 2014 (CTL) and a version operational from 2014 (NEW). As discussed in Bechtold et al. (2014), the latter contains a revision to the convective available energy (CAPE) closure of the convection parametrization, whereby the efficiency with which convection responds to either surface heating or large-scale forcing has been altered. The conclusion from Fig. 11 for all regions is that CTL produced a diurnal cycle that is in phase with the surface heating and therefore precedes the observed afternoon peak by roughly 6h. Since then progress has been made and NEW which is still the operational version in 2019 reasonably reproduces the daytime evolution of precipitation, although the afternoon precipitation still occurs 1–2h too early. The main remaining systematic error in NEW is the underestimation of nighttime precipitation.

Recently, a lightning parametrization has been developed for the IFS (Lopez, 2016) that is based on the output from the convection scheme such as the CAPE, the convective cloud base height and the frozen water content. In Fig. 12 we have taken advantage of available lightning data from ground stations and GOES-16 satellite data and have repeated the analysis of Fig. 11 by comparing the observed normalized intensity of cloud to ground and intracloud lightning strikes from the different available data sources with that produced by the IFS during the summer season of 2018 over the continental United States east of the Rocky Mountains. Both, Figs. 11 and 12 consistently suggest that the main remaining model error is an underestimation of summer-time nocturnal convection with a too rapid decline in convective activity in the early evening hours.

5.2 Mesoscale convective systems over the Sahel

As an example of the difficulties, we can face in representing propagating mesoscale convective systems including squall lines we have selected the forecasts for the August 12, 2017 over northern Africa. This is admittedly an example of limited predictability and an unsuccessful operational forecast. On the Meteosat-10 3-hourly sequence of infrared satellite images from 15 to 21 UTC (Fig. 13A–C) one readily identifies large and deep (blue colors) mesoscale convective systems. These systems form first around 10°N, then later during the day also form around 16°N in a drier environment with generally large convective inhibition. The convective systems tend to grow further during the late evening-early night hours and propagate westward with the mean mid-tropospheric wind. The 3-hourly accumulated rainfall from the TRMM 3B42 version 7 product (Fig. 13D and E) shows, consistently with the satellite images, more spotty rainfall during the day south of 10°N with rainfall intensities of 5–15mm/3h, while larger and more intense systems develop during the evening around 16°N with rainfall intensities of 10–30mm/3h.

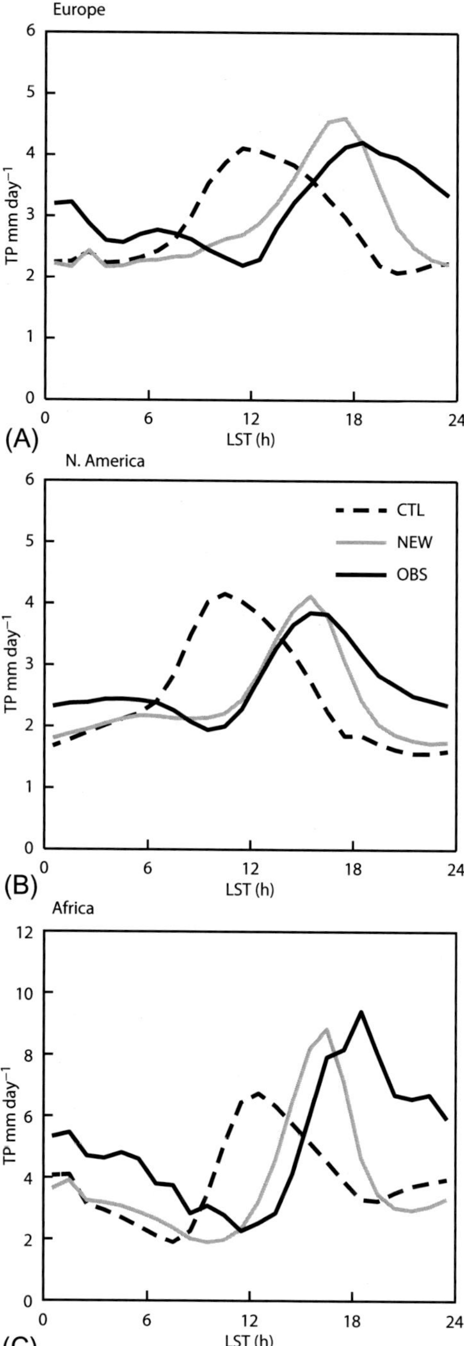

FIG. 11 The diurnal cycle of summer-time (JJA) precipitation (mm) as a function of local solar time over Europe, the United States east of the Rockies and Saharan Africa from the short-range forecasts, the radar observations for central Europe and the United States and the TRMM rainfall climatology over the Sahel.

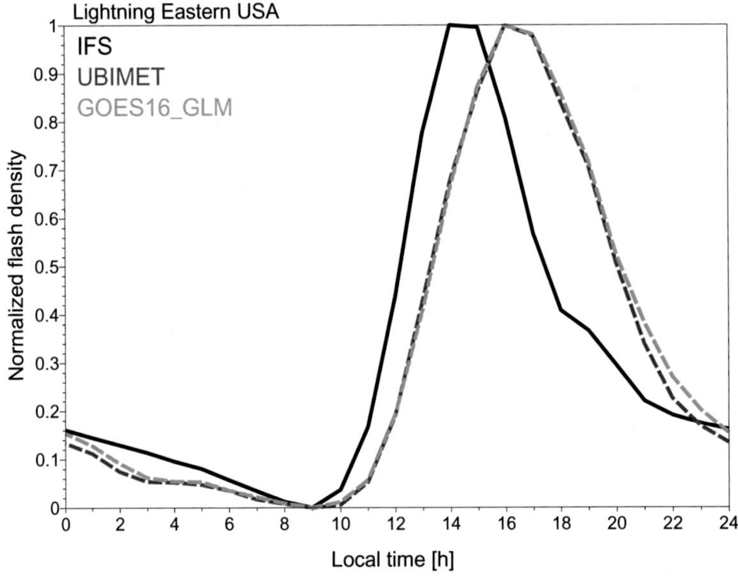

FIG. 12 Same as Fig. 11 but comparing the diurnal cycle of normalized intensity of cloud to ground + intracloud lighting strikes over (A) Europe for summer 2015 and (B) the Eastern United States for summer 2018 as obtained from ground stations and satellite data with the lightning product from the IFS short-range forecasts. The data sources are the Institute for Ubiquitous Meteorology (UBIMET) and GOES-16 satellite data. *Figure courtesy Philippe Lopez, ECMWF.*

The operational high-resolution forecasts with the IFS in 2019 are run at a horizontal resolution of 9 km. Here we have also performed two experimental very high-resolution reforecasts for the August 12, 2017 at 4 km global resolution with the model cycle operational in 2019, starting at August 11, 2017 00 UTC. The first forecast labeled OPER4 uses the standard deep convection parametrization with resolution-dependent convective fluxes (tendencies) (Bechtold et al., 2014), while the second forecast, labeled NO DEEP is run in convection-permitting mode that is without the deep convection parametrization, but retains the shallow convection parametrization. The 3-hourly accumulated rainfall for OPER and NO Deep during day-2 of the forecasts is depicted in Fig. 13G–K, respectively. The rainfall intensities in OPER roughly correspond to that of TRMM 3B42, also OPER can produce systems that become stronger and more organized during the afternoon. However, OPER produces more widespread light rainfall during daytime compared to TRMM and misses the propagating systems north of 10°N. In contrast, NO DEEP produces a too spotty and intense rainfall distribution south of 10°N and off the coast with local maxima up to 50 mm/3 h during the afternoon hours, but succeeds in producing propagating convective systems north of 10°N. However, NO DEEP overestimates the global precipitation by around 10% and produces too spotty a precipitation field over the tropical oceans (not shown). Therefore, we think that the desired model performance is somewhat in between OPER4 and NO DEEP.

Consequently, we are putting a lot of effort in trying to improve the model performance by a revision of the convective closure that also includes a scaled total moisture advection term.

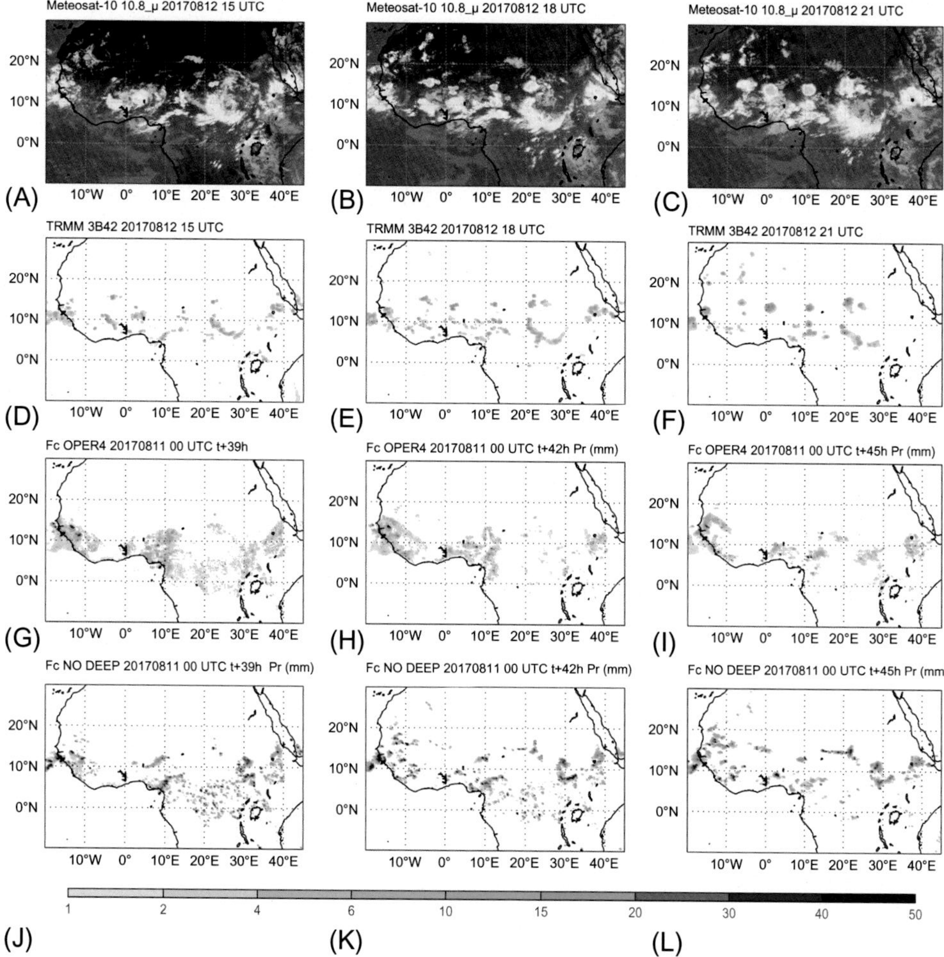

FIG. 13 Evolution of continental convective systems over tropical Africa during September 12, 2017 in 3-hourly slots from 15 to 21 UTC as seen by Meteosat-11 infrared image at 10.9 μm wavelength (A–C), as well as 3 hourly accumulated rainfall (mm) from 12–15, 15–18 and 18–21 UTC from the TRMM 3B42 product (D–F) and from the 4 km IFS reforecasts with (operational version) and without the deep convection scheme. The IFS reforecasts start on September 11, 2017 at 00 UTC and use the model cycle operational in 2019. There is no TRMM 3B42 data East of 25°E at 21 UTC.

As depicted in Fig. 14, the first results are promising as with the revised convective closure the deep convection scheme produces more intense mesoscale systems north of 10°N that now propagate westward during the evening as also proven by more extensive monthly statistics (not shown).

The kinetic energy in the upper troposphere is markedly different for the simulations in Figs. 13 and 14 as illustrated in Fig. 15 by the spectra at 200 hPa as a function of the global wavenumber. The 4 km resolution simulation with deep convection off (black line) (corresponding to Fig. 13J–L) exhibits significantly more energy for wavenumbers >60, that

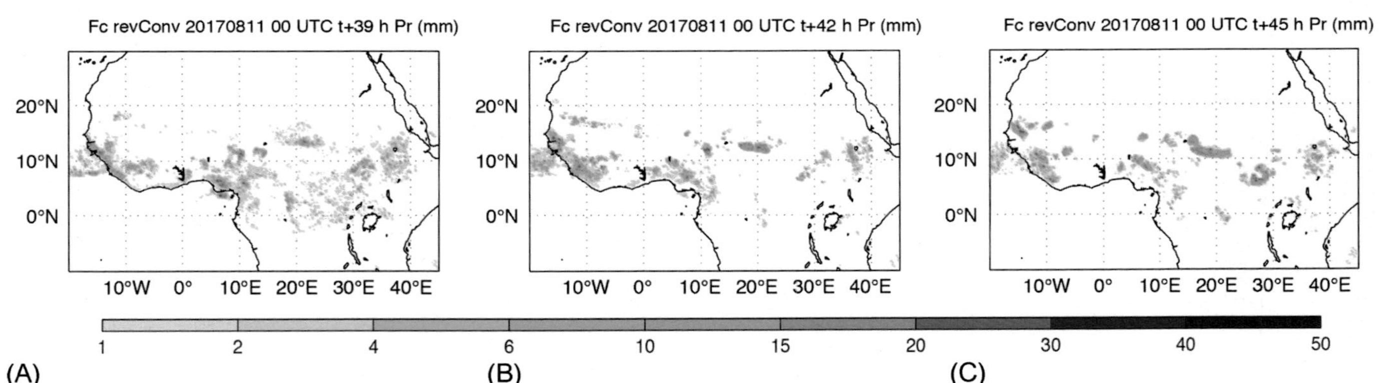

FIG. 14 As in Fig. 13G–I, but with the revised deep convective closure.

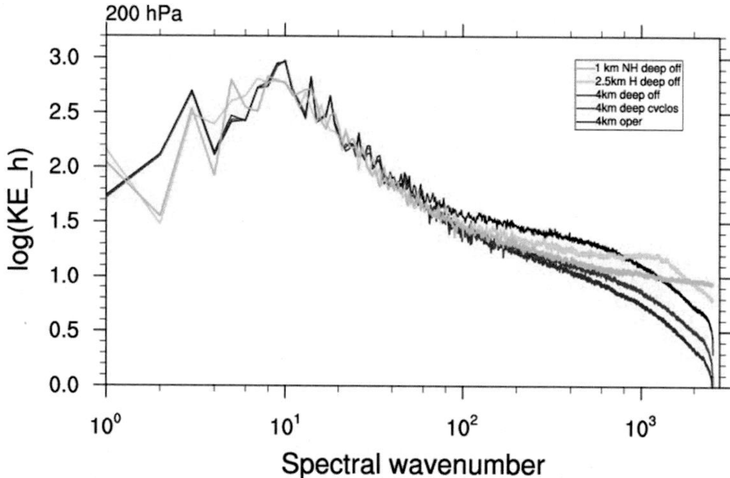

FIG. 15 Global spectra of the kinetic energy at 200 hPa as a function of the global wavenumber for different model configurations: 4 km resolution simulations with the operational model version and with deep convection on (*blue*), deep convection off (*black*) and simulation with deep convection but using an extended convective closure including total moisture convergence (*red*). In addition, spectra from a hydrostatic convection-permitting simulation at 2.5 km and a nonhydrostatic convection-permitting simulation at 1 km have been included.

is, on scales <300 km, than the operational version with deep convection on (blue line) (corresponding to Fig. 13G–I). However, the simulation with the revised convective closure (red line) (corresponding to Fig. 14) increases the kinetic energy in the upper troposphere in the mesoscales with respect to the operational version and also produces a precipitation distribution with a broader tail (not shown) which is a desired and potentially a more realistic feature of the forecasts (Sun et al., 2017). For comparison, we have also added the spectra from a hydrostatic convection-permitting simulation at 2.5 km with the IFS and nonhydrostatic convection-permitting simulation at 1 km resolution. The interesting result here is that with increasing model resolution the spectral energy in the mesoscales significantly reduces with respect to the convection-permitting simulation at 4 km and that at 1 km resolution the kinetic energy spectra have converged to values that are not far off from those obtained with the revised deep convection simulation at 4 km.

Finally, forecasts of convective systems at subsynoptic scales cannot be deterministic and must be probabilistic. There is currently at all NWP centers a strong investment in further improving the respective ensembles systems by including a description of uncertainty of the model physics through stochastic perturbations. At ECMWF we are currently developing a stochastically perturbed parameter scheme for the model physics, where selected parameters or parts of the individual parametrization schemes are given specified standard deviations and are modulated by a global spatially and temporarily varying pattern generator (Ollinaho et al., 2017).

Here, we use a 15-member ensemble where only perturbations to six selected parameters of the convection scheme are applied: the deep and shallow entrainment and detrainment rates, the convective adjustment time, the autoconversion rate from cloud water to precipitation and the convective momentum transport. No initial perturbations haven been applied to the ensemble members and the ensemble is run for August 2017 with the operational model cycle in 2019 and at the operational horizontal resolution of the ensemble system that is 16km. Similar to Fig. 13 we display in Fig. 15 the standard deviation of the 3-hourly accumulated precipitation at 18 UTC and 21 UTC during August 2017 from the IFS ensemble (Fig. 15A and B) and compare it to the corresponding standard deviations from the TRMM 3B42 product (Fig. 15C and D). Although the model data has 15 times more samples than the observations we consider both as broadly comparable. Indeed, in both the model and the observations the monthly mean standard deviations for 3-hourly precipitation vary between 1 and 5mm mainly in a band between 5 and 15°N, although the observations have some precipitation even north of 20°N. The monthly mean variability gives a picture that is quite consistent with the individual day in Fig. 13 in that the precipitation is most variable and/or uncertain north of 10°N and at the West African coast. However, Fig. 16A and B indicates more variability/uncertainty of precipitation in the model north of 10°N at 18 UTC than at 21 UTC, while there is some hint in the observations of larger uncertainty at 21 UTC.

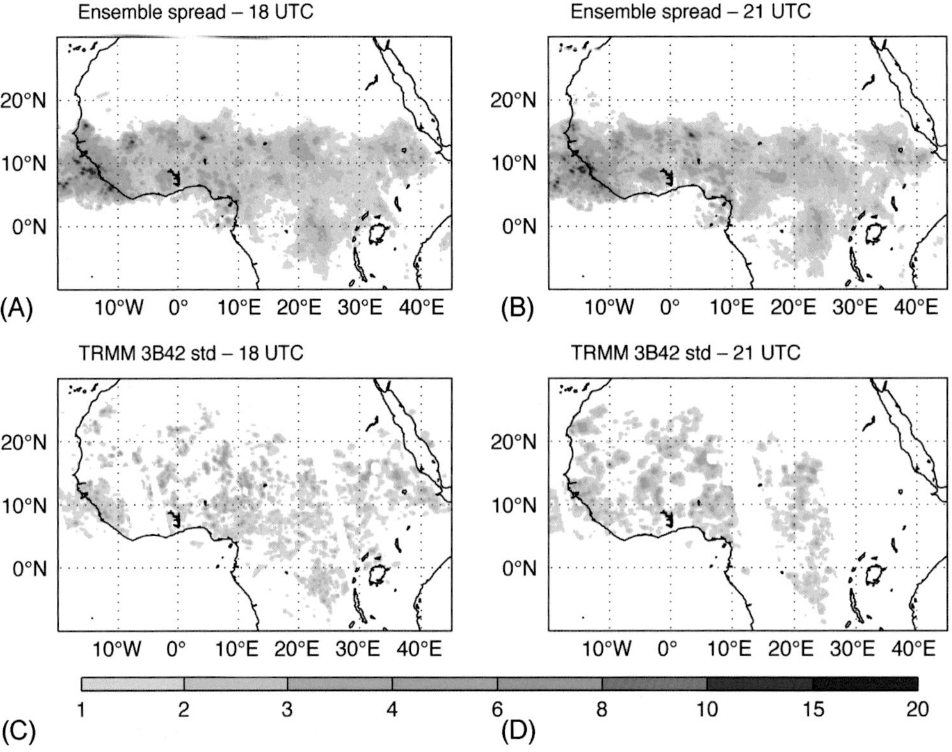

FIG. 16 Ensemble standard deviation of 3-hourly accumulated precipitation (mm) for (A) 18 UTC and (B) 21 UTC during August 2016 from an IFS 15-member ensemble at 16km resolution and corresponding rainfall standard deviation from the TRMM 3B42. (C) and (D) Note the limited data coverage (white stripes) in (C) and (D), e.g., between 10°–15°E and East of 25°E.

We have used this study on mesoscale convective systems to illustrate that their forecasts must be probabilistic but that the probabilistic forecast cannot in general compensate for systematic errors in the propagation and diurnal cycle of these systems and errors in the large-scale meridional circulation.

Clearly, a realistic convection-permitting run cannot be done at 4 km resolution but requires resolutions of 2 km or higher. But can we correct for the systematic errors by further improving the deep convection parametrization? The current diagnostic IFS convection parametrization is coupled to the large-scale dynamics via the convective tendencies for temperature, moisture, momentum, and cloud variables. It assumes that all convective subsidence occurs locally on the subgrid-scale. Furthermore, the parametrization does not include convective memory and horizontal advection. Malardel and Bechtold (2019) have therefore explored a direct coupling of the convection to the model dynamics whereby the divergence of the convective mass flux is passed to the continuity equation of the dynamical core. The convective compensating subsidence is then no longer subgrid-scale but is performed by the grid-scale advection scheme of the dynamical core of the model. The authors showed improved convective organization and more intense squall line with this configuration for horizontal resolutions higher than 4 km, but little effect for lower resolutions. We, therefore, think that the next step forward should be the inclusion of advective effects in the convection parametrization which is not without considerable numerical and theoretical difficulties (Gerard, 2015).

6 Boundary-layer clouds and the radiation budget

Finally, in this last section is discussed what we think is the challenge for the moist boundary-layer that is getting the low-level cloud amount and the shortwave radiation budget right. This is of primary importance for accurate 2-m temperature forecasts and accurate longer-range predictions with a coupled ocean-atmosphere system. The importance of the moist convective boundary-layer is displayed in Fig. 17 where is plotted the annual mean frequency of shallow convection in the IFS. This diagnostic of the shallow convective boundary-layer is certainly model dependent, but we think that Fig. 17 provides realistic statistics as the IFS shallow convection scheme is active in both cumulus and stratocumulus topped boundary-layers. Indeed, the main information from Fig. 16 is that shallow convection is ubiquitous over the oceanic regions and it occurs quasipersistently with a frequency of over 80% in the subtropical oceanic high-pressure areas.

In Fig. 18 are displayed the IFS (operational model version from 2019) annual mean errors in the net shortwave radiation at the top of the atmosphere as deduced from multiannual coupled integrations verified against the Clouds and Earth's Radiant Energy System (CERES) Energy Balanced and Filled (EBAF) dataset (Kato et al., 2018). These errors have a distinct spatial distribution and can be summarized as: the tropical anticyclonic regions with shallow cumulus clouds (Fig. 17) are 5–15 W/m^2 too reflective, while the stratocumulus regions over the cold currents off the West coasts of the contents are not reflective enough, with errors that do exceed 30 W/m^2. Other regions that stick out are the Southern Ocean and West Pacific Kuroshio regions that are also not reflective enough, as well as the equatorial

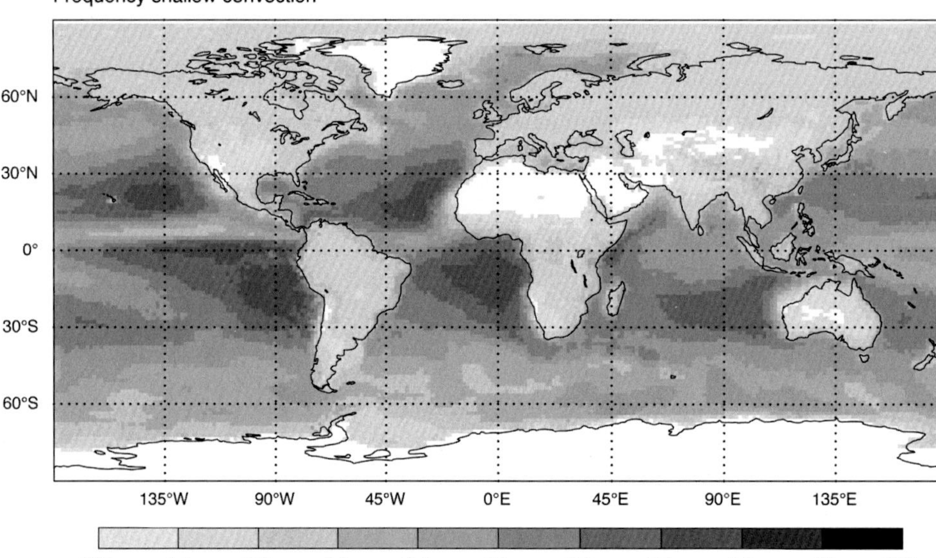

FIG. 17 Annual mean frequency of occurrence of shallow convection that is a convective cloudy boundary-layer as diagnosed by the IFS convection scheme in multiannual integrations.

West Pacific. The error in the tropical West Pacific only occurs in coupled simulations and is due to a model dry bias and a lack of convection resulting from feedback between the equatorial winds, the sea surface temperature, and the convection. Li et al. (2013) have evaluated the global models that have participated in the 20th century Coupled Model Intercomparison Project Phase 5 (CMIP5) and Phase 3 (CMIP3) simulations and found a very similar distribution of the shortwave radiation errors as in the IFS, i.e., the trade cumulus regions are too reflective, and the stratocumulus regions and the Southern Ocean regions are not reflective enough.

It can be readily shown that these errors are related to the model's representation of boundary-layer clouds and the questions are then are these errors linked to errors in the cloud cover or errors in the total condensate content or errors in the phase of the condensate (liquid vs ice). To address these questions, Ahlgrimm et al. (2018) have evaluated the model against CERES Moderate-resolution Imaging Spectroradiometer (MODIS) retrievals along a transect from the Southern California coast to Hawaii that encompasses a cloudiness transition from stratocumulus to trade cumulus clouds. The results for the total column liquid water content are displayed in Fig. 19 and show that the model strongly underestimates the liquid water content near the coast, while the liquid water content in the trade cumulus regions is slightly overestimated. However, the model errors in-cloud cover (not shown) are small. The conclusion, therefore, is that the overestimation in outgoing shortwave radiation in the trade cumulus regions is due to an overestimation of the liquid water content of the clouds, while the liquid water content in the stratocumulus clouds are underestimated, resulting in insufficient upwelling shortwave radiation.

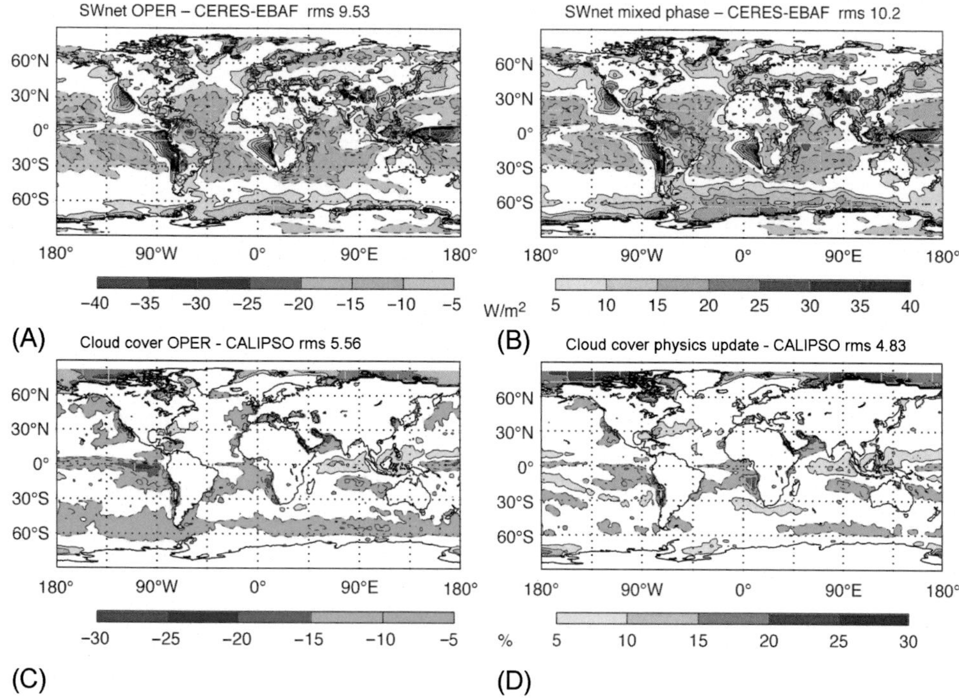

FIG. 18 Cloud and radiation evaluation from multiannual coupled integrations with the IFS. (A and B) Difference in the top of atmosphere net shortwave radiation (W/m^2) between the model and the Clouds and the Clouds and the Earth's Radiant Energy System (CERES) Energy Balanced and Filled (EBAF) product for (A) the operational version in 2019 and (B) an experimental version where we have reverted the developments to the mixed-phase microphysics in the convection. Negative (positive) values correspond to excessive (insufficient) outgoing shortwave radiation. (C and D) differences in total cloud cover over open water between the model and the Cloud-Aerosol Lidar and Infrared Pathfinder Satellite Observations (CALIPSO) climatology for (C) the operational version in 2019 and (D) an experimental physics upgrade with a complete revision to the convective cloudy boundary-layer and the cloud scheme. Significant differences are denoted by the hatched areas.

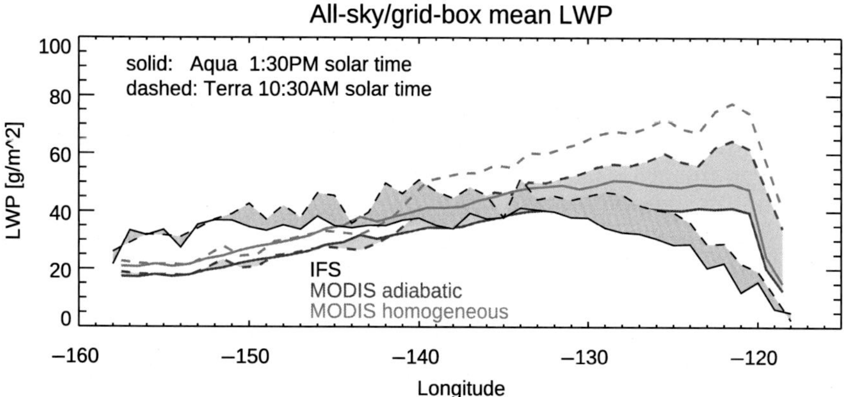

FIG. 19 Mean total column liquid water path (g/m^2) for summer 2013 along a transect from California to Hawaii [33°N,118°W–21°N,158°W] from the IFS and from different retrieval methods using Clouds and the Earth's Radiant Energy System/Moderate-resolution Imaging Spectroradiometer, CERES/MODIS data. *Solid and dashed lines* are for different solar times.

The situation in the Southern Oceans is different. Indeed, many studies have been devoted to this region and found that the shortwave radiation errors in global models in these regions are due to an overestimation of the cloud ice content to the detriment of the cloud liquid water content (Hyder et al., 2018). The latter has been observed to occur at the top of cold shallow convective boundary-layer clouds and in cumulus congestus during cold air outbreaks and increases the shortwave reflectivity of these clouds with respect to ice clouds. A lot of progress has been made in the IFS in recent years in improving the microphysics of these clouds. This is documented in Fig. 18B where we have reverted the recent changes to the convective microphysics, i.e., we use a temperature transition from 0 to −23°C for the cloud glaciation (instead of −38°C for the operational version in 2019, Fig. 18A), we assume that shallow convective clouds are in the mixed phase (and not purely in the liquid phase) and we do not detrain the precipitating species rain and snow to the stratiform cloud scheme. The resulting annual mean shortwave model errors in Fig. 18B are up to $10\,W/m^2$ larger compared to the control model in Fig. 18A over the Southern Oceans, over the Kuroshio region off North-eastern Asia and Greenland.

Finally, a lot of efforts have also been made in recent years in improving the representation of the stratocumulus clouds. In our group, these developments have taken years and are still not operational as it involves both changes to the turbulent and convective transport and the representation of boundary-layer clouds in general (either statistical or prognostic). The final goal is some universal representation of both stratocumulus and cumulus clouds. While sophisticated statistical schemes for the boundary-layer turbulence and clouds are now used in some global models (Thayer-Calder et al., 2018), we decided to further develop the traditional mass flux and eddy diffusion approaches that are widely used in global NWP models and which was originally introduced in the IFS by Köhler et al. (2011). The idea is that the classical mass flux shallow convection scheme performs most of the transport in the cloudy part of the convective layer, while the diffusion scheme does all the dry turbulent mixing while also contributing to the quasi well-mixed stratocumulus boundary-layer. Importantly, both schemes are comprehensively coupled to the prognostic cloud scheme via the respective source terms. The revised model, therefore, can in principle represent all types of boundary-layer clouds, the cumulus type or positively skewed distributions via the mass flux source and the Gaussian and negatively skewed distributions as in the stratocumulus and cumulus to stratocumulus transition regimes via the contributions from the mass flux and the vertical diffusion, plus the contribution from the relative humidity dependent quasi-Gaussian subgrid condensation in the clear sky part of the grid. However, we needed to introduce a criterion based on the boundary-layer inversion strength to distinguish stratocumulus, where the turbulent mixing extends to the cloud top, from other cloud types where the turbulent mixing is limited to near cloud base (by using a stronger include entrainment rate in the test parcel that determines the mixing height). The still somewhat preliminary results of all these developments are shown in Fig. 18C and D for the total cloud cover against the Cloud-Aerosol Lidar and Infrared Pathfinder Satellite Observations (CALIPSO) climatology. The physics upgrade improves the total cloud cover distribution by reducing the negative biases in the East Pacific stratocumulus and transition regions and in the storm tracks. Overall, the new model is simpler, has better physical consistency, and has more sensitivity in addressing the characteristic cloud and low-level errors. An operational implementation of the moist physics upgrade is planned for 2021.

7 Conclusions

All subgrid moist physical processes are important for the forecast performance and individual process is relevant for a particular aspect of the forecast system. Therefore, our choice of moist physical processes and their coupling to the atmospheric flow has been somewhat arbitrary. Nevertheless, we have chosen to focus on processes where the physics-dynamics interaction is strong, which are crucially important in terms of potential forecast improvements and understanding and which remain challenging at future even higher O (1–5 km) resolutions for both the model and the data assimilation: (i) the coupling of deep convection with the larger-scale circulation, (ii) vertically propagating gravity waves and their impact in the stratosphere, (iii) the diurnal cycle of convection and the propagation of mesoscale convective systems, and (iv) the cloudy convective boundary-layer and its impact on the global shortwave radiation budget. We used the European Centre of Medium-range Weather Forecast (ECMWF) Integrated Forecast System (IFS) with its "high-resolution" forecasts at 9 km horizontal resolution as a benchmark for global NWP, and also explore 4 km resolution reforecasts with and without the deep convection parametrization. Furthermore, we have also explored ensemble simulations at 16 km horizontal resolution using stochastically perturbed parameters in the convection parametrization.

1. It is shown that the IFS realistically reproduces the large-scale convectively coupled waves and heating profiles in the tropics and that getting these perturbations right is of fundamental importance for tropical predictions in general and for middle latitude predictions on the medium and subseasonal time range. The model's ability to realistically represent the large-scale tropical modes critically depends on its ability to represent over the tropical oceans the observed variability in the shallow, congestus, and deep convective heating profiles. However, there are still challenges in improving the analysis of these convective structures as the successful assimilation of satellite data (mainly microwave data) critically depends on a successful simulation of the spatial distribution and mixed-phase microphysics of these clouds. Microwave satellite data have now become the most important data source in NWP (Geer et al., 2018). Also, there is potential to further improve the forecasts of the 30–60 days Madden Julian oscillation beyond the current predictability of around 30 days. Improving the Madden Julian oscillation is also of fundamental importance in improving the middle latitude predictions beyond the medium-range via the well-established teleconnections between the tropics and the middle latitudes (Vitart and Molteni, 2010b).
2. In the high frequency or high phase speed part of the atmospheric wave spectrum, large-amplitude gravity waves of orographic or convective origin or from dynamical adjustments in the jet regions have a large impact on the stratospheric circulation and via the stratosphere-troposphere interaction also on the troposphere. Overall the model is able to represent the different gravity waves as generated by orography, convection and dynamical sources. However, selected temperature soundings over Chile document profiles where the model struggles to represent the amplitude and phase of these waves.

   Orographic or nonorographic gravity wave parametrization schemes cannot resolve the wave structure and only represent the mean effect on the flow in a simplified manner, e.g., by considering purely vertically propagating waves while observed gravities follow an oblique path (Preusse et al., 2014). We suggest that increasing the horizontal resolution to 4 km, while likely also increasing the vertical resolution allows to more realistically represent the resolved stratospheric gravity wave spectrum in the mesoscales. However, the diffusive/dissipative processes in the model have to be carefully evaluated and the overall gravity wave activity carefully monitored as a function of resolution.

3. As discussed in Lillo and Parsons (2017) and Rodwell et al. (2013) accurately forecasting mesoscale convective systems is fundamental in NWP as forecast errors strongly project on to the large-scale flow and therefore into the medium-range, where these systems interact with the upper-level jet stream. There are still considerable problems in representing nighttime organized propagating mesoscale convective systems with parametrized convection and that quasiindependently of model resolution. In contrast, convection-permitting simulations at 4 km can more realistically represent the occurrence of these systems and their propagation, although they significantly overestimate the intensity of convection and consequently the upper-tropospheric kinetic energy. A convective ensemble system using, e.g., stochastically perturbed convection parameters is clearly needed to represent the uncertainty in forecasting the timing and triggering of mesoscale convective systems, but it cannot correct for the systematic model errors in the representation of moist convection.

Significant progress in representing mesoscale convective systems at resolutions around 5 km, which is the envisaged resolution of the ECMWF ensemble system in 2025, can only be made by understanding the fundamental differences between parametrized and convection-permitting simulations. The probably most fundamental difference is that in convection-permitting simulations condensational heating occurs at grid-scale supersaturation generated by grid-scale lifting. Parametrized convection is generally diagnosed before the grid-column becomes saturated and the grid-scale vertical motion results from the quasigeostrophic adjustment to the subgrid-scale heating. In nonequilibrium convection (Bechtold et al., 2014), the large-scale motion field is shifted in time with respect to the heating. Other differences include the lack of memory or horizontal advection in many convection parametrizations and the environmental subsidence. In the parametrization, the sinking motion that is the response to the subgrid updraught mass flux occurs on the subgrid by construction, while in nature it occurs on larger scales surrounding the convective clouds. Pan and Randall (1998) and Gerard (2015) have already developed a framework to include convective memory and horizontal advection in the parametrization, while Malardel and Bechtold (2019) have included the compensating subsidence in the model dynamics through the divergence of the convective mass flux. Here we have shown that with an extended deep convective closure that includes, e.g., the total moisture convergence it is in principle possible to more realistically represent mesoscale convective systems, their propagation and consequently the upper-tropospheric energy in the mesoscales. However, clearly more theoretical work and detailed model evaluations are needed.

4. Finally, representing the different convective boundary-layer cloud regimes and their radiative effects in coupled ocean-atmosphere systems remains challenging. Considerable progress has and can still be made through an adequate description of the mixed-phase microphysics, the local turbulent and nonlocal moist convective moist mixing, and through an adequate coupling of the turbulent mixing scheme and a subgrid cloud scheme. We decided to pursue an approach where the shallow convection scheme does most of the transport in the cloudy part of the boundary-layer and where the diagnostic shallow convection scheme and the diagnostic turbulent diffusion scheme are coupled to a prognostic subgrid cloud scheme via their respective sources. In contrast to a statistical approach, which requires assumptions on the probability distribution functions of the boundary-layer turbulence and cloudiness, the chosen approach can consistently represent positively skewed cumulus boundary-layers and Gaussian or negatively skewed stratocumulus or transition boundary-layers. The largest sensitivity of the boundary-layer scheme is to the height to which the turbulent diffusive mixing is applied. Optimal results are obtained if turbulent diffusion is limited to around cloud base in cumulus type boundary-layers but extends up to cloud top that is the main inversion in stratocumulus type boundary-layers. The problem of representing diffusive mixing in the cloudy boundary-layer is somewhat mitigated when using a prognostic higher-order turbulent approach via the turbulent kinetic energy or including further prognostic turbulent moments. However, diffusion schemes in contrast to mass flux schemes tend to produce well-mixed boundary-layers. In the prognostic approach, the sensitivity to a specified mixing height is replaced by a sensitivity to the buoyant turbulent production term. We think there is still a long future for a combined approach of nonlocal cloudy mass flux and turbulent diffusion schemes.

Acknowledgments

In addition to the colleagues already acknowledged who have provided figures, my special gratitude goes to my colleagues Gabriella Szépszo for proofreading and advice, Anabel Bowen for professional figure editing, and to my team Irina Sandu, Richard Forbes and Maike Ahlgrimm for our work on the new boundary-layer cloud scheme.

References

Ahlgrimm, M., Forbes, R.M., Hogan, R.J., Sandu, I., 2018. Understanding global model systematic shortwave radiation errors in subtropical marine boundary layer cloud regimes. J. Adv. Mod. Earth Syst. 10, 2042–2060.

Alexander, M.J., Geller, M., McLandress, C., Polavarapu, S., Preusse, P., Sassi, F., Sato, K., Eckermann, S., Ern, M., Hertzog, A., Kawatani, Y., Pulido, M., Shaw, T.A., Sigmond, M., Vincent, R., Watanabe, S., 2010. Recent developments in gravity-wave effects in climate models and the global distribution of gravity-wave momentum flux from observations and models. Q. J. Roy. Meteorol. Soc. 136, 1103–1124.

Arakawa, A., Schubert, W., 1974. Interaction of a cumulus ensemble with the large-scale environment. Part I. J. Atmos. Sci. 31, 674–701.

Bacmeister, J.T., Eckermann, S.D., Newman, P.A., Lait, L., Chan, K.R., Loewenstein, M., Proffitt, M.H., Gary, B.L., 1996. Stratospheric horizontal wavenumber spectra of winds, potential temperature, and atmospheric tracers observed by high-altitude aircraft. J. Geophys. Res. 101, 9441–9470. https://doi.org/10.1029/95JD03835.

Bechtold, P., Köhler, M., Jung, T., Leutbecher, M., Rodwell, M., Vitart, F., Balsamo, G., 2008. Advances in predicting atmospheric variability with the ECMWF model, from synoptic to decadal time-scales. Q. J. Roy. Meteorol. Soc. 134, 1337–1351.

Bechtold, P., Semane, N., Lopez, P., Chaboureau, J.-P., Beljaars, A., Bormann, N., 2014. Representing equilibrium and non-equilibrium convection in large-scale models. J. Atmos. Sci. 7, 734–753.

Beljaars, A., Balsamo, G., Bechtold, P., Bozzo, A., Forbes, R., Hogan, R.J., Köhler, M., Morcrette, J.-J., Tompkins, A.M., Viterbo, P., Wedi, N., 2018. The numerics of physical parametrization in the ECMWF model. Front. Earth Sci 6, 1–18. https://doi.org/10.3389/feart.2018.00137.

Birch, C.E., Parker, D.J., Marsham, J.H., Copsey, D., Garcia-Carreras, L., 2014. A seamless assessment of the role of convection in the water cycle of the West African Monsoon. J. Geophys. Res. 119, 2890–2912. https://doi.org/10.1002/2013JD020887.

Brousseau, P., Seity, Y., Ricard, D., Léger, J., 2016. Improvement of the forecast of convective activity from the AROME-France system. Q. J. Roy. Meteorol. Soc. 142, 2231–2243.

Dias, J., Gehne, M., Kiladis, G.N., Sakaeda, N., Bechtold, P., Haiden, T., 2018. Equatorial waves and the skill of NCEP and ECMWF numerical weather prediction systems. Mon. Weather Rev. 146, 1763–1784.

Geer, A.J., Bauer, P., O'Dell, C.W., 2009. A revised cloud overlap scheme for fast microwave radiative transfer in rain and cloud. J. Appl. Meteorol. Clim. 48, 2257–2270.

Geer, A.J., et al., 2018. All-sky satellite data assimilation at operational weather forecasting centres. Q. J. Roy. Meteorol. Soc. 144, 1191–1217.

Geerts, B., Parsons, D., Ziegler, C.L., Weckwerth, T.M., Biggerstaff, M.I., Clark, R.D., Coniglio, M.C., Demoz, B.B., Ferrare, R.A., Gallus Jr., W.A., Haghi, K., K., 2017. The 2015 plains elevated convection at night field project. Bull. Am. Meteorol. Soc. 98 (2017), 767–786.

Gerard, L., 2015. Bulk mass-flux perturbation formulation for a unified approach of deep convection at high resolution. Mon. Weather Rev. 143, 4038–4063.

Gregory, D., Miller, M., 1989. A numerical study of the parameterization of deep tropical convection. Q. J. Roy. Meteorol. Soc. 115, 1209–1241.

Guichard, F., Couvreux, F., 2017. A short review of numerical cloud-resolving models. Tellus A 69, 1–36.

Herman, M.J., Fuchs, Ž., Raymond, D., Bechtold, P., 2016. Convectively coupled Kelvin waves: from linear theory to global models. J. Atmos. Sci. 73, 407–428.

Hirons, L.C., Inness, P., Vitart, F., Bechtold, P., 2013. Understanding advances in the simulation of intraseasonal variability in the ECMWF model. Part II: The application of process-based diagnostics. Q. J. Roy. Meteorol. Soc. 139, 1427–1444. https://doi.org/10.1002/qj.2059.

Hyder, P., Edwards, J.M., Allan, R.P., Hewitt, H.T., Bracegirdle, T.J., Gregory, J.M., Wood, R.A., Meijers, A.J.S., Mulcahy, J., Field, P., Furtado, K., Bodas-Salcedo, A., Williams, K.D., Copsey, D., Josey, S.A., Liu, C., Roberts, C.D., Sanchez, C., Ridley, J., Thorpe, L., Hardiman, S.C., Mayer, M., Berry, D.I., Belcher, S.E., 2018. Critical Southern Ocean climate model biases traced to atmospheric model cloud errors. Critical Southern Ocean climate model biases traced to atmospheric model cloud errors. Nat. Commun. 9, 3625.

Kato, S., Rose, F.G., Rutan, D.A., Thorsen, T.J., Loeba, N.G., Doellinga, D.R., Huang, X., Smith, W.L., Su, W., Ham, S.-H., 2018. Surface irradiances of edition 4.0 Clouds and the Earth's Radiant Energy System (CERES) Energy Balanced and Filled (EBAF) data product. J. Climate 31, 4501–4537.

Kim, J.-E., Zhang, C., Kiladis, G.N., Bechtold, P., 2018. Heating and moistening of the MJO during DYNAMO in ECMWF reforecasts. J. Atmos. Sci. 75, 1429–1452.

Köhler, M., Ahlgrimm, M., Beljaars, A., 2011. Unified treatment of dry convective and stratocumulus-topped boundary layers in the ECMWF model. Q. J. Roy. Meteorol. Soc. 137, 43–57.

Kurowski, M.J., 2018. Shallow-to-deep transition of continental moist convection: cold pools, surface fluxes, and mesoscale organization. J. Atmos. Sci. 75, 4071–4090.

Li, J.-L., Waliser, D., Stephens, G., Lee, S., L'Ecuyer, T., Kato, S., Loeb, N., Ma, H.-Y., 2013. Characterizing and understanding radiation budget biases in CMIP3/CMIP5 GCMs, contemporary GCM, and reanalysis. J. Geophys. Res. Atmos. 118, 8166–8184. https://doi.org/10.1002/jgrd.50378.

Lillo, P., Parsons, D.B., 2017. Investigating the dynamics of error growth in ECMWF medium-range forecast busts. Q. J. Roy. Meteorol. Soc. 143, 1211.

Lopez, P., 2016. A lightning parameterization for the ECMWF Integrated Forecasting System. Mon. Weather Rev. 144, 3057–3075.

Madden, R.A., Julian, P.R., 1971. Detection of a 40-50 day oscillation in the zonal wind in the tropical Pacific. J. Atmos. Sci. 28, 702–708.

Malardel, S., Bechtold, P., 2019. The coupling of deep convection with the resolved flow via the divergence of mass flux in the IFS. Q. J. Roy. Meteorol. Soc. 722, 1832–1845.

Malardel, S., Wedi, N.P., 2016. How does subgrid-scal parametrisation influence non-linear spectral energy fluxes in global NWP models. J. Geophys. Res. 121, 5395–5410. https://doi.org/10.1002/2015JD023970.

Matsuno, T., 1966. Quasi-geostrophic motions in the equatorial area. J. Meteorol. Soc. Japan 44, 25–43.

Ollinaho, P., Lock, S.-J., Leutbecher, M., Bechtold, P., Beljaars, A., Bozzo, A., Forbes, R., Haiden, T., Hogan, R., Sandu, I., 2017. Towards process-level representation of model uncertainties: stochastically perturbed parametrisations in the ECMWF ensemble. Q. J. Roy. Meteorol. Soc. 143, 408–422. https://doi.org/10.1002/qj.2931.

Pan, D.-M., Randall, D.A., 1998. A cumulus parametrization with a prognostic closure. Q. J. Roy. Meteorol. Soc. 124, 949–981.

Preusse, P., Ern, M., Bechtold, P., Eckermann, S.D., Kalisch, S., Trinh, Q.T., Riese, M., 2014. Characteristics of gravity waves resolved by ECMWF. Atmos. Chem. Phys. 14, 10483–10508. https://doi.org/10.5194/acp-14-10483-2014.

Redelsperger, J.L., Thorncroft, C.D., Diedhou, A., Levbel, T., Parker, D.J., Polcher, J., et al., 2006. Bull. Am. Meteorol. Soc. 86, 1739–1746.

Rodwell, M., Richardson, D.S., Hewson, T.D., Haiden, T., 2010. A new equitable score suitable for verifying precipitation in numerical weather prediction. Q. J. Roy. Meteorol. Soc. 136, 1344–1363.

Rodwell, M., Magnusson, L., Bauer, P., Bechtold, P., Bonavita, M., Cardinali, C., Diamantakis, M., 2013. Characteristics of occasional poor medium-range weather forecasts for Europe. Bull. Am. Meteorol. Soc. 1393–1405. https://doi.org/10.1175/BAMS-D-12-00099.1.

Satoh, M., Noda, A.T., Seiki, T., Chen, Y.-W., Kodama, C., Yamada, Y., Kuba, N., Sato, Y., 2018. Toward reduction of the uncertainties in climate sensitivity due to cloud processes using a global non-hydrostatic atmospheric model. Prog. Earth Planet. Sci. 5 (1), 67.

Shin, H.H., Dudhia, J., 2016. Evaluation of PBL Parameterizations in WRF at subkilometer grid spacings: turbulence statistics in the dry convective boundary layer. Mon. Weather Rev. 144, 1161–1177.

Shutts, G., 2008. The forcing of large-scale waves in an explicit simulation of deep tropical convection. Dyn. Atmos. Oceans 45, 1–25.

Stelten, S., Gallus Jr., W.A., 2017. Pristine nocturnal convective initiation: a climatology and preliminary examination of predictability. Weather Forecast. 32 (4), 1613–1635.

Stephan, C.C., Strube, C., Klocke, D., Ern, M., Hoffmann, L., Preusse, P., Schmidt, H., 2019. Gravity waves in global high-resolution simulations with explicit and parameterized convection. J. Geophys. Res. Atmos. 124, 4446–4459. https://doi.org/10.1029/2018JD030073.

Sun, Y.Q., Rotunno, R., Zhang, F., 2017. Contributions of moist convection and internal gravity waves to building the atmospheric -5/3 kinetic energy spectrum. J. Atmos. Sci. 74, 185–201.

Tan, Z., Kaul, C.M., Pressel, K.G., Cohen, Y., Schneider, T., Teixeira, J., 2018. An extended eddy-diffusivity mass-flux scheme for unified representation of subgrid-scale turbulence and convection. J. Adv. Mod. Earth Syst. 10, 770–800.

Thayer-Calder, K., Gettleman, A., Craig, C., Goldhaber, S., Bogenschutz, P.A., Chen, C.-C., Morrison, H., Höft, J., Raut, E., Griffin, B.M., Weber, J.K., Larsson, V.E., Wyant, M.C., Wang, M., Guo, Z., Ghan, S.J., 2018. A unified parameterization of clouds and turbulence using CLUBB and subcolumns in the Community Atmosphere Model. Geosci. Model. Dev. 8, 3801–3821. https://doi.org/10.5194/gmd-8-3801-2015.

The DYAMOND Initiative, n.d. https://www.esiwace.eu/services/diamond

Vitart, F., Molteni, F., 2010a. Simulation of the Madden–Julian Oscillation and its teleconnections in the ECMWF forecast system. Q. J. Roy. Meteorol. Soc. 136, 842–855.

Vitart, F., Molteni, F., 2010b. Simulation of the MJO and its teleconnections in the ECMWF forecast system. Q. J. Roy. Meteorol. Soc. 136, 842–855.

Vitart, F., et al., 2017. The Subseasonal to Seasonal (S2S) prediction project database. Bull. Am. Meteorol. Soc. 98 (1), 163–173. https://doi.org/10.1175/BAMS-D-16-0017.1.

Vizy, E.K., Cook, K.H., 2017. Mesoscale convective systems and nocturnal rainfall over the West African Sahel: role of the inter-tropical front. Climate Dynam. 50, 587–614.

Wheeler, M.C., Hendon, H.H., 2004. An all-season real-time multivariate MJO index: development of an index for monitoring and prediction. Mon. Weather Rev. 132, 1917–1932.

Wheeler, M., Kiladis, G.N., 1999. Convectively coupled equatorial waves: analysis of clouds and temperature in the wavenumber-frequency domain. J. Atmos. Sci. 56, 374–399.

White, B., Gryspeerdt, E., Stier, P., Morrison, H., Thompson, G., Kipling, Z., 2017. Uncertainty from the choice of microphysics scheme in convection-permitting models significantly exceeds aerosol effects. Atmos. Chem. Phys. 17, 12145–12175.

Williamson, D.L., 2013. The effect of time steps and time-scales on parametrization suites. Q. J. Roy. Meteorol. Soc. 139, 548–560.

Williamson, L., et al., 2013. The Aqua Planet Experiment (APE): response to changed meridional SST profile. J. Meteorol. Soc. Japan 91A, 57–89.

Wu, D., Dong, X., Xi, B., Feng, Z., Kennedy, A., Mullendore, G., Gilmore, M., Tao, W.-K., 2013. Impacts of microphysical scheme on convective and stratiform characteristics in two high precipitation squall line events. J. Geophys. Res. 118, 11119–11135.

Yanai, M., Esbensen, S., Chu, J., 1973. Determination of bulk properties of tropical cloud clusters from large-scale heat and moisture budgets. J. Atmos. Sci. 30, 611–627.

Žagar, N., Jelić, D., Blaauw, M., Bechtold, P., 2017. Energy spectra and inertia-gravity waves in global analyses. J. Atmos. Sci. 74, 2447–2466. https://doi.org/10.1175/JAS-D-16-0341.1.

Zuidema, P., Torri, G., Mueller, C., Chandra, A., 2017. A survey of precipitation-induced atmospheric cold pools over oceans and their interactions with the large-scale environment. Surv. Geophys. 38, 1283–1305.

CHAPTER

6

Ensemble data assimilation for estimating analysis uncertainty

Roland Potthast[a,b]

[a]Deutscher Wetterdienst (DWD), Offenbach, Germany [b]Applied Mathematics, University of Reading, Reading, United Kingdom

1 State estimation and state uncertainty

1.1 Basic setup for data assimilation

Forecasting weather by numerical systems is first based on some *numerical model*. Clearly, without a deep process understanding and the simulation of processes based on partial differential equations, we would not have *numerical weather prediction* (NWP) and we would not be able to achieve the forecasting power of state-of-the-art NWP systems. If we collect all dynamic variables of the model into some state variable $x \in \mathbb{R}^n$ in an n-dimensional state space $\mathbb{R}^n$, the model M calculates the change of the state x from time $s \in \mathbb{R}$ to time $t \in \mathbb{R}$, $t > s$, such that

$$x_t = M_{t,s}(x_s). \tag{1}$$

Here, as model systems we will look at examples from either the three-dimensional (3D) Lorenz63 model, the Lorenz96 model with $N \in \mathbb{N}$ variables of a coupled oscillator or a two-dimensional (2D) *simple oscillator* defined by

$$\begin{aligned} \frac{\partial x_1}{\partial t} &= x_2 - \overbrace{c_0 x_1 \cdot (|x_1| > c_1)}^{\textit{damping}} \\ \frac{\partial x_2}{\partial t} &= -\underbrace{\frac{x_1}{p \cdot (0.5 + \|x\|)}}_{\textit{oscillation speed}} \end{aligned} \tag{2}$$

https://doi.org/10.1016/B978-0-12-815491-5.00006-9

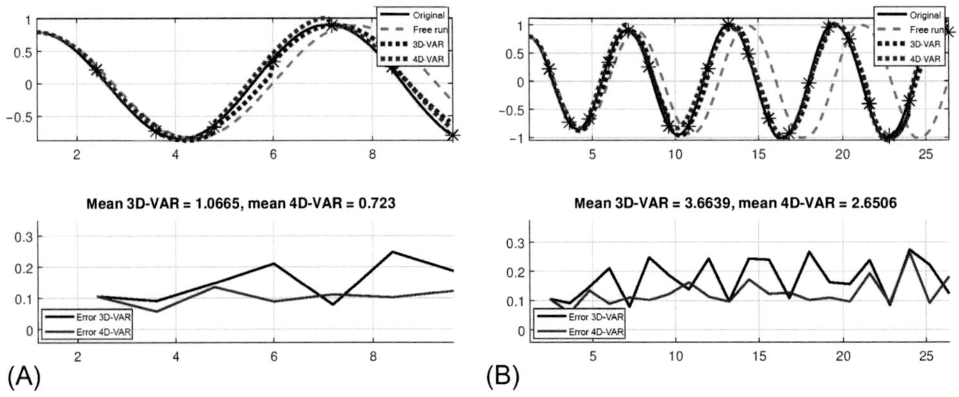

FIG. 1 The principle of data assimilation, using the observation (indicated by *black* "*") to pull the trajectory to the true trajectory. We show the result for 3D-VAR and 4D-VAR applied to an oscillating system defined by Eq. (2) for $P > 0$. Here, the observation is the value x_1 of the system only, where the free variables are the value and the speed $x_2 = dx_1/dt$ (i.e., the temporal derivative). Our nature run (*black*) is carried out with $P = 1$, the "numerical" system employs $P = 1.4$. 4D-Var is minimizing with respect to both values (x_1, x_2), whereas 3D-VAR can only adapt the observed value x_1. Note that the free run (the *dashed gray line*; without assimilation) is desynchronized shortly after several oscillations.

with some constants $c_0 = 0.1$ and $c_1 = 1.2$. Note that here x_1 corresponds to the location of some object and x_2 describes its speed with some damping if x_1 becomes larger than c_1. Here, the oscillation frequency p is made dependent on the *total energy* term $E = |x_1|^2 + |\frac{\partial x_1}{\partial t}|^2 = ||x||^2$.

The second key component of NWP systems is the *atmospheric state estimation*, that is, the estimation of the values of atmospheric variables x_s at each single grid point of the numerical model for the initial time s of each of its forecasting steps. Usually we do not *directly* measure some atmospheric variable at one of the model grid points. Many *remote sensing* techniques do not measure any variable at all, but complex functions of several variables or integrated values of such complex functions. In all these cases, the measurement itself needs to be simulated, often with techniques not much different from the physical components of the model itself. Typical examples are the measurements of satellite radiances, which are simulated by sophisticated fast *radiative transport models* such as the RTTOV code maintained by the NWP-SAF consortium of EUMETSAT. We usually call the simulation of measurements the *observation operator* H, which maps the atmospheric state $x \in \mathbb{R}^n$ onto the collection of measurement $y \in \mathbb{R}^m$, that is,

$$y = H(x), \quad x \in \mathbb{R}^n, \quad y \in \mathbb{R}^m. \tag{3}$$

Often, the full observation operator H consists of several special observation operators, for example, for RADAR, conventional measurements, radiances, or radio occultations.

It is well-known today, compare, for example, Dalety (1993), Kalnay (2003), Lewis et al. (2006), and Nakamura and Potthast (2015), that state estimation cannot be carried out with sufficient quality when measurements of one point in time only are used. All modern data assimilation system collects information overtime and passes it based on the numerical model

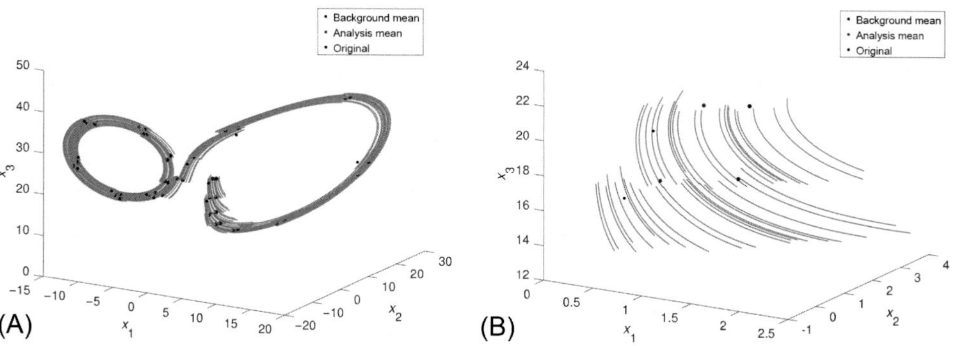

FIG. 2 Lorenz63 dynamical system assimilation cycle, with background mean at the analysis times in *blue dots*, analysis mean as *red dots*, truth as *black dots*, and the ensemble evaluation between the analysis times as *gray lines*. Here, we just observe the x_1 component and employ the ensemble Kalman filter for assimilation.

itself, integrating both observations and modeling information into the combined state estimate. Usually, we collect all m observations at some given point s in time into an observation vector $y \in \mathbb{R}^m$, where $\mathbb{R}^m$ is called the *observation space*. Then, the data assimilation calculates an atmospheric analysis $x_s^{(a)}$ for time $s \in \mathbb{R}$ given some *first guess* or *background* atmospheric state $x_s^{(b)}$ at time s and observations $y \in \mathbb{R}^m$ with observations from the interval

$$I_s := [s - \Delta s_1, s + \Delta s_2], \quad \Delta s_1, \Delta s_2 \geq 0, \tag{4}$$

where Δs_1 and Δs_2 define the time interval from which observations are used to calculate the current analysis at $s \in \mathbb{R}$.

1. The analysis $x_{s_k}^{(a)}$ at time s_k can be calculated based on some initial first guess $x_{s_k}^{(b)}$.
2. Then, the analysis state $x_{s_k}^{(a)}$ is propagated to the next analysis time s_{k+1}, leading to a background state $x_{s_{k+1}}^{(b)}$.

The two steps are *cycled*, that is, we always carry out an analysis step first, then a propagation step. We call the analysis step CORE and the propagation step MORE. Proper data assimilation will usually need several CORE and MORE steps. For the core assimilation step, let us look at three important scenarios:

(A) For *3D variational data assimilation (3D-VAR)*, the values of Δs_1 and Δs_2 are usually chosen small and observations are treated as being valid for the time s.

(B) For *4D variational data assimilation (4D-VAR)*, the values are chosen as $\Delta s_1 = 0$ and Δs_2 relatively large, for example, $s_2 = 12h$ for global NWP, and the full trajectory $x_\tau, \tau \in [s, s + \Delta s_2]$ is fit to the observations $y \in \mathbb{R}^m$ each at the correct point in time.

(C) For the *4D ensemble Kalman filter*, the values are chosen as $\Delta s_1 > 0$ of medium size and $\Delta s_2 = 0$, for example, $s_1 = 1h$ for convective-scale NWP, and the trajectory takes into account the correct values of the trajectory over the interval $[s - \Delta s_1, s]$ to calculate the analysis x_s at time $s \in \mathbb{R}$.

In scenario (A) basically *one point in time* is used for state estimation. In this case, the temporal development of states and transport of information overtime is carried out through the *first guess* or *background*

$$x_{s_k}^{(b)} = M_{s_{k-1},s_k}\left(x_{s_{k-1}}^{(a)}\right) \tag{5}$$

from the previous analysis time s_{k-1} to the current analysis time s_k with temporal index $k \in \mathbb{N}$. Scenarios (B) and (C) use information about the *temporal correlations* implicitly when analysis is calculated, either by some minimization method based on the *tangent linear model* in (B) or by an ensemble of states calculated over the full interval and exploited in the analysis of the *ensemble Kalman filter* in (C).

The numerical model can be seen as an *approximation* to the true atmospheric dynamics. Also, observations provide an *approximate value* of either some atmospheric variable in some point in space and time, or of some integrated derived quantity which is modeled itself based on basic atmospheric variables and underlying processes. Clearly, neither the model nor the observation establishes some *true* state, both come with inherent errors and uncertainty. Traditional views such as "observations are the truth" are strongly neglecting most of the measurement uncertainty, they are neglecting the deep problem of representativity of observations with respect to the scales a model can resolve, and they are usually generating large problems. Indeed, the uncertainty includes

- the uncertainty of the measured value itself based on the *design and errors* of the *measuring device*,
- the uncertainty of *definition* of the *measuring process* and *measuring setup*,
- its *representativity* with respect to the atmospheric scales under consideration, and
- further *practical limitations* of the real measurement, the transformation, decoding, transmission, and storage of the observation values.

Traditionally, in engineering and many applied sciences, the *uncertainty* of some state estimate is measured in terms of *error bars*, which are

(I) either understood in some probabilistic sense as the *variance* of the variable under consideration, or

(II) they are understood in the sense of some *quantile* or *confidence interval*, which provides an interval of values to contain a given percentage $p \in (0, 100)$ of the given data.

However, in the framework of *nonlinear* dynamical systems, usually the distributions of states or observations will not be *Gaussian*, and for estimating the *uncertainty* of some quantity, we need to go to more complex *distributions* of each of the variables under consideration. Thus, we will need to consider

(III) *uncertainty estimates* which provide an approximation to the full nonlinear or non-Gaussian *distribution* of a set of variables.

The full task (III) will lead us to *nonlinear* estimators and *particle filters* in Sections 2.3 and 3. Here, we will first explore the classical approaches to data assimilation and then discuss the estimation of the variance and covariance of states in the sense of approach (I) in Section 2.1.

For proper estimation, we need knowledge not only about the *variance* of all variables, but also about their statistical dependencies. In the easiest possible case, this is the *covariance* of their distributions, defined as $B := \mathbb{E}[(x-\overline{x})(x-\overline{x})^T] \in \mathbb{R}^{n \times n}$, defined as the estimation

$$B_{i,j} := \int_{\mathbb{R}^n} (x_i - \overline{x}_i)(x_j - \overline{x}_j)\, d\mu(x), \quad i,j=1,\ldots,n, \tag{6}$$

where $\mu(x)$ is the probability density of the value $x \in \mathbb{R}^n$ and $\overline{x}$ is the mean of x given by

$$\overline{x} := \int_{\mathbb{R}} x\, d\mu(x). \tag{7}$$

An example for a covariance estimation is displayed in Fig. 3. Note that the figure shows that dependent on the current dynamics of your system, the knowledge about x_1 would lead to different increments of x_3 at different times—something a static covariance cannot provide.

Classical data assimilation methods have been working with the approach (I) for decades, estimating the uncertainty of some model *first guess* $x_s^{(b)}$ by a *Gaussian distribution*

$$p_B(x) := \frac{1}{\sqrt{(2\pi)^n \det(B)}} e^{-\frac{1}{2}\left(x - x^{(b)}\right)^T B^{-1}\left(x - x^{(b)}\right)}, \quad x \in \mathbb{R}^n \tag{8}$$

and the uncertainty induced by the measurement error by a second Gaussian distribution

$$p_R(x) := c e^{-\frac{1}{2}(y - H(x))^T R^{-1}(y - H(x))}, \quad x \in \mathbb{R}^n, \tag{9}$$

with norming constant c, the measurement $y \in \mathbb{R}^m$ and the *measurement error covariance matrix* $R \in \mathbb{R}^{m \times m}$. In the case of *nonlinear* observation operators, usually their derivative $\mathbf{H} := dH(x)/dx$ is employed within an iterative algorithm.

Both for didactic purposes as well as for the development of algorithms is often instructive, to consider a *low-dimensional example* and carry out all calculations explicitly to gain detailed insight into the role of all parameters and components of a method under consideration. Here, let us consider the case where $x \in \mathbb{R}^2$ and where $y \in \mathbb{R}^1$, that is, where we measure a scalar value to determine two variables x_1 and x_2 at a given point in time. When the observation operator H

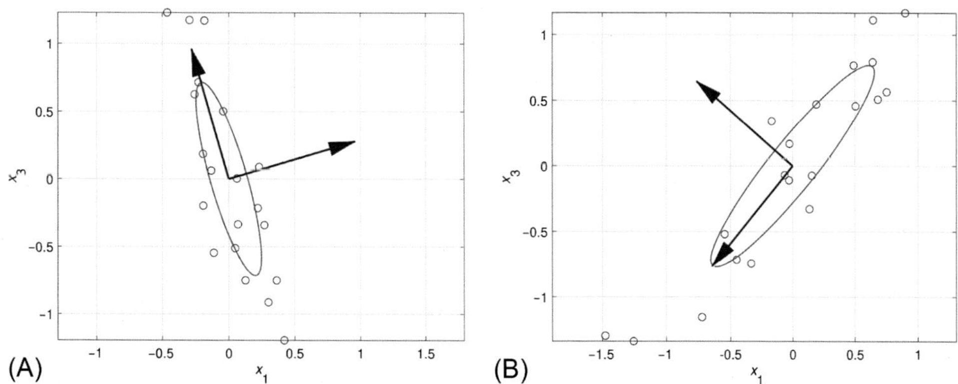

FIG. 3 Covariance estimation for two different time steps of the assimilation shown in Fig. 2. We display the ensemble by *circles*, here looking at the x_1–x_3-plane only, the principle vectors of the ensemble by *blue arrows* and the ellipse reflecting both the standard deviation and covariance information by a *red line*.

is *linear*, it must be of the form $H=(h_1,h_2)\in\mathbb{R}^{1\times 2}$, with $h_1,h_2\in\mathbb{R}$, such that the simulated observation is

$$H(x)=h_1x_1+h_2x_2,\quad x\in\mathbb{R}^2. \tag{10}$$

Let us stay with the case where $h_1=1$ and $h_2=0$, that is, where we measure the first variable, and do not know the second variable. In our case, the error covariance matrix is just a scalar $R=r\in\mathbb{R}$, which reflects the variance of a Gaussian distribution of the possible true of value of x_1 given the measurement y. In this case, the measurement $y\in\mathbb{R}$ leads to the uncertainty estimate

$$p_R(x)=ce^{-\frac{1}{2}\frac{(x_1-y)^2}{r}},\quad x\in\mathbb{R}^2. \tag{11}$$

If we had no further information, we would estimate x_1 to be equal to y and we would have *no information* on x_2. But we would also like to use knowledge from the past. We now first discuss the classical approach.

Usually, data assimilation exploits the knowledge from the past by using the *first guess* and its corresponding uncertainty given by Eq. (8). The proper framework to combine the first guess and the observation is the *Bayesian framework*, which is the sum of techniques and methods based on the elementary probabilistic formula

$$p(x|y)=cp(x)\cdot p(y|x),\quad x\in\mathbb{R}^n,\quad y\in\mathbb{R}^m, \tag{12}$$

where $p(x)$ is the *prior distribution* of the state before the measurement y is given and $p(y|x)$ is the *conditional distribution* of the measurements y given the state x. In our case, $p(x)$ is given by the background uncertainty $p_B(x)$ defined in Eq. (8) and $p(y|x)$ is given by the measurement uncertainty $p_R(x)$ defined in Eq. (9). The so-called *posterior distribution* $p(x|y)$ of the states $x\in\mathbb{R}^n$ based on the measurements $y\in\mathbb{R}^m$ is given by Eq. (12).

1.2 3D-VAR and the Kalman filter

In the case where both $p(x)$ and $p(y|x)$ are *Gaussian distributions* as in the case earlier, the product $p(x|y)$ in Eq. (12) is again a Gaussian distribution. It is completely determined by its *mean* $x^{(a)}$ and *covariance matrix* $B^{(a)}$, which with a linearization $\mathbf{H}$ of $H(x)$ around the state $x^{(b)}$ can be approximated by

$$x^{(a)} := x^{(b)}+B\mathbf{H}^T(R+\mathbf{H}B\mathbf{H}^T)^{-1}(y-H(x^{(b)})) \tag{13}$$

and

$$B^{(a)} := (I-K\mathbf{H})B \tag{14}$$

with the *Kalman gain matrix* K given by

$$K := B\mathbf{H}^T(R+\mathbf{H}B\mathbf{H}^T)^{-1}. \tag{15}$$

We note that the analysis $x^{(a)}$ is the minimizer of the quadratic functional

$$J_{lin}(x) := \frac{1}{2}\left(x-x^{(b)}\right)^T B^{-1}\left(x-x^{(b)}\right)+\frac{1}{2}(y-\mathbf{H}x)^T R^{-1}(y-\mathbf{H}x), \tag{16}$$

$x \in \mathbb{R}^n$, which is the exponent of the product of the terms $p_B(x)$ and $p_R(x)$, with $\mathbf{H}$ being the linearization of the observation operators $H(x)$. Often, the minimization of the full nonlinear functional

$$J(x) := \frac{1}{2}\left(x - x^{(b)}\right)^T B^{-1}\left(x - x^{(b)}\right) + \frac{1}{2}(y - H(x))^T R^{-1}(y - H(x)), \tag{17}$$

$x \in \mathbb{R}^n$ is carried out by a conjugate gradient method, where the minimization of $J_{lin}(x)$ is the *inner loop* and the determination of $\mathbf{H}$ from $H(x)$ is carried out in an *outer loop*. For an explicit calculation for the derivations earlier, we refer to Chapter 5 of Nakamura and Potthast (2015). Here, let us explore how to move from 3D-VAR to a Kalman filter.

Calculating an analysis in each time step of a cycled data assimilation system based on the minimization of Eq. (17) in the case where the covariance matrix B is fixed is known as 3D-VAR. For linear systems, it can be calculated by Eq. (13). An example of applying 3D-VAR to the oscillator (2) is shown in Fig. 1.

1.2.1 *Determining a statistical covariance matrix* B

For practical systems, usually the covariance matrix B is derived from a time series of first guess and analysis states—or more general from the difference of two forecasts with different lead times, usually calculated within some analysis cycle based on an easier or earlier version of B. This process of estimating B can be iterated: after the update of B, a new time series is calculated, leading to a better covariance estimator B. Since calculating this so-called *reanalysis* of the whole system over a longer time period is very costly in terms of computational time and costs, an update of B is carried out only occasionally for typical operational systems.

1.3 Two-dimensional example

For the simple 2D system given around Eq. (10), with measurements y of x_1 (i.e., for $h_1 = 1$ and $h_2 = 0$), the 3D-VAR or Kalman filter update (13) is given by

$$\begin{pmatrix} x_1^{(a)} \\ x_2^{(a)} \end{pmatrix} = \begin{pmatrix} x_1^{(b)} \\ x_2^{(b)} \end{pmatrix} + \begin{pmatrix} b_{11} & b_{12} \\ b_{21} & b_{22} \end{pmatrix} \begin{pmatrix} 1 \\ 0 \end{pmatrix} \frac{1}{r + b_{11}} \left(y - x_1^{(b)}\right) \tag{18}$$

$$= \begin{pmatrix} x_1^{(b)} \\ x_2^{(b)} \end{pmatrix} + \begin{pmatrix} b_{11} \\ b_{21} \end{pmatrix} \frac{1}{r + b_{11}} \left(y - x_1^{(b)}\right). \tag{19}$$

This means that $x_1^{(a)}$ is given by a convex combination

$$x_1^{(a)} = x_1^{(b)} + \frac{b_1 1}{r + b_{11}} \left(y - x_1^{(b)}\right) \tag{20}$$

$$= \frac{r + b_{11}}{r + b_{11}} x_1^{(b)} + \frac{b_1 1}{r + b_{11}} \left(y - x_1^{(b)}\right) \tag{21}$$

$$= \frac{r}{r + b_{11}} x_1^{(b)} + \frac{b_{11}}{r + b_{11}} y \tag{22}$$

$$= \alpha x_1^{(b)} + (1 - \alpha) y \tag{23}$$

with $\alpha = r/(r + b_{11})$. The analysis of the second variable x_2 is given by

$$x_2^{(a)} = x_2^{(b)} + \frac{b_{21}}{r + b_{11}} \left(y - x_1^{(b)} \right). \tag{24}$$

Here, the increment

$$\delta x_2 := x_2^{(a)} - x_2^{(b)} = \frac{b_{21}}{b_{11}} \left(\frac{b_{11}}{r + b_{11}} \left(y - x_1^{(b)} \right) \right) \tag{25}$$

depends on the increment

$$\delta x_1 := \frac{b_{11}}{r + b_{11}} \left(y - x_1^{(b)} \right) \tag{26}$$

scaled with a factor b_{21}/b_{11}, that is, it depends on the covariance b_{21} between the variables x_1 and x_2. The smaller the measurement error is, the smaller the influence of the background state $x^{(b)}$. If the correlation between x_1 and x_2 is strong, then also x_2 receives a strong increment. If the correlation is zero, there is no increment of x_2 at all and x_2 will stay as given by $x_2^{(b)}$.

1.4 Analysis uncertainty and its propagation

We need to discuss the uncertainty of the current state for the *analysis*. Clearly, the *analysis uncertainty* is usually much smaller than the first-guess uncertainty, since the use of observations feeds information into the system and decreases the covariance of the distribution. The *posterior covariance matrix* $B^{(a)}$ can be defined by representing $J(x)$ of Eq. (16) in the form

$$J(x) = c\left(x - x^{(a)}\right)^T \left(B^{(a)}\right)^{-1} \left(x - x^{(a)}\right) \tag{27}$$

with some constant c, which after some patient algebraic derivations leads to Eq. (14), compare Lemma 5.4.2. of Nakamura and Potthast (2015).

Calculating the analysis covariance matrix $B^{(a)}$ based on the prior covariance matrix $B = B_k$ at time t_k as described by Eq. (14) and then propagating it to the next time step t_{k+1} to use it as new prior covariance matrix $B = B_{k+1}$ is known as *Kalman filter*. The full Kalman filter needs to calculate the $n \times n$-dimensional matrix $B^{(a)}$ given the matrix B in each analysis step. Also, it needs to propagate this matrix, which for linear model dynamics $\mathbf{M}$ can be calculated as $B_{k+1} = \mathbf{M} B_k^{(a)} \mathbf{M}^T$. For high-dimensional systems, this is not feasible and approximation methods need to be employed. The *ensemble Kalman filter* can be viewed as a low-dimensional approximation to this full update. We will discuss it in more detail in Section 2.1, since it is also be used to estimate the *uncertainty* of the analysis in one joint assimilation step.

1.5 F4D-VAR

We can also understand *4D variational data assimilation* (4D-VAR) as minimizer of a functional of the form (17). To this end, we combine the model propagation with the observation operator by

$$H^{(4d)}(x_s) := \begin{pmatrix} H(M_{s,t_1}(x_s)) \\ H(M_{s,t_2}(x_s)) \\ \vdots \\ H(M_{s,t_N}(x_s)) \end{pmatrix} \tag{28}$$

and

$$y^{(4d)} := \begin{pmatrix} y_{t_1} \\ y_{t_2} \\ \vdots \\ y_{t_N} \end{pmatrix}, \quad R^{(4d)} := \begin{pmatrix} R_{t_1} & 0 & \dots & 0 \\ 0 & R_{t_2} & 0 & \vdots \\ \vdots & & \ddots & \\ 0 & \dots & 0 & R_{t_N} \end{pmatrix} \tag{29}$$

with measurement times $t_1,\dots,t_N$, $N \in \mathbb{N}$, and model propagations M_{s,t_ξ} from time s to t_ξ, $\xi = 1,\dots,N$ and measurement error correlations $R^{(4d)}$ which are uncorrelated between different measurement times. Then, the functional which searches for the best trajectory for times $t_1,\dots,t_N$ becomes

$$\begin{aligned} J^{(4d)}(x_s) :=& \frac{1}{2}\left(x_s - x_s^{(b)}\right)^T B^{-1}\left(x_s - x_s^{(b)}\right) \\ &+ \frac{1}{2}\left(y^{(4d)} - H^{(4d)}(x_s)\right)^T \left(R^{(4d)}\right)^{-1} \left(y^{(4d)} - H^{(4d)}(x_s)\right) \\ =& \frac{1}{2}\left(x_s - x_s^{(b)}\right)^T B^{-1}\left(x_s - x_s^{(b)}\right) \\ &+ \sum_{\xi=1}^{N} \frac{1}{2}\left(y_{t_\xi} - H(M_{s,t_\xi}(x_s))\right)^T (R_{t_\xi})^{-1}\left(y_{t_\xi} - H(M_{s,t_\xi}(x_s))\right). \end{aligned} \tag{30}$$

Finding a minimizer of $J^{(4d)}(x_s)$ can be carried out by an efficient and well-conditioned *gradient method* based on the derivatives of $H^{(4d)}$ with respect to the initial state x_s. With the derivative $\mathbf{H}$ of H, the gradient of the nonlinear function $f(x) = \frac{1}{2}(y - H(x))^T A(y - H(x))$ with respect to $x \in \mathbb{R}^n$ is calculated by

$$\begin{aligned} \frac{df(x)}{dx_\xi} &= \frac{d}{dx_\xi}\left(\frac{1}{2}\sum_{j,k=1}^{m}(y - H(x))_j A_{j,k}(y - H(x))_k\right) \\ &= -\frac{1}{2}\sum_{j,k=1}^{m}\left(\mathbf{H}_{j\xi}A_{j,k}(y - H(x))_k + (y - H(x))_j A_{jk}\mathbf{H}_{j\xi}\right) \\ &= -\frac{1}{2}\left((\mathbf{H}^T A(y - H(x)))_\xi + ((y - H(x))^T A\mathbf{H})_\xi^T\right), \end{aligned} \tag{31}$$

for $\xi = 1,\dots,n$, such that the full gradient in the case of symmetric A is given by

$$\nabla f(x) = -\mathbf{H}^T A(y - H(x)). \tag{32}$$

By the chain rule, differentiation of $H(M_{s,t_\xi})$ yields

$$\begin{aligned} \frac{d}{dx_s}\left[H(M_{s,t_\xi})\right] &= \left.\frac{dH(x)}{dx}\right|_{M_{s,t_\xi}(x)} \cdot \frac{dM_{s,t_\xi}}{dx_s} \\ &= \mathbf{H}|_{M(x,t_\xi)} \cdot \mathbf{M}_{s,t_\xi} \end{aligned} \tag{33}$$

with the *derivative* $\mathbf{M}_{s,t_\xi}$ of M_{s,t_ξ}, also known as *tangent linear model*. With this the gradient of Eq. (30) is given by

$$\nabla_{x_s} = -B^{-1}\left(x_s - x_s^{(b)}\right) - \sum_{\xi=1}^{N} \mathbf{H}^T \mathbf{M}_{s,t_\xi}^T (R_{t_\xi})^{-1} \left(y_{t_\xi} - H(M_{s,t_\xi}(x_s))\right). \tag{34}$$

Here, we do not want to go into details of how to carry out the calculation of Eq. (34) efficiently, we refer to Nakamura and Potthast (2015), Chapter 5, or the classical book (Kalnay, 2003).

Finally, note that in the earlier form, 4D-VAR does not estimate the *uncertainty* of a state, but calculates a best estimate for a Gaussian *background distribution* given by $x_s^{(b)}$ and the covariance matrix B, and Gaussian error distributions R_{t_ξ} for $\xi = 1,\ldots,N$. The method does, however, take into account the *nonlinear trajectory* $M_{s,t}(x_s^{(b)})$ for $t \in I_s = [s, t_N]$. We will discuss this in more detail in Section 2.3 and demonstrate the potential of ensembles of 4D-VAR.

2 Ensembles of states for uncertainty estimation

For uncertainty estimation, you need more than just one state which fits best to your data. You want to understand the uncertainty of this estimate and as far as possible calculate the distribution of the uncertainty.

2.1 Ensemble Kalman filters

The basic idea of an *ensemble Kalman filter* (EnKF) is to employ an ensemble of states to estimate the covariance matrix B. Let $x^{(\ell)} \in \mathbb{R}^n$ for $\ell = 1,\ldots,L$ be an ensemble of states. The standard *stochastic estimator* for the covariance matrix is given by

$$B := \frac{1}{L-1}\sum_{\ell=1}^{L}\left(x^{(\ell)} - \overline{x}\right)\left(x^{(\ell)} - \overline{x}\right)^T, \tag{35}$$

where the *mean* $\overline{x} \in \mathbb{R}^n$ of the states is calculated by

$$\overline{x} := \frac{1}{L}\sum_{\ell=1}^{L} x^{(\ell)}. \tag{36}$$

If we define the full ensemble matrix $X^{(full)}$ by

$$X^{(full)} := \left(x^{(1)},\ldots,x^{(L)}\right) \in \mathbb{R}^{n\times L}, \tag{37}$$

we obtain the matrix of ensemble differences to the mean $\overline{x}$ and its normalized version by

$$X := X^{(full)} - \underbrace{(\overline{x},\ldots,\overline{x})}_{L\text{ times}}, \quad Q := \frac{1}{\sqrt{L-1}}X. \tag{38}$$

Then, the B matrix estimator (35) can be written in the form

$$B = \frac{1}{L-1}XX^T = QQ^T. \tag{39}$$

This means that we can use an ensemble of states to estimate their variance and covariance based on Eq. (35) or (39).

To carry out *uncertainty estimation* in a high-dimensional framework, we first need to investigate the quality of the estimator (35). It is well known from classical statistics that the convergence of the estimator (35) obeys a rate of $L^{-1/2}$, that is, the error for the covariance decays with the square root $\sqrt{L}$ of the ensemble size. For many practical systems, this is, however, a very low convergence order. If you can afford 40 or 50 or 100 ensemble members, the best you can hope for is an error of the approximate size of 10%, which of course depends on the constants and practical situation. An important observation is that if you carry out the estimate with a large dimension $n = 10^8$ of your state space, usually you end up with a lot of *spurious correlations* over long distances, which are far from being physically plausible. The weather is usually not strongly correlated over distances more than 600–1000 km. If you distribute measurements at one place with 10% error over the whole globe, this will immediately lead to useless estimates and forecasts.

A key idea which has become very popular over the past decades is to damp out spurious correlations by *localization*. If you know that your correlations should go down with distance to some given location of a variable, you could just enforce this behavior by taking the Schur product

$$C \circ B := \left(c_{jk} \cdot B_{jk}\right)_{j,k=1,\ldots,n} \tag{40}$$

of your B estimator with some *localization matrix* C of the form

$$C := \left(f_\rho(|p_j - p_k|)\right)_{j,k=1,\ldots,n}, \tag{41}$$

where p_j is the location of the component x_j of the state x and where f_ρ is a scalar sufficiently smooth function with

$$\begin{aligned} &f_\rho(0) = 1, \\ &f_\rho(s) \in (0,1) \quad \text{for } s \in (0,\rho), \\ &f_\rho(s) = 0 \quad \text{for } s \geq \rho. \end{aligned} \tag{42}$$

This approach is known as *localization in state space*. The constant ρ is known as *localization radius*. Typical functions which are popular in the community are the polynomials suggested by Gaspari and Cohn (1999). An alternative is to use a Gaussian matrix

$$C := \left(e^{-\frac{1}{2}|p_j - p_k|^2/\sigma^2}\right)_{j,k=1,\ldots,n}, \tag{43}$$

which is positive for all entries, but effectively has exponentially small entries for $|p_j - p_k| \gg \sigma$, which practically can be neglected.

The analysis step of the *ensemble Kalman filter* with *localization in state space* carries out a state estimate based on Eq. (13), but with covariance matrix calculated by $C \circ QQ^T$.

A second important question is how to generate an ensemble estimating the current uncertainty, which can be used to propagate the uncertainty through time (carrying out the more step) and calculate the *background uncertainty* or *prior* for the subsequent core analysis step.

Today, two basic approaches are used. The first carries out L independent assimilation steps based on randomly perturbed observations y. The second, known as *square root filter*, determines the posterior ensemble explicitly based on the formula (14) in ensemble space.

Let us study the ensemble generation based on *perturbed observations* first. Consider some mapping $f: \Omega \rightarrow V$, $y \mapsto x = f(y)$ from a set Ω into some set V. Canonically, a probability distribution $q(x)$ on Ω will be mapped into a probability distribution $p(x)$ on V by

$$p(x) := q(f^{-1}(x)), \quad x \in V. \tag{44}$$

To generate samples $x^{(\ell)}$ from p, we can generate samples $y^{(\ell)}$ from q and map them into $x^{(\ell)} := f(y^{(\ell)})$, $\ell = 1,2,3,\ldots$. Now, for the data assimilation system, if y is a given observation and we consider

$$q(\tilde{y}) := ce^{-\frac{1}{2}(\tilde{y}-y)^T R^{-1}(\tilde{y}-y)}, \quad \tilde{y} \in Y \tag{45}$$

to be the distribution describing the uncertainty of the observation, we can employ a standard *pseudorandom number generator* to generate samples $y^{(\ell)}$, $\ell = 1,\ldots,L$ from q. Then, we calculate

$$x^{(a,\ell)} := K\left(y^{(\ell)} - H(x^{(b,\ell)})\right), \quad \ell = 1,\ldots,L, \tag{46}$$

with the Kalman matrix K defined in Eq. (15). This is a very practical and generic approach, since it can be carried out easily when some analysis method is available. It does not need much coding to implement a new method. However, to generate your ensemble you have to carry out L assimilations, usually with high computational cost.

2.2 Square root filter

A *second* approach is known as *square root filter* (SRF or EnKF-SRF). It does not need to carry out L assimilations, but employs the formula (14) as follows. Let us consider the case of a linear observation operator H and abbreviate $\gamma = 1/(L-1)$. We note that we have $Y^{(b)} = HX^{(b)}$. In this case, using $X = X^{(b)}$ and $Y = Y^{(b)}$ for readability, we insert Eq. (39) into Eq. (14). We note that since by elementary calculations we have

$$(I + \gamma Y^T R^{-1} Y) Y^T = Y^T R^{-1} (R + \gamma Y Y^T), \tag{47}$$

which leads to

$$Y^T (R + \gamma Y Y^T)^{-1} = (I + \gamma Y^T R^{-1} Y)^{-1} Y^T R^{-1}. \tag{48}$$

With $X = X^{(b)}$, we have

$$\begin{aligned} K &= B\mathbf{H}^T(R+\mathbf{H}B\mathbf{H}^T)^{-1} \\ &= \gamma XX^T\mathbf{H}^T(R+\gamma\mathbf{H}XX^T\mathbf{H}^T)^{-1} \\ &= \gamma XY^T(R+\gamma YY^T)^{-1}. \end{aligned} \tag{49}$$

Now, we calculate

$$\begin{aligned} B^{(a)} &= (I-K\mathbf{H})B \\ &=^{(15)} (I-\gamma XX^T\mathbf{H}^T(R+\gamma\mathbf{H}XX^T\mathbf{H}^T)^{-1}\mathbf{H})\gamma XX^T \\ &= (I-\gamma XY^T(R+\gamma YY^T)^{-1}\mathbf{H})\gamma XX^T \\ &=^{(48)} \left(I-\gamma X(I+\gamma Y^TR^{-1}Y)^{-1}Y^TR^{-1}\mathbf{H}\right)\gamma XX^T \\ &= X\left(I-\gamma(I+\gamma Y^TR^{-1}Y)^{-1}Y^TR^{-1}Y\right)\gamma X^T \\ &= X\left((I+\gamma Y^TR^{-1}Y)^{-1}\left[I+\gamma Y^TR^{-1}Y-\gamma Y^TR^{-1}Y\right]\right)\gamma X^T \\ &= X(I+\gamma Y^TR^{-1}Y)^{-1}\gamma X^T \\ &= \gamma X(I+\gamma Y^TR^{-1}Y)^{-1}X^T \\ &= \gamma\left(X(I+\gamma Y^TR^{-1}Y)^{-1/2}\right)\left((I+\gamma Y^TR^{-1}Y)^{-1/2}X^T\right). \end{aligned} \tag{50}$$

According to Eq. (50), the analysis ensemble $X^{(a)}$, which generates the correct posterior covariance in the form

$$B^{(a)} = \gamma X^{(a)}\left(X^{(a)}\right)^T \tag{51}$$

is given by

$$X^{(a)} := X\left(I+\gamma Y^TR^{-1}Y\right)^{-1/2}. \tag{52}$$

The matrix

$$W := \left(I+\gamma Y^TR^{-1}Y\right)^{-1/2} \in \mathbb{R}^{L\times L} \tag{53}$$

has the ensemble dimension L. Calculation of W is usually very efficient, and the full analysis ensemble can be obtained by an ensemble transformation of W.

To make the square root filter work in a high-dimensional system, we also need to introduce *localization*. Here, we might use what is known as *localization in observation space*. The idea is to calculate different transform matrices $W = W(p)$ for each grid point p of some underlying analysis grid.

1. For each point p, we employ a localization matrix C as defined in Eq. (41).
2. Then, the matrix R^{-1} is replaced by $R^{-1} \circ C$, that is, we give less weights to observations, which are further away from the analysis grid point.

3. With this new observation error matrix, the transform matrix $W(p)$ is calculated for each analysis grid point.
4. After this, the matrices $W(p)$ are *interpolated* to the model grid.
5. Finally, in each model grid point, the resulting analysis ensemble members $x^{(a,\ \ell)}$ are calculated.

The shift of the ensemble mean of the *square root filter* is carried out as in Eq. (13), with B given by Eq. (35). Using the same algebra as in Eq. (50), we obtain the form

$$\begin{aligned} x^{(a)} &:= x^{(b)} + K\left(y - H\left(x^{(b)}\right)\right) \\ &= x^{(b)} + X^{(b)}\gamma\left(I + \gamma YTR^{-1}Y\right)^{-1} YTR^{-1}\left(y - H\left(x^{(b)}\right)\right) \end{aligned} \tag{54}$$

for the mean of the ensemble, such that the full analysis ensemble can be calculated by

$$x^{(a,\ell)} = x^{(a)} + X^{(a)} \cdot e_\ell, \quad \ell = 1, \ldots, L. \tag{55}$$

An example for the ensemble Kalman square root filter used for Lorenz 63 is shown in Fig. 2.

2.3 Ensembles of 4D state estimation

The transform of a probability distribution (44) is not limited to linear mappings f or to any particular type of probability distribution $q(x)$. It is a very well-known technique, for example, for ensemble reanalysis systems or to generate an ensemble of initial states when a 4D-VAR system is available for data assimilation. In this case, the generation of the EnKF ensemble (46) is replaced by

$$x^{(a,\ell)} := A_{4D}\left(x^{(b)}, y^{(4d,\ell)} - H^{(4d)}\left(x^{(b)}\right)\right), \quad \ell = 1, \ldots, L, \tag{56}$$

where A_{4D} denotes the 4D-Var analysis operator depending on the first guess $x^{(b)}$ and observation increments $y^{(4d)} - H^{(4d)}(x^{(b)})$. Here, $y^{(4d,\ \ell)}$ is the set of perturbed observations over the full 4D analysis interval I_s.

In contrast to sequential methods, which move from time step to time step and calculate samples of the underlying probability distribution, the 4D sampling is a so-called *smoother*, since it calculates a distribution given knowledge about the future evolution of the system. Thus, in contrast to sequential methods, it has additional knowledge to calculate the samples at present time. Further, it employs detailed knowledge of the evolution of the system and its derivatives to carry out the sampling process. These are advantages, which sequential methods cannot fully replace. However, an ensemble of 4D-Vars (En-4D-Var) is expensive and needs a lot of approximations to be able to efficiently carry out the minimization.

Let us study the result of this type of ensemble data assimilation when a full interval of observations is used. We employ the oscillator (2) to sample at time t_0 given the knowledge of the observation at time t_1 and t_2. Here, to visualize that we can obtain good reconstruction of a distribution from non-Gaussian distributions,

- we chose an initial value $x(t_0) = (0.8, 0)$ and constructed a set of points $x^{(true,\ \ell)}(t_0)$ on a circle surrounding the point, compare Fig. 4.
- We calculated observations $y^{(\ell)}(t_\xi)$, $\xi = 1, 2$, at points t_1 and t_2 for all trajectories $x^{(\ell)}(t)$, $t \in [t_0, t_2]$ starting at the above points.
- Then, we carried out a 4D-VAR minimization by a gradient method to reconstruct the analysis points $x^{(a,\ \ell)}(t_0)$ from $\left\{y^{(\ell)}(t_1), y^{(\ell)}(t_2)\right\}$.

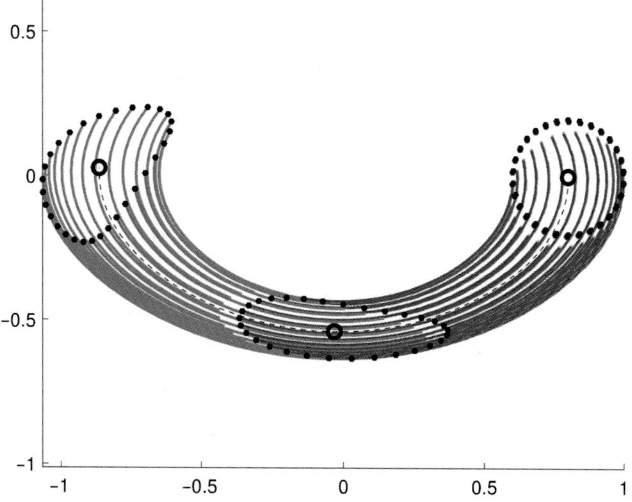

FIG. 4 We show the 4D sampling of a non-Gaussian distribution of the simple oscillator model (2). Here, we used the observation values $x_1(t_2)$ and $x_1(t_1)$—that is, the x_1 coordinates of the *black dots*—to recover the full state $x(t_0)$ by Eq. (56). The analysis results $x^{(a,\ \ell)}$ are shown as *blue dots*. For illustration purposes, we chose the observations from trajectories arising from a circle around the original value $x_{true}(t_0)$, visualizing a Gaussian distribution at t_0.

- The analysis points are shown in Fig. 4 as blue dots, the curves starting in the analysis points are shown in pink color for $[t_0, t_1]$ and in gray for $]t_1, t_2]$.

We observe in Fig. 4 that the 4D smoother 4D-Var can map the distributions at t_1 and t_2 back to a symmetric distribution at t_0 (the blue dots). In the same way, it will map the Gaussian model error at t_ξ, $\xi = 1, 2$ into some non-Gaussian distribution at t_0 when drawing from observations. An example of such a mapping can be found in Fig. 6, where one propagation step is studied within the framework of a *particle smoother*.

3 Particle filters in high dimensions

Let us go back to the Bayes formula (12) and consider some nonlinear dynamical system M on a state space $\mathbb{R}^n$, given by the forcing function $F(x, p)$ for $x \in \mathbb{R}^n$ and parameter setting p within the evolution equation

$$\frac{dx}{dt} = F(x,p), \quad x \in \mathbb{R}^n, \quad t \in \mathbb{R}^+ \tag{57}$$

with initial condition $x(0) = x_0 \in \mathbb{R}^n$. Assume that at time t_0 we can measure $y_0 = x_0 + \mathcal{N}(0,R)$, that is, we have normally distributed measurement error with covariance R as described by Eq. (9) in the case where H is the identity operator. If we have no further knowledge, then according to Eq. (12) the distribution our knowledge at time t_0 for the possible distribution of the true state is given by

$$p_0(x) = c e^{-\frac{1}{2}(y_0 - x)^T R^{-1} (y_0 - x)}, \quad x \in \mathbb{R}^n, \tag{58}$$

with norming constant c, which is a Gaussian distribution around the mean y_0.

Let us now propagate this distribution through time from time t_0 to the point $t_1 = t_0 + \delta t$. What is the distribution of the uncertainty at time t_1? First, if the model M is linear, that is, given by a matrix $\mathbf{M} \in \mathbb{R}^{n \times n}$, then we obtain the distribution

$$\begin{aligned} p_1(x) &= ce^{-\frac{1}{2}(y_0 - \mathbf{M}^{-1}x)^T R^{-1}(y_0 - \mathbf{M}^{-1}x)}, \\ &= ce^{-\frac{1}{2}(\mathbf{M}y_0 - x)^T ((\mathbf{M}^{-1})^T R^{-1} \mathbf{M}^{-1})(\mathbf{M}y_0 - x)}, \end{aligned} \tag{59}$$

with a weight propagation as in Nakamura and Potthast (2015, Eq. 5.4.27). Note that Eq. (59) is again a Gaussian distribution with mean $\mathbf{M}y_0$ and covariance

$$\begin{aligned} \widetilde{R} &:= ((\mathbf{M}^{-1})^T R^{-1} \mathbf{M}^{-1})^{-1} \\ &= \mathbf{M}R\mathbf{M}^T. \end{aligned} \tag{60}$$

In general, the model is *not* linear, and the distribution p_1 at time t_1 will not be a Gaussian. In the case of a model which consists of advection with vector $a(x,t) \in \mathbb{R}^m$ plus diffusion with matrix coefficient $b(x,t) \in \mathbb{R}^{n \times n}$, the propagation of a distribution is given by the *Fokker-Planck equation*

$$\frac{\partial p(x,t)}{\partial t} = -\sum_{j=1}^{n} \frac{\partial}{\partial x_j}(a_j(x,t)p(x,t)) + \frac{1}{2}\sum_{i,j=1}^{n} \frac{\partial}{\partial x_i}\frac{\partial}{\partial x_j}(b_{i,j}(x,t)p(x,t)), \tag{61}$$

for $x \in \mathbb{R}^n, t \in \mathbb{R}^+$. A full integration of Fokker-Planck or its generalizations is usually time consuming and not feasible for any high-dimensional problem. However, ensemble methods can be employed to calculate low-dimensional approximations to the distribution $p_1(x)$. We approximate $p_0(x)$ by an ensemble $\{x^{(\ell)},\ \ell = 1,\ldots,L\}$ of states in the form

$$p_{0,L} := c\sum_{\ell=1}^{L} \delta\left(x - x^{(\ell)}\right) \tag{62}$$

with δ-functions $\delta(x - x^{(\ell)})$ at points $x^{(\ell)}$. Then, the propagation of the states $x^{(\ell)}$ for $\ell = 1, \ldots, L$ is carried out by an application of the nonlinear model M to calculate the approximation

$$p_{1,L}^{(b)}(x) := c\sum_{\ell=1}^{L} \delta\left(x - M\left(x^{(\ell)}\right)\right), \quad x \in \mathbb{R}^n. \tag{63}$$

When δ-functions are used, the previous method is usually called *particle approximation*. Often, the term *ensemble approximation* is used when more specific basis functions are calculated from the ensemble distributions, for example, we can

1. draw L ensemble members $x_0^{(\ell)}$ from the distribution $p_0(x)$,
2. propagate the ensemble members through time to calculate

$$x_1^{(b,\ell)} := M\left(x_0^{(\ell)}\right), \quad \ell = 1,\ldots,L, \tag{64}$$

3. approximate the propagated distribution $p_1(x)$ by a superposition

$$p_{1,L}(x) := \sum_{\xi=1}^{K} \phi_{\xi}\left(x - x^{(\xi)}\right) \tag{65}$$

with basis functions $\varphi_{\xi}(x)$, $x \in \mathbb{R}^n$, now centered at new nodes $x^{(\xi)}$ for $\xi = 1,\ldots,K$.
The choices for the approximation (65) offers different options, which lead to well-known assimilation methods as follows:

1. Assume we choose the number of nodes to be $K = 1$, the only node $x^{(1)}$ to be the mean (36) and B_1 to be the covariance (35) both estimated from the ensemble $\{x^{(b,\ell)},\ \ell = 1,\ldots,L\}$. Then, the method leads to the *ensemble Kalman filter* described in Section 2.1.
2. In the case where $K = L$ and $\varphi_{B_\xi}(x) = \delta(x)$ is a δ-function, we have set up the input for the *classical* or *bootstrap particle filter*. With the localized adaptive particle filter (LAPF) of Potthast et al. (2019), we will describe a version of this particle filter which works in high-dimensional systems in Section 3.1.
3. If we choose $K = L$ and employ Gaussian basis functions $\phi_{B_\xi}(x)$, $x \in \mathbb{R}^n$, with covariance matrix B, we obtain a *Gaussian mixture filter*. Many versions of Gaussian mixture filters have been suggested in the literature, see, for example, Hoteit et al. (2008). Here, we describe the local mixture coefficient particle filter (LMCPF) of Walter et al. (2020), which has recently been shown to be able to provide high quality in high-dimensional systems.

Clearly, the propagation of the distribution is just one of several crucial ingredients of the assimilation method. We need to describe the assimilation step next and discuss stability and further properties of the filters, which arise from the choices we make.

Before we continue, let us demonstrate the deformation of a Gaussian distribution through the application of the nonlinear oscillator described in Eq. (2). Overtime, even if you would start with a Gaussian distribution. In Fig. 5, it is visualized by points sampling a level curve of the distribution. The level curve is distorted by the application of the model, here indicated by the trajectories for each of the sampling points. The non-Gaussianity increases with time between two assimilation steps.

We can measure the non-Gaussianity by any metric, which calculates the distance of the distribution with respect to a Gaussian distribution. The *reference* Gaussian distribution could be either

1. the Gaussian distribution with the same mean μ and covariance B than the distribution under consideration, or
2. the Gaussian distribution which fits best to the current distribution under the metric under consideration.

Here, let us briefly look at the first reference. A Gaussian distribution is characterized by its first and second moments, the mean μ and the standard deviation σ. We also know that higher moments are determined by the first and second moment, in particular

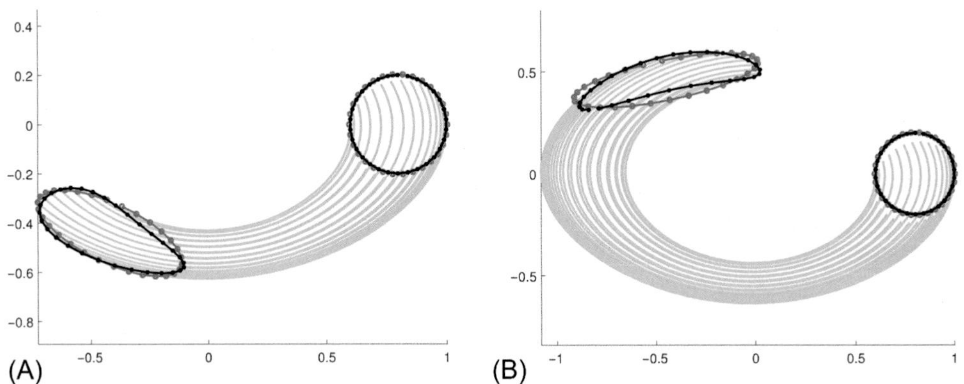

FIG. 5 We demonstrate the deformation of some Gaussian prior during the model propagation based on the simple oscillator model (2). The *black dots* show the ensembles at t_0 and t_1, the *gray circles* are the Gaussian estimates based on a singular value decomposition of the covariance matrix of distributions given by the *black points*. We display two different analysis time intervals $dt = 2$ and $dt = 4$.

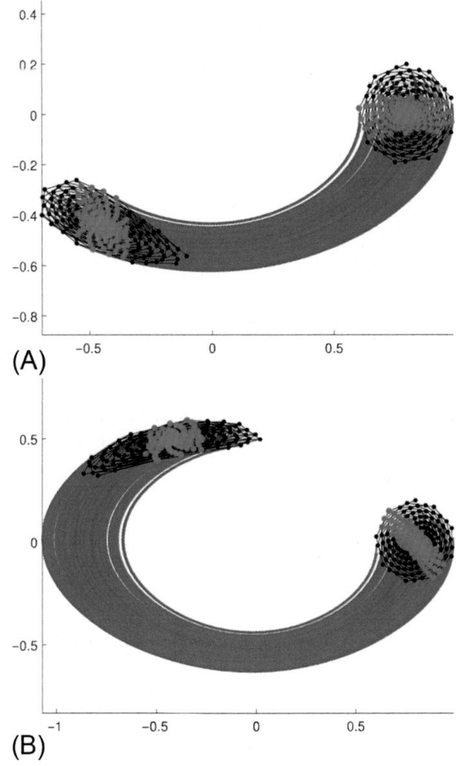

FIG. 6 We demonstrate the prior (*black*) and posterior distribution (*green*) for a classical particle filter at time t_1 (around the point (− 0.4, −0.4)) of the simple oscillator model (2), as well as the particles which survive for a particle smoother as the green points from the original draw at t_0 (around the point (0.8, 0)). The *left image* shows the result for $dt = 2$ and the *right-hand side* for $dt = 4$.

- *Skewness* or *third moment*

$$Sk := \mathbb{E}\left\{\left(\frac{x-\mu}{\sigma}\right)^3\right\} = 0. \tag{66}$$

- *Kurtosis* or *fourth moment*

$$Ku := \mathbb{E}\left\{\left(\frac{x-\mu}{\sigma}\right)^4\right\} = 3. \tag{67}$$

Possible *indices* to measure non-Gaussianity can be the derivation of the third and fourth moment of your current distribution from these values, for example,

$$I_1 := |Sk|,\ I_2 := |Ku-3|,\ \text{ or } I_3 := \sqrt{I_1^2 + I_2^2}. \tag{68}$$

We close with the remark that for *small ensemble size*, the stochastic indicators need to be treated with care, since the probability to observe deviations from a Gaussian distribution even if you have a Gaussian distribution are high for low sampling size.

3.1 The localized adaptive particle filter (LAPF)

The core idea of a *particle filter* is to use the approximation (63) for the prior and then carry out a Bayesian assimilation step as described in Eq. (12) to calculate the posterior distribution $p(x|y_k)$ with the observations y_k at time t_k. Here, we use the standard notation

$$x_k^{(b,\ell)} := M\left(x_{k-1}^{(a,\ell)}\right),\quad \ell = 1,\ldots,L \tag{69}$$

for the prior ensemble at time t_k. Given the distribution $p(y_k|x)$ as in Eq. (9), applying Bayes theorem (12) the posterior distribution is given by

$$p(x|y_k) = c\sum_{\ell=1}^{L} e^{-\frac{1}{2}\left(y_k - H\left(x_k^{(b,\ell)}\right)\right)^T R^{-1}\left(y_k - H\left(x_k^{(b,\ell)}\right)\right)}\delta\left(x - x_k^{(b,\ell)}\right) \tag{70}$$

with norming constant c.

There are several important points to take into account when you try to apply such a particle filter to *high-dimensional* and *real-world* systems. Here, we first introduce the *localized adaptive particle filter* suggested by Potthast et al. (2019). The filter belongs to the class of *ensemble transform filters*, that is, it calculates an increment in ensemble space spanned by the ensemble differences. Let us use the notation

$$X^{(b)} := \left(x^{(b,1)},\ldots,x^{(b,L)}\right) \in \mathbb{R}^{n\times L} \tag{71}$$

for the matrix of *prior states* and denote the matrix of *analysis states* by

$$X^{(a)} := \left(x^{(a,1)},\ldots,x^{(a,L)}\right) \in \mathbb{R}^{n\times L}. \tag{72}$$

The row of jth components is denoted by $X_j^{(b)}$ and $X_j^{(a)}$, respectively. The *localized* filter for each component $j = 1,\ldots,n$ calculates an update matrix $W^{(j)} \in \mathbb{R}^{L\times L}$ such that

$$X_j^{(a)} = X_j^{(b)} W^{(j)}, \quad j=1,\ldots,n. \tag{73}$$

This means that the analysis state is calculated as a linear combination of the prior ensemble members.

The calculation of the matrix $W^{(j)}$ is carried out based on localized weights

$$w_\ell := ce^{-\frac{1}{2}\left(y_k - H\left(x_k^{(b,\ell)}\right)\right)^T R_{loc}^{-1}\left(y_k - H\left(x_k^{(b,\ell)}\right)\right)}, \quad \ell=1,\ldots,L, \tag{74}$$

where localization R_{loc}^{-1} is achieved by multiplication of the entries of R^{-1} by a function depending on the distance of a variable to a given localization point $p \in \mathbb{R}^3$ in 3D space, and where c is a norming constant such that

$$\sum_{\ell=1}^{L} w_\ell = L. \tag{75}$$

Fig. 6 shows some particle distribution at time t_0 (around the point (0.8, 0)) and the propagated distribution at time t_1 (e.g., around the point (− 0.4, −0.4)), both as black dots connected with lines. We now calculate the weights w_ℓ, $\ell = 1,\ldots,L$ defined in Eq. (74). The particle filter now resamples from this distribution by taking all points which obey $w_\ell \geq 1$.

The points are shown in green color, both for time t_1 as well as their origin at t_0. For this particular example, with choosing $R = 50$, we have 110 out of 200 original particles to survive the resampling step.

- The *particle filter* takes the green points at time t_1 as a basis for further processing.
- The *particle smoother* would take the green points at time t_0 for further action.

We call this selection the *classical resampling step*. We define accumulated weights by

$$w_{ac,0} := 0, \quad w_{ac,\ell} := w_{ac,\ell-1} + w_\ell, \quad \ell=1,\ldots,L. \tag{76}$$

Taken a random draw $r_\ell \in U[0, 1]$, the uniform measure between 0 and 1, and $R_\ell := \ell - 1 + r_\ell$, the matrix $\breve{W}$ is defined by

$$\breve{W}_{i,\ell} := \begin{cases} 1, & \text{if } R_\ell \in (w_{ac,i-1}, w_{ac,i}], \\ 0, & \text{otherwise,} \end{cases} \tag{77}$$

for $i, \ell = 1,\ldots,L$. As an example, consider three particles, where the relative weight of the first is one and the others are zero. Then, $w_{ac,\,1-3} = (1, 1, 1)$. Assume we obtain the craws $R_{1-3} = (0.5, 1.2, 2.7)$. Then, the matrix $\breve{W}$ is given by

$$\breve{W} = \begin{pmatrix} 1 & 1 & 1 \\ 0 & 0 & 0 \\ 0 & 0 & 0 \end{pmatrix}, \tag{78}$$

selecting the first particle three times when used in the ensemble analysis Eq. (73). However, we are not yet complete with our particle filter.

The core next step of a particle filter is *rejuvenation*. We want to draw from the distribution to get back to L distinct particles for further propagation.

Clearly, the selection of particles based on the weights (74) will lead to a decrease of the spread of the ensemble of particles under consideration. This is very natural and it is a desired effect, since by measuring you put information into your system, which decreases the uncertainty.

But often there is a lot of uncertainty in the system, which is not captured by the prior ensemble. The *true* system is not the *model* system. For practical systems, we need to get back to a reasonable uncertainty in the posterior, that is, in the *rejuvenation step*. Here, the observation statistics need to be taken into account.

In general, based on $o - b = (o - t) + (t - b)$, the observation minus background statistics obeys the equation

$$(o-b)(o-b)^T = (o-t)(o-t)^T + (t-b)(t-b)^T + (o-t)(t-b)^T + (t-b)(o-t)^T, \tag{79}$$

where o is the observation values, b is the model equivalents in observation space, and t is the true system in observation space. If the measurement errors are statistically independent of the difference between truth and background, the expectation of $(o - t) \cdot (t-b)^T$ and its transpose is zero, and we obtain

$$\mathbb{E}[(o-b)(o-b)^T] = \mathbb{E}[(o-t)(o-t)^T] + \mathbb{E}[(t-b)(t-b)^T]. \tag{80}$$

Recall that $\mathbb{E}[(o-t)(o-t)^T] = R$ is the variance of the observation error and, if the ensemble reflects the true uncertainty, the value $\mathbb{E}[(t-b)(t-b)^T]$ should be the covariance matrix of the ensemble. This provides a tool to correct the observed covariance of the ensemble by an introduction of a multiplicative factor ρ into Eq. (80), leading to

$$\begin{aligned}\mathbb{E}[(o-b)^2] &= \mathbb{E}[(o-t)^2] + \rho\mathbb{E}[(t-b)^2] \\ &= R + \rho\mathbb{E}[HBH^T].\end{aligned} \tag{81}$$

For a larger set of observations and high-dimensional systems, we calculate ρ for each point in space, collecting observations with their weight according to the localization and taking the trace of the matrices which appear in Eq. (81), such that

$$\rho := \frac{\mathbb{E}[(o-b)^T(o-b)] - \mathrm{Tr}(R)}{\mathrm{Tr}(\mathbb{E}[HBH^T])}. \tag{82}$$

The basic idea of the LAPF is to carry out *rejuvenation* by drawing from a Gaussian around each remaining ensemble member after the classical resampling step. This random draw is carried out in a particular way, to keep the spatial structure of the particles intact as much as possible.

1. First, a global random matrix $N_{rand} \in \mathbb{R}^{L \times L}$ is chosen, where each component is drawn from a unit normal distribution.
2. For each analysis point in space $\mathbb{R}^3$, the matrix is scaled proportional to the local multiplication factor ρ, which reflects in what way the ensemble spread should be increased or decreased. In Potthast et al. (2019), the choice was taken linear in the form

$$\sigma(\rho) := \begin{cases} c_0, & \rho < \rho_0, \\ c_0 + (c_1 - c_0) \cdot \dfrac{\rho - \rho_0}{\rho_1 - \rho_0}, & \rho_0 \leq \rho \leq \rho_1, \\ c_1, & \rho > \rho_1, \end{cases} \tag{83}$$

where c_0, c_1, ρ_0, ρ_1 are used as tuning constants.

3. Finally, rejuvenation is carried out by definition of the final transformation matrix

$$W(p) := \breve{W} + \sigma(\rho(p))N_{rand}. \tag{84}$$

We call this last step (84) *modulated rejuvenation*, since the spread of the classical particle filter posterior ensemble is enhanced based on the local observation minus background statistics consistent with basic statistics, while at the same time the change carried out to each particle is a continuous function based on the given ensemble perturbations.

3.2 Localized mixture coefficient particle filter (LMCPF)

Data assimilation methods calculate initial conditions for models, which are built to approximate the true dynamics of some system. But the numerical model is not able to fully simulate the truth, and therefore its short-term forecasts will be different from the true state of the model. This applies to each single forecast, but it also applies to a whole ensemble of forecasts $x^{(\ell)}$, $\ell = 1, \ldots, L$. Here, we need to be aware of one of the underlying challenges of *ensemble data assimilation*:

1. The spread and statistics of the ensemble at some time t reflects the uncertainty at the initial time t_0 propagated through the model dynamics to time t.
2. The ensemble does not know differences to the true dynamics. This means that the ensemble spread and statistics miss an important part of the forecast uncertainty: the model error.
3. There are further errors, which are not properly captured by the ensemble, in particular the sampling error which comes from the finite number of ensemble members which are used in real-world systems.

In ensemble filters like the ensemble Kalman filter (in any of its many variants), we employ covariance inflation to model the uncertainty of the model error. This can be used in the form of

1. *Multiplicative covariance inflation*, where the ensemble spread is increased after the analysis step by a multiplicative factor.
2. *Additive covariance inflation*, where we add some random perturbation to the analysis ensemble members, either before the analysis step or after the analysis is calculated.
3. Different relaxation methods implicitly add spread by pulling the analysis ensemble toward the prior ensemble, for example, *relaxation to prior spread* (RTPS) or *relaxation to prior perturbations* (RTPP).

Often, the size of the inflation is calculated adaptively based on the estimate Eq. (82). The *adaptive rejuvenation* (84) of the LAPF can be understood as an additive inflation method carried out in ensemble space.

In variational methods as well as the ensemble Kalman filter, in the analysis step, the model states are drawn toward the observations. The strength of this move is determined by the relative size of the observation error covariance matrix R in comparison to the model uncertainty covariance matrix B.

For the classical particle filter, the ensemble members are understood as δ-functions, such that their individual uncertainty is zero, and the ensemble members as such are *not* moved toward the observations. The filter moves the system toward observations by removing particles, which are further away from the observed states, and it further moves by the random adaptive rejuvenation. But the rejuvenation step does not include any direction, but is merely increasing the spread of the ensemble by a random draw.

For real-world systems, which are strongly influenced by the difference between truth and model system, it is desirable to enable the particle filter to include model error as well. This can be achieved by a change in the interpretation of the individual particle. Instead of using a δ-function, the idea of the *localized mixture coefficient particle filter* employs ideas of *Gaussian mixture filters* by approximating the prior by

$$p(x) := c\sum_{\ell=1}^{L} e^{-\frac{1}{2}\left(x-x^{(\ell)}\right)^T (\gamma B)^{-1}\left(x-x^{(\ell)}\right)}, \quad x\in\mathbb{R}^n, \tag{85}$$

with a norming constant c, where γ is a constant and B is the covariance matrix (35). The posterior $p(x|y)$ defined by Bayes formula (12) is now given by

$$p(x|y) = c\sum_{\ell=1}^{L} e^{-\frac{1}{2}\left(y_k - H\left(x_k^{(b,\ell)}\right)\right)^T R^{-1}\left(y_k - H\left(x_k^{(b,\ell)}\right)\right)} \cdot e^{-\frac{1}{2}\left(x-x^{(\ell)}\right)^T (\gamma B)^{-1}\left(x-x^{(\ell)}\right)}, \tag{86}$$

$x\in\mathbb{R}^n$. For each of the terms $\ell=1,\dots,L$ of Eq. (86), the product of the two Gaussians can be carried out analytically as derived for the Kalman filter in Eqs. (52)–(55). All other elements of this filter are chosen as for the LAPF described earlier, that is, the resampling step to select which terms in Eq. (86) are chosen with what weight, and the adaptive rejuvenation step by drawing from the posterior with some adaptive additive Gaussian distribution to achieve a posterior ensemble spread, which fits the observation minus first-guess statistics.

Finally carry out some tests for the localized filters LETKF, LAPF, and LMCPF for the Lorenz96 model of N coupled oscillators

$$\frac{\partial x_\xi}{\partial t} = (x_{\xi+1} - x_{\xi-2})x_{\xi-1} - x_\xi + F, \tag{87}$$

where we assume that $x_{-1} = x_{N-1}$, $x_0 = x_N$, and $x_{N+1} = x_1$. We consider the nodes x_ξ to be located on a circle of radius $r = 1$ and base the localization of the spatial distance $d_{\xi,\eta}$ of the variables x_ξ and x_η with a localization function as defined in Eq. (42) given by

$$f(d_{\xi,\eta}) := c\frac{\rho}{d_\xi}, \quad \xi,\eta = 1,\dots,N, \tag{88}$$

with $c = 0.1$ and $\rho = 6\pi/N$. In our first test, visualized in Fig. 7, we chose $F_{true} = 8$ for the nature run, that is, the simulation of the truth and the observations, and $F = 10$ for the model system, that is, as the parameter employed within the data assimilation cycle and for forecasts. We have started the simulation with a state $x_\xi = F_{true}$ at $t_0 = 0$ and chose a time step $h = 0.01$ for the fourth-order Runge-Kutta method for time integration. Here, we have chosen to

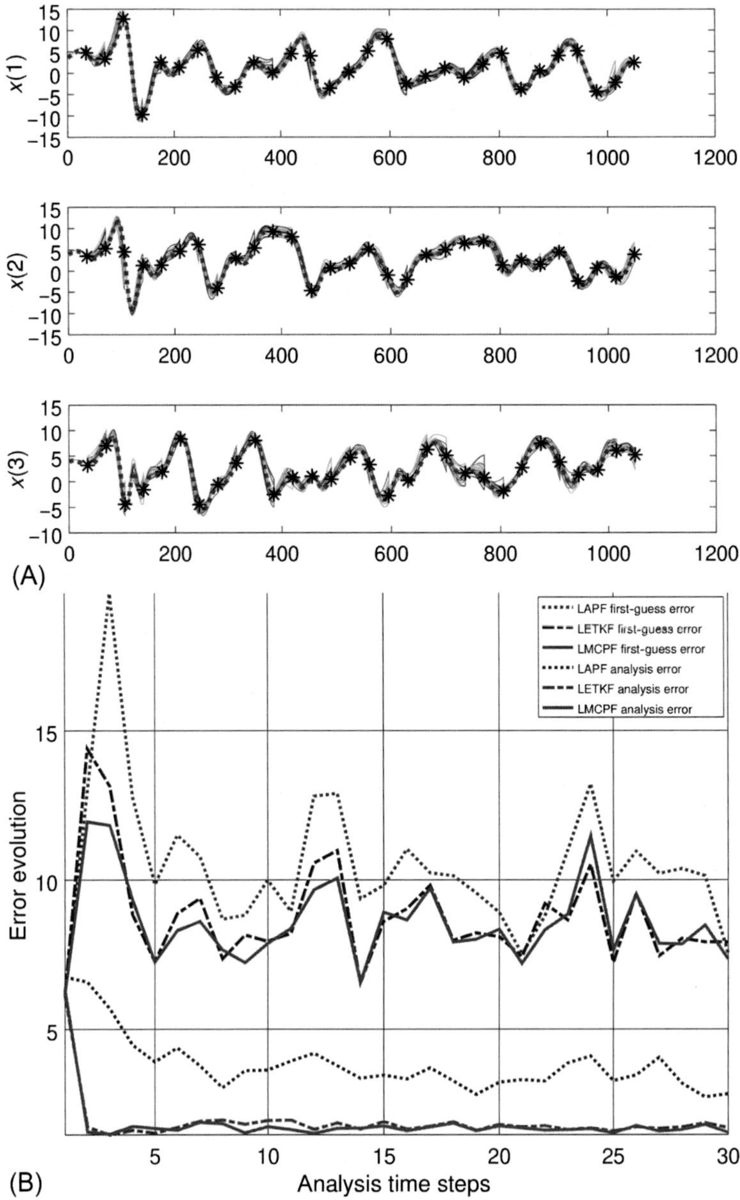

FIG. 7 In (A), we show the trajectories for the variables x_ξ, $\xi = 1, 2, 3$ of the Lorenz96 model with $N = 100$ coupled variables run over 1000 time steps with an analysis each 25 time steps. The development of the analysis error and first-guess error for the three filters LAPF, LETKF, and LMCPF is shown in (B). Here, we took the classical choice $F_{true} = 8$ and for assimilation $F = 10$, compare also Fig. 8 with another choice.

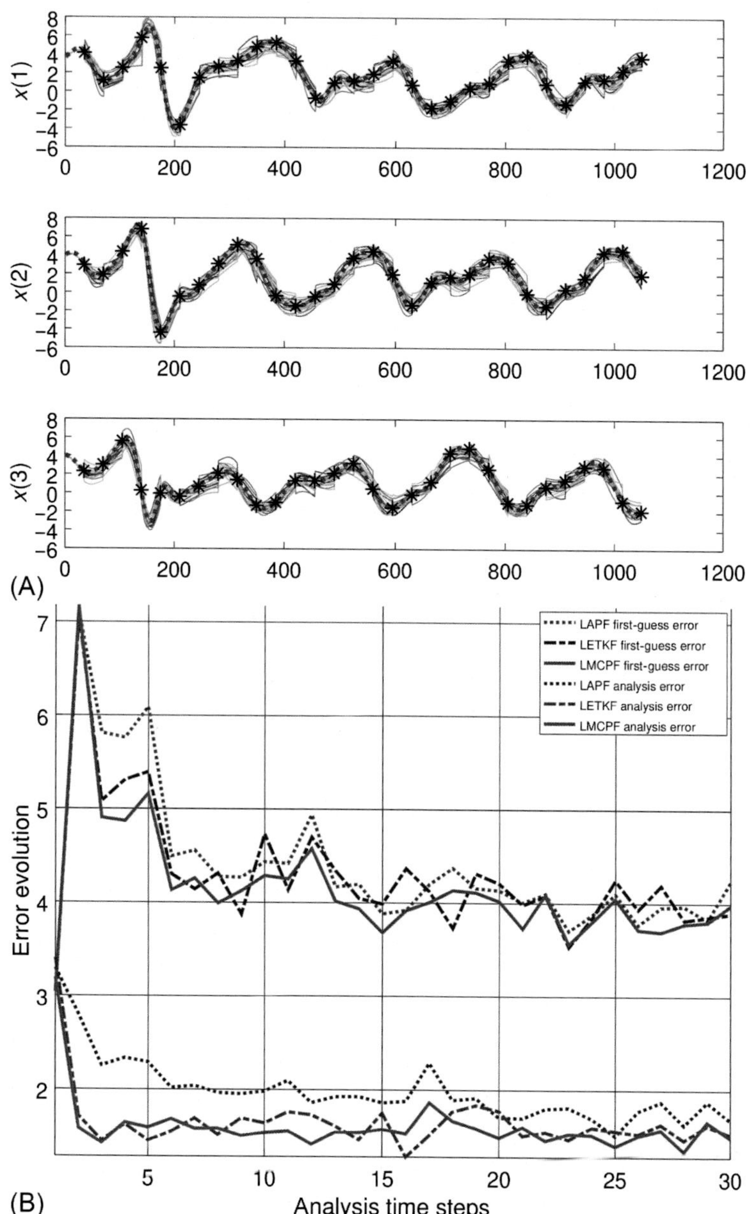

FIG. 8 In (A), we show the trajectories for the variables x_ξ, $\xi = 1, 2, 3$ of the Lorenz96 model with $N = 100$ coupled variables run over 1000 time steps with an analysis each 25 time steps. The development of the analysis error and first-guess error for the three filters LAPF, LETKF, and LMCPF is shown in (B). In contrast to Fig. 7, here we choose $F_{true} = 4$ and $F = 3$ for assimilation.

observe at each node with an observation error with standard deviation $\sigma = 0.1$. Another test is carried out with $F_{true} = 4$ and $F = 3$ for assimilation and forecasting, visualized in Fig. 8.

The localized ensemble and particle filters all run stably for high-dimensional systems, as shown in Potthast et al. (2019) and Walter et al. (2020) within the operational framework of NWP. Here, we show results for dimension $N = 100$ of the state space and $L = 20$ ensemble members. Figs. 7 and 8 show the evolution of some trajectories in (A), the evolution of the first guess and analysis errors in (B). We observe that usually all filters are able to synchronize the assimilation cycle with the nature run. Overall, the LMCPF shows the least first-guess errors, the LETKF is next and the LAPF often has difficulties to pull the ensemble strongly enough to the nature run. This is consistent with the theory and setup of the method, since the difference between LAPF and LMCPF is the shift based on additional model error, such that the LMCPF pulls the particles closer to the observations even if the ensemble does not catch the difference.

References

Dalety, R., 1993. Atmospheric Data Analysis. Cambridge Atmospheric and Space Science SeriesCambridge University Press. p. 472. http://books.google.de/bookss?id=RHM6pTMRTHwC.

Gaspari, G., Cohn, S.E., 1999. Construction of correlation functions in two and three dimensions. Q. J. R. Meteorol. Soc. 125 (554), 723–757. https://doi.org/10.1002/qj.49712555417.

Hoteit, I., Pham, D.T., Triantafyllou, G., Korres, G., 2008. A new approximate solution of the optimal nonlinear filter for data assimilation in meteorology and oceanography. Mon. Weather Rev. 136, 317–334. https://doi.org/10.1175/2007MWR1927.1.

Kalnay, E., 2003. Atmospheric Modeling, Data Assimilation and Predictability. Cambridge University Press. http://books.google.de/books?id=Uqc7zC7NULMC.

Lewis, J.M., Lakshmivarahan, S., Dhall, S., 2006. Dynamic Data Assimilation: A Least Squares Approach. Dynamic Data Assimilation: A Least Squares Approach, vol. 13. Cambridge University Press. http://books.google.co.in/books?id=MkydCKWlm_QC.

Nakamura, G., Potthast, R., 2015. Inverse Modeling. IOP Publishing. https://doi.org/10.1088/978-0-7503-1218-9.

Potthast, R., Walter, A., Rhodin, A., 2019. A localized adaptive particle filter within an operational NWP framework. Mon. Weather Rev. 147 (1), 345–362. https://doi.org/10.1175/MWR-D-18-0028.1.

Walter, A., Schenk, N., van Leeuwen, P.J., Potthast, R., 2020. Particle filtering and Gaussian mixtures—on a localized mixture coefficients particle filter (LMCPF) for global NWP. (under review).

CHAPTER

7

Subgrid turbulence mixing

Xu Zhang

Shanghai Typhoon Institute, China Meteorological Administration, Key Laboratory of Numerical Modeling for Tropical Cyclone, China Meteorological Administration, Shanghai, China

1 Introduction

The spatial scales of turbulent flows in the atmosphere range from boundary layer depth z_i to a few millimeters (smallest dissipative eddies). In the modeling community of atmospheric turbulence, it is unfeasible to resolve all scales of atmospheric motion with high Reynolds number (*Re*), because it is beyond the capabilities of computers today or in the foreseeable future. In lieu of simulating all scales of turbulent motion, high wave numbers and high-frequency fluctuations are removed from the solution by filtering (or averaging) the Navier–Stokes equation, which makes the problem manageable with available computing resources. The filtering operation separates the fields formally into filtered-scale (resolved) and subfilter-scale (unresolved) components, where the former is usually referred to as grid scale (resolved) component and the latter as subgrid scale (SGS) component. Since the SGS motions are not explicitly calculated by the numerical dynamics, the parameterization problem is introduced to represent the effect of SGS motions on the resolved motions. The purpose of subgrid mixing parameterization in the numerical atmospheric modeling is to account for the small-scale turbulent mixings that are cut off from a solution of the dynamical equations. Subgrid turbulent mixing parameterizations use resolvable fields to represent the effect of SGS motions on the resolved motions.

Historically, based on the filter scale Δ and the energy-containing turbulence scale l_e, two types of approaches for modeling atmospheric turbulence were developed. In mesoscale numerical weather prediction (NWP) or general circulation models (GCMs), $\Delta >> l_e$, which is referred to as the mesoscale limit. In this limit, none of the turbulence is resolved, and all the turbulent motions are SGS. The planetary boundary layer (PBL) is assumed to be horizontally homogeneous, so conventional one-dimensional (1D) PBL schemes are usually used to represent SGS turbulent motions. The other approach for simulating atmospheric turbulence is the three-dimensional (3D) large eddy simulation (LES), where $\Delta << l_e$ (which is referred to

https://doi.org/10.1016/B978-0-12-815491-5.00007-0

as to the LES limit). In the LES limit, the large eddies are explicitly resolved, and the 3D SGS model is usually to represent the effect of small eddies, i.e., to drain energy from the large eddies. The most widely used LES SGS models include the 3D Smagorinsky–Lilly (Smagorinsky, 1963; Lilly, 1967) and Deardorff's 3D turbulent kinetic energy (TKE)-based models (Deardorff, 1980).

Lilly's (1967) derivations of the filtered equations and the governing equation for SGS stress provided the basis for simulating the turbulent flows and developing the turbulence parameterization schemes. Using the Reynolds averaging that is thought to describe the ensemble mean statistical moments, the transport equation for SGS stress is given by

$$\begin{aligned}\frac{\partial\overline{u_i'u_j'}}{\partial t}+\overline{u}_k\frac{\partial\overline{u_i'u_j'}}{\partial x_k}=&\left(-\overline{u_i'u_k'}\frac{\partial\overline{u}_j}{\partial x_k}-\overline{u_j'u_k'}\frac{\partial\overline{u}_i}{\partial x_k}\right)-\frac{\partial\overline{u_i'u_j'u_k'}}{\partial x_k}+\frac{g}{\theta_v}\left(\delta_{j3}\overline{u_i'\theta_v'}+\delta_{i3}\overline{u_j'\theta_v'}\right)\\&-\frac{1}{\rho_0}\left[\frac{\partial\overline{p'u'_j}}{\partial x_i}+\frac{\partial\overline{p'u'_i}}{\partial x_j}-\overline{p'\left(\frac{\partial u_i'}{\partial x_j}+\frac{\partial u_j'}{\partial x_i}\right)}\right]-2\nu\overline{\frac{\partial u_i'}{\partial x_k}\frac{\partial u_j'}{\partial x_k}}\end{aligned} \tag{1}$$

Here, u_i are the velocity components, θ_v is the virtual potential temperature, ρ_0 is the reference density, p is the pressure, and ν is the kinematic viscosity. The overbars denote the quantities that are explicitly computed (resolved) by model dynamics, and primes denote deviations from the resolved quantities. The second term on the left-hand side is the advection of momentum flux by the mean wind. On the right-hand side, the terms represent (in order): production of momentum flux by the mean wind shears, turbulent transport, buoyant production, pressure transport and pressure destruction, and dissipation. Note that the Coriolis force has been neglected.

The transport equation of SGS heat flux is written as follows:

$$\frac{\partial\overline{u_i'\theta'}}{\partial t}+\overline{u}_j\frac{\partial\overline{u_i'\theta'}}{\partial x_j}=\left(-\overline{\theta'u_j'}\frac{\partial\overline{u}_i}{\partial x_j}-\overline{u_i'u_j'}\frac{\partial\overline{\theta}}{\partial x_j}\right)-\frac{\partial\overline{u_i'u_j'\theta'}}{\partial x_j}+\delta_{i3}\frac{g}{\overline{\theta_v}}\overline{\theta'\theta'_v}-\frac{1}{\rho_0}\left(\frac{\partial\overline{p'\theta'}}{\partial x_i}-\overline{p'\frac{\partial\theta'}{\partial x_i}}\right). \tag{2}$$

The second term on the left-hand side is the advection of heat flux by the mean wind. On the right-hand side, the terms represent (in order): production of heat flux by the mean wind shears and the thermal gradient, turbulent transport, buoyant production, pressure transport and destruction, and dissipation. Note that the molecular destruction terms are absent under the assumption of local isotropy (Wyngaard et al., 1971).

It should be emphasized that the form of the Reynolds-averaged equations does not depend on the filter scale Δ. The Reynolds-averaged equations (1) and (2) can be applied in both LES and mesoscale modeling, but the corresponding closure methods may be different. The closure (or parameterization) problem essentially depends on the scale of filtering, i.e., the grid size of the numerical model. Uncertainties in subgrid mixing parameterizations result from the different closure assumptions.

To facilitate the closure to these equations, Mellor (1973) and Mellor and Yamada (1982) proposed the Level 4 model using closure assumptions for unknown second- and third-order terms. Following Mellor and Yamada (1974), the Level 4 model is then simplified to the Level 3 model, in which the differential equations for SGS stress and flux are reduced to algebraic equations. For the sake of simplicity in the discussion, the algebraic equation of the deviatoric SGS stress ($i \neq j$) in Level 3 model is considered here

$$\overline{u_i'u_j'} = -\frac{3l_1}{\sqrt{2e}}\left[\left(\overline{u_i'u_k'} - 2C_1 e\delta_{ki}\right)\frac{\partial \overline{u}_j}{\partial x_k} + \left(\overline{u_k'u_j'} - 2C_1 e\delta_{kj}\right)\frac{\partial \overline{u}_i}{\partial x_k} - \frac{g}{\theta_v}\left(\delta_{j3}\overline{u_i'\theta_v'} + \delta_{i3}\overline{u_j'\theta_v'}\right)\right]. \tag{3}$$

The algebraic equation of SGS heat flux in Level 3 model is written as

$$\overline{u_i'\theta'} = -\frac{3l_2}{\sqrt{2e}}\left(\overline{\theta' u_j'}\frac{\partial \overline{u}_i}{\partial x_j} + \overline{u_i'u_j'}\frac{\partial \overline{\theta}}{\partial x_j} - \delta_{i3}\frac{g}{\theta_v}\overline{\theta'\theta'_v}\right). \tag{4}$$

Here, l_1 and l_2 are length scales, which are proportional to the master mixing length scale l; C_1 is a nondimensional constant, and $e=\overline{u_i'2}/2$, the turbulent kinetic energy. The first two terms on the right of Eqs. (3) and (4) are flux productions by the local gradient of the mean state, while the third term defines the flux production by buoyancy. Suggested by Wyngaard (2004), $l_1/\sqrt{2e}$ and $l_2/\sqrt{2e}$ can be thought of as time scales of unresolved turbulence.

In the convective boundary layer (CBL), organized structures due to the buoyancy dominate the vertical turbulent flux in a nonlocal way (Shin and Hong, 2013; Hellsten and Zilitinkevich, 2013). The flux productions by buoyancy in Eqs. (3) and (4) can be regarded as the nonlocal effects. From the view of local and nonlocal way, the Level 3 model of SGS stress and flux Eqs. (3) and (4) can be expressed by conventional eddy- diffusivity models including the nonlocal effects as follows:

$$\overline{u_i'u_j'} = -K_{ij}^M\left(\frac{\partial \overline{u}_i}{\partial x_j} + \frac{\partial \overline{u}_j}{\partial x_i}\right) + \overline{u_i'u_j'}^{NL}, \tag{5}$$

$$\overline{u_i'\theta'} = -K_{ij}^H\frac{\partial \overline{\theta}}{\partial x_j} + \delta_{i3}\overline{u_i'\theta'}^{NL}, \tag{6}$$

where superscripts M, H, and NL indicate momentum, heat, and nonlocal, respectively. Thus, the flux can be decomposed into two components: the local flux due to the gradient of mean fields and the nonlocal flux due to the buoyant-production terms. Note that the nonlocal terms are only retained in the vertical. In the Level 3 model, the nonlocal effect is directly related to the buoyancy. Wyngaard (2004) emphasized that the eddy diffusivity is a second-order tensor rather than a scalar.

Wyngaard (2004) assessed the importance of buoyant-production terms (nonlocal terms) at different grid size based on the ratio of buoyant- and gradient-production terms in turbulent stress conservation equations, the ratio becomes a scale-dependent turbulent Richardson number:

$$Ri(\Delta) = \frac{\text{buoyant production term}}{\text{gradient production term}} \sim \left(\frac{l_e}{\Delta}\right)^{-2/3} \tag{7}$$

The relative importance of the buoyancy production term (nonlocal effect) decreases with increasing l_e/Δ, that is, in the mesoscale limit ($\Delta >> l_e$) the buoyant-production term (nonlocal effect) is very important, while in the LES limit ($\Delta << l_e$), the direct effects of buoyancy on the SGS turbulence budgets are small. In mesoscale simulations, all nonlocal flux is contained in the SGS component. In other words, SGS nonlocal flux is saturated as grid spacings approaching the mesoscale limit and no longer changes for coarser spacings.

According to Wyngaard (2004), the role of nonlocal buoyancy flux becomes more important as the grid size increases. In the mesoscale limit, the model cannot maintain any turbulence and the inclusion of the nonlocal terms is necessary. Then if one makes the boundary-layer approximation and retains the term involving the scalar gradient in the direction of the flux, the Level 3 model of SGS stress and flux in Eqs. (5) and (6) is reduced to

$$\overline{w'u'} = -K_{Mv}\frac{\partial \overline{u}}{\partial z} + \overline{w'u'}^{NL}, \tag{8}$$

$$\overline{w'\theta'} = -K_{Hv}\frac{\partial \overline{\theta}}{\partial z} + \overline{w'\theta'}^{NL}, \tag{9}$$

which are the popular forms in the 1D PBL scheme. K_{Mv} and K_{Hv} are vertical eddy diffusivity for momentum and heat, respectively. The eddy diffusivity is usually approximated as a scalar rather than a tensor. The vertical eddy diffusivity takes the form of $K_{Mv} \sim le^{1/2}$, where l is the vertical mixing length, and $K_{Hv} = K_{Mv}P_r^{-1}$. In regards to the nonlocal term, there are many definitions based on different assumptions, especially for $\overline{w'\theta'}^{NL}$ (Troen and Mahrt, 1986; Deardorff, 1972; Holtslag and Moeng, 1991; Siebesma et al., 2007; Noh et al., 2003; Shin and Hong, 2015), which will be discussed later. In the LES limit, two assumptions are made: (1) the grid size is well within the inertial subrange; (2) the grid size is much smaller than the scale of the energy-containing eddies to be simulated ($\Delta < < l_e$) (Bryan et al., 2003). The buoyancy production terms in Eqs. (5) and (6) are neglected. That is, the nonlocal mixing is explicitly calculated, and only small-scale eddies are parameterized, thus the local eddy-diffusivity formulation is suitable for the LES closures.

Different formulations of horizontal mixing parameterizations are adopted in LES and mesoscale simulations. In mesoscale simulations with horizontal grid sizes Δ considerably larger than the scale of the energy-containing eddies ($\Delta < < l_e$), the boundary layer is assumed to be horizontally homogeneous. It is common practice in mesoscale simulations to treat the vertical and horizontal mixings separately. Conventional 1D PBL schemes are usually used to represent vertical mixing. Horizontal mixing is usually treated by the simple eddy diffusivity formulation and uses the deformation-based diffusivity coefficients. SGS horizontal fluxes can be expressed by eddy-diffusivity formulations,

$$\overline{u'v'} = -K_{Mh}\left(\frac{\partial \overline{u}}{\partial y} + \frac{\partial \overline{v}}{\partial x}\right), \tag{10}$$

$$\overline{u'\theta'} = -K_{Hh}\frac{\partial \overline{\theta}}{\partial x}, \tag{11}$$

where K_{Mh} and K_{Hh} are horizontal eddy diffusivities for momentum and heat, respectively. In the Weather Research and Forecasting (WRF) model, K_{Mh} and K_{Hh} are calculated by 2D Smagorinsky first-order closure based on the horizontal resolvable-scale velocity deformation. In practical applications, horizontal mixing requirements are largely a function of numerical constraints such as computational stability and small-scale noise (Deardorff, 1985; Jablonowski and Williamson, 2011). The distinction between the numerical and physical reasons for horizontal mixing is often obscure in NWP models. The horizontal mixing is one of the main sources of uncertainty in NWP models. In the LES limit, the calculation of horizontal diffusivity is based on the 1.5-order TKE closure, which is consistent with the vertical mixing.

In addition, the SGS turbulent closures typically used in the mesoscale models and LES are of the same form—an eddy-diffusivity model with $K \sim le^{1/2}$ and of the same expression for the TKE dissipation rate with $\varepsilon \sim e^{3/2}/l$. It is very important to represent the master mixing length in both limits. The mixing length expression is one of the main different aspects between the LES and mesoscale simulation closures.

As mentioned above, in NWP models, nonlocal flux (including heat and momentum), turbulent mixing length, and horizontal mixing parameterization are three main sources of uncertainty with regard to subgrid mixing parameterization. The uncertainties of subgrid mixing parameterizations are discussed from the three issues in this chapter.

As the resolution of NWP models steadily increases, such that grid size becomes comparable to the typical size of the energy-containing eddies ($\Delta \sim l_e$), the grid sizes fall within a "Terra Incognita" (Wyngaard, 2004) that is also referred to as the gray zone. In this regime, neither traditional PBL schemes which are designed to parameterize all PBL turbulence, nor LES SGS models which are designed to resolve the energy-containing eddies, perform appropriately. The conventional formulations of the three aspects (nonlocal flux, mixing length, and horizontal mixing) that were originally developed for relatively coarse model grids should be reexamined in the context of high-resolution modeling. The reformulations may bring new uncertainties. Meanwhile, the failure of the statistical assumptions in high-resolution model adds the extra level of uncertainty in the subgrid mixing parameterizations.

In the last couple of decades, the representations of turbulence in NWP models have been mostly devoted to the development of PBL parameterization. The above-mentioned issues of subgrid turbulence mixing have been comparatively less studied in the context of moist convection. Based on the fact that energetic eddies associated with dry PBL turbulence and moist convection have different scales, NWP models with O (10 km) grid spacing usually contain separate subgrid schemes for dry PBL turbulence and moist convection. However, in cloud models with O(1 km) grid spacing, there is no spectral gap between resolvable and SGS motions in a convective cloud system (Moeng et al., 2010). Currently, most cloud modeling at kilometer-scale resolution still relies on the traditional subgrid mixing parameterization approaches developed for PBL turbulence in mesoscale models or LES-type turbulence schemes. The weaknesses of convectional PBL turbulence parameterizations and LES-type turbulence schemes in the kilometer-scale cloud models have been pointed out (Bryan et al., 2003; Moeng et al., 2010; Verrelle et al., 2017; Shi et al., 2019). The parameterization of subgrid turbulence mixing in convective cloud modeling presents more uncertainty than that in the dry PBL.

2 Nonlocal flux

Large eddies associated with the rise of warm air parcels transport heat from hot to cold regardless of the local gradient of the background environment. When large eddies exist, the local eddy-diffusivity method often fails owing to upgradient, or countergradient, fluxes. This is perhaps the most important criticism of local eddy-diffusivity method—that the heat and momentum transport in the boundary layer during the daytime is mostly accomplished by the largest eddies, and that these eddies are more representative of the properties of the entire boundary layer than the local conditions at one vertical level. Nonlocal-closure methods are motivated by the recognition that much of the mixing in the boundary layer can be associated with large eddies. In order to overcome the deficiencies, nonlocal mixing schemes have been developed. The local-closure approximation to the vertical heat flux can incorporate a correction term to the local gradient that incorporates the effects of the large eddies.

The popular format for the SGS heat flux in the mesoscale limit can be expressed by the sum of eddy-diffusivity component (local) and nonlocal component as shown in Eq. (9). The nonlocal heat flux is usually written as

$$\overline{w'\theta'}^{NL} = K_{Hv}\gamma \tag{12}$$

The countergradient term γ quantifies the effect of nonlocal transport. However, there is no universal definition of the nonlocal heat flux.

Troen and Mahrt (1986) defined the countergradient term using the surface heat flux and PBL height. In the Medium-Range Forecast (MRF) (Troen and Mahrt, 1986) and Yonsei University (YSU) (Hong et al., 2006) PBL parameterization schemes, the countergradient term γ is given by

$$\gamma = b\frac{\overline{(w'\theta')_0}}{w_s h} \tag{13}$$

where $\overline{(w'\theta')_0}$ is the surface heat flux, b is a coefficient of proportionality, w_s is the mixed-layer velocity scale, and h is the height of the PBL. The formulations of nonlocal flux in these schemes use results from LES.

Deardorff (1972) and Holtslag and Moeng (1991) attempted to formally justify the countergradient term γ based on turbulent heat flux equation [Eq. (2)]. The equation for the time rate of change of vertical heat flux is

$$\frac{\partial\left(\overline{w'\theta'}\right)}{\partial t} = -\overline{w'^2}\frac{\partial\bar{\theta}}{\partial z} - \frac{\partial\left(\overline{w'^2\theta'}\right)}{\partial z} + \left(\frac{\overline{\theta'^2}}{\theta_0}\right)g - \frac{1}{\bar{\rho}}\left(\overline{\theta'\frac{\partial p'}{\partial z}}\right) \tag{14}$$

By neglecting the turbulent transport term (the second term on the right of equation (14)), Deardorff (1972) proposed a theoretical expression of the countergradient term:

$$\gamma = \frac{g}{\theta_0}\frac{\overline{\theta'^2}}{\overline{w'^2}} \tag{15}$$

which indicates the countergradient heat flux comes from the buoyancy production term.

On the other hand, based on the different closure assumptions applied to the buoyancy flux budgets, Holtslag and Moeng (1991) proposed a countergradient term with different expressions from Deardorff's. They presented the countergradient term as

$$\gamma = b\frac{w_*^2\theta_*}{\overline{w'^2}h} \tag{16}$$

where w_* and θ_* are the convective velocity scale and the convective temperature scale, respectively. The countergradient term results from the third-moment turbulent transport effect. Note that the physical interpretation of countergradient terms in Holtslag and Moeng (1991) is different from Deardorff's results.

Zilitinkevich et al. (1999) argued that the third-moment turbulent transports (fluxes of fluxes) are largely responsible for the nonlocal nature of turbulent motions, and proposed a turbulent advection plus diffusion parameterization for the third-order moments. Using the proposed parameterization can result in a nonlocal integral closure, in which Deardorff's expression for the countergradient term becomes a particular case of the integral closure.

In addition, the nonlocal flux caused by strong organized updrafts can be represented by the top-hat mass-flux model that is initially used in most convection parameterizations (Siebesma et al., 2007). Based on a scale separation assumption between strong organized updrafts and the remaining turbulence, Siebesma et al. (2007) proposed an eddy-diffusivity mass-flux (EDMF) approach for the CBL, in which the nonlocal flux is described explicitly by a mass flux term, whereas the remaining turbulent part is represented by an eddy-diffusivity approach. The nonlocal heat flux in EDMF is usually represented as

$$\overline{w'\theta'}^{NL} = M(\theta_u - \overline{\theta}) \tag{17}$$

where the subscript u refers to the updraft properties and M denotes the updraft mass flux.

According to the top-hat representation methods of Siebesma and Cuijpers (1995) and Siebesma et al. (2007), when applied to the potential temperature θ, the total turbulent heat flux can be decomposed into three terms:

$$\overline{w'\theta'} = a\overline{w'\theta'}^{c} + (1-a)\overline{w'\theta'}^{e} + a(1-a)(w_c - w_e)(\theta_c - \theta_e), \tag{18}$$

where a is the areal fraction of convective updrafts in grid boxes. The overbar indexed c (e) denotes an average of the perturbations from the convective updraft (environmental) average within the area-designated convective (environment). The subscript c (e) refers to an average over the convective updraft (environmental) area. The first term is the turbulent heat flux in the convective updraft; the second describes the environmental turbulence; while the third term describes the organized turbulence term. In the dry CBL, the first two terms can be considered to be the local heat flux due to small eddies, while the third term can be understood to be nonlocal heat flux induced by organized structures (i.e., large eddies). Therefore, the local (L) and nonlocal (NL) heat flux are separated as

$$\overline{w'\theta'}^{NL} = a(1-a)(w_c - w_e)(\theta_c - \theta_e), \tag{19}$$

$$\overline{w'\theta'}^{L} = a\overline{w'\theta'}^{c} + (1-a)\overline{w'\theta'}^{e}. \tag{20}$$

However, the quantification of nonlocal and local flux presents large uncertainties, which may result from sampling methods, analyzed data, and boundary layer types. Based on various sampling and filtering methods, many studies tried to quantify the contribution of the organized structures to the total flux. Lenschow and Stephens (1980) analyzed the organized thermal structures over a relatively warm ocean surface using aircraft observations and a conditional sampling method using a threshold value of humidity as an indicator of thermals. They estimated that the organized thermal structures account for about 50% of the heat flux and about 60% of the moisture flux in the surface layer. Greenhut and Khalsa (1982, 1987) also studied the convective structures in the marine boundary layer from aircraft observations using a conditional sampling method based on the vertical velocity as an indicator function. They showed that the contribution of the organized structures to the total fluxes is approximately 85–90%. Using mast measurements and conditional sampling, Wilczak (1984) showed that the large eddies are responsible for essentially all of the heat and momentum fluxes in the surface layer. Using different filtering techniques, Hellsten and Zilitinkevich (2013) estimated that roughly 85% of the vertical heat flux and about 72% of the kinetic energy is carried by the large eddies in the mixed layer in two CBL LES simulations. Note that there is no unique and strict way to separate the organized structures from the environmental turbulence as the two kinds of motion are strongly intertwined. In reality, the nonlocal transports drive the local mixing through turbulence cascade, thus there is no spectral gap separating them. Such separation between local and nonlocal flux can only be done approximately.

A key issue for the flux decomposition is the method used to sample organized structures in LES. The quantification and determination of nonlocal flux have large uncertainty, which comes from the different conditional sampling methods, and their associated parameter. For example, the conditional sampling method is used in Siebesma et al. (2007) to identify the organized structures and distinguish the nonlocal transport from the total transport:

$$w(x, y, z) > w_{p\%}(z). \tag{21}$$

The p-percentile velocity $w_{p\%}(z)$ is defined as the vertical velocity for which exactly p% of the grid points exceed $w_{p\%}(z)$ at height z. Here, p is the parameter with some arbitrariness, which is likely to affect the results. Couvreux et al. (2010) used a conditional sampling based on the combination of a passive tracer emitted at the surface and thermodynamic variables with a lifetime of typically 15 min to characterize organized structures. Hellsten and Zilitinkevich (2013) applied three different filtering methods (short-time averaging, Fourier filtering, and proper orthogonal decomposition) to quantify the roles and the relative importance of the organized structures in the budget of heat flux, while the quantification is still somewhat sensitive to the filter width. Chinita et al. (2018) developed a method based on the joint probability density function (PDF) of vertical velocity and a scalar (e.g. total water mixing ratio or potential temperature) to detect organized motions in the atmospheric boundary layer. The overarching goal of the conditional sampling and filtering method is to guide the development and evaluation of boundary layer and convection parameterizations. However, it must also be kept in mind that the various conditional sampling and filtering techniques involve parameter thresholds that have no unique optimal values. They just have to be set more or less arbitrarily and are likely to influence the results (Schmidt and Schumann, 1989).

According to the discussions in the introduction, the importance of the SGS nonlocal flux depends on the grid size of the model. Wyngaard (2004) theoretically analyzed the importance of nonlocal flux at different grid sizes. The magnitude of the SGS nonlocal flux decreases as the grid size decreases, which means the nonlocal flux is explicitly represented by the model dynamics. Using Eqs. (19) and (21), vertical profiles of the SGS nonlocal heat flux in an idealized CBL for different grid sizes are diagnosed from the LES data and are shown in Fig. 1.

Shin and Hong (2013, 2015) presented a vertical profile of nonlocal heat flux based on the top-hat representation of subgrid-scale variables and conditional sampling approach using the LES data. Shin and Hong (2015) replaced the traditional nonlocal term (Troen and Mahrt,

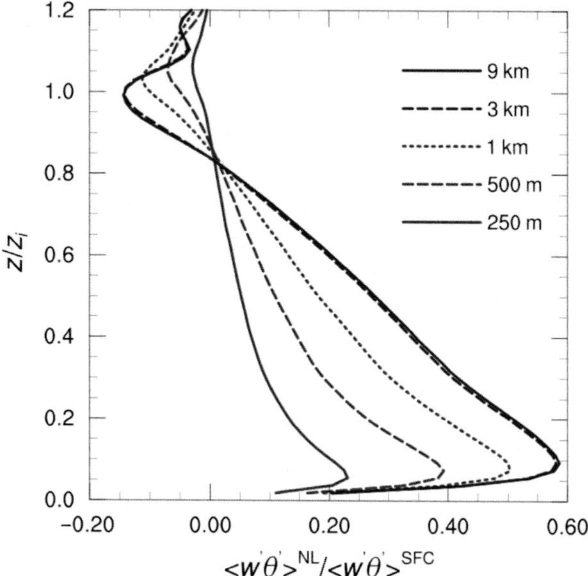

FIG. 1 Vertical profiles of horizontally averaged SGS nonlocal heat flux at different resolutions (9 km, 3 km, 1 km, 500 m, and 250 m) for CBL case, normalized by the surface heat flux. The vertical axis is height, z, normalized by the PBL depth, z_i. *From Zhang, X., Bao, J., Chen, B., Grell, E., 2018. A three-dimensional scale-adaptive turbulent kinetic energy scheme in the WRF-ARW model. Mon. Wea. Rev. 146 (7), 2023–2045.*

1986; Noh et al., 2003) in the YSU scheme (Hong et al., 2006) with the prescribed nonlocal heat flux profile that is diagnosed from the LES data, multiplying the corresponding grid-size dependency functions with the local and nonlocal profiles, to modify the YSU scheme to the gray zone. As the grid size increases to the mesoscale limit, the grid boxes are large enough to contain turbulence samples with statistical significance, thus the vertical profile of the SGS nonlocal heat flux converges to a single profile (compare profiles for 3 km and 9 km in Fig. 1). In the mesoscale limit, the effect of nonlocal flux is necessary to be accurately represented in the subgrid mixing parameterization. When the grid size approaches the LES limit, the models are able to resolve the nonlocal flux induced by energetic large eddies. In the gray zone resolutions, the nonlocal flux is partly resolved. The loss of statistical significance adds extra uncertainty and stochasticity to the subgrid mixing parameterizations at gray-zone resolutions.

In recent studies about the scale-dependent subgrid mixing parameterizations (Boutle et al., 2014; Shin and Hong, 2015; Ito et al., 2015; Zhang et al., 2018; Efstathiou and Plant, 2018), partition functions were pragmatically introduced in parameterizations to account for gradual variations of the SGS nonlocal flux across the gray zone (Fig. 2). At gray zone resolutions, model dynamics would better resolve the large eddies responsible for nonlocal heat flux. The SGS nonlocal flux term in most current scale-dependent parameterizations is downweighted by the partition function $P(\Delta/z_i)$ (Fig. 2):

$$\overline{w'\theta'}^{\Delta,NL} = \overline{w'\theta'}^{NL} P(\Delta/z_i), \tag{22}$$

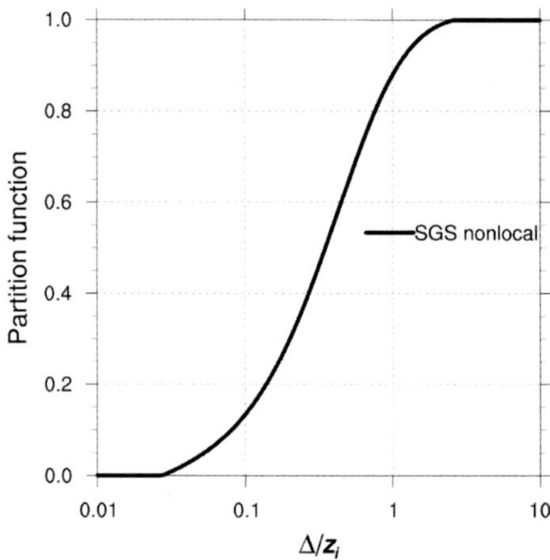

FIG. 2 Partition function $P(\Delta/z_i)$ for SGS nonlocal heat flux derived from LES data for the CBL case. *From Zhang, X., Bao, J., Chen, B., Grell, E., 2018. A three-dimensional scale-adaptive turbulent kinetic energy scheme in the WRF-ARW model. Mon. Wea. Rev. 146 (7), 2023–2045..*

where $\overline{w'\theta'}^{NL}$ is the converged SGS heat flux as the grid size approaches the mesoscale limit (profile for 9 km in Fig. 1), $\overline{w'\theta'}^{\Delta,NL}$ is the SGS heat flux for a particular gray zone grid size Δ. The introduction of partition function adds further the uncertainty to the subgrid mixing parameterizations.

Honnert et al. (2011) determined the partition function form by analyzing the resolved and subgrid parts of turbulent fluxes at various grid sizes using the LES data of five dry and cloudy CBLs. Although the partition function is defined to be 0 (1) at the LES limit (mesoscale limit), there are some degrees of uncertainties in the partition function $P(\Delta/z_i)$. Shin and Hong (2013) discussed the effects of stability on the partition functions. The partition functions show some discrepancies in terms of different variables and boundary layer types (Honnert et al., 2011).

Besides the nonlocal heat flux, the nonlocal effect of momentum flux in the subgrid mixing parameterization presents some uncertainty (Frech and Mahrt, 1995; Brown and Grant, 1997). The mixing of momentum is more complicated than that of heat since it is a vector and is affected by both baroclinity and pressure fluctuations. Frech and Mahrt (1995) proposed a formulation for the nonlocal mixing of momentum. Unfortunately, the magnitude of nonlocal momentum mixing is a function of baroclinity that cannot be adequately determined from existing data.

Previous work mostly focuses on improving the representation of nonlocal flux in the simulations of the PBL. But, relatively few studies have been devoted to the parameterization of nonlocal flux in deep moist convection. The nonlocal turbulent flux in deep convection has been represented by the cumulus convection parameterizations based on mass flux approaches in GCMs (Arakawa and Schubert, 1974). With the increasing application of kilometer-scale cloud models for NWP, the ensemble-mean closure concept used in the conventional climate models is no longer valid. Atmospheric moist convection has a continuous, wide range of scales with no spectral gap between resolvable and the SGS motions. Moeng et al. (2010) showed that the energy spectral peak of deep convection lies in the range of a typical cloud model grid cutoff. The parameterization of SGS nonlocal flux associated with deep convection in kilometer-scale cloud models still is not addressed. Recently, some studies tried to include the nonlocal turbulent flux associated with the deep convection in the subgrid turbulence mixing scheme (Moeng, 2014; Verrelle et al., 2017; Shi et al., 2019). Moeng (2014) proposed a turbulence closure scheme for nonlocal turbulent flux inside the convective cloud in terms of the horizontal gradients of resolved variables. The deficiencies and uncertainties of subgrid mixing parameterization inside convective clouds still exist (Guichard and Couvreux, 2017).

Note that the above representations of nonlocal flux including mass-flux approach and prescribed profile using conditional sampling are built on the top-hat PDF distribution, which is equivalent to two-delta-function PDF. Based on top-hat distribution, the motions are only partitioned into two components: updrafts and the environment surrounding the updrafts. In reality, however, the distinction between updraft and environment is not clear. The top-hat PDF cannot fully describe the statistics of the subgrid-scale motions and remove the subplume-scale variability. Recently, EDMF multi-plume schemes were developed to consider the subplume-scale variability (e.g., Suselj et al., 2019). The multiple plumes represent the skewed part of PDF. Golaz et al. (2002) proposed an assumed PDF scheme to account for the nonlocal transport in PBL and shallow convection, in which the two-delta-function PDF is replaced with a two-Gaussian-function PDF. Bogenschutz et al. (2010) evaluated several PDFs using the LES results from different cloud regimes for a wide range of horizontal grid sizes. They concluded that every PDF has its own strengths and weaknesses, and it does not appear that either one is resolutely better than the others. None of the PDF satisfactorily works across all the cloud regimes for all grid sizes.

3 Mixing length

The SGS turbulent mixing closures typically used in the mesoscale models and LES are of the same scaling expression with $K \sim le^{1/2}$ and the TKE dissipation rate with $\varepsilon \sim e^{3/2}/l$, but with different mixing length scale formulations. The mixing length expression is one of the main different aspects between the LES and mesoscale simulation closures. A major obstacle facing the development of eddy-diffusivity closures based on a balance equation for TKE is the specification of the mixing length. The mixing length is a measure of the ability of turbulence to cause mixing. The mixing length has to be specified, and the appropriate specification is inevitably dependent on the geometry of the flow. For the complex atmospheric flow, the specification of mixing length requires a large measure of guesswork (Pope, 2000).

In the surface layer, a linear relationship between the mixing length and height has been established for a long time (e.g., Von Karman, 1930; Prandtl, 1932). Blackadar (1962) suggested the following formulation for the momentum mixing length in a neutral atmosphere,

$$\frac{1}{l} = \frac{1}{kz} + \frac{1}{\lambda}, \tag{23}$$

where k is the Von Karman constant, and λ is an asymptotic value often taken as constant and an adjustable parameter. In NWP models, Blackadar's (1962) mixing length formulation is often still used (e.g., ECMWF, 2016) for wind, temperature, and specific humidity. The variant forms of Blackadar's mixing length and more complex formulations have been designed for boundary-layer and mesoscale models (e.g., Yu, 1977; Bodin, 1979; Therry and Lacarrère, 1983; Nakanishi and Niino, 2009; Bogenschuz and Krueger, 2013).

For instance, the mixing length scale in the boundary layer should be highly correlated with the distance for the ground, structure, and strength of the turbulence, and local thermal stability. Considering these effects, Nakanishi and Niino (2009) proposed a length scale formulation in Mellor-Yamada Level-3 (MYNN) PBL scheme. The mixing length scale l is determined by a harmonic average of three length scales as

$$\frac{1}{l}=\frac{1}{L_S}+\frac{1}{L_T}+\frac{1}{L_B} \tag{24}$$

where L_S is the mixing length in the surface layer controlled by the effects of wall and stability, L_T the mixing length depending on the turbulent structure of the PBL (Mellor and Yamada, 1982), and L_B the mixing length limited by the thermal stability. The L_S, L_T, and L_B are given by

$$L_S=\alpha_1 kz \tag{25}$$

$$L_T=\alpha_2\frac{\int_0^\infty e^{1/2}z dz}{\int_0^\infty e^{1/2}dz} \tag{26}$$

$$L_B=\begin{cases}\left[\alpha_3 e^{1/2}+\alpha_4 e^{1/2}(q_c/L_T N)^{1/2}\right]/N, & \partial\bar{\theta}/\partial z>0\\ \infty, & \partial\bar{\theta}/\partial z\le 0\end{cases} \tag{27}$$

where α_1, α_2, α_3, and α_4 are empirical constants, k the von Karman constant, N is the Brunt-Väisälä frequency, and $q_c \equiv [(g/\theta_0)Q_0 L_T]^{1/3}$ a velocity scale similar to the convective velocity w_*.

There exist many other definitions of mixing length in the PBL parameterizations in mesoscale NWP models. Teixeira and Cheinet (2004) suggested a simple mixing length formulation that relates to the square root of TKE and a time scale. Thus the closure problem is shifted from determining a length scale to the determining of a time scale. Bougeault and Lacarrère (1989) suggested that the length scale for turbulent eddies is primarily determined by the resistance to vertical displacements due to static stability. They proposed a parameterization of mixing length that is determined as a function of the stability profile of the adjacent levels. The algorithms rely on the computation of the maximum vertical displacement allowed, for a parcel of air having the mean kinetic energy of the level as initial TKE. The maximum vertical upward (l_{up}) and downward (l_{down}) displacements are given by,

$$\int_z^{z+l_{up}}\frac{g}{\theta_0}(\theta(z')-\theta(z))dz'=e(z) \tag{28}$$

$$\int_{z-l_{down}}^{z}\frac{g}{\theta_0}(\theta(z)-\theta(z'))dz'=e(z)$$

respectively. Then the mixing length is defined by

$$l = \min\left(l_{up}, l_{down}\right) \text{ or } l = \left(l_{up} l_{down}\right)^{1/2} \tag{29}$$

The main advantage of the method is to consider the effect of remote stability.

Note that although in some situations the above formulations lead to more realistic results, there is no such thing as a mixing length formulation that is robust, flexible, and simple enough to allow for a realistic simulation of the variety of flows that occur in the atmosphere. Several formulations give results considerably different from each other, which may result from a variety of subgrid mixing closures and the lack of data.

In the LES limit where the grid size lies within the inertial subrange, it is assumed that the most energetic parameterized eddies are just a little smaller than the grid size, hence the length scale is usually set to the vertical grid size. Deardorff's mixing length is commonly applied, which is given as the minimum of the scale corrected by stability and the grid size

$$l = \begin{cases} \min\left[0.76 e^{1/2} \left|\dfrac{g}{\theta}\dfrac{\partial \theta}{\partial z}\right|^{-1/2}, \Delta s\right] & \text{for } N^2 > 0 \\ \Delta s & \text{for } N^2 \le 0 \end{cases}, \tag{30}$$

where $\Delta s = (\Delta x \Delta y \Delta z)^{1/3}$.

The above mixing length formulations are designed for the mesoscale and LES limits. Since the specification of mixing length depends on the geometry of the flow, they may be inappropriate for the flows at gray-zone resolutions. In practical applications, the blending approaches are usually proposed to determine the mixing length at the gray-zone resolutions. Many studies blended the mixing length scale from the LES to the mesoscale limit using the partition function $P(\Delta/z_i)$ (Boutle et al., 2014; Zhang et al., 2018; Efstathiou and Plant, 2018).

The above-mentioned formulations of mixing length are mainly used for the simulations of the dry CBL. The mixing length inside deep convective clouds has been explored very little in comparison. Machado and Chaboureau (2015) showed that the simulations of cloud organizations with a 3D turbulence scheme were highly sensitive to in-cloud mixing length parameterization. The adjustment of the mixing length inside the cloud can result in a better description of cloud organizations. Hanley et al. (2015) investigated the sensitivity of storm morphology to the mixing length used in the subgrid turbulence scheme. They found an increase in the number of small storms with a decrease in subgrid mixing length. But, the specification of mixing length may physically depend on the different cloud regimes, the subgrid mixing parameterization formulations, and the model's horizontal resolution.

The mixing length is one of the main sources of uncertainty in the subgrid mixing parameterizations, particularly at the gray-zone resolutions. A scale-dependent and flow-adaptive mixing length are desired for the simulations of various and complex atmospheric motions. The performance of the subgrid mixing parameterization is very sensitive to the mixing length formulations. In designing the subgrid mixing scheme, one can choose the appropriate formulation of mixing length according to the specific closure methods and practical performance.

4 Subgrid horizontal mixing parameterizations

Compared to subgrid vertical mixing, horizontal turbulent mixing receives much less attention. Based on the horizontal homogeneity assumption, most PBL parameterizations deal with the effects of mixing in the vertical direction only. However, as the resolution of NWP models steadily increases, when the grid size falls within the gray zone where the local circulation is partly resolved, or due to thermal contrast (such as sea-breeze circulation) and topographies, the role of the horizontal mixing becomes more significant. Based on CBL cases, Honnert and Masson (2014) suggested a critical horizontal resolution at which 1D PBL schemes of NWP models must be replaced by 3D turbulence schemes. Muñoz-Esparza et al. (2016) discussed the limitations of 1D PBL schemes in reproducing mountain-wave flow where the assumption of horizontal homogeneity is violated and suggested the importance of horizontal mixing. For the deep convective systems, Verrelle et al. (2014) pointed out that the difference between 1D and 3D turbulence scheme becomes perceptible at the resolution of 2 km, and suggested that a 3D turbulence scheme is necessary for a good representation of deep convection at kilometric resolutions.

In mesoscale simulations, anisotropic turbulence implies that the contributions from horizontal turbulent fluxes become negligible when compared to the dominating vertical transports. The boundary-layer approximation that imposes horizontal homogeneity of variables and fluxes is made, which results in the neglect of all horizontal turbulent fluxes. Therefore, it is common practice in mesoscale simulations to treat the vertical and horizontal mixing separately. The conventional 1D PBL schemes are usually used to represent vertical mixing, while horizontal mixing is usually handled by the deformation-based eddy diffusivity formulation (Smagorinsky closure). Horizontal mixing largely depends on numerical reasons such as computational stability and small-scale noise (Deardorff, 1985; Jablonowski and Williamson, 2011). So, physical or energetic inconsistencies may exist between vertical and horizontal mixing.

Current parameterizations of horizontal mixing in NWP models are often based on the eddy-diffusivity formulation, where the horizontal flux is represented by the product of a horizontal diffusivity K_h and the horizontal gradient of the respective variable [Eqs. (10) and (11)]. Such simple closure of the horizontal mixing may be inappropriate at all scales because the closure implies that the SGS mixing process is always diffusive. Wyngaard (2004) pointed out that neglecting the production by the tilting caused by the mean velocity gradient in Eq. (4) may result in the underestimation of horizontal SGS flux, and that neglecting the tilting term is not inappropriate in both LES and mesoscale limits. Including the tilting term in the horizontal mixing parameterization can allow backscatter, that is, the transfer of energy from smaller to larger scales.

Current NWP models often use the 2D Smagorinsky first-order closure to parameterize K_h, which is defined as

$$K_h = c_s^2 l_h^2 \left[0.25(D_{11} - D_{22})^2 + D_{12}^2\right]^{1/2}, \tag{31}$$

where c_s is a dimensionless constant. The horizontal mixing length is defined as $l_h = (\Delta x \Delta y)^{1/2}$. And $D_{ij} = \partial \overline{U}_i / \partial x_j + \partial \overline{U}_j / \partial x_i$ is the deformation tensor of the resolved flow. The 2D Smagorinsky closure was initially used in the GCMs. Byun and Schere (2006) showed that the Smagorinsky closure may not be rational to use in a mesoscale model to predict pollutant diffusion. Zhou et al. (2017) also showed that the limitation of the Smagorinsky closure to simulate the CBL eddies.

Another parameterization of K_h is based on the 1.5-order TKE closure, which is commonly used in LES closure, the K_h is defined as

$$K_h = c_k l_h e^{1/2}, \tag{32}$$

where c_k is a dimensionless constant. Since K_h increases with the horizontal grid size, it is likely to overestimate the magnitude of horizontal diffusivity in mesoscale simulations (Zhou et al., 2017). Takemi and Rotunno (2003) examined the effects of 1.5-order TKE closure on the idealized simulation of a squall line at 1-km grid size. They pointed out that applying the 1.5-order TKE closure to convection-resolving simulations is inappropriate because the 1-km grid size is probably well beyond the inertial subrange.

The horizontal mixing has a strong effect on turbulence statistics. An example of an idealized CBL is given to show the role of horizontal mixing and the sensitivity of turbulence statistics to horizontal mixing length based on the 1.5-order TKE closure. First, a benchmark LES was performed over a 30 km × 30 km domain with a horizontal grid size of 50 m. The LES was driven by surface heat flux ($Q_0 = 0.24\,\mathrm{K\,m\,s^{-1}}$ or about $273\,\mathrm{W\,m^{-2}}$) and geostrophic wind in the x direction ($U_g = 10\,\mathrm{m\,s^{-1}}$). The Deardorff's model (Deardorff, 1980) is selected for SGS turbulence parameterization. Time integration is performed with a timestep of 0.2 s for a 4 h period. The initial sounding of the potential temperature is

$$\theta = \begin{cases} 300\mathrm{K} & : 0 < z \le 925\mathrm{m} \\ 300\mathrm{K} + (z - 925\mathrm{m}) \times 0.0536\mathrm{K\,m^{-1}} & : 925\mathrm{m} < z \le 1075\mathrm{m} \\ 308.05\mathrm{K} + (z - 1075\mathrm{m}) \times 0.003\mathrm{K\,m^{-1}} & : z > 1075\mathrm{m} \end{cases} \tag{33}$$

The LES output was used to investigate the performance of the horizontal mixing based on 1.5-order TKE closure. The coarse-graining approach can be used as a tool to examine model behavior with varying grid sizes (Honnert et al., 2011; Shin and Hong, 2015). In the coarse-graining method, successive horizontal spatial means are applied to the original LES fields to derive lower resolution fields, which are considered to be the "truth" for the lower resolution.

The 3DTKE scheme (Zhang et al., 2018) was used in the sensitivity experiments. In the scheme, the horizontal eddy diffusivity is set by $K_h = c_k l_h e^{1/2}$, where $l_h = (\Delta x \Delta y)^{1/2}$, $c_k = 0.1$. A sensitivity test was carried out by increasing and decreasing the horizontal mixing length by a scale factor. Three scale factors were considered: one 10 times smaller ($0.1 l_h$), one 2 times smaller ($0.5 l_h$), and the other two times larger ($2 l_h$). We found that the horizontal mixing length has a significant influence on the resolved vertical velocity variance ($<w'^2>$) and TKE. Fig. 3 shows the sensitivities of $<w'^2>$ to the horizontal mixing length at the resolutions of 250 m, 500 m, and 1 km. As the horizontal mixing length decreases, the profiles of resolved $<w'^2>$ get close to the coarse-grained profiles. While l_h multiplied by a constant 0.1, the resolved $<w'^2>$ profile is closest to the coarse-grained profile. At the relatively coarse resolution (1 km), the 1.5-order TKE closure obviously overestimates the horizontal mixing because l_h increases with the horizontal grid size. Therefore, it is deduced that the horizontal mixing length in 1.5-order TKE formulation should be

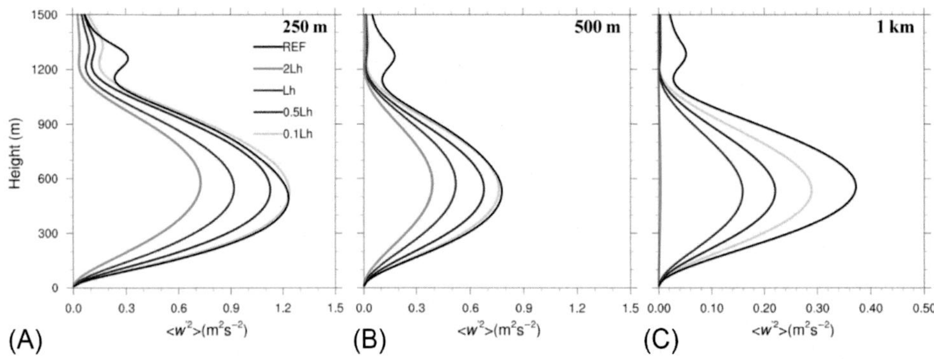

FIG. 3 Vertical profiles of resolved vertical velocity variance ($<w'^2>$) for the idealized CBL simulations using the 3DTKE scheme at resolutions of (A) 250m, (B) 500m, and (C) 1km, and with different horizontal mixing lengths (*green*: $2l_h$; *red*: l_h; *blue*: $0.5l_h$; *yellow*: $0.1l_h$). And $l_h = \sqrt{\Delta x \Delta y}$. The *black lines* show the corresponding coarse-grained fields from the LES reference (REF) data.

saturated as the grid size approaches mesoscale and no longer varies for coarser grid sizes, as suggested by Cuxart et al. (2000) and Wyngaard (2004).

To illustrate the importance of horizontal mixing in resolved TKE, Fig. 4 presents a plot of the resolved TKE at $z/z_i = 0.5$ simulated by the 3DTKE scheme for different horizontal mixing lengths as a function of Δ/z_i. The resolved TKE with the horizontal mixing length $0.1l_h$ is

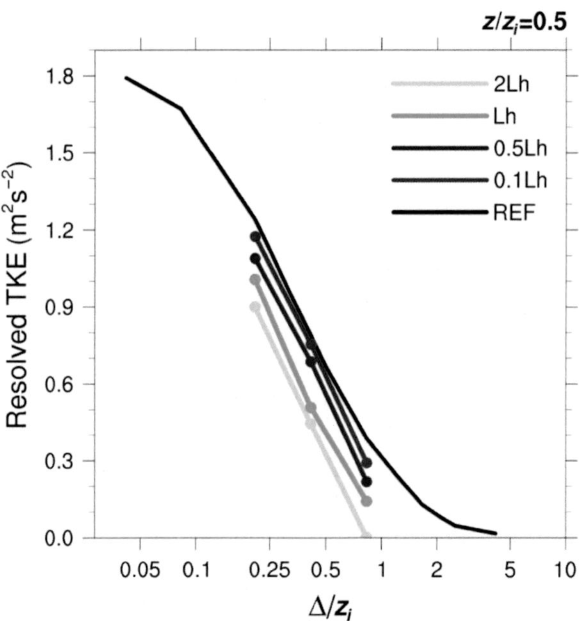

FIG. 4 Resolved TKE at $0.5z_i$ simulated by the 3DTKE scheme for different horizontal mixing lengths at resolutions of 250m, 500m, and 1km. The *black line* shows coarse-grained fields from the LES reference (REF) data as a function of Δ/z_i.

closest to the coarse-grained data, which implies that the 1.5-order TKE closure with l_h produces too much horizontal mixing. The sensitivity test using these adjustments shows the strong effect of horizontal mixing on the turbulence statistics. The simulations with a 3D turbulence scheme are highly sensitive to the mixing length parameterization.

Since the 2D Smagorinsky closure was originally designed for the GCM, it accounts for the horizontal mixing by mesoscale or large-scale circulations only. The 1.5-order TKE closure is commonly used in LES. The applications of 2D Smagorinsky closure and 1.5-order TKE closure may not be justified when the grid size is in the gray zone (Ito et al., 2014; Bryan et al., 2003). In the gray zone, there are no proper horizontal mixing parameterizations in numerical models yet. Zhang et al. (2018) proposed a pragmatic and heuristic blending approach for the transition of horizontal mixing across the gray zone:

$$K_h = K_D + K_T = P(\Delta x/z_i)(c_s l_h)^2 D/\sqrt{2} + [1 - P(\Delta x/z_i)]c_k l_h e^{1/2}, \tag{34}$$

where K_D is the diffusivity based on the deformation (i.e., Smagorinsky-type formula), K_T is the diffusivity based on the TKE. And $D=(D_{ij}D_{ij})^{1/2}$. Using the partition function $P(\Delta/z_i)$, this approach blends the Smagorinsky-Lilly closure based on the deformation for mesoscale simulation, with a TKE-based subgrid mixing scheme for LES. When at the mesoscale limit $P(\Delta/z_i)=1$ (Fig. 2), the horizontal diffusivity based on the Smagorinsky—Lilly formula is obtained. At the LES limit $P(\Delta/z_i)=0$, the horizontal eddy diffusivity is calculated by TKE, which retains the formulation in the original LES subgrid model (consistent with the vertical eddy diffusivity). The analogous approaches which include the TKE term in the deformation formula have been proposed (Lilly, 1962; Xu, 1988; Janjic, 1990). Janjic (1990) proposed a modified Smagorinsky-type parameterization for the horizontal mixing including the presence of TKE term. The TKE term is included to account for the effects of horizontal mixing due to physical processes (e.g., entrainment and detrainment).

The implementation of the horizontal mixing in NWP models is somewhat controversial. Some horizontal mixing processes are used for purely numerical reasons to prevent the model from becoming unstable (Xue, 2000; Langhans et al., 2012). Others are meant to mimic SGS turbulence processes. In practice, it is difficult to evaluate their individual effects. It is not always clear to what extent the horizontal mixing is needed to alleviate numerical problems. How to quantify the horizontal mixing resulted from the physical process, and to distinguish it from numerical diffusion pose challenges for the NWP community. It is difficult to quantify the horizontal mixing due to the physical processes in real weather forecasting. The parameterization of horizontal mixing is still one of the most uncertain aspects in NWP models.

The above results show that horizontal mixing has a significant effect on the turbulence statistics in dry CBL. Recently, more and more studies have addressed the importance of horizontal mixing in more realistic cases, and have attempted to evaluate and quantify the horizontal mixing based more on a physical basis. Verrelle et al. (2014) and Machado and Chaboureau (2015) suggested that horizontal mixing was important for the cloud size, lifetime, and organization. Bryan and Rotunno (2009) addressed that the maximum intensity of the simulated tropical cyclones was very sensitive to the horizontal mixing length in an axisymmetric numerical model. Rotunno and Bryan (2012) took a closer look at the effects of horizontal mixing on the nearly steady axisymmetric numerical simulations of hurricanes and concluded that horizontal mixing was the most important control factor on the maximum simulated hurricane intensity. Using a numerical model with higher vertical resolution and a more realistic microphysics scheme, Bryan (2012) confirmed the above findings. Honnert (2016) calculated the horizontal mixing length from LES of neutral and convective cases in the gray-zone resolutions. But, there is no theory for how to quantify the intensity of horizontal mixing, and few observations are available to constrain the adjustable settings (e.g., horizontal mixing length). Hanley et al. (2015) investigated the sensitivities of the cloud morphology and statistical properties to the horizontal mixing and evaluated it against the Met Office rainfall radar network. Evaluating and developing the horizontal mixing parameterization using the observations may be the future direction. The horizontal mixing is strongly related to the vertical, especially for the gray zone, one must consider both in a more coherent way. Consistency of vertical and horizontal mixing needs to be studied more in the future.

5 Discussions

Parameterizations of subgrid turbulence mixing contain closure assumptions and related parameters with inherent uncertainty. The uncertainty comes from not only uncertain closure coefficients but also the assumptions inherent in the mathematical formulation of the subgrid turbulence mixing parameterizations. Furthermore, the scenario in which a subgrid mixing parameterization is applied can also introduce uncertainties. It is important to note that, while all subgrid mixing parameterizations are calibrated to fit a set of observational or LES data, this does prove their predictive capability.

The chapter gives a review of uncertainty in subgrid turbulence mixing parameterizations with an emphasis on the nonlocal flux, mixing length, and horizontal mixing. The differences in various subgrid mixing parameterizations are reflected mainly in the above three aspects. In some higher-order parameterizations, the nonlocal flux and mixing length can be determined by prognostic equations for higher-order moments. But, there is no guarantee that the higher the order of turbulence closure, the better the results of subgrid mixing parameterization, because the higher-order closure may bring more uncertainties.

In addition to the above three aspects, there are many uncertainties in subgrid mixing parameterizations, such as Prandtl number, PBL height, and entrainment. The turbulence Prandtl number under unstable conditions is found to be strongly dependent on stability in the surface layer (Businger et al., 1971). In the mixed layer above the surface layer, the Prandtl number is not very well-defined. Troen and Mahrt (1986) defined the Prandtl number value at the top of the surface layer, and the value is assumed to remain invariant over the whole boundary layer. Noh et al. (2003) suggested that Prandtl number varies within the PBL according to the local stability, and has a presumed profile. There is no universal formulation of the Prandtl number. In the K-profile closures of PBL parameterization (e.g., Troen and Mahrt, 1986; Hong et al., 2006), the performances of parameterization rely heavily on the determination of PBL height that is often difficult to define precisely and also hard to obtain accurately. Different algorithms for the calculation of the PBL height in models often lead to significantly different results. Moreover, there is no single, universally accepted definition of the entrainment zone (Brooks and Fowler, 2012). Entrainment is poorly represented in numerical models and its parameterization is highly questionable (Fedorovich et al., 2004). A good representation of entrainment is also essential in the description of cloud processes and organization (Lebo and Morrison, 2015; Tompkins and Semie, 2017). Few studies have been devoted to the effect of turbulence mixing on cloud organization (Machado and Chaboureau, 2015).

The most common way to evaluate and improve subgrid mixing parameterization is to use LES. However, the statistical convergence of deep moist convection simulations with decreased grid spacing has not been well characterized. Bryan et al. (2003) found a lack of convergence of statistical properties of the deep convection with horizontal grid size decreased to 125 m, and suggested that grid spacing with approximately 100 times smaller than the energetic large-eddy scale may be necessary for statistical convergence of deep convection characteristics. Khairoutdinov et al. (2009) found a convergence of statistical properties for a tropical convective system between the LES benchmark and the 200 m grid spacing run. Based on different definitions of statistical convergence, some other studies showed that convergence can be achieved as the grid spacing is reduced to 200–500 m (Fiori et al., 2010, 2011; Langhans et al., 2012; Ricard et al., 2013; Verrelle et al., 2014).

From a statistical point of view, subgrid mixing parameterization relies on the statistical significance assumption, which results from the "law of large numbers." If the number of turbulent eddies within the model grid box is large, then the average outcome of subgrid-scale processes will be uniquely determined by the known large-scale quantities. As the resolution of the NWP model increases, the decrease of the number of eddies in the grid box leads to the loss of statistical significance of the ensemble average, which adds an extra level of uncertainty in the subgrid mixing parameterizations.

Because each grid box is not large enough to contain a robust number of turbulent eddies in the gray zone, the "law of large numbers" underlying deterministic parameterizations is no longer valid. Hence a stochastic approach becomes essential. Developing stochastic physics parameterizations presents a future direction for high-resolution NWP. The stochastic approach would contain the statistical information about the uncertainty in the subgrid-scale processes for given grid size, and provide a way to represent and quantify the uncertainty of subgrid mixing parameterization. Plant and Craig (2008) proposed a stochastic parameterization for deep convection to represent the uncertainties that arise naturally from the limited random sampling of the plume ensemble in each grid box. Sakradzija et al. (2016) developed a stochastic parameterization of shallow convection based on the EDMF framework. The parameterization provides the distribution of all possible subgrid states for a given grid-scale flow, instead of producing an ensemble average response, and thus provides a way to quantify the uncertainty of subgrid convection in NWP models. Also, the notion of stochastic parameterization can be applied in the subgrid turbulent mixing parameterization.

In fact, due to the aforementioned modeling uncertainties, the validity of a single deterministic prediction is simply unknown. To circumvent this problem, a predictive simulation with quantified uncertainty and continuous description of the space of possible solutions should be made. In addition to the initial condition error, NWP models are sensitive to errors associated with the subgrid physical parameterizations as well. The uncertainty due to the parameterizations of SGS physical processes is known to play a crucial role in the predictability of a system (Palmer, 2001). Some research has been done to develop formulations to insert SGS physical parameterization uncertainty to ensemble prediction systems. Buizza et al. (1999) imposed a stochastic term to the tendencies from physical parameterizations which increases the spread of the ensemble and improves the skill of the probabilistic prediction of precipitation in the European Center for Medium-Range Weather Forecasts Ensemble Prediction System. Teixeira and Reynolds (2008) implemented a stochastic convection scheme in the Navy Operational Global Atmospheric Prediction System ensemble and presented a general methodology to incorporate a stochastic component into a subgrid mixing parameterization. The development of stochastic parameterizations in the context of ensemble prediction could improve the ensemble spread and provide probabilistic forecasting of the variable of interest.

Acknowledgments

This work is supported by the National Key Research and Development Program of China (2017YFC150190X) and the National Science Foundation of China (Grants 41975133).

References

Arakawa, A., Schubert, W.H., 1974. Interaction of a cumulus cloud ensemble with the large-scale environment, Part I. J. Atmos. Sci. 31, 674–701.

Blackadar, A., 1962. The vertical distribution of wind and turbulent exchange in a neutral atmosphere. J. Geophys. Res. 67, 3095–3102.

Bodin, S., 1979. A Predictive Numerical Model of the Atmospheric Boundary Layer Based on the Turbulent Energy Equation. SMHI Report No. 13, Norrkoping, Sweden.

Bogenschutz, P.A., Krueger, S.K., Khairoutdinov, M., 2010. Assumed probability density functions for shallow and deep convection. J. Adv. Model. Earth Syst.. 2 (10) 24 pp.

Bogenschuz, P.A., Krueger, S.K., 2013. A simplified PDF parameterization of subgrid-scale clouds and turbulence for cloud-resolving models. J. Adv. Model. Earth Syst. 5, 195–211.

Bougeault, P., Lacarrère, P., 1989. Parameterization of orography-induced turbulence in a mesobeta-scale model. Mon. Wea. Rev. 117, 1872–1890.

Boutle, I.A., Eyre, J.E.J., Lock, A.P., 2014. Seamless stratocumulus simulation across the turbulent gray zone. Mon. Wea. Rev. 142, 1655–1668.

Brooks, I.M., Fowler, A.M., 2012. An evaluation of boundary layer depth, inversion and entrainment parameters by large eddy simulation. Bound.-Layer Meteor. 142, 245–263.

Brown, A.R., Grant, A.L.M., 1997. Non-local mixing of momentum in the convective boundary layer. Bound.-Layer Meteor. 84, 1–22.

Bryan, G.H., 2012. Effects of surface exchange coefficients and turbulence length scales on the intensity and structure of numerically simulated hurricanes. Mon. Wea. Rev. 140, 1125–1143.

Bryan, G.H., Rotunno, R., 2009. The maximum intensity of tropical cyclones in axisymmetric numerical model simulations. Mon. Wea. Rev. 137, 1770–1789.

Bryan, G.H., Wyngaard, J.C., Fritsch, J.M., 2003. Resolution requirements for the simulation of deep moist convection. Mon. Wea. Rev. 131, 2394–2416.

Buizza, R., Miller, M., Palmer, T.N., 1999. Stochastic representation of model uncertainties in the ECMWF ensemble prediction system. Quart. J. Roy. Meteor. Soc. 125, 2887–2908.

Businger, J.A., Wyngaard, J.C., Izumi, Y., Bradley, E.F., 1971. Flux-profile relationships in the atmospheric surface layer. J. Atmos. Sci. 28, 181–189.

Byun, D., Schere, K.L., 2006. Review of the governing equations, computational algorithms, and other components of the models-3 community multiscale air quality (CMAQ) modeling system. Appl. Mech. Rev. 59, 51–77. https://doi.org/10.1115/1.2128636.

Chinita, M.J., Matheou, G., Teixeira, J., 2018. A joint probability density-based decomposition of turbulence in the atmospheric boundary layer. Mon. Wea. Rev. 146, 503–523.

Couvreux, F., Hourdin, F., Rio, C., 2010. Resolved versus parameterized boundary-layer plumes. Part I: A parameterization oriented conditional sampling in large-eddy simulations. Bound.-Layer Meteor. 134, 441–458.

Cuxart, J., Bougeault, P., Redelsperger, J.-L., 2000. A turbulence scheme allowing for mesoscale and large-eddy simulations. Quart. J. Roy. Meteor. Soc. 126, 1–30.

Deardorff, J.W., 1972. Theoretical expression for the counter-gradient vertical heat flux. J. Geophys. Res. 77, 5900–5904.

Deardorff, J.W., 1980. Stratocumulus-capped mixed layers derived from a three-dimensional model. Bound.-Layer Meteor. 18, 495–527.

Deardorff, J.W., 1985. Sub-grid-scale turbulence modeling. Adv. Geophys. 28, 337–343.

ECMWF, cited, 2016. IFS documentation. Part IV: Physical processes. ECMWF Meteorological Bulletin, Reading, United Kingdom.

Efstathiou, G.A., Plant, R.S., 2018. A dynamic extension of the pragmatic blending scheme for scale-dependent sub-grid mixing Quart J. Roy. Meteor. Soc. 145 (719), 884–892

Fedorovich, E., Conzemius, R., Mironov, D., 2004. Convective entrainment into a shear-free, linearly stratified atmosphere: bulk models reevaluated through large eddy simulations. J. Atmos. Sci. 61, 281–295.

Fiori, E., Parodi, A., Siccardi, F., 2010. Turbulence closure parameterization and grid spacing effects in simulated supercells. J. Atmos. Sci. 67, 3870–3890.

Fiori, E., Parodi, A., Siccardi, F., 2011. Uncertainty in prediction of deep moist convection processes: turbulence parameterizations, microphysics and grid-scale effects. Atmos. Res. 100, 447–456.

Frech, M., Mahrt, L., 1995. A two-scale mixing formulation for the atmospheric boundary layer. Bound.-Layer Meteor. 73, 91–104.

Golaz, J.-C., Larson, V.E., Cotton, W.R., 2002. A PDF-based model for boundary layer clouds. Part I: Method and model description. J. Atmos. Sci. 59, 3540–3551.

Greenhut, G.K., Khalsa, S.J.S., 1982. Updraft and downdraft events in the atmospheric boundary layer over the equatorial Pacific Ocean. J. Atmos. Sci. 39, 1803–1818.

Greenhut, G.K., Khalsa, S.J.S., 1987. Convective elements in the marine atmospheric boundary layer. Part I: Conditional sample statistics. J. Climate Appl. Meteor. 26, 813–822.

Guichard, F., Couvreux, F., 2017. A short review of numerical cloud-resolving models. Tellus A.. 69.

Hanley, K.E., Plant, R.S., Stein, T.H.M., Hogan, R.J., Nicol, J.C., Lean, H.W., Halliwell, C., Clark, P.A., 2015. Mixing-length controls on high-resolution simulations of convective storms. Quart. J. Roy. Meteor. Soc. 141, 272–284.

Hellsten, A., Zilitinkevich, S., 2013. Role of convective structures and background turbulence in the dry convective boundary layer. Bound.-Layer Meteor. 149, 323–353.

Holtslag, A.A.M., Moeng, C.-H., 1991. Eddy diffusivity and counter gradient transport in the convective atmospheric boundary layer. J. Atmos. Sci. 48, 1690–1698.

Hong, S.-Y., Noh, Y., Dudhia, J., 2006. A new vertical diffusion package with an explicit treatment of entrainment processes. Mon. Wea. Rev. 134, 2318–2341.

Honnert, R., 2016. Representation of the grey zone of turbulence in the atmospheric boundary layer. Adv. Sci. Res. 13, 63–67.

Honnert, R., Masson, V., 2014. What is the smallest physically acceptable scale for 1D turbulence schemes? Front. Earth Sci. 2, 27. https://doi.org/10.3389/feart.2014.00027.

Honnert, R., Masson, V., Couvreux, F., 2011. A diagnostic for evaluating the representation of turbulence in atmospheric models at the kilometric scale. J. Atmos. Sci. 68, 3112–3131.

Ito, J., Niino, H., Nakanishi, M., 2014. Horizontal turbulent diffusion in a convective mixed layer. J. Fluid Mech. 758, 553–564. https://doi.org/10.1017/jfm.2014.545.

Ito, J., Niino, H., Nakanishi, M., Moeng, C.-H., 2015. An extension of Mellor-Yamada model to the terra incognita zone for dry convective mixed layers in the free convection regime. Bound.-Layer Meteor. 157, 23–43.

Jablonowski, C., Williamson, D., 2011. The pros and cons of diffusion, filters and fixers in atmospheric general circulation models. In: Lauritzen, P.H. et al., (Ed.), Numerical Techniques for Global Atmospheric Models. In: Lecture Notes in Computational Science and Engineering, vol. 80. Springer, pp. 381–493.

Janjic, Z.I., 1990. The step-mountain coordinate: physical package. Mon. Wea. Rev. 118, 1429–1443.

Khairoutdinov, M.F., Krueger, S.K., Moeng, C.-H., Bogenschutz, P.A., Randall, D.A., 2009. Large-eddy simulation of maritime deep tropical convection. J. Adv. Model. Earth Syst.. 1(15)https://doi.org/10.3894/JAMES.2009.1.15.

Langhans, W., Schmidli, J., Schär, C., 2012. Mesoscale impacts of explicit numerical diffusion in a convection-permitting model. Mon. Wea. Rev. 140, 226–244.

Lebo, Z.J., Morrison, H., 2015. Effects of horizontal and vertical grid spacing on mixing in simulated squall lines and implications for convective strength and structure. Mon. Wea. Rev. 143, 4355–4375.

Lenschow, D.H., Stephens, P.L., 1980. The role of thermals in the convective boundary layer. Bound.-Layer Meteor. 19, 509–532.

Lilly, D.K., 1962. On the numerical simulation of buoyant convection. Tellus 14, 148–172.

Lilly, D.K., 1967. The representation of small-scale turbulence in numerical simulation experiments. In: Proc. IBM Scientific Computing Symp. on Environmental Sciences, Yorktown Heights, NY, Thomas J, Watson Research Center, IBMpp. 195–210.

Machado, L.A., Chaboureau, J.-P., 2015. Effect of turbulence parameterization on assessment of cloud organization. Mon. Wea. Rev. 143, 3246–3262.

Mellor, G.L., 1973. Analytic prediction of the properties of stratified planetary surface layers. J. Atmos. Sci. 30, 1061–1069.

Mellor, G.L., Yamada, T., 1974. A hierarchy of turbulence closure models for planetary boundary layers. J. Atmos. Sci. 31, 1791–1806.

Mellor, G.L., Yamada, T., 1982. Development of a turbulence closure model for geophysical fluid problems. Rev. Geophys. Space Phys. 20, 851–875.

Moeng, C.-H., 2014. A closure for updraft-downdraft representation of subgrid-scale fluxes in cloud-resolving models. Mon. Wea. Rev. 142, 703–715.

Moeng, C.-H., Sullivan, P.P., Khairoutdinov, M.F., Randall, D.A., 2010. A mixed scheme for subgrid-scale fluxes in cloud-resolving models. J. Atmos. Sci. 67, 3692–3705.

Muñoz-Esparza, D., Sauer, J., Linn, R., Kosovic, B., 2016. Limitations of one-dimensional mesoscale PBL parameterizations in reproducing mountain-wave flows. J. Atmos. Sci.. https://doi.org/10.1175/JAS-D-15-0304.1.

Nakanishi, M., Niino, H., 2009. Development of an improved turbulence closure model for the atmospheric boundary layer. J. Meteor. Soc. Jpn. 87, 895–912.

Noh, Y., Cheon, W.-G., Hong, S.-Y., Raasch, S., 2003. Improvement of the K-profile model for the planetary boundary layer based on large eddy simulation data. Bound.-Layer Meteor. 107, 401–427.

Palmer, T.N., 2001. A nonlinear dynamical perspective on model error: a proposal for non-local stochastic-dynamic parameterization in weather and climate prediction models. Quart. J. Roy. Meteor. Soc. 127, 279–304.

Plant, R., Craig, G., 2008. A stochastic parameterization for deep convection based on equilibrium statistics. J. Atmos. Sci. 65, 87–105.

Pope, S.B., 2000. Turbulent Flows. Cambridge University Press 771 pp.

Prandtl, L., 1932. Zur turbulenten Stromung in Rohren und langs platen. Ergbn. Aerodyn. Versuchsanstalt Gottingen B4, 18–29.

Ricard, D., Lac, C., Riette, S., Legrand, R., Mary, A., 2013. Kinetic energy spectra characteristics of two convection-permitting limited-area models AROME and Meso-NH. Quart. J. Roy. Meteor. Soc. 139, 1327–1341.

Rotunno, R., Bryan, G.H., 2012. Effects of parameterized diffusion on simulated hurricanes. J. Atmos. Sci. 69, 2284–2299.

Sakradzija, M., Seifert, A., Dipankar, A., 2016. A stochastic scale-aware parameterization of shallow cumulus convection across the convective gray zone. J. Adv. Model. Earth Syst. 8, 786–812.

Schmidt, H., Schumann, U., 1989. Coherent structure of the convective planetary boundary layer. J. Fluid Mech. 200, 511–562.

Shi, X., Chow, F.K., Street, R.L., Bryan, G.H., 2019. Key elements of turbulence closures for simulating deep convection at kilometer-scale resolution. J. Adv. Model. Earth Syst. 11, 818–838.

Shin, H.H., Hong, S.-Y., 2013. Analysis of resolved and parameterized vertical transports in convective boundary layers at gray-zone resolutions. J. Atmos. Sci. 70, 3248–3261.

Shin, H.H., Hong, S.-Y., 2015. Representation of the subgrid-scale turbulent transport in convective boundary layers at gray-zone resolutions. Mon. Wea. Rev. 143, 250–271.

Siebesma, A.P., Cuijpers, J.W.M., 1995. Evaluation of parametric assumptions for shallow cumulus convection. J. Atmos. Sci. 52, 650–666.

Siebesma, A.P., Soares, P.M.M., Teixeira, J., 2007. A combined eddy-diffusivity mass-flux approach for the convective boundary layer. J. Atmos. Sci. 64, 1230–1248.

Smagorinsky, J., 1963. General circulation experiments with the primitive equations: I. The basic experiment. Mon. Wea. Rev. 91, 99–164.

Suselj, K., Kurowski, M.J., Teixeira, J., 2019. A unified Eddy-diffusivity/mass-flux approach for modeling atmospheric convection. J. Atmos. Sci. 76, 2505–2537.

Takemi, T., Rotunno, R., 2003. The effects of subgrid model mixing and numerical filtering in simulations of mesoscale cloud systems. Mon. Wea. Rev. 131, 2085–2101.

Teixeira, J., Cheinet, S., 2004. A simple mixing length formulation for the eddy-diffusivity parameterization of dry convection. Bound.-Layer Meteor. 110, 435–453.

Teixeira, J., Reynolds, C.A., 2008. Stochastic nature of physical parameterizations in ensemble prediction: a stochastic convection approach. Mon. Wea. Rev. 136, 483–496.

Therry, G., Lacarrère, P., 1983. Improving the eddy kinetic energy model for planetary boundary layer description. Bound.-LayerMeteor. 25, 63–88.

Tompkins, A.M., Semie, A.G., 2017. Organization of tropical convection in low vertical wind shears: role of updraft entrainment. J. Adv. Model. Earth Syst. 9, 1046–1068.

Troen, I., Mahrt, L., 1986. A simple model of the atmospheric boundary layer: sensitivity to surface evaporation. Bound.-Layer Meteor. 37, 129–148.

Verrelle, A., Ricard, D., Lac, C., 2014. Sensitivity of high resolution idealized simulations of thunderstorms to horizontal resolution and turbulence parameterization. Quart. J. Roy. Meteor. Soc. 141, 433–448.

Verrelle, A., Ricard, D., Lac, C., 2017. Evaluation and improvement of turbulence parameterization inside deep convective clouds at kilometer-scale resolution. Mon. Wea. Rev. 145, 3947–3967.

Von Karman, T., 1930. Mechanische Ahnlichkeit und Turbulenz. Nach. Ges. Wiss. Gottingen, Math.-Phys. Kl. 58–76.

Wilczak, J., 1984. Large-scale eddies in the unstably stratified atmospheric surface layer. Part I: Velocity and temperature structure. J. Atmos. Sci. 15, 3537–3550.

Wyngaard, J.C., 2004. Toward numerical modeling in the "terra incognita" J. Atmos. Sci. 61, 1816–1826.

Wyngaard, J.C., Cote, O.R., Izumi, Y., 1971. Local free convection, similarity, and the budgets of shear stress and heat flux. J. Atmos. Sci. 28, 1171–1182.

Xu, Q., 1988. A formula for eddy viscosity in the presence of moist symmetric instability. J. Atmos. Sci. 45, 5–8.

Xue, M., 2000. High-order monotonic numerical diffusion and smoothing. Mon. Wea. Rev. 128, 2853–2864.

Yu, T.W., 1977. A comparative study on parameterization of vertical turbulent exchange. Mon. Wea. Rev. 105, 55–66.

Zhang, X., Bao, J., Chen, B., Grell, E., 2018. A three-dimensional scale-adaptive turbulent kinetic energy scheme in the WRF-ARW model. Mon. Wea. Rev. 146 (7), 2023–2045.

Zhou, B.W., Zhu, K.F., Xue, M., 2017. A physically based horizontal subgrid-scale turbulent mixing parameterization for the convective boundary layer. J. Atmos. Sci. 74, 2657–2674. https://doi.org/10.1175/JAS-D-16-0324.1.

Zilitinkevich, S., Gryanik, V.M., Lykossov, V.N., Mironov, D.V., 1999. Third-order transport and nonlocal turbulence closures for convective boundary layers. J. Atmos. Sci. 56, 3463–3477.

CHAPTER

8

Uncertainties in the surface layer physics parameterizations

Haiqin Li[a,b] *and Jian-Wen Bao*[c]

[a]Cooperative Institute for Research in Environmental Sciences, University of Colorado Boulder, Boulder, CO, United States [b]NOAA/Global Systems Laboratory, Boulder, CO, United States [c]NOAA/Physical Sciences Laboratory, Boulder, CO, United States

1 Introduction

While the interaction between the atmosphere and the surfaces of the ocean, land, and sea ice involves processes on a wide range of spatial and temporal scales, it is the turbulent processes transferring momentum, heat, moisture, gaseous chemicals, and aerosols across the air and the underneath surface. In numerical weather prediction (NWP) and climate models, it is necessary to parameterize these processes using the model-resolvable variables that drive and influence them. Despite the fact that some of the parameterized processes may be more important than others in NWP models, it is a minimal requirement for an NWP model to have the parameterizations of surface momentum, sensible, and latent heat fluxes across the interface between the atmosphere and the underneath surface. Such parameterizations are commonly formulated based on the so-called Monin–Obukhov (M–O) similarity theory in which many of the physical processes involved in the turbulent transfer of momentum, sensible heat, or latent heat flux are embedded within a transfer coefficient for the particular flux. Like most physics parameterizations in the NWP models, the parameterizations of turbulent transfer processes at the interface between the atmosphere and the underneath surface are deduced from observations, and they are approximations to the processes occurring in nature. This chapter provides a brief overview of the major uncertainties in the parameterizations of turbulent transfer processes used in NWP and climate models.

Uncertainties in Numerical Weather Prediction
https://doi.org/10.1016/B978-0-12-815491-5.00008-2

2 Atmosphere-ocean interaction

Ocean and sea ice play an important role in numerical weather prediction models on various temporal and spatial scales. The advantage of coupling ocean includes the effects such as heat storage and the poleward heat transport in the ocean (Bryan, 1991; Zanna et al., 2019), cold sea surface temperature due to coastal upwelling (Li et al., 2012, 2014), and variability in upper ocean mixed layer (Kara et al., 2003). The ocean and sea ice have been included in ECMWF's Integrated Forecasting System (IFS), which significantly improve sea-surface temperature forecast in Europe, and the NWP forecast of near-surface air temperature is further improved (Mogensen et al., 2017). Ocean variability is a strong source for seasonal forecast and climate projection (Troccoli, 2010). In the ocean-atmosphere coupled NWP and climate models, ocean and atmosphere components exchange energy, freshwater, and momentum at the air-sea interface. However, the uncertainties in ocean-atmosphere coupling are inevitable.

2.1 Air-sea fluxes uncertainties

One kind of widely used flux product for stand-alone ocean modeling is the atmospheric reanalysis, such as the NCEP/NCAR reanalysis (R1, Kalnay et al., 1996), NCEP/DOE reanalysis (R2, Kanamitsu et al., 2002), Climate Forecast System Reanalysis (CFSR, Saha et al., 2010), ERA-Interim (Dee et al., 2011), Modern-Era Retrospective analysis for Research and Applications (MERRA, Reinecker et al., 2011). The net downward shortwave radiation (SW), net upward longwave radiation (LW), latent heat flux (LH), sensible heat flux (SH), evaporation (E), precipitation (P), and surface wind stress (τ_x, τ_y) are the common variables from these reanalysis products for ocean modeling. However, turbulent heat flux and surface wind stress are computed from individual bulk parameterizations schemes used in different reanalysis (Yu, 2019). Another kind of widely used surface flux produced for ocean modeling is Satellite- and ship-based products (Yu, 2019). Such as the satellite-based flux products of the Objectively Analyzed Air-Sea Fluxes 1-degree flux analyses (OAFlux-1x1, Yu and Weller, 2007), OAFlux-HR3 (Yu and Jin, 2014, 2018), OAFlux-HR4 (Yu and Jin, 2014, 2018), and the ship-based flux climatology by the National Oceanography Center (NOC) (Berry and Kent, 2011). Turbulent heat flux, evaporation, and wind stress are computed from the COARE v3 bulk flux algorithm for OAFlux-1x1 while the COARE v4 algorithm is used for OAFlux-HR product. The latent heat flux and sensible heat flux coefficients are identical in COAREv3, as the transfer of heat and mass is assumed similar. However, the latent and sensible heat flux are calculated with different formulae in COAREv4 and directly validated with campaign flux measurements. Yu (2019) compared the zonal averages of the annual mean LH + SH from these products and found they differ most at low and middle latitudes. The change of the COARE algorithms resulted in the differences between OAFlux products, particularly that COARE v4 products have stronger $LH + SH$ at all latitudes.

Pillar et al. (2018) assessed the impact of atmospheric reanalysis uncertainty on the simulated ocean state. They used a stand-alone ocean model and fully explored the simulated Atlantic meridional overturning uncertainty at 25°N forced by different atmospheric reanalysis fluxes. The overturning uncertainties with different reanalysis fluxes can exceed 4 Sverdrup

(Sv, $1Sv = 106\ m^3\ s^{-1}$), and it is larger than the overturning ensemble mean anomaly of the 15-year model simulation. Wild et al. (2015) evaluated the energy budgets in 43 Coupled Model Inter-comparison Project Phase 5 (CMIP5) climate models. Compared with Clouds' and the Earth's Radiant Energy System (CERES) satellite observations, the CMIP5 multi-model mean Top-Of-Atmosphere (TOA) solar and thermal fluxes closely match observations when averaged over the ocean, and there are more substantial biases in individual models. However, there is a significant spread in the ocean surface budgets of the CMIP5 models. The majority of the CMIP5 models tend to overestimate the downward solar radiation over the ocean. Huber and Zanna (2017) assessed the impact of uncertainties in air-sea fluxes and ocean model parameters on the Atlantic Meridional Overturning Circulation (AMOC) and ocean heat uptake (OHU). They used the simulated atmospheric forcing from CMIP5 climate models to drive a stand-alone ocean model. Generally speaking, the ocean model adequately reproduces the mean AMOC and the zonally integrated OHU. They found that the ensemble spread in AMOC strength, and Atlantic OHU from different air-sea fluxes is as large as twice of the uncertainty related to vertical and mesoscale eddy diffusivities. Their study demonstrates that air-sea fluxes are one of the key sources of uncertainty of model biases in climate simulations.

2.2 Oceanic and atmospheric parameterization uncertainties

The oceanic vertical mixing involves sub-grid scale turbulence and should be parameterized (Large et al., 1994). Richards et al. (2009) studied the impact of vertical mixing parameterization in a regional ocean-atmosphere coupled model. The changing of vertical diffusion in the ocean component significantly affects the surface ocean temperature, currents, and even the atmospheric fields of surface wind, clouds, and precipitation. They also found that increasing/decreasing ocean mixing has a negative/positive effect on the zonal and meridional asymmetries in the eastern tropical Pacific Ocean. The near-surface heat storage and water mass transformations are also impacted by the changing of vertical mixing.

In the ocean-atmosphere coupled model, the heat flux, momentum, and freshwater flux are from the atmospheric component. The physics of NWP and climate models, including the cloud-radiation, microphysics, convection, and planetary boundary layer are parameterized. All these atmospheric parameterizations can result in uncertainties in ocean-atmosphere coupling. Although the coastal upwelling of California current system is well resolved in a 10-km high-resolution reanalysis ocean-atmosphere coupled downscaling, offshore warm SST bias is caused by the deficiency of the atmospheric physics parameterizations (Li et al., 2012). Less resolved clouds result in surplus net surface heat flux, which further causes the warm SST bias. Actually, the offshore warm SST bias along the coastal upwelling zone is a common issue in contemporary CMIP5 coupled models (Wang et al., 2014). Satoh et al. (2018) used a 7-km nonhydrostatic global model, which has a much higher resolution than conventionally general circulation models (GCMs) and without a cumulus parameterization scheme, to explicitly represent cloud processes by cloud microphysics schemes. When mesoscale convective systems are well resolved, the clouds are more realistically simulated. The ocean-atmosphere coupling would benefit from this kind of improved atmospheric physics.

Bao et al. (2002) evaluated the uncertainties in the parameterizations of air-sea interaction physics in the high wind speed case (>30 m/s) of hurricane development over the Gulf of

Mexico with a regional atmosphere-ocean-wave coupled model. They pointed out that the uncertainties in the parameterizations of air-sea interaction physics are greater for high-wind conditions than the low-wind ones. Their results indicate that there is great sensitivity in the simulated hurricane intensity to the parameterization of wave-age-dependent roughness lengths required in the calculation of momentum and enthalpy fluxes across the air-sea interface. They emphasized that there are still theoretical and practical challenges as to how to migrate the uncertainties. In addition, there are also great uncertainties in the parameterization of the effect of sea spray on air-sea momentum and enthalpy fluxes. They concluded that further theoretical and observational studies are needed to reduce the air-sea fluxes uncertainties under high wind conditions.

3 Atmosphere-land interaction

Another key component in the interaction between the atmosphere and the underneath surface is the land surface. The land surface governs the "partitioning of available energy at the surface between sensible and latent heat, and it controls the partitioning of available water between evaporation and runoff (Pitman, 2003)." The first-generation Land Surface Model (LSM) is a kind of simplest bucket model (Manabe, 1969). The "Manabe bucket model" applied a globally uniform soil depth and water-holding capacity. A soil water threshold limits the evaporation. When a soil water content limit is reached, the extra precipitation is treated as runoff. This kind of bucket model is a key step to implement LSM in weather and climate models. After then, the LSM develops more and more complicated to include the snow physics, hydrological cycle, vegetation parameters, carbon cycle (Chen and Dudhia, 2001; Dai et al., 2003; Niu et al., 2011), etc. However, there are still lots of uncertainties from the atmosphere-land interactions.

3.1 Land surface characteristics

Land cover and soil classification are the commonly used surface characteristics in LSMs. The land component of NOAA GFS is the NOAH LSM (Chen and Dudhia, 2001), which applies the IGBP-MODIS 20-type 1 km land classification, STASGO-FAO 19-type 1-km soil classification, 1/8-deg NESDIS/AVHRR green vegetation fraction by leaf area index (LAI). The Tiled ECMWF Scheme for Surface Exchanges over Land (TESSEL, Balsamo et al., 2009) is used operationally in the ECMWF IFS as a land surface component, and its vegetation characteristic is based on Biosphere-Atmosphere Transfer Scheme (BATS) model (Dickinson et al., 1993). The Community Land Model (CLM) is used in the Community Earth System Model (CESM). The land surface of CLM is presented by five primary sub-grid land cover types (land units: glacier, lake, wetland, urban, vegetated) in each grid cell. The vegetation portion of a grid cell is further divided into patches of plant functional types (PFT), each with its own leaf and stem area index and canopy height. Each sub-grid land cover type and PFT patch is a separate column for energy and water calculations.

Dai et al. (2019a) pointed out that different representation of canopy structure is a kind of essential uncertainty source of land surface modeling. Li et al. (2018) replaced the multi-year averaged Green vegetation fraction (GVF) with near real-time Moderate Resolution Imaging Spectroradiometer (MODIS) GVF data in the NOAH LSM, and performed the offline LSM simulation and WRF NWP forecast. Their results show that the near real-time GVF improves the NWP forecast of near-surface temperature, humidity, and precipitation. The root-mean-square-error (RMSE) of soil moisture from the offline LSM simulation is also reduced. Li et al. (2017) modeled the MODIS land cover type uncertainties over Urumqi, China with Regional Atmospheric Modeling System (RAMS). The model simulation is significantly affected by the land cover categorical uncertainties, for example, the domain averaged latent heat flux difference is as large as 4.32 W/m^2.

Soil plays an important role in land surface processes of water, energy, carbon balance, and biogeochemical cycle, but still remains not well-described for Earth System modeling (Dai et al., 2019b). Currently, in the global modeling framework, the only generally accepted global soil-type map is the FAO-UNESCO Soil Map of the World (SMW; FAO, 1981; Dai et al., 2019b). Most of the LSMs apply constant soil depth to bedrock. For example, NOAH LSM uses 4-layers with a total depth of 2m, while CLM uses 10-layers with constant depths. Actually, this is an unrealistic representation of global soil properties (Dai et al., 2019b).

The soil moisture controls the surface energy and water fluxes, which are coupled to the atmosphere (Santanello et al., 2019). The soil moisture is not only determined by the atmospheric forcing of precipitation, but also by the LSM hydraulic properties. Since the LSM simulated soil moisture is model-dependent, the direct adoption of soil moisture from one model as an initial condition to another model can result in inconsistency (Koster et al., 2009). The soil moisture initial condition is not only very important for the NWP and seasonal forecast, but also necessary for the ecosystem, carbon cycle models, and crop models. Koster et al. (2009) showed that the LSM consistent initialization experiments have better seasonal forecast skill than the LSM inconsistent initialization runs. Santanello et al. (2019) executed a suite of NWP forecasts over the U.S. Southern Great Plains with different soil moisture initialization approaches. It is shown the soil moisture are diverse among different initialization approaches and also impacts the NWP forecasts. Liang and Guo (2003) perturbed three soil parameters and two vegetation parameters to evaluate the sensitivities of evapotranspiration, runoff, turbulent heat flux, and soil moisture simulated by ten land surface models. The impact of soil parameters on surface water and energy partitioning is stronger than the vegetation parameters from their results.

3.2 Atmospheric forcing and LSM physics parameterization uncertainties

The atmosphere component provides the atmospheric forcing fields of surface downward shortwave radiation, surface downward longwave radiation, surface pressure, surface air temperature, precipitation, etc. The accuracy of the atmospheric forcing effects the results of land surface modeling. Gelati et al. (2018) assessed the impact of atmospheric forcing data uncertainties on the LSM simulations. Four atmospheric forcing datasets were used in their land surface modeling. They found that precipitation is the most uncertain forcing field among the datasets while the surface air temperature, shortwave and longwave radiation are most consistent. The largest precipitation uncertainty resulted in a large difference in the simulated river discharge.

Kato et al. (2007) executed a series of sensitivity studies with three state-of-the-art LSMs over four Coordinated Enhanced Observing Period (CEOP) sites. Their studies revealed that different model physics caused the most diversity. They concluded that beyond model physics, sensible heat flux is most sensitive to radiation while both evaporation and soil moisture are most sensitive to precipitation. Their results indicate that continued improving of the LSMs themselves is essential for accurate land surface modeling.

4 Summary

Through a brief literature review, it is shown that the uncertainties in the energy flux parameterizations between the atmosphere and the underneath surface have a few important attributes. It is illustrated that there exists a great sensitivity in coupled NWP and climate models to these attributes. It has been shown by previous studies that although progress has been made in the parameterizations of physical property fluxes across the atmosphere and the underneath surface based on observations, there are still theoretical and practical problems of how to formulate the parameterizations, and caution should be exercised in using a particular parameterization in a coupled NWP model. In particular, great uncertainties remain in the current parameterization of air-sea interaction under high winds. All these point out the need for further research involving both theory and observations to reduce the uncertainties in numerical simulation of the interaction between the atmosphere and the underneath surface.

References

Balsamo, G., Viterbo, P., Beljaars, A., Hurk, B.V.D., Hirschi, M., Betts, A.K., Scipal, K., 2009. A revised hydrology for the ECMWF model: verification from field site to terrestrial water storage and impact in the integrated forecast system. J. Hydro. Meteor. 10, 623–643.

Bao, J.-W., Michelson, S.A., Wilczak, J.M., Fairall, C.W., 2002. Storm simulations using a regional coupled atmosphere-ocean modeling system. In: Perrie, W. (Ed.), In: Chapter 4 in Atmosphere-Ocean Interactions, Advances in Fluid Mechanics, vol. 1. WIT Press, Boston, pp. 115–153.

Berry, D.I., Kent, E.C., 2011. Air-sea fluxes from ICOADS: the construction of a new gridded dataset with uncertainty estimates. Int. J. Climatol. 31, 987–1001.

Bryan, K., 1991. Poleward heat transport in the ocean. Tellus A 43AB, 104–115.

Chen, F., Dudhia, J., 2001. Coupling an advanced land surface-hydrology model with Penn State-NCAR MM5 modeling system. Part I: Model implementation and sensitity. Mon. Wea. Rev. 129, 569–585.

Dai, Y., et al., 2019a. Different representations of canopy structure—a large source of uncertainty in global land surface modeling. Agric. Forest Meteorol. 269-270, 119–135.

Dai, Y., et al., 2019b. A review of the global soil property maps for Earth System models. Soil 5, 137–158.

Dai, Y., Zeng, X., Dickinson, R.E., Baker, I., Bonan, G.B., Bosilovich, M.G., et al., 2003. The common land model. Bull. Am. Meteorol. Soc. 84 (4), 1013–1023.

Dee, D.P., et al., 2011. The ERA-interim reanalysis: configuration and performance of the data assimilation system. Q. J. R. Meteorol. Soc. 137, 553–597.

Dickinson, R.E., Henderson-Sellers, A., Kennedy, P.J., 1993. Biosphere atmosphere transfer scheme (BATS) version 1e as coupled to the NCAR community climate model. NCAR Tech. Note. TN-387+STR, 72 pp.

FAO, 1981. Soil Map of the World. Vol. 110. UNESCO, Paris, France.

Gelati, E., et al., 2018. Hydrological assessment of atmospheric forcing uncertainty in the Euro-Mediterranean area using a land surface model. Hydrol. Earth Syst. Sci. 22, 2091–2115.

Huber, M.B., Zanna, L., 2017. Drivers of uncertainty in simulated ocean circulation and heat uptake. Geophys. Res. Lett. 44, 1402–1413. https://doi.org/10.1002/2016GL071587.

Kalnay, E., Kanamitsu, M., et al., 1996. The NMC/NCAR 40-year reanalysis project. Bull. Am. Meteorol. Soc. 77, 437–471.

Kanamitsu, M., et al., 2002. NCEP-DOE AMIP-II reanalysis (R-2). Bull. Am. Meteorol. Soc. 83, 1631–1643.

Kara, A.B., Rochford, P.A., Hurlburt, H.E., 2003. Mixed layer depth variability over the global ocean. J. Geophys. Res. 108 (C3), 3079.

Kato, H., Rodell, M., Beyrich, F., Cleugh, H., Gorsel, E.V., Liu, H., Meyers, T.P., 2007. Sensitivity of land surface simulations to model physics, land characteristics, and forcing, at four CEOP sites. J. Meteorol. Soc. Jpn 85A, 187–204.

Koster, D.R., Guo, Z., Yang, R., Dirmeyer, P.A., Mitchell, K., Puma, M.J., 2009. On the nature of soil moisture in land surface models. J. Climate 22, 4322–4335.

Large, W.G., McWilliams, J.C., Doney, S.C., 1994. Oceanic vertical mixing: a review and a model with a nonlocal boundary layer parameterization. Rev. Geophys. 32 (4), 363–403.

Li, H., Kanamitsu, M., Hong, S.-Y., 2012. California reanalysis downscaling at 10 km using an ocean-atmosphere coupled regional model system. J. Geophys. Res. 117, D12118. https://doi.org/10.1029/2011JD017372.

Li, H., Kanamitsu, M., Hong, S.-Y., Yoshimura, K., Cayan, D.R., Misra, V., 2014. A high-resolution ocean-atmosphere coupled downscaling of a present climate over California. Clim. Dyn. 42, 701–714. https://doi.org/10.1007/s00382-013-1670-7.

Li, X., Messina, J.P., Mooree, N.J., et al., 2017. MODIS land cover uncertainty in regional climate simulations. Clim. Dyn. 49, 4097.

Li, F., Zhan, X., Hain, C.R., Liu, J., 2018. Impact of using near real-time green vegetation fraction in noah land surface model of NOAA NCEP on numerical weather predictions. Adv. Meteorol. 2018, 9256396.

Liang, X., Guo, J., 2003. Intercomparison of land surface parameterization schemes: sensitivity of surface energy and water fluxes to model land characteristics. J. Hydrol. 279, 182–209.

Manabe, S., 1969. Climate and Ocean circulation: 1, the atmospheric circulation and the hydrology of the Earth's surface. Mon. Wea. Rev. 97, 739–805.

Mogensen, K.S., Hewson, T., Keeley, S., Magnusson, L., 2017. Effects of ocean coupling on weather forecasts. Eur. State Clim. 2017.

Niu, G., et al., 2011. The community NOAH land surface model with multiparameterization options (NOAH-MP): 1. Model description and evaluation with local-scale measurements. J. Geophys. Res. 116, D12109. https://doi.org/10.1029/2010JD015139.

Pillar, H.R., et al., 2018. Impacts of atmospheric reanalysis uncertainty on Atlantic overturning estimes at 25deg N. J. Clim. 31, 8719–8744. https://doi.org/10.1175/JCLI-D-18-0241.1.

Pitman, A.J., 2003. The evolution of, and revolution in, land surface schemes designed for climate models. Int. J. Climatol. 23, 479–510.

Reinecker, M.M., et al., 2011. MERRA: NASA's modern-era retrospective analysis for research and application. J. Clim. 24, 3624–3648.

Richards, K.J., Xie, S., Miyama, T., 2009. Vertical mixing in the Ocean and its impact on the coupled Ocean-Atmosphere system in the Eastern Tropical Pacific. J. Clim. 22, 3703–3719.

Saha, S., et al., 2010. The NCEP climate forecast system reanalysis. Bull. Am. Meteorol. Soc. 91, 1015–1057.

Santanello, J.A., Lawston, J.P., Kumar, S., Dennis, E., 2019. Understanding the impacts of soil moisture initial condition on NWP in the context of land-atmosphere coupling. J. Hydrometeor. 20, 793–819.

Satoh, M., et al., 2018. Toward reduction of the uncertainties in climate sensitivity due to cloud processes using a global non-hydrostatic atmospheric model. Progr. Earth Planet. Sci. 5, 67.

Troccoli, A., 2010. Seasonal climate forecasting. Meteorol. Appl. 17, 251–268.

Wang, C., Zhang, L., Lee, S., Xin, L., Mechoso, C.R., 2014. A global perspective on CMIP5 climate model biases. Nat. Clim. Change 4, 201–205.

Wild, M., et al., 2015. The energy balance over land oceans: an assessment based on direct observations and CMIP5 climate models. Clim. Dyn. 44, 3393–3429.

Yu, L., 2019. Global air-sea fluxes of heat, fresh water, and momentum: energy budget closure and unanswered questions. Annu. Rev. Mar. Sci. 11 (2), 27–48.

Yu, L., Jin, X., 2014. Insights on the OAFlux ocean surface vector wind analysis merged from scatterometers and passive microwave radiometers (1987 onward). J. Geophys. Res. Oceans 119, 5244–5269.

Yu, L., Jin, X., 2018. Retrieving near-surface air humidity and temperature using a regime-dependent regression model. Remote Sens. Environ. 215, 199–216.

Yu, L., Weller, R.A., 2007. Objectively analyzed air-sea heat Fluxex (OAFlux) for the global ocean. Bull. Am. Meteorol. Soc. 88, 527–539.

Zanna, L., Khatiwala, S., Gregory, J.M., Ison, J., Heimbach, P., 2019. Global reconstruction of historical ocean heat storage and transport. PNAS 116 (4), 1126–1131.

CHAPTER

9

Radiation

Kristian Pagh Nielsen[a], Laura Rontu[b], and Emily Gleeson[c]

[a]Department of Research and Development, Danish Meteorological Institute, Copenhagen, Denmark, [b]Meteorological Research, Finnish Meteorological Institute, Helsinki, Finland, [c]Climate, Research and Applications Division, Met Eireann, Dublin, Ireland

Abbreviations

AOD	aerosol optical depth
diff	subscript for diffuse
dir	subscript for direct
$F_{LW}^{\downarrow}$ **and LW↓**	downward LW radiation flux or irradiance
$F_{LW}^{\uparrow}$ **and LW↑**	upward LW radiation flux or irradiance
$F_{SW}^{\downarrow}$ **and SW↓**	downward SW radiation
$F_{SW}^{\uparrow}$	upward SW radiation
g	asymmetry parameter
ice	subscript for ice crystals
IOP	inherent optical properties
k	mass extinction coefficient
lev	subscript for atmospheric model level
liq	subscript for liquid droplets
LW	longwave, thermal radiation
NWP	numerical weather prediction
oro	subscript for orographic
surf	subscript for surface
SW	shortwave, solar radiation
TOA	top of atmosphere
ω	single-scattering albedo

1 Introduction

Radiation from the Sun is, by far, the most important energy source to the Earth's atmosphere. In effect, it powers the weather and climate system and indirectly provides the fuel for hurricanes, tornadoes, thunderstorms, and other weather systems and keeps the

Uncertainties in Numerical Weather Prediction
https://doi.org/10.1016/B978-0-12-815491-5.00009-4

hydrological cycle going. Radiation calculations in weather and climate models are computationally intensive. Therefore, a trade-off is required between accuracy and computational efficiency. When speaking about weather models, "radiation" always refers to electromagnetic radiation and is considered in two major parts: shortwave (SW) radiation, which is the radiation from the Sun and longwave (LW) radiation, which is thermal radiation emitted from the surface of the Earth, and from clouds, aerosols, and gases. Radiation is scattered and absorbed by gases, aerosols, and cloud particles. SW radiation is also reflected by the Earth's surface, and LW radiation is emitted from the Earth's surface and from clouds, aerosols, and gases. The role of a radiation scheme in a meteorological model is to compute:

1. the incoming SW irradiance at the top of the atmosphere (TOA),
2. the radiative heating rates on the atmospheric levels in the meteorological model,
3. the direct SW and diffuse SW and LW irradiances reaching, and leaving the Earth's surface, and
4. the diffuse SW and LW irradiances that escape from the highest model level.

Due to the different nature of SW and LW radiation sources, that is, the direct solar irradiance at the top of the atmosphere and the LW thermal emissions from the surface and model levels, SW and LW radiation calculations are done independently in separate schemes. Both the SW and LW radiation computations can be further subdivided into bands of the electromagnetic spectrum. For instance, the widely used Rapid Radiative Transfer Model (RRTM) includes 14 SW spectral bands (Mlawer and Clough, 1997) and 16 LW spectral bands (Mlawer et al., 1997). These are used for the cloud and aerosol optical properties. For the gas radiative transfer, a finer division of the spectral bands into a total of 112 SW and 140 LW g-points is used (Hogan and Bozzo, 2018). See also Section 3.4 on gas optical properties.

SW and LW irradiances are radiative fluxes given in units of W m^{-2}. In weather models, these are computed on a horizontal plane and considered to be positive in the downwards direction toward the surface and negative in the upwards direction. The atmospheric radiative heating rate on a full level of the model is related to the SW and LW fluxes as given in Eq. (1).

$$\left(\frac{\partial T}{\partial t}\rho c_p\right)_{\text{radiation}} = F^{\downarrow}_{\text{SW,dir}} + F^{\downarrow}_{\text{SW, diff}} - F^{\uparrow}_{\text{SW, diff}} + F^{\downarrow}_{\text{LW}} - F^{\uparrow}_{\text{LW}}, \tag{1}$$

where $F^{\downarrow}$ denotes downward irradiances and $F^{\uparrow}$ denotes upward irradiances on the model level. Separating the irradiances into upward and downward components is referred to as the two-stream approximation, the modified Schuster-Schwarzschild approximation (Schuster, 1905; Schwarzschild, 1906; Sagan and Pollack, 1967), or the Delta-Eddington approximation (Joseph et al., 1976). Most models today use the latter version, but here we will refer to this using the generic term: "The two-stream approximation." This is discussed in Section 3.6. In Eq. (1), the SW irradiances are divided into three components: the downward direct SW irradiance ($F^{\downarrow}_{\text{SW, dir}}$), the downward diffuse SW irradiance ($F^{\downarrow}_{\text{SW, diff}}$), and the upward diffuse SW irradiance ($F^{\uparrow}_{\text{SW, diff}}$). The downward and upward LW irradiances ($F^{\downarrow}_{\text{LW}}$ and $F^{\uparrow}_{\text{LW}}$) are always diffuse. The sum of the terms on the right-hand side of Eq. (1) is referred to as the net irradiance, and the sum of the SW and LW terms separately are referred to as the net SW and the net LW irradiances, respectively.

At the TOA, Eq. (1) reduces to

$$F_{\text{net,TOA rad}} = F^{\downarrow}_{\text{SW,dir}} - F^{\uparrow}_{\text{SW, diff}} - F^{\uparrow}_{\text{LW}}. \tag{2}$$

At the surface, the total net flux is given as

$$\begin{aligned} F_{\text{net,surf}} &= \sum_i \left(F^{\downarrow}_{\text{SW,dir},\,i} + F^{\downarrow}_{\text{SW,diff},\,i} - F^{\uparrow}_{\text{SW,diff},\,i} \right) \\ &+ \sum_j \left(F^{\downarrow}_{\text{LW},j} - F^{\uparrow}_{\text{LW},j} \right), \end{aligned} \tag{3}$$

where

$$F^{\uparrow}_{\text{SW,diff},i} = F^{\downarrow}_{\text{SW,dir},i}\, \alpha_{\text{dir},i}(\mu_0) + F^{\downarrow}_{\text{SW,diff},i}\, \alpha_{\text{diff},i}, \tag{4}$$

$$F^{\uparrow}_{\text{LW},j} = \epsilon_j \sigma_{\text{SB}} T^4_{\text{skin}}. \tag{5}$$

Here, i is the index of the ith SW spectral band, $\mu_0 \equiv \cos(\theta_0)$ is the cosine of the zenith angle of the direct solar beam (θ_0), $\alpha_{\text{dir},i}(\mu_0)$ is the spectral direct albedo, which depends on the zenith angle of the direct solar beam, $\alpha_{\text{diff},i}$ is the spectral diffuse albedo, j is the index of the jth LW spectral band, ϵ_j is the spectral emissivity, σ_{SB} is the Stefan-Boltzmann constant, and T_{skin} is the surface skin temperature. Notice here that spectral rather than broadband fluxes $F^{\downarrow}_{\text{SW,dir},i}$, $F^{\downarrow}_{\text{SW,diff},i}$, and $F^{\downarrow}_{\text{LW},j}$ must be used for the computation, because the albedos and emissivities vary as a function of the electromagnetic wavelength. The fluxes in each of the right-hand terms in Eqs. (1), (2) implicitly include sums over the spectral bands.

2 External uncertainties

The primary inputs to the two-stream radiation solver in a numerical weather prediction (NWP) model are the optical properties of clouds, aerosols, gases, and the surface. The input variables available, on the other hand, are mostly in the form of specific mass concentrations that first need to be converted into optical properties. We will refer to the uncertainties in this conversion and in the radiative transfer computations as internal uncertainties. These are discussed in detail in Section 3. The external uncertainties are due to errors in the input cloud, aerosol, and gas mixing ratios, which will be discussed further in this section.

2.1 Clouds

Cloud fraction, and liquid/ice water mass mixing ratios are provided on model levels by the cloud scheme. These are often incorrect with clouds being in grid columns where they should not be and clouds not being in grid columns where they should be. For cumulus clouds in particular (including cumulonimbus), it is impossible to always get the cloud positions correct in hourly or daily forecasts from highly resolved weather models. This can cause local errors. The frequency of the clouds should ideally be correct on larger scales.

Sometimes, however, the cloud forecasts are also wrong on larger scales. In fact, forecasting clouds correctly is the biggest challenge in current NWP forecasting and climate projections. Since clouds are very sensitive to errors in other model variables, the errors tend to accumulate in the form of cloud cover errors.

In addition to cloud liquid and ice water mass mixing ratios, some models include mass mixing ratios of precipitation species such as rain, snow, and graupel as input to the radiation scheme. This, of course, requires that the model has these species available as model-level variables. Snow and other forms of frozen precipitation, in particular, affect the LW irradiance at the surface (Ivarsson, K.-I., 2016, personal communication).

In Figs. 1 and 2, the impact of cloud errors on surface SW and LW irradiances at a specific location is illustrated using data from the ERA5 reanalysis dataset (Hersbach and Dee, 2016). The data are of hourly resolution and are shown for the year 2017. The location shown is typical of oceanic mid-latitude climates. Specifically, the figures illustrate the relative impact clouds can have on the surface irradiances by showing what these irradiances are, and what they are relative to the irradiances under clear-sky conditions. An interesting feature is that the relative impact patterns of the clouds on the SW and LW net surface irradiances are quite similar (Figs. 1 and 2, bottom panels), with the main difference being that the SW irradiance is only nonzero during daylight hours.

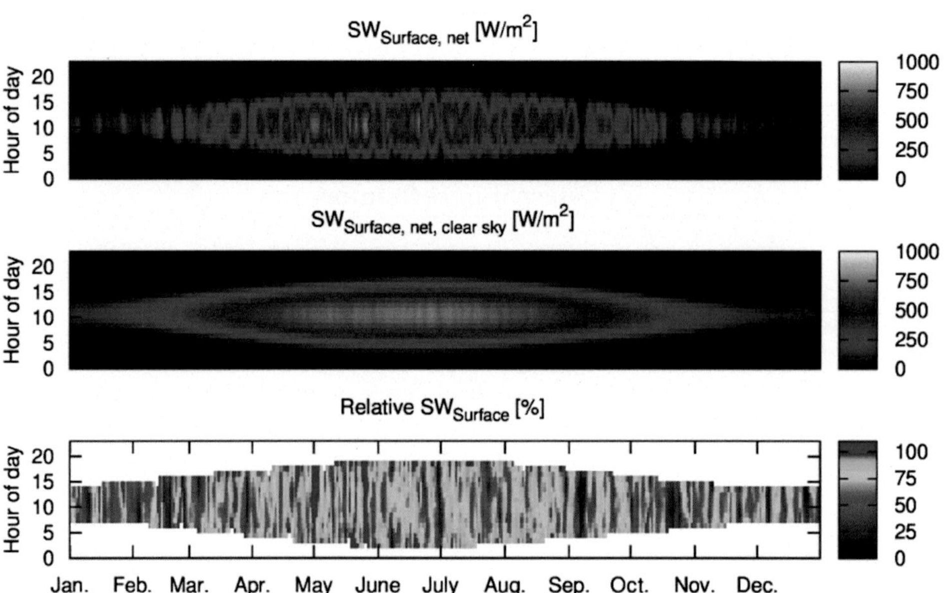

FIG. 1 The radiative effect of clouds illustrated using surface SW irradiances from the ERA5 reanalysis dataset for the year 2017. *Each panel* shows the day of the year on the horizontal axis and the hour of the day on the *vertical axis*. The *top panel* shows the net SW irradiance. The *middle panel* shows the clear-sky net SW irradiance, that is, the net SW irradiance as it would be without clouds. The *lower panel* shows the relative percentage difference caused by the presence of clouds. The data are from the ERA5 grid point nearest to the WMO global radiation station 618800 in Sjælsmark, Denmark (12.41°E, 55.88°N).

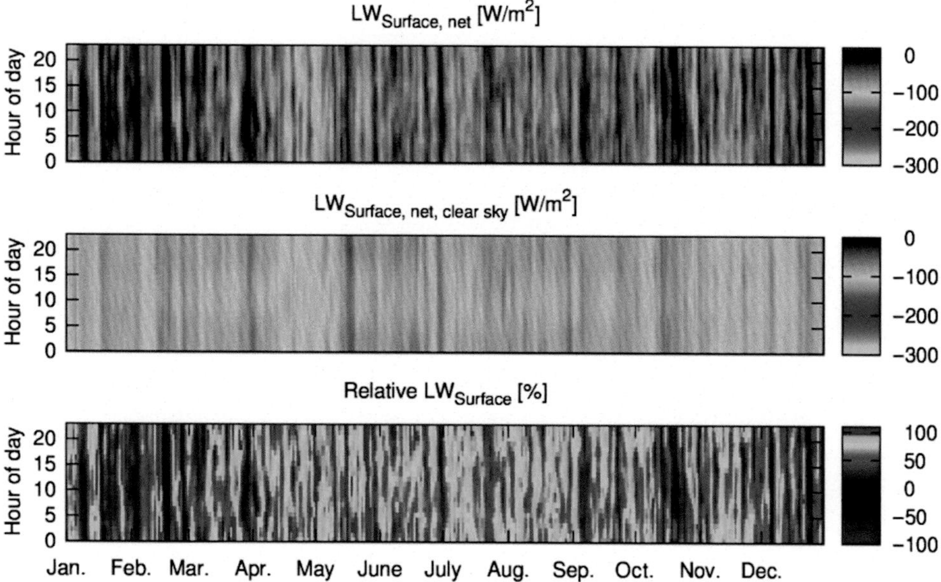

FIG. 2 The same as Fig. 1 but shows surface LW irradiances.

Fig. 1 shows that the clouds have a varying effect on surface SW irradiance ranging from only slightly attenuating to extinguishing almost all of the SW irradiance. The cloud forcing depends on the integrated cloud water load and the internal cloud properties, as detailed in Section 3.2. Fig. 2 illustrates the effect of clouds on the surface LW irradiance, where the clouds reduce the loss of irradiance from the surface. Clouds do not need to be very optically thick to become opaque to LW irradiance. When the cloud-base temperature is the same as the surface skin temperature, the net LW irradiance becomes zero. In a few cases, the clouds even cause a net LW heating of the surface, as shown by the red color in the top panel of Fig. 2. For this to happen clouds, with a cloud-base temperature higher than the surface skin temperature, need to be advected over the location.

Most of the SW irradiance that the clouds prevent from reaching the surface is reflected out of the atmosphere at the TOA as can be seen in Fig. 3. In summer during daytime, the part of this SW irradiance absorbed by the water absorption bands is large enough to cause net heating within the atmosphere, or more specifically within the clouds. During nighttime in winter, clouds increase the net irradiance at the surface. This increase corresponds to a decrease in the net irradiance summed over all atmospheric model levels. The LW irradiance from a cloud is emitted from the top and base of the cloud, while the net LW irradiance within a cloud quickly becomes zero as the cloud becomes optically thicker.

Overall, this section has illustrated the large uncertainty induced locally from incorrect forecasts of clouds. Clouds are able to block much of the SW irradiance from reaching the surface during daylight hours, to trap the LW irradiance at the surface, and to limit the LW irradiance that is emitted to space.

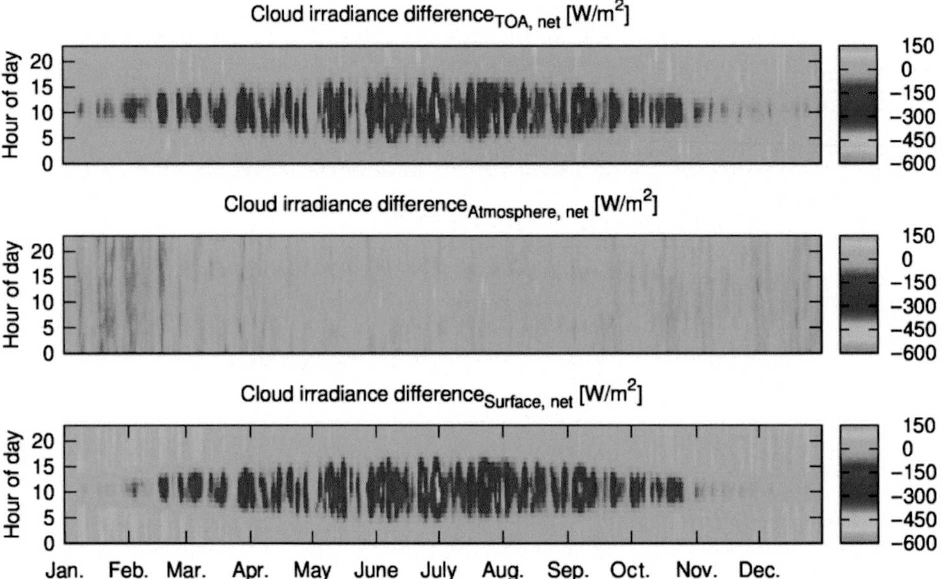

FIG. 3 The net radiative effect of clouds illustrated using net irradiance differences at the TOA (*upper panel*), the sum of net irradiance on atmospheric model levels (*middle panel*), and the surface (*lower panel*). The net irradiance differences are between total irradiance and clear-sky irradiance output from the ERA5 reanalysis dataset. The net irradiances are the sums of the SW and LW net irradiances. *Each panel* shows the day of the year on the horizontal axis and the hour of the day on the vertical axis. The data are from the ERA5 grid point nearest to the WMO global radiation station 618800 in Sjælsmark, Denmark (12.41°E, 55.88°N) for the year 2017.

2.2 Aerosols

In this section, we focus on the direct radiative effect of aerosols and discuss uncertainties related to aerosol input, that is, to the aerosol load and aerosol inherent optical properties (IOPs).

Aerosols are small liquid or solid particles suspended in the air. Examples include sea salt from sea spray, desert dust, sulfates from anthropogenic sources, and carbonates from biomass burning. Most types of aerosols have only a short atmospheric lifetime. They, therefore, mainly occur in regions near the sources of such aerosols. Since desert dust and biomass burning aerosols are highly variable in space and time, they are a challenge to include in NWP models.

The exception to this spatiotemporal aerosol intermittency issue is volcanic stratospheric sulfates that can be spread around the globe over months following very large volcanic eruptions. While the mineral particles from a volcanic eruption quickly sediment from the atmosphere, droplets of sulfuric acid that reach into the stratosphere can stay suspended for years. This occurs a couple of times each century with the last example being the Pinatubo eruption in 1991. Using sulfate deposition on the Antarctica and Greenland ice sheets, the global sulfate aerosol load caused by Pinatubo has been estimated to be in the range 0.01–0.02 g m^{-2} (Sigl et al., 2015). This caused a global cooling in the lower troposphere of 0.5°C (Dutton and Christy, 1992).

Most regional NWP models use global monthly climatologies of a few aerosol species as input to the radiative transfer calculations. Commonly used climatologies include the Tegen climatology (Tegen et al., 1997) and the newer Copernicus Atmospheric Monitoring Service (CAMS) aerosol climatology (Inness et al., 2019; Bozzo et al., 2020). The global climatologies come at relatively coarse horizontal resolutions and their interpolation to fine-resolution NWP model grids leads to inaccuracies due to smoothing. For example, the sea-salt load may be overestimated and spread too far inland by the interpolations. CAMS also provides near real-time forecasts of three-dimensional (3D) aerosol mass mixing ratios (MMRs) for 15 species. Advanced data assimilation, using information about emission sources and conventional and space-born observations, is used to constrain the model processes (Rémy et al., 2019) in the CAMS near real-time aerosol forecasts. The spatial and temporal resolution of the near real-time data are higher than the climatological data. Regional NWP models may benefit from use of such external aerosol data, instead of applying computationally demanding aerosol modeling and data assimilation internally.

Measured aerosol mass mixing ratios can differ significantly from climatological mean values during an influx of dust from a far-away desert, when large-scale wildfires occur, or after a volcanic eruption. The impact of an increased aerosol load on radiation is often significant. For SW radiation, the local impact can be comparable to the impact of clouds. A sensitivity study by Gleeson et al. (2016) for a severe wildfire episode in Russia in August 2010 showed that excluding the direct effect of aerosols resulted in an error of over 15% in the surface SW irradiance during the middle of the day. Similarly, Rontu et al. (2019) reported differences of up to 20% in the simulated local SW irradiance at noon when comparing climatological to near real-time aerosol input in a Saharan dust intrusion case. Specifically, the input was aerosol optical depths (AODs) at a wavelength of 550 nm. On the other hand, replacement of the coarse-resolution climatological total column AOD dataset with another dataset led to only minor changes in SW irradiance. A single-column model study (Rontu et al., 2020) confirmed that the largest differences in radiative fluxes and heating rates are due to different aerosol loads.

Uncertainties related to the ways in which aerosol MMRs and IOPs are converted to aerosol optical properties during an NWP model forecast are discussed in Section 3.3.

2.3 Aerosol-cloud interactions

In addition to the direct radiative effects of aerosols, they have indirect and semidirect effects on clouds (Hansen et al., 1997). The direct effect of aerosols involves their absorption and scattering of SW and LW radiation. Indirect effects of aerosols include their effect on cloud microphysics, and the formation and lifetime of clouds. The indirect effects are primarily due to the impact of aerosols on cloud albedo—clouds with more condensate or an increased number of small liquid droplets, instead of fewer larger droplets, reflect more SW radiation.

The absorption of SW radiation by aerosols within and around clouds can have an impact on air temperature and humidity and hence on atmospheric stability. This effect is known as the semidirect effect of aerosols.

The direct radiative effect of aerosols on weather and climate is much better quantified than the cloud-related indirect effects (Toll et al., 2019). Many NWP models use simplified formulations of cloud particle effective size (e.g., the HARMONIE-AROME model, Bengtsson et al., 2017), which is the key parameter used for estimation of the indirect radiative effect of aerosols. Ideally, cloud particle effective size should be input to the radiation scheme of the NWP model, for instance, in the form of cloud liquid droplet effective radius and cloud ice particle equivalent radius. In most radiation schemes, however, these are computed by parameterizations within the schemes.

Cloud liquid droplet effective radius $r_{e,liq}$ is often parameterized following (Martin et al., 1994) as:

$$r_{e,liq} = \left(\frac{L_{liq}}{c\rho_w N_{\text{TOT}}}\right)^{1/3} \quad [\mu\text{m}], \tag{6}$$

where L_{liq} is the cloud liquid water concentration, c is a parameterization variable, ρ_w is the density of water, and N_{TOT} is the cloud droplet number concentration. Martin et al. (1994) found that c has different values for maritime and continental air masses. For stratocumulus clouds, N_{TOT} depends on the number of cloud condensation nuclei (CCN). CCN are complex aerosol combinations that enable cloud droplet formation (Dunne et al., 2016).

There are sufficient CCN in the air for clouds to form when the relative humidity is slightly higher than 100%. Things are more complicated with respect to cloud ice nucleation. Homogeneous cloud ice nucleation, that is, nucleation without ice nuclei, occurs at temperatures lower than −37°C. This means that most cloud ice particles form from ice nuclei and liquid cloud droplets. If sufficient ice nuclei are present, this can occur at only a few degrees below zero, but lower temperatures are mostly required (Sahyoun et al., 2016).

Cloud ice equivalent radius $r_{e,\,ice}$ has been parameterized by Sun and Rikus (1999) and Sun (2001) as

$$\begin{aligned} r_{e,ice} &= 3\sqrt{\frac{3}{8}}(1.2351 + 0.0105T_C) \\ &\quad \times \left(45.8966L_{ice}^{0.2214} + 0.7957L_{ice}^{0.2535}(T_C + 190.0)\right) \quad [\mu\text{m}], \end{aligned} \tag{7}$$

where L_{ice} is the cloud ice water concentration and T_C is the temperature in Celsius. In Figs. 4 and 5, this parameterization is plotted to illustrate the range of values of $r_{e,ice}$. The temperature is included in the parameterization of $r_{e,ice}$, unlike in the parameterization of $r_{e,liq}$. The reason is that cloud ice particles can have many different shapes depending on the temperature, while cloud liquid droplets are always spherical.

2.4 Gases

Input variables to the gas optics calculations are given in the form of gas MMRs. The most important gas for radiation calculations is water vapor. This is given in the form of specific humidity, q, which is computed at each model full level. In weather models, the other gases are either assumed to be constant or, in the case of the greenhouse gases (CO_2, CH_4, N_2O, CFC-11, CFC-12, etc.), given based on specific climate scenarios that evolve year by year. Not using up-to-date mixing ratios for these greenhouse gases is a source of error in the model. The greenhouse gases are often assumed to be uniformly mixed and thus have the same concentration in each model grid box. In Fig. 6, an example of the

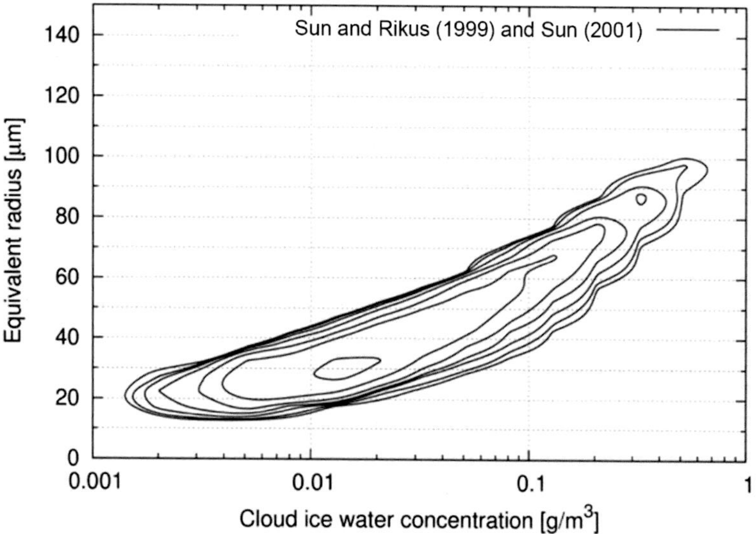

FIG. 4 Density plot of $r_{e,\,ice}$ for a large sample of realistic atmospheric conditions as a function of the cloud ice water concentration. For each contour in the plot, the density increases twofold.

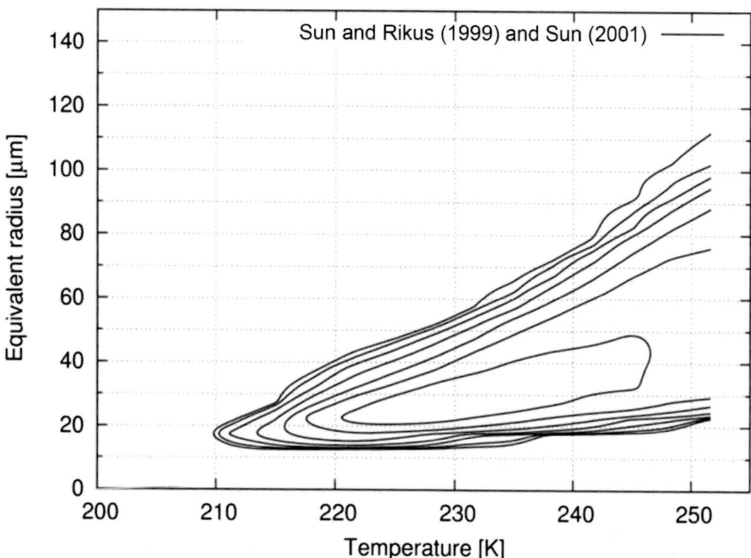

FIG. 5 Density plot of $r_{e,\,ice}$ for a large sample of realistic atmospheric conditions as a function of the temperature. For each contour in the plot, the density increases twofold.

climatological evolution of CO_2 assumed in an NWP model, as well as local measurements, are shown. Here, the B1, A2, and A1B scenarios are from Nakicenovic et al. (2000). Even though they were made in the year 2000, they capture the real trend well. The newer Representative Concentration Pathway 8.5 (RCP8.5) scenario from Meinshausen et al. (2011) is also

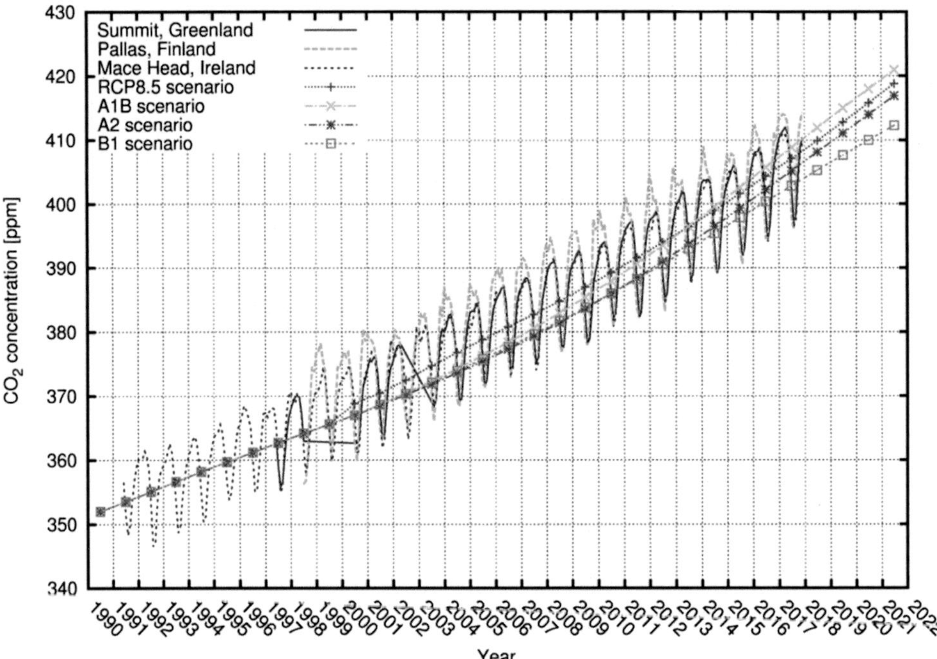

FIG. 6 A Comparison of modeled CO_2 molecule fractions and measured CO_2 molecule fraction. The *red*, *green*, and *blue curves* show the measurements in Greenland, Finland, and Ireland, respectively. The climate scenarios B2, A2, A1B, and RCP8.5 are shown with the *gray*, *orange*, *cyan*, and *magenta curves*, respectively.

shown. The model scenarios here do not include the intraannual variation of CO_2. In Fig. 7, similar data are shown for CH_4. Here, it can be seen that the older scenarios (B1, A2, and A1B) overestimate the growth in CH_4 concentration in the first decade of the 21st century. All of the scenarios have concentrations that are approximately 5% too low for the historical periods. This is due to the limitation of the assumption that CH_4 is a well-mixed greenhouse gas. In fact, the concentration of CH_4 is approximately 10% higher at northern latitudes compared to southern latitudes. Most schemes also use monthly climatologies of atmospheric O_3 that vary as a function of latitude and altitude.

The impact of integrated water vapor on the spectrally differentiated SW↓ at the surface can be seen in Fig. 8. The first spectrum shown is for a very dry atmosphere containing a water vapor load of 2 kg m^{-2} (red line), while the two other spectra are for water vapor loads of 20 kg m^{-2} (green line) and 40 kg m^{-2} (blue line), respectively. It is clear from this figure that water vapor absorption bands are the main atmospheric SW absorption bands. The daily variability of the clear-sky surface net SW irradiance, seen in the middle panel of Fig. 1, is due to variations in integrated water vapor load. The other bands are the Fraunhofer lines (absorption by gases in the solar atmosphere), O_3, O_2, and CO_2. The O_2 A absorption band is clearly seen at a wavelength of 762 nm, while the small wiggle in the spectrum around a wavelength of 2000 nm is due to CO_2 absorption.

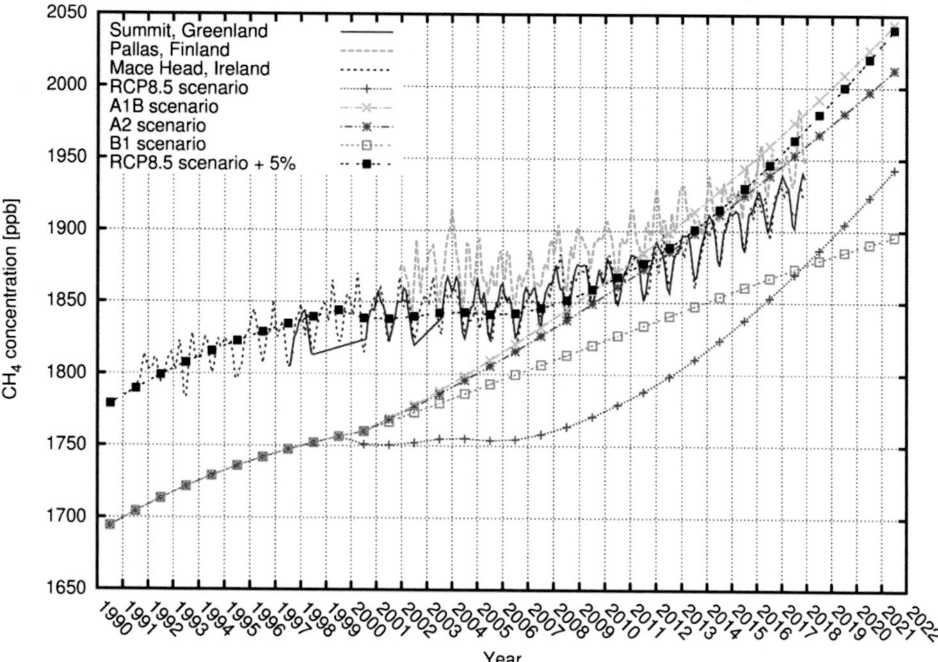

FIG. 7 A comparison of modeled CH_4 molecule fractions and measured CH_4 molecule fraction. The *red, green,* and *blue curves* show the measurements in Greenland, Finland, and Ireland, respectively. The climate scenarios B2, A2, A1B, and RCP8.5 are shown with the *gray, orange, cyan,* and *magenta curves,* respectively. The *black curve* shows the RCP8.5 scenario for methane multiplied by 1.05.

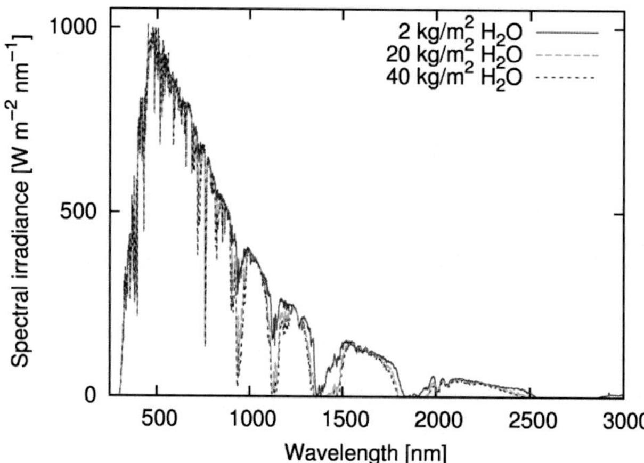

FIG. 8 Typical spectra of SW↓ at the surface shown for varying atmospheric integrated water vapor loads. These were computed using 1 nm spectral resolution in the DISORT radiative transfer solver (Stamnes et al., 1988) as applied in the libRadtran script library (Mayer and Kylling, 2005).

O_3 is a very important gas for ultraviolet spectral irradiances, but for total SW↓, its overall effect is small. Reducing the ozone layer from 500 to 100 DU causes a reduction in surface SW↓ of 2.3% (Nielsen et al., 2014). If a reasonable ozone climatology is used, the uncertainty in the surface SW↓ is less than 1%. For heating in the stratosphere, which occurs due to O_3 SW absorption, it is more important to use correct O_3 mixing ratios. The stratospheric heating rates are not important for the accuracy of day-to-day weather forecasts, run using limited area models, but they are important for weekly weather forecasts run with global weather models.

Fig. 9 is similar to Fig. 8 but illustrates how LW irradiances are affected by varying integrated water vapor. The top panel shows LW↑ at the TOA in red, green, and blue, while the cyan color shows LW↑ at the surface. The spectrally integrated difference between the cyan area and each of the other colored areas is defined as the greenhouse effect. It shows that increasing the integrated water vapor increases the greenhouse effect. The spectral fingerprint of the other greenhouse gasses can also be seen. The CO_2 absorption band around 15 μm and the O_3 absorption band around 9.6 μm are the most prominent.

In the lower panel of Fig. 9, the corresponding surface LW↓ radiative effect of varying the integrated water vapor is illustrated. The cyan color again shows LW↑ at the surface. The red, green, and blue colors show LW↓ at the surface for integrated water vapor loads of 2, 20, and

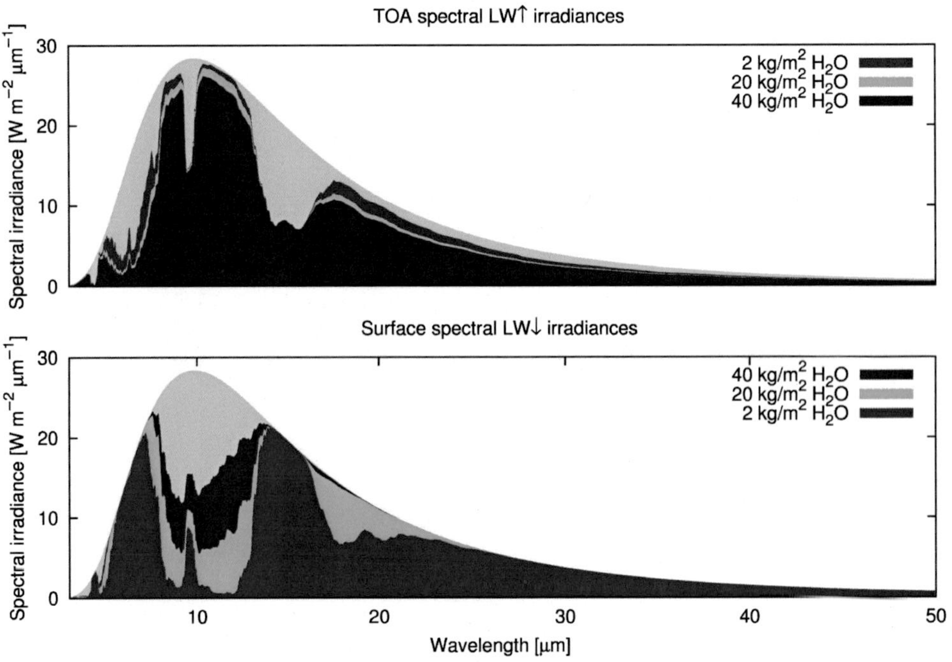

FIG. 9 *Top panel*: Typical spectra of LW↑ at the TOA. *Bottom panel*: Typical spectra of LW↓ at the surface. In both panels, the *cyan-colored filled area* shows LW↑ at the surface, that is, the Planck function. The *red*, *green*, and *blue colors* illustrate the LW spectral irradiances for water vapor loads of 2, 20, and 40 kg m^{-2}, respectively. The spectra are computed using the LOWTRAN spectral bands (Pierluissi and Peng, 1985; Ricchiazzi et al., 1998) with the DISORT radiative transfer solver (Stamnes et al., 1988) applied in the libRadtran script library (Mayer and Kylling, 2005).

40 kg m^{-2}, respectively. The sensitivity of the LW↓ surface irradiances with respect to the integrated water vapor is high. Here, a specific case is shown, and it must be remembered that T_{skin} also varies during the year. If the emissivity is assumed to be 1.0, the effect of increasing T_{skin} in Eq. (5) from 273 to 293 K is an increase in LW↑ of 103 W m^{-2} or a relative increase of 33%.

2.5 Surface

The atmospheric radiative transfer calculations use surface properties as the lower boundary condition. The SW albedo and LW emissivity depend on the properties of the underlying surface (e.g., whether it comprises of vegetation, soil, water, snow and ice, etc.). In an NWP model, the spectral albedo and emissivity are required at every model integration time-step and are provided by the basic physiography description in the model, surface data assimilation, and physical parameterizations. At the surface-atmosphere interface average values of spectral albedo and emissivity, weighted by the fraction of the surface types present in each grid box, are usually used by the atmospheric radiation parameterizations. To illustrate, grid averages can be compared to a bird's-eye view of the surface from well above tree tops, similar to the view from a satellite where the minor details are smoothed out.

SW albedo, or the relationship between the diffuse reflection of the solar radiative flux to the incoming solar flux is defined separately for direct ($\alpha_{dir,i,\ surf}(\mu_0)$) and diffuse ($\alpha_{diff,i,\ surf}$) irradiances at the surface level. Albedo depends on the properties of the surface, μ_0 and electromagnetic wavelength. Typically, values of $\alpha_{diff,surf}$ in the visible part of the spectrum vary from the 0.07 for sea surfaces to over 0.9 for freshly fallen snow. For example, due to the albedo effect the uncertainties in determining the snow and ice cover extent in an NWP model may lead to an order of magnitude inaccuracy in the estimated reflected solar radiation flux $F^{\uparrow}_{SW,diff,i,\ surf}$. For snow-free surfaces, the value of $\alpha_{dir,i,\ surf}(\mu_0)$ typically differs from the value of $\alpha_{diff,surf}$ by between −0.02 (Sun at zenith) and +0.08 (Sun on the horizon) (Yang et al., 2008). Photosynthetically active plants absorb solar radiation in the visible spectral range (small albedo), while they reflect most of the solar irradiance (large albedo) in the near-infrared spectral range. The opposite is true for fresh snow.

Emissivity is the effectiveness of the Earth's surface at emitting thermal radiation. It is less variable than albedo over typical wavelengths in the terrestrial infrared range in NWP models. For example, according to satellite measurements of radiation within the atmospheric window 8–14 μm in the zenith direction (i.e., vertically upwards), the surface broadband emissivity varies from approximately 0.8 over dry sandy deserts to values of 0.95–0.99 for other surfaces such as different types of vegetation, ice or snow. Local emissivity variations might typically lead to differences of up to 10 W m^{-2} in outgoing surface LW irradiance. In NWP models, it is considered sufficient to use constant emissivity values derived from ground-based or satellite measurements.

In most NWP models, a terrain-following vertical coordinate is used. It is based on grid-average surface elevation obtained by upscaling or smoothing fine-resolution terrain elevation data to the grid scale of the NWP model. Statistical subgrid scale variables required by the parameterizations of orographic effects on radiation (Section 4.4) are also derived from fine-resolution terrain elevation data.

3 Internal uncertainties

By internal uncertainties, we mean uncertainties that are not affected by errors or uncertainties in the main input variables to the radiation scheme. The top of the atmosphere solar irradiance is considered to be an internal computation as are the computations of optical properties and the two-stream radiative transfer calculations.

3.1 Top-of-atmosphere solar irradiance

The solar constant is the normal surface solar irradiance at the average distance of the Earth from the Sun. Based on satellite measurements, the current best estimate of the solar constant is 1361 W m^{-2} with a value approximately 0.1% larger at the maximum of the solar cycle relative to the value at the minimum of the cycle (Kopp and Lean, 2011; Coddington et al., 2016). During the past 400 years, variations in the solar constant have also been estimated to be within 0.1% of 1361 W m^{-2} (Kopp et al., 2016). Until recent years, the solar constant was assumed to be 1367 W m^{-2} (Kurucz, 1992). If this value is used rather than current value, a relative difference of +0.5% occurs in modeled SW irradiances.

When the TOA solar irradiance is computed, the seasonal variation in the Sun-Earth distance is accounted for, and the solar zenith angle at the TOA is calculated. Several different algorithms are used to compute the solar angles at the TOA and have accuracies of $\pm 0.01\%$. Some of these approximations are only valid for a little more than a decade, while comprehensive computations that are valid for millennia (Reda and Andreas, 2008) are not used in weather models. For applications, such as various types of concentrating solar energy, the errors in solar angles computed in weather models can be too high.

3.2 Cloud inherent optical properties

The IOPs determine how a cloud interacts with radiation. These can be computed from the cloud water mixing ratio and effective particle size. The IOPs can be defined as the mass extinction coefficient (k), the single-scattering albedo (ω), and the asymmetry factor (g). All of these vary as a function of wavelength. The much-used RRTM scheme (Mlawer and Clough, 1997; Mlawer et al., 1997) has 14 SW spectral bands and 16 LW spectral bands for which an equivalent number of spectrally averaged IOPs can be computed. This is done for both cloud liquid droplets and cloud ice particles. In addition, the clouds have a cloud fraction in the range 0–1. Within the cloudy fraction of a grid box, the cloud is assumed to be homogeneous.

The mass extinction coefficient is given in units of m^2 kg^{-1}. When this is multiplied by the cloud water load in units of kg^{-1} m^2, the dimensionless quantity of cloud optical thickness $\tau_{i,\text{cloud}}$ is obtained. For a specific model level, lev, and spectral band, i, the direct solar beam extinction through the level can be computed as

$$F^{\downarrow}_{\text{SW,dir},i,lev} = \exp\left(-\tau_{i,lev\ \text{cloud}}/\mu_0\right) F^{\downarrow}_{\text{SW,dir},i,lev-1}, \tag{8}$$

where $F^{\downarrow}_{\text{SW,dir},\, i,\, lev-1}$ is the direct solar beam above the model level, $F^{\downarrow}_{\text{SW,dir},\, i,\, lev}$ is the direct solar beam below the model level, $\tau_{i,\, lev\ \text{cloud}}$ is the model-level spectral band cloud optical thickness, and μ_0 is the cosine solar zenith angle.

The single-scattering albedo ω is the ratio of scattering to total extinction and varies from 0 for to 1. Here, all extinction events, as defined by the mass extinction coefficient, are due to absorption if $\omega = 0$, and all extinction events are due to scattering if $\omega = 1$. Scattering means that radiance, for instance, the direct solar beam, is diffracted, refracted, or reflected in another direction.

The final IOP is the asymmetry factor g. It can vary from -1 to 1 and is the main variable in the Henyey-Greenstein normalized volume scattering function (Henyey and Greenstein, 1940)

$$p_{\mathrm{HG}}(\Theta) \equiv \frac{1-g^2}{(1+g^2-2g\cos\Theta)^{3/2}}. \tag{9}$$

Here, Θ is the angle relative to the direction of the incident beam that is scattered. p_{HG} is a purely mathematical construction that does not represent a physically realistic volume scattering function. Even a single cloud droplet creates a complex diffraction pattern which depends on the ratio between the droplet radius and the wavelength of the incident beam. In a liquid cloud droplets of many sizes exist, and the incident beams are considered in broad spectral bands in NWP radiation schemes. This makes p_{HG} an overall good approximation that is also mathematically convenient, as will be explained further in Section 3.6. The shape of p_{HG} as a function of g is such that it varies smoothly with Θ and has a peak that becomes increasingly pronounced as g approaches 1. For $g = 1$, all scattering is in the forward direction. For $g = 0$, the scattering is isotropic.

3.2.1 *Liquid water clouds*

If liquid cloud droplets are assumed to be spherical, their optical properties can be accurately calculated from Mie theory (Mie, 1908; Wiscombe, 1980). It was shown by Nielsen et al. (2014) that many cloud liquid optical property computations that are used in NWP and climate models are inaccurate. Thus, negative errors of the order of tens of W m^{-2} were found for clouds with cloud water loads of the order of 0.1 kg m^2 when the Slingo liquid cloud optical property scheme (Slingo, 1989) was used. The Nielsen liquid cloud optical property scheme for the RRTM SW spectral bands is given in Eqs. (10)–(12) and Tables 1 and 2.

$$k_i = a_i r_{\mathrm{e,liq}}^{b_i}, \tag{10}$$

$$\omega_i = c_i + d_i r_{\mathrm{e,liq}}, \tag{11}$$

$$g_i = e_i + f_i r_{\mathrm{e,liq}} + h_i \exp\left(j_i r_{\mathrm{e,liq}}\right). \tag{12}$$

Here, the first RRTM SW spectral band is omitted because UV radiation with wavelengths shorter than 270 nm does not reach the atmospheric clouds. RRTM SW spectral band 14, which covers the wavelength range 3.846–12.195 μm, is omitted because this spectral band has almost no SW irradiance.

The LW cloud liquid optical properties can be parameterized in a similar manner. This has been done by Lindner and Li (2000).

3.2.2 *Ice clouds*

For ice clouds things are more challenging. The ice cloud particles can be made up of crystals of varying shapes and sizes, or be amorphous. For known shapes the optical properties

TABLE 1 Coefficients used for the Nielsen liquid cloud optical parameterization.

i	2	3	4	5	6	7
[μm]	0.2632–0.3448	0.3448–0.4415	0.4415–0.6250	0.6250–0.7782	0.7782–1.242	1.242–1.299
a_i	1.614	1.633	1.649	1.677	1.719	1.788
b_i	−1.018	−1.021	−1.022	−1.026	−1.032	−1.042
c_i	1.000	1.000	1.000	1.000	0.99998	0.9999
d_i	0.000	0.000	-4×10^{-8}	-1×10^{-6}	-2.7×10^{-5}	-9.7×10^{-5}
e_i	0.861	0.869	0.874	0.870	0.869	0.870
f_i	2.5×10^{-4}	1.1×10^{-4}	4.0×10^{-5}	2.3×10^{-4}	2.3×10^{-4}	2.3×10^{-4}
h_i	−0.050	−0.064	−0.060	−0.067	−0.088	−0.095
j_i	−0.30	−0.25	−0.19	−0.19	−0.19	−0.15

Note: *The columns show the wavelength bands (i) 2–7.*

TABLE 2 Coefficients used for the Nielsen liquid cloud optical parameterization.

i	8	9	10	11	12	13
[μm]	1.299–1.626	1.626–1.942	1.942–2.150	2.150–2.500	2.500–3.077	3.077–3.846
a_i	1.869	1.845	1.740	1.832	1.870	2.098
b_i	−1.054	−1.049	−1.028	−1.043	−1.049	−1.080
c_i	0.9985	0.9993	0.988	0.993	0.525	0.775
d_i	-9×10^{-4}	-6×10^{-4}	-3.1×10^{-3}	-2.2×10^{-3}	1.1×10^{-3}	-4.8×10^{-3}
e_i	0.867	0.863	0.863	0.862	0.944	0.871
MMR f_i	4.5×10^{-4}	5.1×10^{-4}	1.0×10^{-3}	1.0×10^{-3}	2.8×10^{-4}	1.7×10^{-3}
h_i	−0.092	−0.17	−0.78	−0.78	−0.22	−0.48
j_i	−0.15	−0.20	−0.40	0.40	−0.40	−0.30

Note: *The columns show the wavelength bands (i) 8–13.*

can be calculated, so the challenge is in knowing the cloud ice particle shapes. Much-used ice cloud optical property parameterizations have been made by Fu (1996) and Fu et al. (1998) for the SW and LW spectral regions, respectively.

3.3 Aerosol optical properties

The aerosol optical properties, needed by the radiation scheme during the model forecast, can be derived from the aerosol MMRs and the atmospheric humidity at each time step for each location in the model's 3D grid (see Section 2.2 for the aerosol input). Aerosol IOPs can be introduced to NWP models as external information. For example, for the

Integrated Forecasting System (IFS) of the European Center for Medium-range Weather Forecasting (ECMWF), the spectral k_i, ω_i, and g_i for each aerosol type are derived using Mie theory (Mie, 1908) and assuming a log-normal size distributions of spherical particles (Bozzo et al., 2020). Refractive indices of the aerosol particles of different chemical origin, that are applied in the calculations, are based on laboratory measurements. For hydrophilic aerosol species hygroscopic growth is accounted for in the calculation of the optical properties. The IOPs are thus based on theoretical and laboratory studies and contain inaccuracies. However, they most likely play a minor role compared to deficiencies in our knowledge of the loads of the different species and their vertical and horizontal distributions in the atmosphere.

In NWP models, the approaches vary depending both on the properties of the aerosol input—the species that are included, whether it is total-column or 3D distributions, MMRs or 550 nm AODs and on the properties of the radiation scheme—the number of spectral intervals in the SW and LW, radiative transfer assumptions, etc. When 550 nm AODs are supplied as input, wavelength-dependent AOD scaling factors are provided as IOPs rather than k_i.

The radiation schemes operate on the model's 3D grid. If total-column (vertically integrated) MMRs or 550 nm AODs are given as input, exponential profiles can be assumed to distribute the different species of tropospheric and stratospheric aerosol on to the model levels. For example, Bozzo et al. (2020) suggest using a constant scale height of 3 km for dust and 2 km for all other species (sea salt, organic matter, black carbon, and sulfates) in cases where real 3D distributions are unavailable. (The scale height indicates the height of the maximum aerosol load in the atmosphere, whose general scale height is approximately 8.3 km.) In reality, the vertical distributions of different aerosol species vary from case to case. This influences the distribution of the optical properties and leads to uncertainties in radiation fluxes and atmospheric radiative heating. For example, aerosol SW transmission differences of up to 10% were found between experiments with different vertical distributions of the same total mass of desert dust aerosol (Rontu et al., 2020).

The spectral absorption and scattering properties of the different aerosol species vary because of their chemical composition and size distributions. For example, sensitivity experiments by Rontu et al. (2020) demonstrated that a total-column 550 nm AOD of 0.5 could be due to either 37 μg of black carbon or almost 1 g of sea-salt particles suspended in air. The SW transmission due to this amount of black carbon was 0.56 but was 0.95 for sea salt, despite both having the same 550 nm AOD values. Black carbon is known to be highly absorbing and sea salt or desert dust is strongly scattering in the SW part of the spectrum. The typical size of sea salt or desert dust particles is at least one order of magnitude greater than that of black carbon particles. An uncertainty in the estimated concentration of the species has, therefore, different impacts. An inaccuracy of 1 μg in the mass of black carbon may influence the radiative transfer more than 10- or 100-fold times an inaccuracy in the estimation of the mass of coarser particles.

Generally, the impacts of aerosols on SW radiation dominate LW impacts. For example, in the sensitivity study by Rontu et al. (2020), LW transmission was 0.88 for desert dust and 0.93 for sea salt. For all other species, the LW transmission was close to unity, that is, they were transparent to LW radiation. LW scattering is not always included in the radiation schemes of NWP models. This causes inaccuracy in cases where coarse aerosol particles play a significant role. The role of LW scattering was found to be significant for dust, sea salt, and organic matter, for which ω was estimated to vary from 0.34 to 0.44 (Rontu et al., 2020). In these experiments, a value of 0.5 for the 550 nm AOD was assumed. The results are sensitive to the assumed size distribution of the species.

Uncertainties in aerosol transmission due to spectral assumptions in the radiation schemes may be minor. For example, Gleeson et al. (2016) showed that the differences in surface SW↓ between two broadband schemes and a spectrally more detailed scheme were small when all schemes used observation-based optical properties of organic aerosol.

Some aerosol parameterizations, particularly older schemes, assume a fixed value of relative humidity. Gleeson et al. (2016) tested the influence of relative humidity on land aerosols in the HLRADIA broadband radiation scheme using 550 nm AOD input. Under normal pollution (AOD at 550 nm of 0.1), increasing the relative humidity from 0 to 1 increased the surface SW↓ by 1.5%. On the other hand, under heavy pollution (AOD at 550 nm of 1), an increase in relative humidity from 0 to 1 resulted in an increase in surface SW↓ of 12%.

The optical properties of an aerosol mixture on the 3D grid can be calculated using the spectral resolution of the radiation scheme as suggested by Rontu et al. (2020). In Rontu et al. (2020), the relative humidity available in the NWP model during the forecast run was used to select the IOP values for hydrophilic aerosol species. This resulted in a run-time AOD, ω, and g that allowed the radiation schemes to calculate the spectral transmittance and absorptivity of the aerosol mixture at each grid point, without knowing the detailed properties of each aerosol species. The magnitude of the uncertainty introduced by the combination of the optical properties of different species to the optical properties of an aerosol mixture has not yet been estimated.

3.4 Gas optical properties

Many radiation schemes in weather and climate models use the correlated k-distribution (CKD) method to treat gaseous absorption (Fu and Liou, 1992; Mlawer et al., 1997; Lacis and Oinas, 1991). This method has been around since the 1980s and involves dividing the SW and LW spectra into a number of bands that have similar spectral properties (i.e., monochromatically valid subintervals) and reordering the gaseous absorption coefficients, k, within each band as a function of a new coordinate, called g. Note that k and g in this context are different to the optical properties described in Section 3.2! In this coordinate system, 0 represents least absorbing and 1 most absorbing. The reordered absorption spectrum in g space is then discretized into N classes or g-points and pseudo-monochromatic radiative transfer calculations are performed on each discrete class for each gas and atmospheric layer.

Thus, the correlated-k method approximates the integrations over hundreds of thousands of spectral lines by N "monochromatically valid" radiative transfer calculations, where N ranges from tens to hundreds. Larger values of N mean greater accuracy but a higher computational cost. Nevertheless, the correlation-k method is capable of achieving an accuracy comparable to that of current line-by-line radiative transfer models (LBLRTM) while substantially reducing the number of radiative transfer calculations required. It should be noted that line-by-line models such as HITRAN (Gordon et al., 2017), MODTRAN (Berk and Hawes, 2017), and others are continually being improved.

The state-of-the-art rapid radiative transfer model (RRTM) scheme is used by many NWP models for the computation of the optical properties of gases. Calculations made using LBLRTMs are used to determine the *k*-distributions for use in the CKD approach. The RRTMG (RRTM for General circulation model applications) setup commonly used has 112 SW *g*-points and 140 LW *g*-points. The newer RRTMGP (RRTM for General circulation model applications—Parallel; Pincus et al., 2019) setup will be capable of using twice as many *g*-points.

The term "correlated" refers to the assumption that over a given spectral range, absorption coefficients with similar values at a level in the atmosphere are also spectrally correlated at other levels in a vertically inhomogeneous atmosphere. The assumption of correlation is not always fulfilled but CKD methods are still sufficient for describing total transmission due to atmospheric gaseous absorption. An alternative approach involves using an uncorrelated *k*-distribution (Doppler et al., 2014), which has the added advantage of being able to model layer, as well as total, transmission. This method can account for different line shapes making it more accurate.

Overall, the uncertainty in the gas optics calculations due to the use of the correlated-*k* approach is orders of magnitude smaller than other uncertainties in the radiation calculations. The correlated-*k* method is needed when accurate stratospheric SW heating rates or greenhouse gas forcing estimates are essential. For limited area modeling applications, in which accurate radiative heating rates on model levels are not important, simplified gas optics schemes can be applied. One example is forecasting potentially available solar energy at the surface for the day ahead. In that case differences in the clear-sky SW↓ are mainly due to uncertainties related to the estimation of aerosol optical depth and water vapor content (Sections 2.2–2.4) rather than due to gas optics assumptions.

3.5 Surface-radiation coupling

Atmospheric radiation parameterizations provide surface-level downwelling solar direct ($F^{\downarrow}_{SW,dir,i,surf}$) and diffuse ($F^{\downarrow}_{SW,diff,i,surf}$) and terrestrial ($F^{\downarrow}_{LW,j,surf}$) radiation fluxes. At the surface, SW↓ consists of direct and diffuse components. In NWP models, these are estimated diagnostically. Clear-sky diffuse radiation is calculated by taking the solar elevation and atmospheric turbidity (aerosol content) into account.

SW albedo and LW emissivity (Section 2.5) are needed for the calculation of net fluxes (Eq. 3). Uncertainties in downwelling fluxes and surface radiative properties lead to inaccuracies in the forecast surface energy balance, which influences the surface temperature.

The temperatures of soil, sea, lake, snow, and ice surfaces, along with emissivities determine the outgoing LW radiation flux ($F^{\uparrow}_{LW,j,surf}$), which is a source term in the atmospheric radiation parameterizations. An uncertainty of 2 K in T_{surf} corresponds to a difference of approximately 10 W m^{-2} in $F^{\uparrow}_{LW,j,j,surf}$.

Surface-atmosphere coupling for radiation may be organized as follows:

1. Call the atmospheric radiation parameterizations assuming that the surface is flat and homogeneous in each grid box.
2. Modify the grid-average downward SW and LW spectral fluxes to account for topographic effects.

3. Pass the modified SW and LW spectral fluxes to the surface schemes (often separate parameterizations for soil, lake, sea, snow, town, etc.) for use as their upper boundary condition. For example, solar radiation may penetrate below the surface of a lake or interact with vegetation.
4. Obtain updated upwelling and net SW and LW spectral fluxes for inclusion in the model output; store the grid box-averaged real-time albedo, emissivity, and T_{surf} for the next time step of the atmospheric radiation calculations.

In this interactive process simplifications, such as averaging of subgrid information and making assumptions about the spectral properties of direct and diffuse radiation take place. If everything is done consistently, significant information is not expected to be lost and the uncertainty of the forecast should not increase during the model's time integration.

Surface-atmosphere radiation interactions related to topography variations are discussed in more detail in Section 4.4.

3.6 Radiative transfer approximations

Almost all NWP model radiation schemes use a variant of the two-stream approximation (Schuster, 1905; Schwarzschild, 1906) to compute the radiative transfer. The spectral fluxes are computed from the IOPs. First, the average IOPs are computed from the cloud, aerosol, and gas IOPs. The optical thicknesses are summed, the respective single-scattering albedos are weighted using the optical thicknesses and averaged, and the asymmetry factors are averaged weighted using the product of the optical thicknesses and single-scattering albedos.

Because the two-stream approximation only includes the upward and downward directions, detailed angular scattering cannot be represented. To use the IOPs within the two-stream approximation, so-called reduced IOPs (Joseph et al., 1976) must be computed

$$\tau' = (1-\omega g^2)\tau, \tag{13}$$

$$\omega' = \frac{(1-g^2)\omega}{1-\omega g^2}. \tag{14}$$

Here, both the reduced optical thickness (τ') and the reduced single-scattering albedo (ω') become relatively smaller than the unreduced values as g approaches 1, which corresponds to the volume scattering distribution function becoming more forward peaked. Different two-stream approximation methods exist. This approach is called the Delta-Eddington scaling method (Joseph et al., 1976).

In Fig. 10, this and other two-stream approximation methods (Fouquart and Bonnel, 1980; Thomas and Stamnes, 2002) are compared to the DISORT radiative transfer model (Stamnes et al., 1988, 2000), which can resolve the angular variability of scattered fluxes. The error in the Delta-Eddington approximation of Joseph et al. (1976) can be seen to be up to 5%. The error can be even larger in the forecast of $SW \downarrow_{dir}$. This issue has been discussed in detail by Räisänen and Lindfors (2019).

As will be discussed further in Section 4 the NWP model assumption that all radiative fluxes and heating rates are computed in independent columns is a shortcoming. To resolve this, full 3D radiative transfer is needed—for instance, with a Monte Carlo model (e.g., Hestenes et al., 2007). The SPARTACUS (Speedy Algorithm for Radiative Transfer through Cloud Sides) radiative transfer model (Schäfer et al., 2016; Hogan et al., 2016, 2019) makes

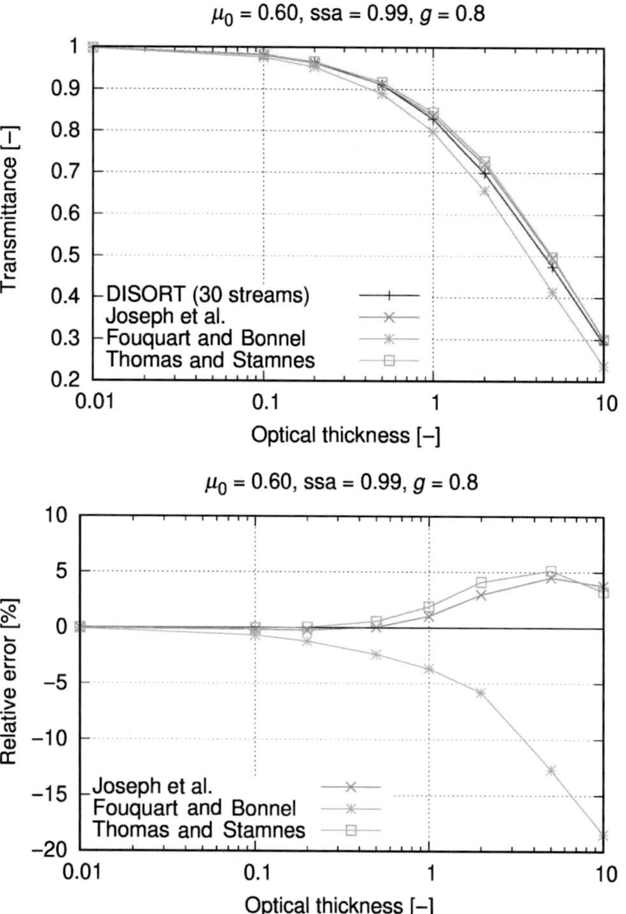

FIG. 10 *Top panel*: Comparison of transmittance for a model grid box with a single-scattering albedo of 0.99 and an asymmetry factor of 0.8 computed with the DISORT radiative transfer scheme (*purple curve*), the (Joseph et al., 1976) Delta-Eddington approximation (*green curve*), the (Fouquart and Bonnel, 1980) two-stream approximation (*cyan curve*), and the (Thomas and Stamnes, 2002) Schuster-Schwarzchild approximation (*orange curve*). The cosine of the solar zenith angle (μ_0) is set to 0.60. *Bottom panel*: The same as above, but here the relative error relative to the DISORT computation is shown.

it possible to account for subgrid-scale 3D cloud scattering effects. This has been designed for the IFS/ECMWF model, but is still computationally too expensive to be used for operational weather forecasting (Hogan and Bozzo, 2018).

4 Subgrid assumptions

4.1 Discretization

Detailed radiative transfer calculations are computationally demanding. Compromises must often be made between the accuracy of the computations and computational speed, depending on the application. Reductions in the spectral, spatial, and time resolution of the calculations are made in order to save computing time. This results in uncertainties. For example, a climate model needs accurate calculations of spectral gas optics that depend on the composition of the atmosphere spanning from the Earth's surface to the stratosphere but may treat the fast spatial and temporal variations in cloudiness in a simplified way. Using a reduced-resolution horizontal grid (a so-called radiation grid) and a reduced time resolution (i.e., calling the radiation scheme only at selected times), as applied for example in the IFS/ECMWF model (Hogan and Bozzo, 2018), is justified in order to allow increased spectral resolution in the radiation calculations. On the contrary, a mesoscale NWP model focuses on detailed short-range local weather forecasting. In such models, it is possible, and also necessary, to take spatial and temporal variations in cloud-radiation interactions into account. The use of a broadband or single spectral interval (i.e., one SW and one LW radiation band) schemes (e.g., Mašek et al., 2016; Geleyn et al., 2017; Rontu et al., 2017) at each grid point at each time-step may therefore be optimal in mesoscale models.

The horizontal and vertical resolution of NWP models continues to increase toward subkilometer scales. This means that the grid-scale equations of the model dynamics are formulated in a framework, where the details of the surface elevation or near-surface temperature and humidity profiles become more resolved. The concept of model physical parameterizations that account for subgrid-scale processes independently in each grid-column requires rethinking. Microphysical processes related to clouds and radiation will always require explicit parameterizations. However, variables and concepts that depend on the grid resolution, such as cloud fraction, assumptions about cloud overlap in the vertical direction and topographic effects on radiation, are affected. This poses requirements for a consistent, seamless formulation of the changing subgrid- and grid-scale processes across different horizontal and vertical scales.

4.2 Cloud overlap models

When you have two or more partially cloudy grid cells in a column of an NWP model, how the clouds overlap is of concern. Commonly used approaches include maximum-random overlap and exponential overlap. In the maximum-random overlap assumption, vertically adjacent grid boxes are treated as overlapping maximally, while nonadjacent grid boxes are treated as overlapping randomly. A detailed description of the exponential overlap method can be found in Hogan and Bozzo (2018).

4.3 3D cloud effects

Examples of 3D radiative effects of clouds include the side illumination of clouds by SW radiation, which is particularly important for low solar angles, the escape of SW radiation from the sides of clouds, strongest when the sun is overhead as forward scattering leads to more penetration of SW irradiance though the clouds, and SW irradiance reaching the surface via reflections from the top and bottom of nearby clouds. 3D radiative effects of clouds also apply in the LW.

3D radiative effects of clouds are becoming an increasingly important topic as we move to higher-resolution model grids. However, such effects are not employed in most contemporary models. ECMWF's IFS model is the first global model capable of representing such effects with their SPARTACUS radiative transfer solver.

The 3D radiative effect of clouds results in a warming of the surface both due to LW and SW. The effects are a similar order of magnitude to the doubling of CO_2 (Hogan, R., personal communication, 2016) and are shown to be about 4 W m^{-2} at the surface and TOA.

4.4 Topographical 3D effects

Over complex orography the elevation angle of the slope, and its orientation, define the local incident angle of direct SW radiation. The largest impact of topography on radiation fluxes is usually due to this so-called slope effect. Shadowing by nearby mountains may block the arrival of direct SW radiation when the Sun is low enough. This is called the shadow effect. An obscured view of the sky, or the sky-view effect, limits the arrival of diffuse SW radiation to the surface but also limits cooling to space by LW radiation. The resulting radiation balance differences impact on local weather and climate via surface temperature gradients, which may also lead to local thermal circulations. The relative impacts are greater when the solar angle is low, and may be important at high latitudes. The sky-view effect on diffuse SW and LW radiation can be significant at all latitudes and times of the day. For example, maximum local changes in the direct SW radiation flux in NWP experiments over the Caucasian mountains were reported to vary from −200 to +100 W m^{-2} in February while the changes in diffuse radiation were an order of magnitude smaller (Rontu et al., 2016). The aim of including a parameterization of orographic radiation effects in an NWP model is to account for these local differences. Such parameterizations have been developed for several mesoscale NWP models (Müller and Scherer, 2005; Senkova et al., 2007; Helbig and Löwe, 2012; Manners et al., 2012; Rontu et al., 2016).

The net SW radiation flux at the model surface is modified by multiplying the grid-scale components of the flux on a flat terrain by the following factors: slope (δ_{sl}), shadow (δ_{sh}), and sky view (δ_{sv}). These factors are defined as follows, and are slightly modified from those in Senkova et al. (2007):

$$F_{\mathrm{SW,net,oro}} = \sum_i \left(F^{\downarrow}_{\mathrm{SW,dir},i,\mathrm{oro}} + F^{\downarrow}_{\mathrm{SW,diff,oro}} - F^{\uparrow}_{\mathrm{SW,diff},i,\mathrm{oro}} \right). \tag{15}$$

The terrain-modified direct, diffuse upward, and diffuse downward fluxes with subscripts *oro* denoting orographically modified flux, and *surf* referring to unmodified surface flux from Eq. (3) for each spectral interval are given by the following:

$$\begin{aligned}
F^{\downarrow}_{\mathrm{SW,dir},i,\mathrm{oro}} &= \delta_{sl}\delta_{sh}F^{\downarrow}_{\mathrm{SW,dir},i,\,\mathrm{surf}} \\
F^{\uparrow}_{\mathrm{SW,diff},i,\mathrm{oro}} &= \alpha_{\mathrm{dir},i,\,\mathrm{surf}}(\mu_0)F^{\downarrow}_{\mathrm{SW,dir},i,\mathrm{oro}} + \alpha_{\mathrm{diff},i,\,\mathrm{surf}}F^{\downarrow}_{\mathrm{SW,diff},i,\,\mathrm{surf}} \\
F^{\downarrow}_{\mathrm{SW,diffioro}} &= \delta_{sv}F^{\downarrow}_{\mathrm{SW,diffi,\,surf}} + (1-\delta_{sv})F^{\uparrow}_{\mathrm{SW,diffi,oro}}.
\end{aligned}$$

The slope factor δ_{sl} represents the local incidence angle of the Sun's rays on the slope (Kondratyev, 1977), δ_{sh} is the shadow factor related to cast shadows, and δ_{sv} is the sky view, or local horizon angle from the inclined surface, integrated over 360 degrees. The other terms in the expressions are explained in Section 1.

The corresponding modification to the net LW radiation is simple. LW fluxes in valleys are assumed to be balanced in all directions except in the direction of the sky:

$$F_{\mathrm{LW,net,oro}} = \delta_{\mathrm{sv}} F_{\mathrm{LW,net,surf}}. \tag{16}$$

The basic building blocks for the grid-scale factors δ_{sl}, δ_{sh}, and δ_{sv}, which are applied during the model's time integration, are the prescribed slope height, azimuth angles and the directional local horizon angles. These are derived from fine-resolution surface elevation data. Such data are available from a digital elevation model (DEM) such as SRTM (Jarvis et al., 2008), which provides global data up to a resolution of 3 arcseconds. The prescribed angles can be calculated within the grid of the source DEM (Rontu et al., 2016) or within the grid of a specified very high-resolution NWP model (Manners et al., 2012). Calculations within the DEM grid are followed by statistical aggregation for use on the NWP model grid. Here, directional slopes, fractions of slopes, and the mean local horizon in each sector (e.g., 8 sectors each of 45 degrees, centered on each grid square) are calculated. The changing position of the Sun is accounted for by selection of the precalculated parameters in the correct direction at each time-step during the model run. Such a method is universally applicable in NWP or climate models of any resolution. Without aggregation of the orography-related parameters, the fine-resolution slope and horizon angles could be used for downscaling simulated radiation fluxes to specific locations. Such downscaling may be useful for road surface weather or solar energy applications.

By introducing an orographic radiation parameterization, the uncertainties in the surface energy balance due to new interactions and systematic dependencies within the model are potentially increased. Locally, systematic effects on radiation fluxes may be introduced that mostly smooth out in averages over larger area. The parameterizations are constrained by the prescribed topography, the diurnal cycle of solar irradiance (prescribed), and the SW and LW atmospheric radiative transfer (given by the atmospheric radiation parameterizations). The largest uncertainties in the orographic radiation parameterization are related to the estimation of the orography-dependent variables, which are used to modify the grid-scale surface-radiation fluxes during a forecast.

Uncertainties in their derivation are illustrated by an example of the sky-view factor in Rontu et al. (2016). The sky view should represent the visibility of the open sky as seen from a point on an inclined surface (Manners et al., 2012) (a grid box in the terrain-following coordinate of an atmospheric model may also be sloped). There are different ways to do the aggregation and to take the slope into account. In the experiments in Rontu et al. (2016), three different methods resulted in the average δ_{sv} values varying between 0.87 and 0.91 over an area in the Caucasian mountains. According to the single-column experiments at chosen locations, the night-time T_{surf} increased when the sky-view factor decreased, the maximum increase being of the order of 1°C when δ_{sv} decreased from 1 to 0.85. In an NWP model, the sky-view effect on LW radiation is systematic, and depends only on surface elevation variations within and beyond each grid box (beyond, if the directional local horizons are estimated across several grid boxes). Thus, there is a possibility to introduce a systematic error and not just a temporary uncertainty if δ_{sv} is not properly calculated.

In orographic radiation parameterizations (e.g., Senkova et al., 2007, as described earlier), the direct and diffuse SW radiation fluxes are treated independently and differently using diagnostic values from the atmospheric radiation parameterizations as input. Any uncertainty in these diagnostics may be enhanced in the final values of the SW↓ and net SW irradiances, updated by the orographic parameterization, which influence the surface energy balance. The parameterization uses the surface albedo and emissivity, α and ϵ (Eq. 15), and assumes that the values in the grid box are representative of the surroundings. For example, in a valley surrounded by snow-capped mountains, this may lead to inaccuracies, depending on the resolution of the model grid.

Acknowledgment

The authors thank the international HIRLAM consortium for its support that made writing this radiation chapter possible.

References

Bengtsson, L., Andrae, U., Aspelien, T., Batrak, Y., Calvo, J., de Rooy, W., Gleeson, E., Sass, B.H., Homleid, M., Hortal, M., Ivarsson, K.I., Lenderink, G., Niemelä, S., Nielsen, K.P., Onvlee, J., Rontu, L., Samuelsson, P., Santos Muñoz, D., Subias, A., Tijm, S., Toll, V., Yang, X., Køltzow, M.Ø., 2017. The HARMONIE-AROME model configuration in the ALADIN-HIRLAM NWP system. Mon. Weather Rev. 145 (5), 1919–1935.

Berk, A., Hawes, F., 2017. Validation of MODTRAN®6 and its line-by-line algorithm. J. Quant. Spectrosc. Radiat. Transf. 203, 542–556. https://doi.org/10.1016/j.jqsrt.2017.03.004.

Bozzo, A., Benedetti, A., Flemming, J., Kipling, Z., Rémy, S., 2020. An aerosol climatology for global models based on the tropospheric aerosol scheme in the integrated forecasting system of ECMWF. Geosci. Model Dev. 13, 1007–1034. https://doi.org/10.5194/gmd-13-1007-2020.

Coddington, O., Lean, J.L., Pilewskie, P., Snow, M., Lindholm, D., 2016. A solar irradiance climate data record. Bull. Am. Meteorol. Soc. 97 (7), 1265–1282. https://doi.org/10.1175/bams-d-14-00265.1.

Doppler, L., Preusker, R., Bennartz, R., Fischer, J., 2014. k-bin and k-IR: k-distribution methods without correlation approximation for non-fixed instrument response function and extension to the thermal infrared—applications to satellite remote sensing. J. Quant. Spectrosc. Radiat. Transf. 133, 382–395.

Dunne, E.M., Gordon, H., Kürten, A., Almeida, J., Duplissy, J., Williamson, C., Ortega, I.K., Pringle, K.J., Adamov, A., Baltensperger, U., et al., 2016. Global atmospheric particle formation from CERN CLOUD measurements. Science 354 (6316), 1119–1124.

Dutton, E.G., Christy, J.R., 1992. Solar radiative forcing at selected locations and evidence for global lower tropospheric cooling following the eruptions of El Chichón and Pinatubo. Geophys. Res. Lett. 19 (23), 2313–3116.

Fouquart, Y., Bonnel, B., 1980. Computations of solar heating of the Earth's atmosphere—a new parameterization. Beitr. Phys. Atmos. 53, 35–62.

Fu, Q., 1996. An accurate parameterization of the solar radiative properties of cirrus clouds for climate models. J. Clim. 9, 2058–2082.

Fu, Q., Liou, K.N., 1992. On the correlated k-distribution method for radiative transfer in nonhomogeneous atmospheres. J. Atmos. Sci. 49 (22), 2139–2156. https://doi.org/10.1175/1520-0469(1992)049<2139:OTCDMF>2.0.CO;2.

Fu, Q., Yang, P., Sun, W.B., 1998. An accurate parameterization of the infrared radiative properties of cirrus clouds for climate models. J. Clim. 11, 2223–2237.

Geleyn, J.F., Mašek, J., Brožková, R., Kuma, P., Degrauwe, D., Hello, G., Pristov, N., 2017. Single interval longwave radiation scheme based on the net exchanged rate decomposition with bracketing. Q. J. R. Meteorol. Soc. 143 (704), 1313–1335. https://doi.org/10.1002/qj.3006.

Gleeson, E., Toll, V., Nielsen, K.P., Rontu, L., Mašek, J., 2016. Effects of aerosols on clear-sky solar radiation in the ALADIN-HIRLAM NWP system. Atmos. Chem. Phys. 16 (9), 5933–5948. https://doi.org/10.5194/acp-16-5933-2016.

Gordon, I.E., Rothman, L.S., Hill, C., Kochanov, R.V., Tan, Y., Bernath, P.F., Birk, M., Boudon, V., Campargue, A., Chance, K.V., Drouin, B.J., Flaud, J.M., Gamache, R.R., Hodges, J.T., Jacquemart, D., Perevalov, V.I., Perrin, A., Shine, K.P., Smith, M.A.H., Tennyson, J., Toon, G.C., Tran, H., Tyuterev, V.G., Barbe, A., Császár, A.G., Devi, V.M., Furtenbacher, T., Harrison, J.J., Hartmann, J.M., Jolly, A., Johnson, T.J., Karman, T., Kleiner, I., Kyuberis, A.A., Loos, J., Lyulin, O.M., Massie, S.T., Mikhailenko, S.N., Moazzen-Ahmadi, N., Müller, H.S.P., Naumenko, O.V., Nikitin, A.V., Polyansky, O.L., Rey, M., Rotger, M., Sharpe, S.W., Sung, K., Starikova, E., Tashkun, S.A., Auwera, J.V., Wagner, G., Wilzewski, J., Wcisło, P., Yu, S., Zak, E.J., 2017. The HITRAN2016 molecular spectroscopic database. J. Quant. Spectrosc. Radiat. Transf. 203, 3–69. https://doi.org/10.1016/j.jqsrt.2017.06.038.

Hansen, J., Sato, M., Lacis, A., Ruedy, R., Lelieveld, J., 1997. The missing climate forcing [and discussion]. Philos. Trans. Biol. Sci. 352 (1350), 231–240. https://doi.org/10.2307/56565.

Helbig, N., Löwe, H., 2012. Shortwave radiation parameterization scheme for subgrid topography. J. Geophys. Res. Atmos. 117 (D3), 2156–2202. https://doi.org/10.1029/2011JD016465.

Henyey, L.G., Greenstein, J.L., 1940. Diffuse radiation in the galaxy. Ann. Astrophys. 3, 117–137.

Hersbach, H., Dee, D.J.E.N., 2016. Era5 reanalysis is in production. ECMWF Newslett. 147 (7), 5–6.

Hestenes, K., Nielsen, K.P., Zhao, L., Stamnes, J.J., Stamnes, K., 2007. Monte Carlo and discrete-ordinate simulations of spectral radiances in a coupled air-tissue system. Appl. Opt. 46 (12), 2333–2350.

Hogan, R.J., Bozzo, A., 2018. A flexible and efficient radiation scheme for the ECMWF model. J. Adv. Model. Earth Syst. 10. https://doi.org/10.1029/2018MS001364.

Hogan, R.J., Schäfer, S.A., Klinger, C., Chiu, J.C., Mayer, B., 2016. Representing 3-D cloud radiation effects in two-stream schemes: 2. Matrix formulation and broadband evaluation. J. Geophys. Res. Atmos. 121 (14), 8583–8599.

Hogan, R.J., Fielding, M.D., Barker, H.W., Villefranque, N., Schäfer, S.A.K., 2019. Entrapment: an important mechanism to explain the shortwave 3D radiative effect of clouds. J. Atmos. Sci. 76 (7), 2123–2141.

Inness, A., Ades, M., Agustí-Panareda, A., Barré, J., Benedictow, A., Blechschmidt, A.M., Dominguez, J.J., Engelen, R., Eskes, H.F.J., Huijnen, V., Jones, L., Kipling, Z., Massart, S., Parrington, M., Peuch, V.H., Razinger, M., Remy, S., Schulz, M., Suttie, M., 2019. The CAMS reanalysis of atmospheric composition. Atmos. Chem. Phys. 19, 3515–3556. https://doi.org/10.5194/acp-19-3515-201.

Jarvis, A., Reuter, H.I., Nelson, A., Guevara, E., 2008. Hole-filled seamless SRTM for the globe Version 4. In: CGIAR Consortium for Spatial InformationCGIAR-CSI, Washington, DC, pp. 1–9 Available from http://srtm.csi.cgiar.org.

Joseph, J.H., Wiscombe, W.J., Weinman, J.A., 1976. The Delta-Eddington approximation for radiative flux transfer. J. Atmos. Sci. 33, 2452–2459.

Kondratyev, K.Y., 1977. Radiation regime of inclined surfaces. WMO Tech. Note 152, 82 pp.

Kopp, G., Lean, J.L., 2011. A new, lower value of total solar irradiance: evidence and climate significance. Geophys. Res. Lett. 38, L01706. https://doi.org/10.1029/2010GL045777.

Kopp, G., Krivova, N., Wu, C.J., Lean, J.L., 2016. The impact of the revised sunspot record on solar irradiance reconstructions. Solar Phys. 291 (9–10), 2951–2965. https://doi.org/10.1007/s11207-016-0853-x.

Kurucz, R.L., 1992. Synthetic infrared spectra. In: Rabin, D.M., Jefferies, J.T. (Eds.), Infrared Solar Physics, IAU Symp. 154. Kluwer Academic Publishers, Norwell, MA.

Lacis, A.A., Oinas, V., 1991. A description of the correlated k distributed method for modeling nongray gaseous absorption, thermal emission, and multiple scattering in vertically inhomogeneous atmospheres. J. Geophys. Res. 96, 9027–9063. https://doi.org/10.1029/90JD01945.

Lindner, T.H., Li, J., 2000. Parameterization of the optical properties for water clouds in the infrared. J. Clim. 13 (10), 1797–1805.

Manners, J., Vosper, S.B., Roberts, N., 2012. Radiative transfer over resolved topographic features for high-resolution weather prediction. Q. J. R. Meteorol. Soc. 138 (664), 720–733. https://doi.org/10.1002/qj.956.

Martin, G.M., Johnson, D.W., Spice, A., 1994. The measurement and parameterization of effective radius of droplets in warm stratocumulus clouds. J. Atmos. Sci. 51, 1823–1842.

Mašek, J., Geleyn, J.F., Brožková, R., Giot, O., Achom, H.O., Kuma, P., 2016. Single interval shortwave radiation scheme with parameterized optical saturation and spectral overlaps. Q. J. R. Meteorol. Soc. 142 (694), 304–326. https://doi.org/10.1002/qj.2653.

Mayer, B., Kylling, A., 2005. Technical note: the libRadtran software package for radiative transfer calculations—description and examples of use. Atmos. Chem. Phys. Discuss. 5 (2), 1319–1381.

Meinshausen, M., Smith, S.J., Calvin, K., Daniel, J.S., Kainuma, M.L.T., Lamarque, J.-F., Matsumoto, K., Montzka, S.A., Raper, S.C.B., Riahi, K., et al., 2011. The RCP greenhouse gas concentrations and their extensions from 1765 to 2300. Clim. change 109 (1–2), 213.

Mie, G., 1908. Beiträge zur optik trüber medien, speziell kolloidaler metallösungen. Ann. Phys. 25 (3), 377–445.

Mlawer, E.J., Clough, S.A., 1997. On the extension of rapid radiative transfer model to the shortwave region. In: CONF-9603149. Proceedings of the 6th Atmospheric Radiation Measurement (ARM) Science Team Meeting, U.-S. Department of Energy, pp. 223–226.

Mlawer, E.J., Taubman, S.J., Brown, P.D., Iacono, M.J., Clough, S.A., 1997. Radiative transfer for inhomogeneous atmospheres: RRTM, a validated correlated-k model for the longwave. J. Geophys. Res. 102 (D14), 16663–16682.

Müller, M.D., Scherer, D., 2005. A grid- and subgrid-scale radiation parametrization of topographic effects for mesoscale weather forecast models. Mon. Weather Rev. 133, 1431–1442.

Nakicenovic, N., Alcamo, J., Grubler, A., Riahi, K., Roehrl, R.A., Rogner, H.H., Victor, N., 2000. Special Report on Emissions Scenarios (SRES), a Special Report of Working Group III of the Intergovernmental Panel on Climate Change. Cambridge University Press.

Nielsen, K.P., Gleeson, E., Rontu, L., 2014. Radiation sensitivity tests of the HARMONIE 37h1 NWP model. Geosci. Model Dev. 7, 1433–1449.

Pierluissi, J.H., Peng, G.S., 1985. New molecular transmission band models for LOWTRAN. Opt. Eng. 24, 541–547.

Pincus, R., Mlawer, E.J., Delamere, J.S., 2019. Balancing accuracy, efficiency, and flexibility in radiation calculations for dynamical models. J. Adv. Model. Earth Syst. 11 (10), 3074–3089. https://doi.org/10.1029/2019MS001621.

Räisänen, P., Lindfors, A.V., 2019. On the computation of apparent direct solar radiation. J. Atmos. Sci. 76 (9), 2761–2780.

Reda, I., Andreas, A., 2008. Solar position algorithm for solar radiation applications. National Renewable Energy Laboratory, Golden, CO NREL/TP-560-34302 NREL/TP-560-34302.

Rémy, S., Kipling, Z., Flemming, J., Boucher, O., Nabat, P., Michou, M., Bozzo, A., Ades, M., Huijnen, V., Benedetti, A., Engelen, R., Peuch, V.H., Morcrette, J.J., 2019. Description and evaluation of the tropospheric aerosol scheme in the European Centre for Medium-Range Weather Forecasts (ECMWF) Integrated Forecasting System (IFS-AER, Cycle 45R1). Geosci. Model Dev. 12, 4627–4659.

Ricchiazzi, P., Yang, S., Gautier, C., Sowle, D., 1998. SBDART: a research and teaching software tool for plane-parallel radiative transfer in the Earth's atmosphere. Bull. Am. Meteor. Soc. 79 (10), 2101–2114.

Rontu, L., Wastl, C., Niemelä, S., 2016. Influence of the details of topography on weather forecast—evaluation of HARMONIE experiments in the Sochi Olympics domain over the Aucasian mountains. Front. Earth Sci. 4. https://doi.org/10.3389/feart.2016.00013.

Rontu, L., Gleeson, E., Räisänen, P., Nielsen, K.P., Savijärvi, H., Sass, B.H., 2017. The HIRLAM fast radiation scheme for mesoscale numerical weather prediction models. Adv. Sci. Res. 14, 195–215.

Rontu, L., Pietikäinen, J.P., Martin Perez, D., 2019. Renewal of aerosol data for Aladin-Hirlam radiation parametrizations. Adv. Sci. Res. 16, 129–136. https://doi.org/10.5194/asr-16-129-2019.

Rontu, L., Gleeson, E., Martin Perez, D., Pagh Nielsen, K., Toll, V., 2020. Sensitivity of radiative fluxes to aerosols in the Aladin-Hirlam numerical weather prediction system. Atmosphere 11, 205. https://doi.org/10.3390/atmos11020205.

Sagan, C., Pollack, J.B., 1967. Anisotropic nonconservative scattering and the clouds of Venus. J. Geophys. Res. 72 (2), 469–477.

Sahyoun, M., Wex, H., Gosewinkel, U., Šantl-Temkiv, T., Nielsen, N.W., Finster, K., Sørensen, J.H., Stratmann, F., Korsholm, U.S., 2016. On the usage of classical nucleation theory in quantification of the impact of bacterial INP on weather and climate. Atmos. Environ. 139, 230–240.

Schäfer, S.A.K., Hogan, R.J., Klinger, C., Chiu, J.C., Mayer, B., 2016. Representing 3-D cloud radiation effects in two-stream schemes: 1. Longwave considerations and effective cloud edge length. J. Geophys. Res. Atmos. 121 (14), 8567–8582.

Schuster, A., 1905. Radiation through a foggy atmosphere. Astrophys. J. 21 (1), 1–22.

Schwarzschild, K., 1906. Ueber der Gleichgewicht der Sonnenatmosphäre. Nachrichten von der Königlichen Gesellschaft der Wissenschaften zu Göttingen. Math. Phys. Klasse 195, 41–53.

Senkova, A.S., Rontu, L., Savijärvi, H., 2007. Parametrization of orographic effects on surface radiation in HIRLAM. Tellus 59A, 279–291.

Sigl, M., Winstrup, M., McConnell, J.R., Welten, K.C., Plunkett, G., Ludlow, F., Büntgen, U., Caffee, M., Chellman, N., Dahl-Jensen, D., et al., 2015. Timing and climate forcing of volcanic eruptions for the past 2,500 years. Nature 523 (7562), 543–549.

Slingo, A., 1989. A GCM parameterization for the shortwave radiative properties of water clouds. J. Atmos. Sci. 46, 1419–1427.

Stamnes, K., Tsay, S.C., Wiscombe, W., Jayaweera, K., 1988. Numerically stable algorithm for discrete-ordinate-method radiative transfer in multiple scattering and emitting layered media. Appl. Opt. 27, 2502–2509.

Stamnes, K., Tsay, S.C., Laszlo, I., 2000. DISORT, A General-Purpose Fortran Program for Discrete-Ordinate-Method Radiative Transfer in Scattering and Emitting Layered Media: Documentation and Methodology. Stevens Institute of Technology, Hoboken, NJ.

Sun, Z., 2001. Reply to comments by Greg M. McFarquhar on Parametrization of effective sizes of cirrus-cloud particles and its verification against observations. Q. J. R. Meteorol. Soc. 127, 267–271.

Sun, Z., Rikus, L., 1999. Parametrization of effective sizes of cirrus-cloud particles and its verification against observations. Q. J. R. Meteorol. Soc. 125, 3037–3055.

Tegen, I., Hollrig, P., Chin, M., Fung, I., Jacob, D., Penner, J., 1997. Contribution of different aerosol species to the global aerosol extinction optical thickness: estimates from model results. J. Geophys. Res. Atmos. (1984–2012) 102 (D20), 23895–23915.

Thomas, G.E., Stamnes, K., 2002. Radiative Transfer in the Atmosphere and Ocean. Cambridge University Press, New York, NY.

Toll, V., Christensen, M., Quaas, J., Bellouin, N., 2019. Weak average liquid-cloud-water response to anthropogenic aerosols. Nature 572, 51–55. https://doi.org/10.1038/s41586-019-1423-9.

Wiscombe, W.J., 1980. Improved Mie scattering algorithms. Appl. Opt. 19, 1505–1509.

Yang, F., Mitchell, K., Hou, Y., Dai, Y., Zeng, X., Wang, Z., Liang, X., 2008. Dependence of land surface albedo on solar zenith angle: observations and model parameterization. J. Appl. Meteorol. Climatol. 47, 2963–2982. https://doi.org/10.1175/2008JAMC1843.1.

CHAPTER

10

Uncertainties in the parameterization of cloud microphysics: An illustration of the problem

Jian-Wen Bao[a], *Sara Michelson*[a,b], *and Evelyn Grell*[a,b]

[a]NOAA/Physical Sciences Laboratory, Boulder, CO, United States [b]CIRES, University of Colorado Boulder, Boulder, CO, United States

1 Introduction

The parameterization of cloud microphysics is essential to numerical weather prediction (NWP) and climate models because it governs the transport, local change, and thermodynamic effects of hydrometeor particles in clouds. Even in very high-resolution NWP models, microphysical processes occur on a scale too small to be simulated explicitly. These processes vary with dynamical feedback of airflow in both the horizontal and vertical. Despite the fact that, over the past few decades, much progress has been made in the observation, understanding, and parameterization of physical processes that govern the production and evolution of cloud particles, there are still great uncertainties in all the parameterization schemes used in the state-of-the-art NWP and climate models. In the context of this chapter, uncertainties do not necessarily mean errors; they represent what we do not know accurately/precisely. Seifert (2011) provided the following three main reasons for the great uncertainties, which we think still remain true as of the writing of this chapter:

1. "There are still gaps in the empirical and theoretical description of cloud processes like ice nucleation, aggregation and splintering of ice particles, collision rates in turbulent flows, the breakup of drops, etc."
2. "The natural variability of clouds, cloud particles, and aerosol is overwhelmingly large, e.g., the different particle habits (including degrees of riming), the time-spatial structures in clouds, as well as the particle size distributions, etc."

https://doi.org/10.1016/B978-0-12-815491-5.00010-0

3. "The strong nonlinearity and high complexity of cloud processes hinder any rigorous analytic and theoretical approaches."

The literature on the parameterization of cloud microphysics is so vast and so many aspects of the science are still being debated that it is impossible to provide a complete review on the subject in the limited space of a single chapter. Instead, in this chapter, only some of the uncertainties in the parameterization of microphysics are illustrated by using a community NWP model to simulate an idealized 2D convection development case. A review is also provided for relevant and recent results pertaining to the microphysics parameterization schemes that are used in the illustration.

2 A brief history of the development of cloud microphysics schemes within the community WRF model

The community model used in this chapter to illustrate uncertainties in the parameterization of microphysics is the Weather Research and Forecasting (WRF) model (Skamarock et al., 2008). This model was developed in the late 1990s to replace the then widely used mesoscale model (referred to as the MM5 model), a nonhydrostatic extension of the hydrostatic model (referred to as the MM4 model) that had been developed in the late 1970s and 1980s by a collaborative effort (Anthes and Warner, 1978; Anthes et al., 1987). Like the MM5 model, the WRF model consists of four-dimensional data assimilation modules and various physical parameterizations. Its development has been supported by a collaborative partnership led by the National Center for Atmospheric Research and the National Oceanic and Atmospheric Administration.

The complexity of the parameterizations of microphysics in the MM4/MM5 and WRF models has evolved as the user community expanded. At the time when the MM5 model was released to the public in the early 1990s, the model had only two options for microphysical parameterizations. The first was the bulk warm rain scheme with mixing ratios of cloud water and rain water predicted (Hsie et al., 1984), which was based on Kessler's warm rain parameterization (Kessler, 1969). The second was the bulk mixed-phase scheme with cloud content in one predictive field and rain water (for $T > 0°C$) or snow (for $T < 0°C$) in another predictive field (Dudhia, 1989). These two schemes were adapted into the WRF model when it superseded the MM5 model in the late 1990s. It is important to mention that besides the two schemes, Reisner et al. (1998) added three new microphysics parameterization options of increasing complexity to the MM5 model. They were (1) a bulk mixed-phase scheme with predictive fields for cloud water, rain water, snow, and cloud ice, (2) a bulk mixed-phase model with predictive fields cloud water, rain water, snow, and cloud ice, graupel and number concentration of cloud ice, and (3) a bulk mixed-phase model with predictive fields of cloud water, rain water, snow and cloud ice, graupel, number concentrations of cloud ice and snow. These new options were based on previous work in this field by Lin et al. (1983), Rutledge and Hobbs (1983), Murakami (1990), and Ikawa and Saito (1991).

In the early 2000s, as the area of the WRF model applications widened, a bulk scheme with predictive fields of cloud water, cloud ice, rain water, snow and graupel was developed by Chen and Sun (2002) and based on the scheme presented by Lin et al. (1983) was added to the option list of the microphysics parameterization schemes in the WRF model. At about the same time, Hong et al. (2004) added a scheme based on Dudhia (1989) and Rutledge and Hobbs (1983) to the list. In particular, to improve explicit, real-time forecasts of supercooled liquid water and hence improve forecasts of aircraft icing, the most complex one of the three bulk, mixed-phase microphysical parameterization scheme described in Reisner et al. (1998) was adapted, further improved and recoded in the WRF model by Thompson et al. (2004, 2008). During this time period, based on the work by Dudhia (1989), Morrison et al. (2003, 2005) developed another bulk scheme that included predictive fields for mass mixing ratio of cloud water, rain water, cloud ice, and snow and for number mixing ratio of these hydrometeor species except for cloud water. This scheme was added to the WRF physics option list with an extension to include graupel in the hydrometeor category and predictive fields.

Starting in the mid-2000s, as the number of available microphysics parameterization schemes increased, more comparative and competitive schemes were included in the WRF model. By the time Version 3.7 of the model was released in 2015, the user could select from more than 24 bulk schemes (including variants of the same scheme). Their complexity varied from single-moment to two-moment formulations with various numbers of hydrometeor categories, and to the so-called spectral bin formulation (see more detailed explanation in the next section). Since then, more microphysics parameterization schemes have been added to the option list of the WRF model that include more flexibility and improvement in microphysics processes for the evolution of frozen hydrometeors and aerosol-cloud interaction.

From a historical perspective, the evolution of the options for microphysics parameterization schemes in the MM4/5 and WRF models followed the general trend in the literature. In particular, all the microphysics parameterization schemes in the WRF model are some realization of the following governing processes that were summarized by Cotton and Anthes (1989) and Straka (2009):

- The nucleation of droplets on aerosol particles
- Condensation and evaporation of cloud droplets as well as drizzle and raindrops
- The development of a mature raindrop spectrum by collection of other liquid species (including cloud droplets, drizzle, and raindrops themselves)
- The inclusion of breakup of raindrops
- The occurrence of self-collection in the droplet and drop spectra
- Homogeneous freezing of cloud drops into ice crystals
- Primary, heterogeneous ice nucleation mechanisms such as contact freezing, deposition, and immersion freezing nucleation
- Secondary ice nucleation mechanisms such as rime-splintering ice production and mechanical fracturing of ice
- Vapor deposition and sublimation of ice particles
- Riming and density changes of ice particles
- Aggregation of ice crystals to form snow aggregates
- Graupel initiation by freezing of drizzle and subsequent heavy riming
- Graupel initiation by heavy riming of ice crystals
- Freezing of raindrops, with smaller particles becoming graupel particle embryos owing to riming, and larger particles possibly becoming hail embryos
- Graupel and frozen drops becoming hail embryos by collecting rain or heavy riming
- Temperature prediction of ice-water particles

- Density changes in graupel and hail
- Shedding from graupel during wet growth and melting
- Soaking of graupel particles during wet growth and melting
- Melting of frozen particles
- Differential sedimentation of liquid and frozen particles

It is important to point out that the development of individual microphysics parameterization schemes in the WRF model is usually motivated by specific applications and the complexity of each scheme is determined by the theoretical equations that can be derived from the first principles (see more detailed discussion in the next section) and the observations that can be used during the development. This explains, at least partially, the diversity of these schemes and is a root cause for the uncertainties illustrated and discussed in this chapter.

3 Theoretical basis for cloud microphysics parameterization: A perspective of cloud particle population balance

All bulk microphysics parameterization schemes share the same theoretical basis, which can be summarized as the following. Physically speaking, clouds are heterogeneous polydisperse systems, which consist of two-phase hydrometeors of different sizes and mass (Pruppacher and Klett, 1997). The work by Smoluchowski (1916) is considered to be the first one to formally use population balances for polydispersed particle dynamics. From a more general physical point of view, population balances are described by the Boltzmann transport equation, or equivalently, by the Reynolds transport theorem applied to the spectral mass density distribution function.

The evolution of the spectral number density distribution function, $f\left(m, \vec{r}, t\right)$, of a hydrometeor per unit volume with mass, m, is governed by the following spectral kinetic equation of population balance (see, e.g., Beheng, 2010):

$$\frac{\partial}{\partial t}f + \nabla\cdot\left[\vec{v}\left(\vec{r},t\right)f\right] + \frac{\partial}{\partial z}\left[v_{fall}(m,t)f.\right]$$
$$= Nucl\left(\vec{r},t\right) - \frac{\partial}{\partial m}(\dot{m}f) + \left(\frac{\partial f}{\partial t}\right)_{coag} + \left(\frac{\partial f}{\partial t}\right)_{break} + \frac{\partial}{\partial z}k_d\frac{\partial}{\partial z}f \quad (1)$$

where $\frac{\partial}{\partial z}\left[v_{fall}(m,t)f\right]$ is the gain and loss of f due to gravitational sedimentation with fall velocity v_{fall}; $Nucl\left(\vec{r},t\right)$ is the gain of f due to nucleation on aerosols (truncated at smallest mass); $\frac{\partial}{\partial m}(\dot{m}f)$ is the gain of f due to continuous mass deposition at the rate $\dot{m}$; $(\partial f/\partial t)_{coag}$ is the gain and loss of f due to coagulation through collision and coalescence; $(\partial f/\partial t)_{break}$ is the gain and loss of f due to breakup; and $\frac{\partial}{\partial z}k_d\frac{\partial}{\partial z}f$ is the gain and loss of f due to turbulent transport processes that are characterized by turbulent diffusivity k_d.

Instead of the above kinetic equation of population balance, for the convenience of practical applications, the so-called bulk representation of population balance is widely used in weather and climate modeling, in which the method of moments is utilized by multiplying the kinetic equation with various powers of radius or mass, integrating over the size spectrum and splitting the higher moments with various approximations. By using the method of moments with the following definition for moment:

$$M_n\left(\vec{r},t\right)=\int_0^\infty m^n f\left(m,\vec{r},t\right)\mathrm{d}m,\quad n=constant\geq 0, \tag{2}$$

Eq. (1) can be expressed as the following kinetic equation of moment balance:

$$\frac{\partial}{\partial t}M_n\left(\vec{r},t\right)+\nabla\cdot\left[\vec{v}\left(\vec{r},t\right)M_n\left(\vec{r},t\right)\right)]+\frac{\partial}{\partial z}\left[V_n\left(\vec{r},t\right)M_n\left(\vec{r},t\right)\right]$$

$$=\int_0^\infty m^n\left[Nucl\left(\vec{r},t\right)-\frac{\partial}{\partial m}(\dot{m}f)\right]\mathrm{d}m+\left(\frac{\partial}{\partial t}M_n\left(\vec{r},t\right)\right)_{coag+break}+\frac{\partial}{\partial z}k_d\frac{\partial}{\partial z}M_n\left(\vec{r},t\right), \tag{3}$$

where the following definition of the nth moment and m-power-weighted fall velocity is used:

$$V_n\left(\vec{r},t\right)=M_n^{-1}\int_0^\infty m^n v_{fall}(m,t)f\left(m,\vec{r},t\right)\mathrm{d}m. \tag{4}$$

For example, with $n=1$, the single-moment scheme for a hydrometeor species is rendered from Eq. (3). The double-moment scheme for a hydrometeor type can be obtained as two separate equations with $n=0$ and 1.

The right-hand side of Eqs. (1) and (3) includes source and sink processes for hydrometeor population, which are generally termed as birth and death processes in the theory and applications of population dynamics (Ramkrishna, 2000). In particular, Eq. (3) can be interpreted dynamically as a model of the competition of precipitating hydrometeor particles for suspended hydrometeor particles that is analogous to the predator-prey model in population dynamics (Wacker, 1995). For example, in an idealized precipitating shallow boundary layer, Eq. (3) can be reduced to an aerosol-cloud-precipitation system that exhibits predator-prey characteristics of population dynamics with rain acting as the predator and cloud as the prey (Koren and Feingold, 2011). It is important to note that a prominent feature of the predator-prey model is that under a wide range of conditions the interaction between predator and prey processes can render the model solution stable with respect to perturbations of intrinsic input parameters (see, e.g., Jackson, 1991). Although it is difficult, if not impossible, to analyze general stability characteristics of an actual bulk scheme of n-moment ($n>1$) and multiple hydrometeor species, it has become increasingly apparent that such a stability is often manifested as the competition of precipitating hydrometeor particles for suspended hydrometeor particles that buffer the scheme against strong perturbations to intrinsic parameters (see, e.g., Stevens and Feingold, 2009; Koren and Feingold, 2011). This stability with respect to perturbations of intrinsic input parameters has significant implications for modeling cloud and precipitation using the method of moments. First, the existence of the stability makes it possible to effectively reduce the number of degrees of freedom in schemes. Second, it also provides the schemes with a tolerance of parametric uncertainties.

The above theoretical overview provides the basis of mathematical physics for all the bulk microphysics parameterization schemes. There exist great uncertainties in the parameterization of all the processes on the right-hand side of Eqs. (1) and (3) due to uncertainties in (i) the parameterization complexity required for specific applications, (ii) cloud particle properties and size distribution assumptions, and (iii) the interaction of aerosols and clouds. In fact, there is no consensus in the literature on how individual parameterization terms should be mathematically formulated due to technical difficulties in observing these processes in nature and laboratory. In fact, for a given process, there are often multiple possible parameterization formulations as described in Straka (2009). This is the reason why there are more than 24 microphysics scheme options in the WRF model and the number is still increasing. In the next sections, by showing the differences in simulated convection due to the use of different microphysics parameterization schemes, we will demonstrate the major uncertainties in the simulated hydrometeor population characteristics in terms of number concentration, size distributions, and total mass.

4 Uncertainties in the parameterization complexity required for applications

The complex interaction between predator and prey processes in microphysics parameterization has two consequences. On the one hand, the feedback between predator and prey processes makes the system stable to perturbations of intrinsic parameters, which can help make the simulated cloud and precipitation production useful even if some aspects of microphysics parameterizations have significant errors and/or uncertainties. On the other hand, the complexity of the feedback makes it difficult, if not impossible, to attribute errors to individual parameterized processes. Thus, uncertainties inevitably exist in any microphysics parameterization scheme used in weather and climate modeling. In the following, three bulk microphysics schemes, representative of various complexities, will be used to illustrate where the major uncertainties lie.

The WRF model version 4.0.2 is used with the configuration of the standard 2D idealized squall line case. The 2D domain is run at 1 km resolution in the horizontal and 80 vertical model levels. All the simulations are run for 6 h, and the output from the model is written every minute. The standard 2D idealized squall line is initialized with a warm perturbation with a maximum perturbation potential temperature of 3 K prescribed at a height of 1.5 km, which tapers to zero following the cosine squared at a horizontal radius of 4 km and a vertical radius of 1.5 km. The environmental wind shear is set to $0.0048\,s^{-1}$ in the lowest 2.5 km. Open boundary conditions are used in all simulations.

The three bulk microphysics schemes used in the simulations include the WRF single-moment 6-class scheme (Hong and Lim, 2006; the simulation using this scheme is referred to as the WSM6 run hereafter), the one-moment and two-moment hybrid scheme developed by Thompson et al. (2008) (the simulation using this scheme is referred to as the Thompson run hereafter) and the two-moment scheme developed by Morrison et al. (2005, 2009) (the simulation using this scheme is referred to as the Morrison run hereafter). All three schemes predict mass mixing ratios of cloud water, rainwater, cloud ice, snow, and graupel. The Thompson scheme predicts hydrometeor particle number mixing ratio for rainwater and cloud ice, while the Morrison scheme predicts the particle number mixing ratio for all the hydrometeors except for cloud water. Hydrometeor size distributions are assumed to be a generalized gamma function, and simple power laws are used to relate mass and fall speed to the size of each hydrometeor.

We first illustrate the differences in the simulated structural characteristics of the idealized squall line using the different microphysics parameterization schemes. In all three simulations, moist convection starts within the first few minutes of the model simulation as the initial warm bubble rises. As condensation occurs in the rising warm bubble, the release of latent heat leads to the formation of cumulus/cumulonimbus clouds. The development of the squall line is the classical one; i.e., it is triggered by tilted updrafts in cumulus/cumulonimbus clouds, and it is maintained by new convection that forms as the downdraft-induced gust front advances. Cloud ice begins to form after 10 min, and precipitation at the surface starts to accumulate after 20 min. The structural evolution of the simulated squall line during the first 2 h of model simulation is depicted in Fig. 1 by the cross sections of equivalent potential temperature, along cross-section winds and potential temperature perturbation every 30 min for the first 2 h of the simulations. At 30 min into the simulations, the simulated convection is

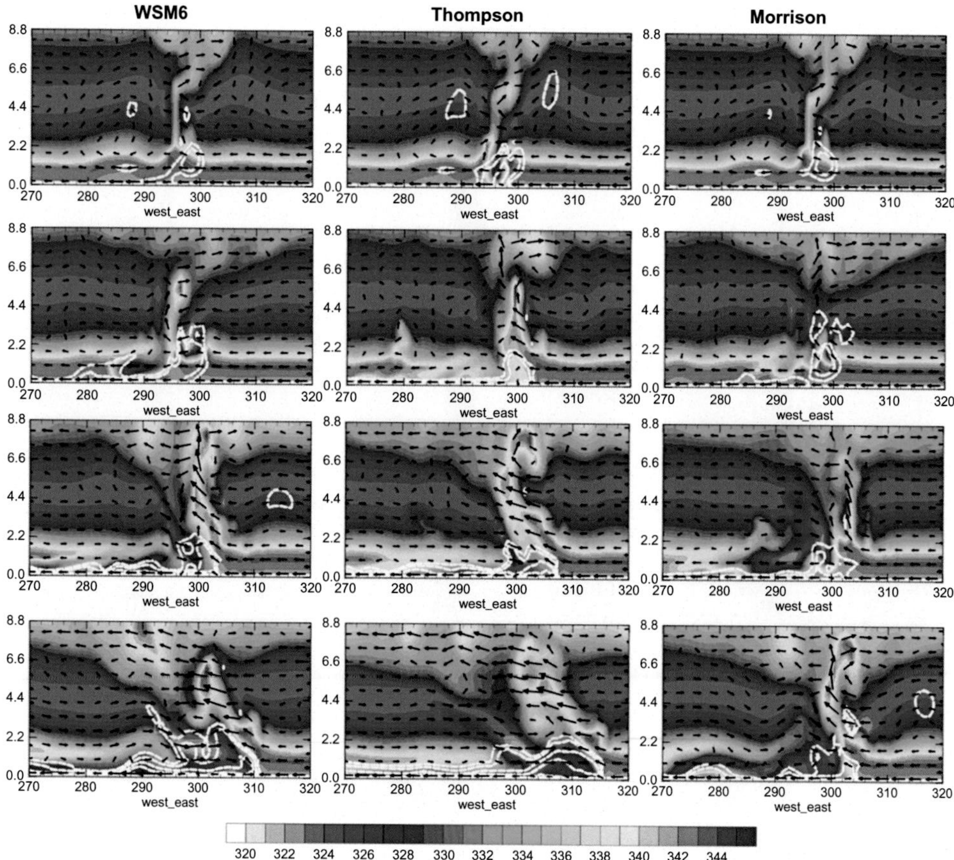

FIG. 1 Cross section of equivalent potential temperature (color-shaded in K), potential temperature perturbation (dashed white lines, contour interval of 2 K beginning with −1 K), along cross-section winds, at 30 min (*top row*), 60 min (*second row*), 90 min (*third row*), and 120 min (*bottom row*) into the simulations. The WSM6 run is in the *left column*, the Thompson run is in the *middle column*, and the Morrison run is in the *right column*.

upright or slightly downshear tilted in all three simulations. By 60 min into the model simulations, the convection in the Thompson run becomes upshear tilted while the convection in both the WSM6 and the Morrison runs is downshear tilted. The transition to upshear-tilted convection in the WSM6 and Morrison runs occurs after 90 min. By 2 h, the convection is upshear tilted in all three runs. The gust front (defined as the leading edge potential temperature perturbation) propagates faster and the cold pool behind the gust front is stronger (in terms of the potential temperature perturbation) in the Thompson run than in either the WSM6 or Morrison runs. These cross sections show that the dependence of the simulated structural evolution of the convection is consistent with previous conceptual models of squall line evolution (Weisman et al., 1988; Lafore and Moncrieff, 1989). That is, when the cold pool is weak, the convection is upright or downshear tilted. When the cold pool becomes stronger, the convection becomes upshear tilted.

Fig. 2 depicts cross sections of the total condensate mass mixing ratio at intervals of 30 min up to 2 h into the simulations. It shows that by 30 min into the simulations, the Thompson run

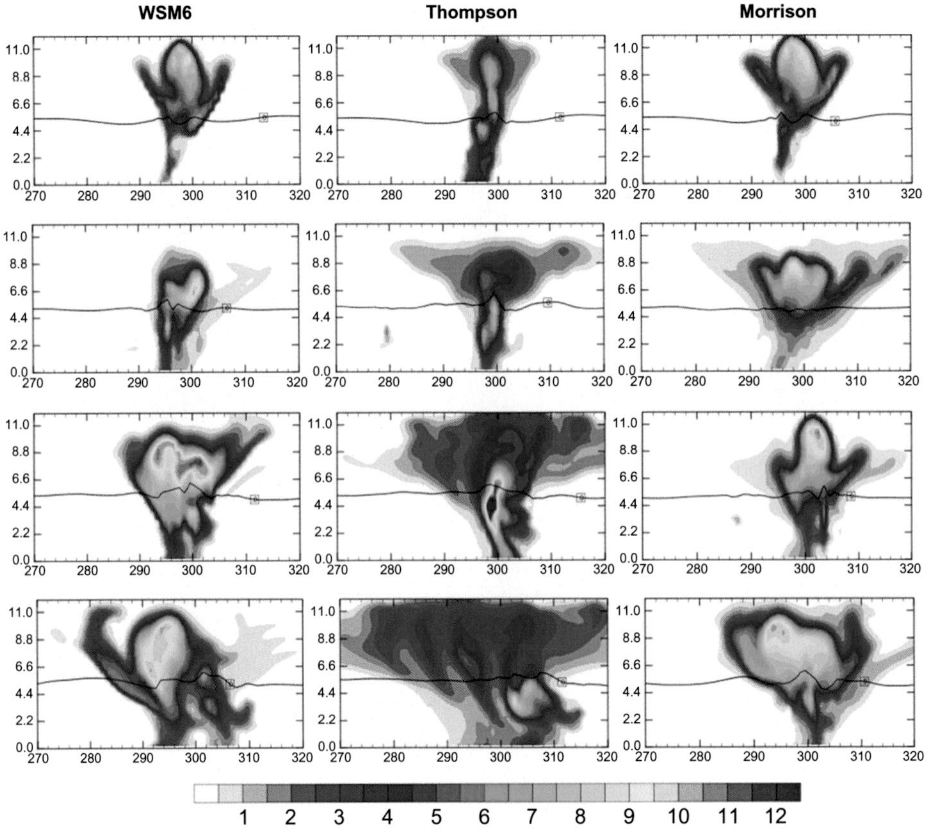

FIG. 2 Cross section of total condensate mass mixing ratio (color-shaded in g kg^{-1}), at 30 min (*top row*), 60 min (*second row*), 90 min (*third row*), and 120 min (*bottom row*) into the simulations. The WSM6 run is in the *left column*, the Thompson run is in the *middle column*, and the Morrison run is in the *right column*. The *black contour* is the 0°C line.

has more total condensate below the freezing level than the other two runs. By 60 min into the simulations, the vertical distribution of the total condensate is distinctly different between the three runs. In the Thompson scheme, the maximum of the total condensate is below the 0°C line while in the WSM6 run it is just above (~6–7 km), and in the Morrison run it is even higher (8–9 km). By 90 min and beyond, the gust front in the Thompson run has propagated the farthest (as seen in Fig. 1) with the most condensate below the freezing level. Since the propagation of the gust front and strength of the cold pool are largely determined by the melting of frozen hydrometeors and rain evaporation, it is consistent that the gust front in the Thompson run has propagated the farthest. Additionally, the shape of the total condensate boundary is different between all three runs at every time suggesting differences in the dynamic fields as a result of the differences in the microphysics schemes.

The quantity and timing of the surface precipitation are also impacted by the differences in the schemes (Fig. 3). At 30 min into the simulations, the Thompson and WSM6 run produce about the same amount of surface precipitation, but the Morrison run produces almost none. At 60 min, the Thompson run has the most precipitation, and the maximum is farther east than the other two runs. In general, precipitation is affected by many factors of the cloud microphysics and environment. Since the same idealized environment is prescribed for all the simulations, different amounts of precipitation depicted in Fig. 3 can only be attributed to the differences between the schemes. In fact, as will be shown later in the analysis and comparison of various microphysics hydrometeor production rates, the differences in the total condensate mass mixing ratio and surface precipitation are due to the differences in the size distributions and the pathways of hydrometeor production. Ultimately, these differences lead to differences in the strength of the cold pool and the structural development of the simulated squall lines.

The differences shown above demonstrate that even in the widely used WRF model, microphysics parameterization schemes of various complexity result in significant differences in the simulated squall line development. There are still substantial uncertainties in the literature about which of these schemes is optimal for the simulation of mixed-phase clouds commonly seen in deep convection. We will further demonstrate next how the uncertainties in certain aspects of the microphysics parameterization lead to the differences in the simulated squall lines.

5 Uncertainties in the parameterized warm rain processes

Bulk microphysics parameterization schemes are expressed as tendency equations that are made up of parameterized microphysical process terms to represent physical production and reduction as well as redistribution via sedimentation for each predicted moment of each hydrometeor species. These parameterized microphysical processes can be grouped into those responsible for the interaction and evolution of liquid hydrometeors and those responsible for the production of frozen hydrometeors, commonly referred to as warm rain processes and cold rain processes, respectively. As shown above, the simulated development of the idealized squall line undergoes three stages: warm bubble rising, mixed-phase cloud production, and formation of the squall line. To illustrate the uncertainties in the parameterized

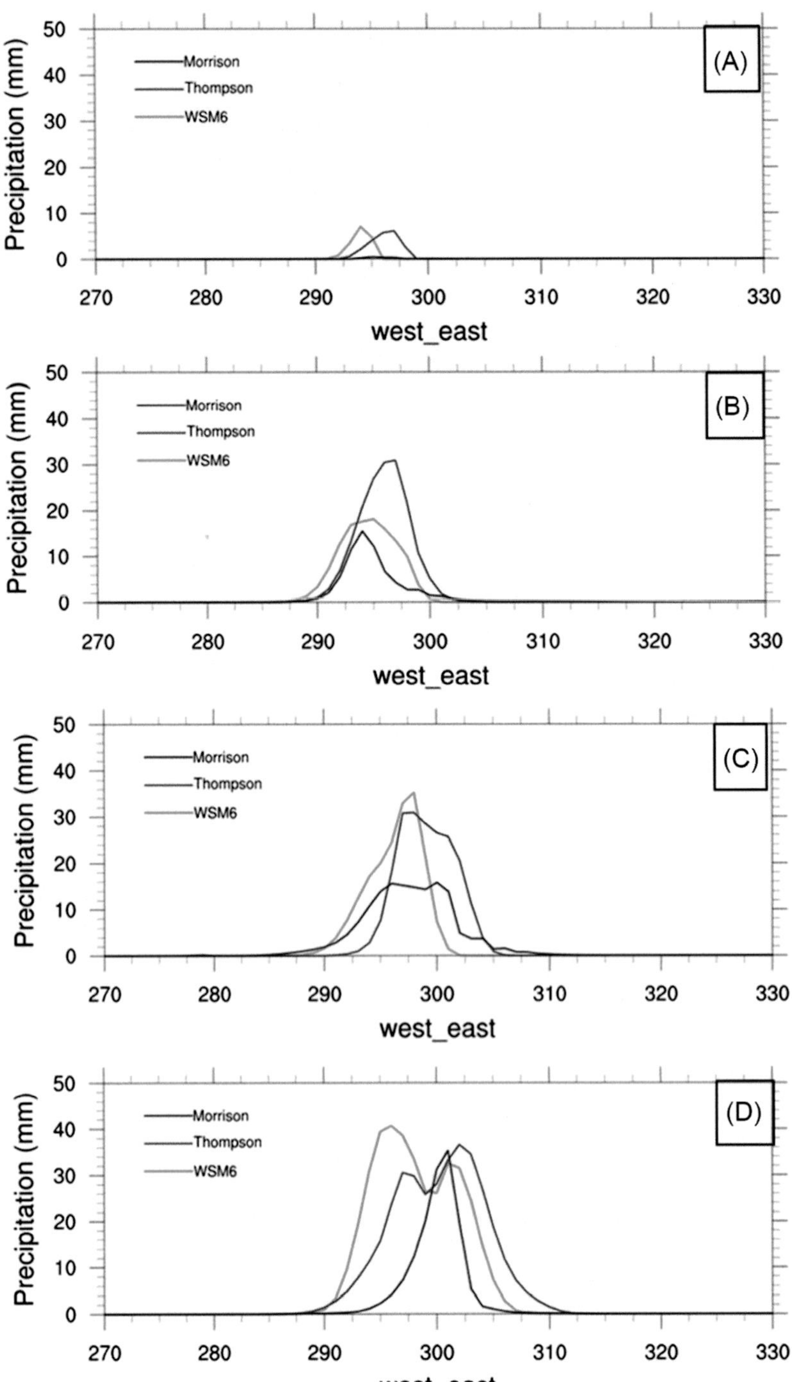

FIG. 3 Along cross section plot of 30-min accumulated precipitation ending at (A) 30, (B) 60, (C) 90, and (D) 120 min into the simulations. The *green lines* are from the WSM6 run, the *red lines* are from the Thompson run, and the *black lines* are from the Morrison run.

microphysical processes associated with each of the stages, we first examine the processes in the tendency equation of cloud water mass mixing ratio, since only warm processes are present during the warm bubble rising and early cumulus formation before the convection becomes deep enough to involve cold rain processes.

Saturation adjustment is used to produce cloud water through condensation in all three schemes used in the simulations shown here. In this treatment of cloud water production, supersaturation with respect to water is adjusted to zero when it occurs, and information on the availability of aerosols is not required. The total number concentration of cloud droplets is given by a prescribed gamma distribution. Fig. 4 shows all the process terms in the

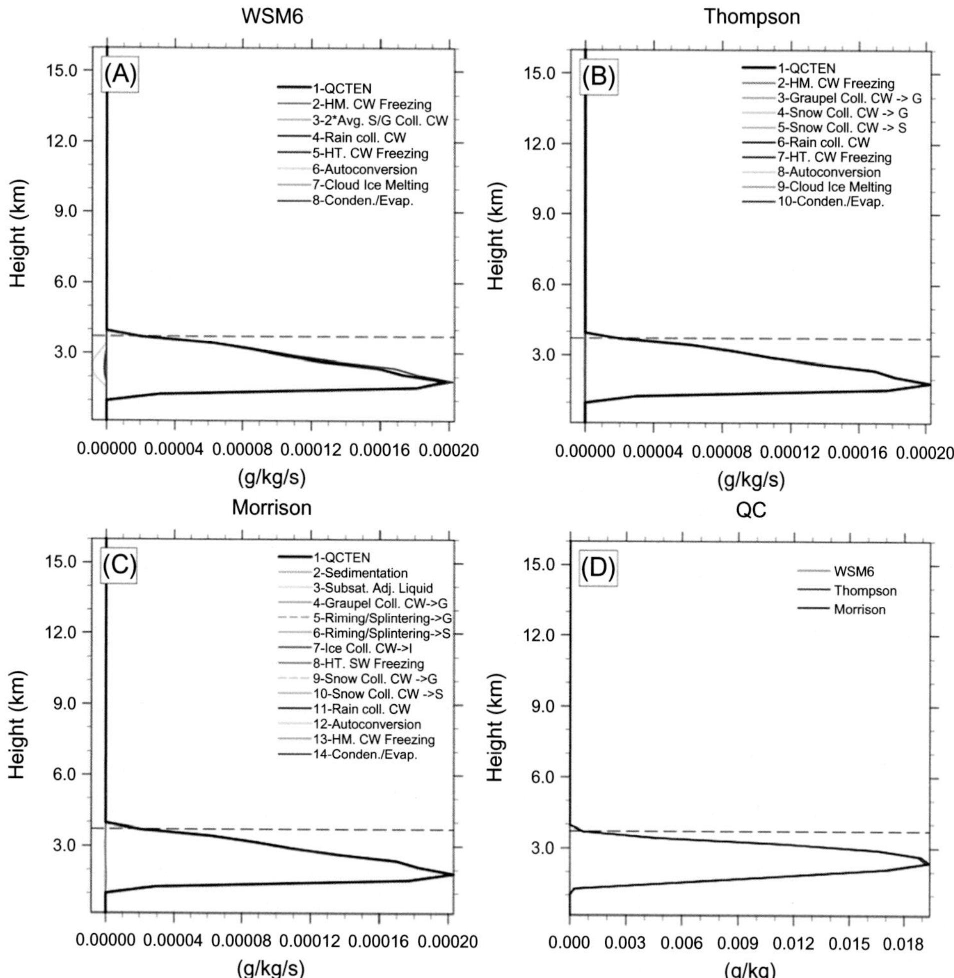

FIG. 4 The vertical profiles of the process terms in the cloud water mass mixing ratio tendency equation averaged over grid points 250–360 at 6 min into the simulations for the (A) WSM6, (B) Thompson, and (C) Morrison runs. Panel (D) is the cloud water mass mixing ratio from the three runs (the *green line* is the WSM6 run, the *red line* is the Thompson run, and the *black line* is the Morrison run). The *black dashed line* is the 0°C line.

tendency equation of cloud water mass mixing ratio averaged over points 250–360 at 6 min into the simulations. Since no frozen processes are initiated by this time, condensation is the dominant process in all three simulations, and the average cloud water profiles are very similar. This result indicates that the production rates of cloud water due to saturation adjustment are numerically very close to each other between the schemes. However, since the gamma distributions for cloud water are prescribed differently, the same mass production rates of cloud water do not lead to the same total number concentrations of cloud droplets between the schemes.

In principle, different total number concentrations of cloud droplets between the schemes will lead to different rates of cloud droplets growing into rain through "autoconversion," a term commonly used to describe the process through which cloud droplets collide and coalescence with each other and grow in size to become raindrops. Although autoconversion describes a physical growth process, it is a result of the separation of liquid water droplets into two categories for the convenience of parameterization: cloud droplets and raindrops. In a typical bulk microphysics parameterization, when cloud droplets become large enough, they are recategorized as raindrops. The "autoconversion" term, therefore, depends on both the parameterization of the collision-coalescence process and on the cloud and raindrop size distributions assumed in the parameterization. Differences between schemes in either of these will lead to differences in the evolution of the cloud and rain mass mixing ratios. In addition, as soon as raindrops form via autoconversion, they begin to collide and collect cloud water. In terms of the predator-prey analogy used earlier, cloud water is the prey, and rain is the predator in this case, as both autoconversion and the collection of cloud droplets by rain are terms that remove cloud water and add to the rainwater content. In fact, by 9 min into the simulations, differences in the cloud water mixing ratio begin to appear (Fig. 5). At this time, the Morrison run has the most cloud water while the WSM6 run has the least. Comparison of the process terms indicates that the greater magnitudes of the autoconversion and collection terms in the WSM6 run correspond to the conversion of more cloud water into rain (not shown). These process terms are comparatively smaller in the Thompson run, and too small to discern in the Morrison run, in which condensation is still the dominant process.

A similar comparison of the averaged process terms in the rainwater mass mixing ratio tendency equation, along with the rainwater mass mixing ratio, time-averaged from 21 to 30 min into the model simulations is shown in Fig. 6. This specific time interval is chosen for temporal averaging because the characteristics of the simulated convection from the three WRF runs start to diverge after 21 min into the simulations as the simulated convection develops further and cold rain processes become more prominent. The dominant pathways for rainwater production in the Thompson run are, in order of magnitude, rain collecting cloud water (Rain Coll. CW in the legend), graupel melting, and enhanced melting of graupel due to rain collecting graupel (Rain Coll. Graupel in the legend). In the Morrison and WSM6 runs, rain collecting cloud water and graupel melting are present, but their magnitudes are different than in the Thompson run. The average vertical profiles (Fig. 6D) indicate that during this time there are large differences in the rainwater mixing ratio. Most prominently, the Thompson run has the most rainwater, particularly above the freezing level. The process terms indicate that above 6 km, there is more rain collecting cloud water in the Thompson run than the other two runs. Also, above the freezing level, the magnitude of rainwater mass mixing ratio in the Thompson run is the greatest among the three, but the magnitude of rainwater

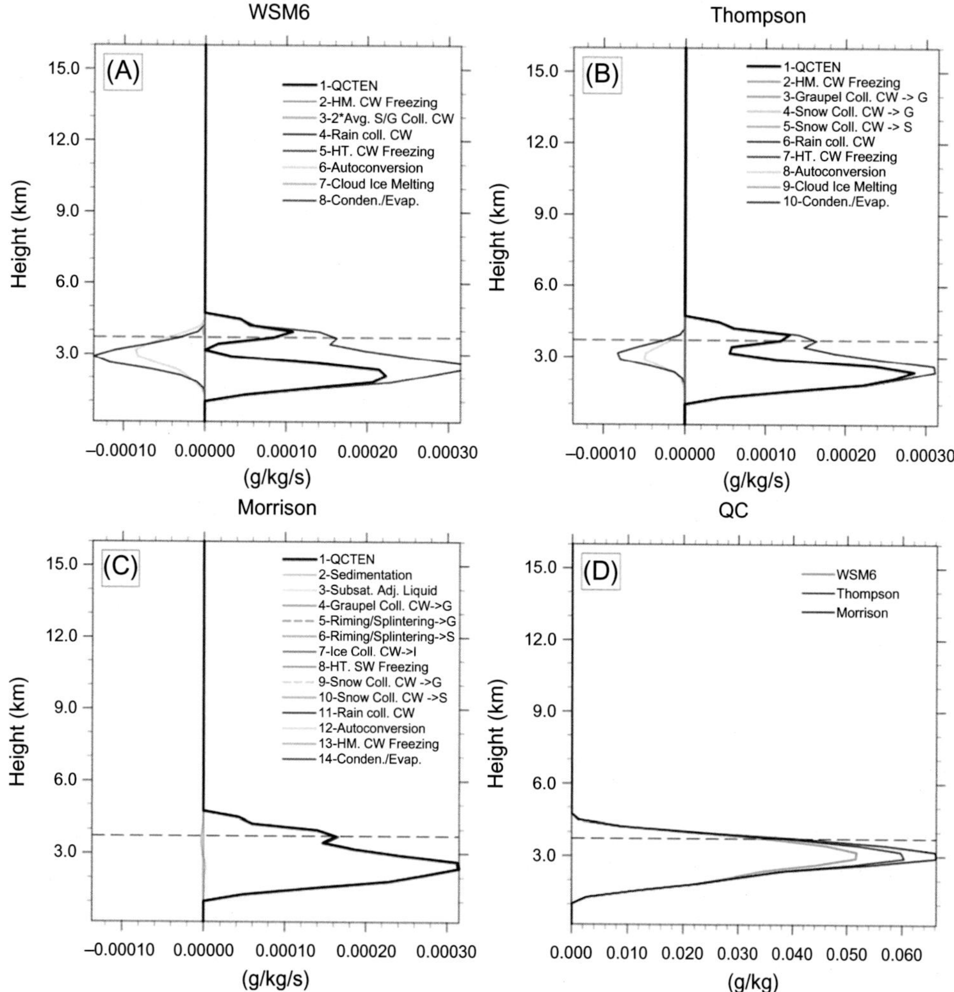

FIG. 5 Same as Fig. 4, except at 9 min.

sedimentation is not, indicating that the rainwater above the freezing level in the Thompson run stays in the air longer than in the other two runs. This is an indication that the size of the raindrops above the freezing level in the Thompson run is smaller than in the other runs. Below the freezing level, the Thompson run also has the most rainwater during this time. In the Thompson run, the peak of the graupel melting is closer to 2 km while the peak of graupel melting in the Morrison and WSM6 runs is around 3 km. This difference suggests that the size and related fallspeed of the graupel in the Thompson scheme is larger than those in the other two schemes. In all the runs, sedimentation, a process of redistribution, not a physical sink/source of rainwater, is significant. The most prominent physical sinks of rainwater are freezing due to collision with frozen hydrometeors above the freezing level (Rain Coll. Ice, Graupel Coll. Rain, Snow Coll. Rain, Ice-Rain Collison → G in the legends) and evaporation

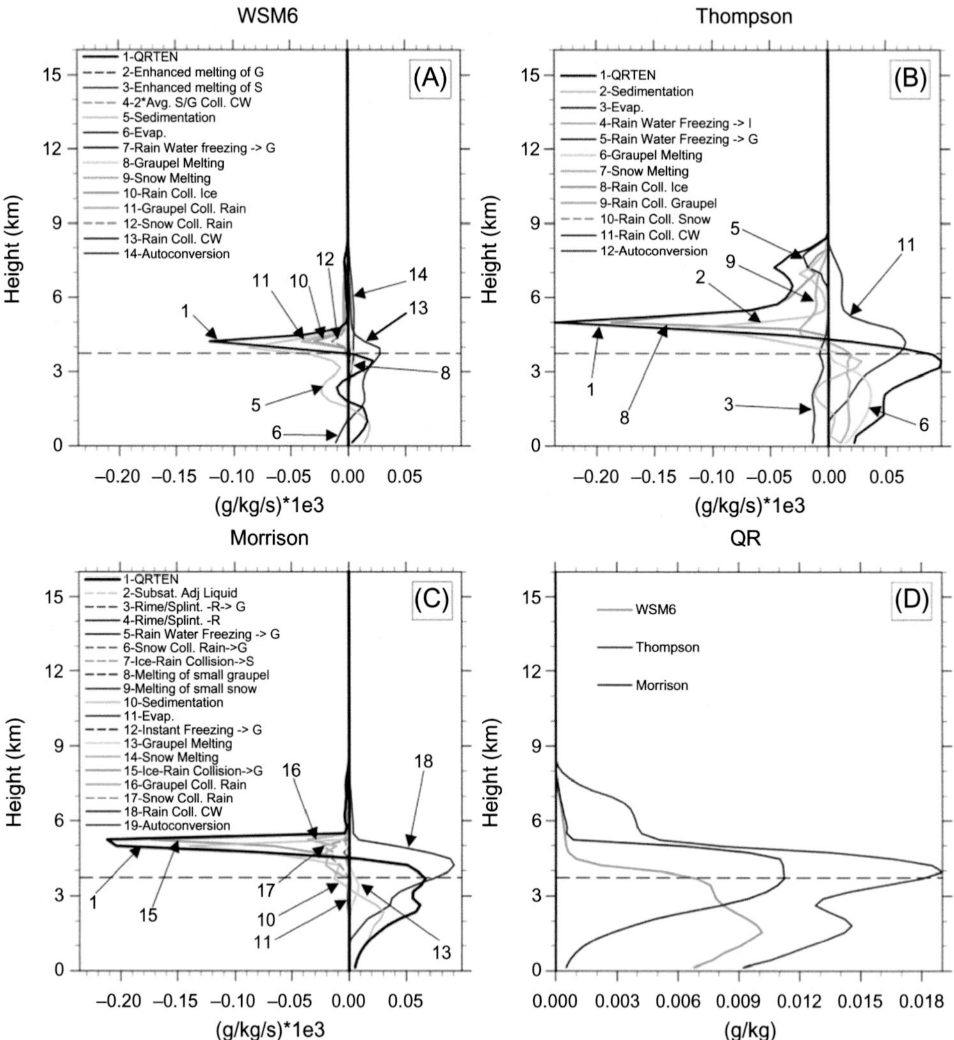

FIG. 6 The domain-averaged vertical profiles of rain water mass mixing ratio averaged from 21 to 30 min into the simulations for the (A) WSM6, (B) Thompson, and (C) Morrison runs. Panel (D) is the rain water mass mixing ratio from the three runs (the *green line* is the WSM6 run, the *red line* is the Thompson run, and the *black line* is the Morrison run). The *black dashed line* is the 0°C line.

below the freezing level in all three runs, with the greatest overall evaporation in the Thompson run and the least in the Morrison run.

It can be summarized that the three schemes have similar prominent pathways to the production of rain water mass that are associated with the processes that prey upon cloud water, but these pathways are significantly different in domain-averaged vertical magnitude distributions. All these indicate that the understanding of these pathways is not complete, leading to uncertainties in the parameterizations of these pathways. We will show in the following section that these uncertainties result from the differences in the simulated size distributions of liquid hydrometeor particles and size-dependent processes between the schemes.

6 Uncertainties in the simulated liquid hydrometeor particle size distributions

Fig. 7 shows the rainwater content vs mean volume diameter (MVD), the particle number concentration of rainwater, and the effective total evaporation area vs MVD over the same area and time period as in Fig. 6. The effective total evaporative area is defined as the multiplication of the square of the MVD and the particle number concentration. This figure demonstrates that the Thompson run has a smaller MVD of raindrops than the other two runs. The MVD of raindrops increases with mixing ratio in the single-moment WSM6 scheme but is limited to below 1250 μm in the Thompson run, which is due to the upper bound imposed on the size of raindrops in terms of a median-volume diameter of 2.5 mm in the scheme. A significant portion of raindrops in the Morrison run have MVDs larger than the upper limit in the Thompson run. The smaller drops in the Thompson run will have more surface area for evaporation, and will fall more slowly, allowing more time to evaporate. This is consistent with the greater evaporation term shown in Fig. 6. Greater evaporation favors the formation of a more intense cold pool behind the gust front. This is one reason why the Thompson run results in a more intense developed squall line than the other two runs during this time period.

Fig. 8 shows the process terms in the tendency equation of rainwater number mixing ratio (Nr) for the Thompson and Morrison runs (since the WSM6 scheme is single-moment, it does not predict number mixing ratio), along with a vertical profile of Nr for the same temporal and areal average as in Fig. 6. Note that the horizontal axes in Fig. 8A and B are different in order to better visualize the processes in the Morrison run. The Thompson run has a much larger rainwater number mixing ratio than the Morrison run, which is consistent with the smaller drops in the Thompson run indicated by Fig. 7. Examination of the process terms shows that in the Thompson run, the term associated with shedding of rainwater in the rain-collecting-graupel term (R_coll_G in the legend) is the largest source of number mixing ratio below the freezing level. This term is much less significant below the freezing level in the Morrison run, although it is a major sink of rainwater above the freezing level in this run. Below the freezing level, the rain production by autoconversion and loss through evaporation (R_Evap in the legend) terms in the Thompson run are much larger than in the Morrison run. The lack of graupel production and the small size and high altitude of the graupel that is produced in the Morrison run at this time means that the graupel falls too slowly to reach below the freezing level, melt, and contribute to the rain production at this time. Since the raindrops are smaller in the Thompson run, it is reasonable that the rainwater number mixing ratio is significantly decreased by evaporation since smaller drops evaporate faster.

In summary, the most important aspect of the differences between the three schemes are the simulated hydrometeor particle size distributions that are governed by process

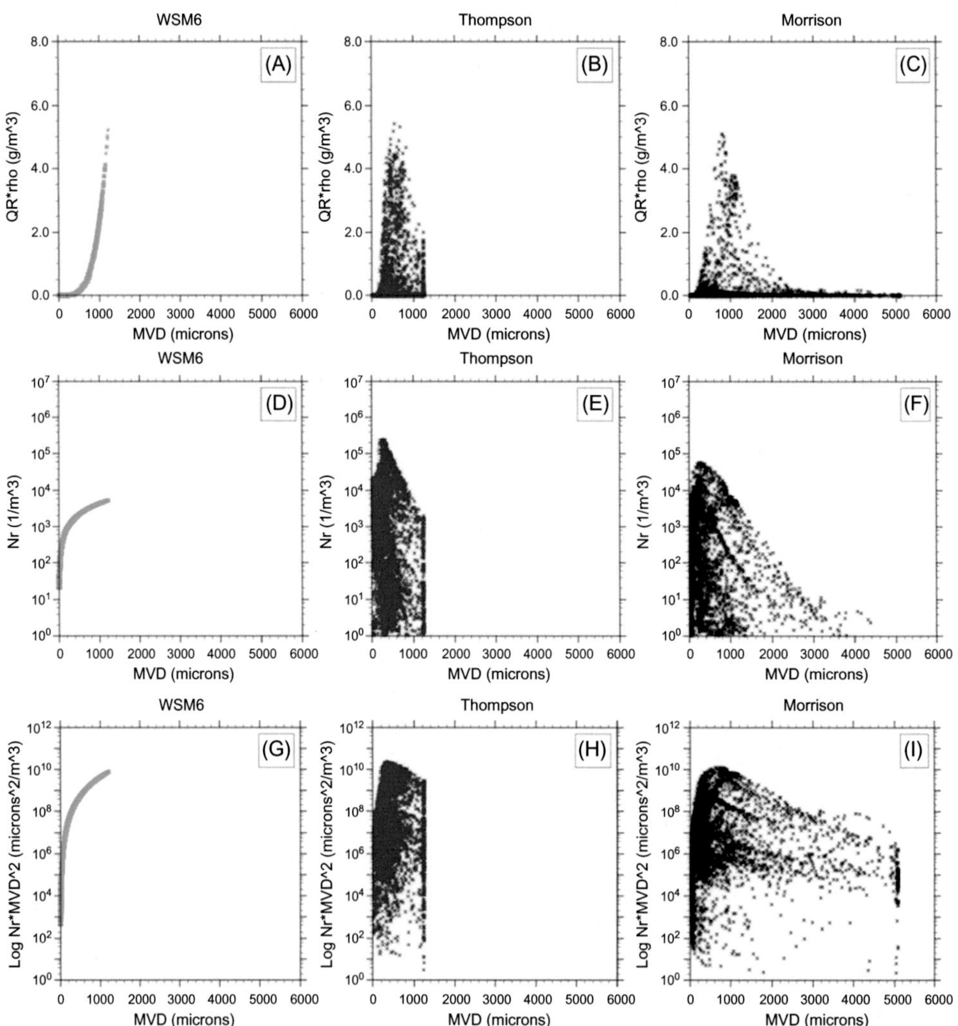

FIG. 7 The rainwater content vs mean volume diameter (MVD, μm) from the (A) WSM6, (B) Thompson, and (C) Morrison runs. The rainwater number concentration (m^{-3}) vs MVD (μm) from the (D) WSM6, (E) Thompson, and (F) Morrison runs. The effective total evaporation area vs MVD from the (G) WSM6, (H) Thompson, and (I) Morrison runs. Data points were sampled over the entire domain from 21 to 30 min into the simulations.

parameterizations along with a priori assumptions about the size distributions. The simulated hydrometeor particle size distributions are critically important because all the pathways to the production of precipitating hydrometeors are physically linked to and therefore numerically dependent on them. Therefore, the uncertainties in the parameterizations of warm rain processes are rooted in the lack of understanding of how to accurately simulate the evolution of hydrometeor particle size distributions.

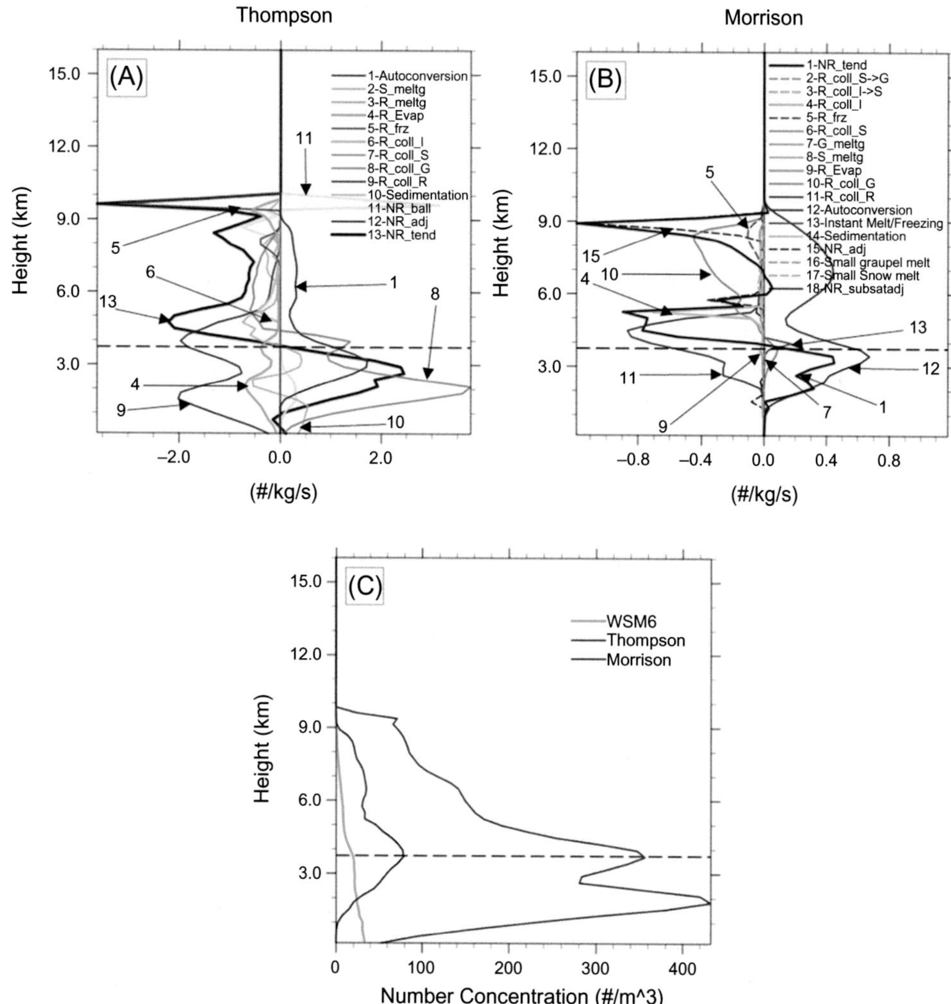

FIG. 8 The domain-averaged vertical profiles of the process terms in the tendency equation of rainwater number mixing ratio averaged from 21 to 30 min into the simulations for the (A) Thompson and (B) Morrison runs (C) is the corresponding average vertical profiles of number concentration. The *black dashed line* is the 0°C line.

7 Uncertainties in the parameterized cold rain processes

Like the warm rain processes described above, there are many uncertainties in the parameterization of the multiple processes that lead to the production and evolution of frozen hydrometeors. To complicate matters further, frozen species come in many different shapes, sizes, and densities, and the assumed size distributions impact the production rates in the parameterization. For instance, the three schemes here each include three frozen species: ice, snow, and graupel, with a specified density for each. An option is included in both the WSM6 and Morrison schemes to make the graupel more hail-like by changing the density and fall speed parameters. As with the liquid species, the assumed size distributions for each hydrometeor type often differs between schemes.

A comparison of the process terms in the tendency equation of cloud ice mass mixing ratio from the three runs reveals that the schemes differ in the processes that dominate the ice formation in the 21–30 min time period (Fig. 9). In the WSM6 run, ice nucleation and

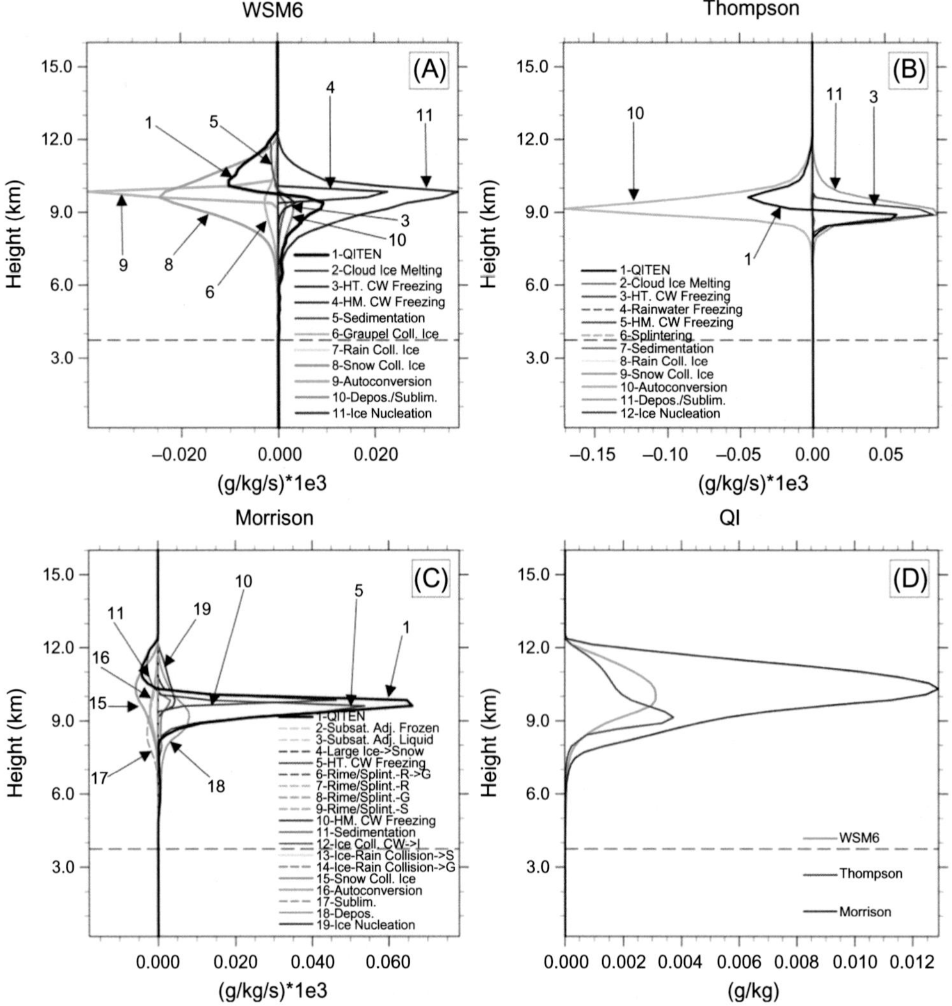

FIG. 9 The domain-averaged vertical profiles of cloud ice mass mixing ratio averaged from 21 to 30 min into the simulations for the (A) WSM6, (B) Thompson, and (C) Morrison runs. Panel (D) is the cloud ice mass mixing ratio from the three runs (the *green line* is the WSM6 run, the *red line* is the Thompson run, and the *black line* is the Morrison run). The *black dashed line* is the 0°C line.

homogeneous freezing of cloud droplets are the processes that contribute most to ice production at this time while in the Thompson run, ice is produced almost entirely by heterogenous freezing and deposition. In the Morrison run, all four of these processes contribute significantly. Ice content decreases due to autoconversion to snow in all three schemes, although the magnitude of this process differs. In the Thompson scheme, autoconversion is the only process of discernible magnitude that leads to a decrease in ice mass, but collection by snow plays a role in both of the other runs. Collection by graupel also preys upon the ice content in the WSM6 run while sublimation accounts for some ice loss in the Morrison run. It is noteworthy that the ice to snow autoconversion term differs between these runs in that they each have their own size threshold for when cloud ice particles are considered large enough to become categorized as snow. This reflects an intrinsic difficulty in the parameterization of hydrometeor conversion from one category to another.

The vertical profiles of the various production terms in the snow mass mixing ratio tendency along with the snow mass mixing ratio that are averaged over the entire domain and the same time period as Fig. 6 are shown in Fig. 10. Overall, the WSM6 and Thompson runs have more snow than the Morrison run. The snow production in the WSM6 run is due to the collection of cloud water by snow and graupel (Avg. S/G coll. CW in the legend), autoconversion of cloud ice to snow, deposition, snow collecting cloud ice, and ice collecting rain (Ice Coll. Rain in the legend). The major pathway for the reduction of snow in the WSM6 run is rain collecting snow (Rain Coll. Snow in the legend) to produce graupel. As in the WSM6 run, the major pathways of snow production in the Thompson run include autoconversion from ice, deposition, and snow collecting cloud water (Snow Coll. CW in the legend). However, the magnitudes differ, most notably the autoconversion in the Thompson run is much larger than in the WSM6 run. Additionally, unlike the WSM6 run, sublimation is the most significant reduction pathway for snow in the Thompson run during this time. The production of snow in the Morrison run is through deposition and snow collecting cloud water like the other two runs, and additionally through snow collecting ice and rain. Autoconversion in the Morrison run is not as significant as it is in the other two runs. Rain collecting snow (as in the WSM6 run) and sublimation (as in the Thompson run) are the major processes that contribute to the reduction of snow in the Morrison run.

The vertical profiles of the graupel mass mixing ratio in Fig. 11 are averaged over the entire domain and the same time period as in previous figures. Overall, the Morrison and WSM6 runs have the most graupel above 7 km while the Thompson run has the least graupel aloft and most below the freezing level. The main pathways for the production of graupel in the WSM6 run are graupel and snow collecting cloud water and rain collecting snow as well as ice collecting rain and graupel collecting rain near the freezing level. The main pathways of graupel reduction in the WSM6 run are melting and sublimation, although their magnitudes are small. The main pathways of graupel production in the Thompson run are graupel collecting cloud water, raindrops freezing, rain collecting ice, and rain collecting graupel when the temperature is below freezing to produce larger graupel. The Thompson run has more freezing of raindrops to form graupel than the other two runs. The dominant pathways of the reduction of graupel in the Thompson run are melting and rain collecting graupel when the temperature is above freezing to produce rain. The main pathways of graupel production in the Morrison run are graupel collecting rain, graupel collecting cloud water, rain collecting snow,

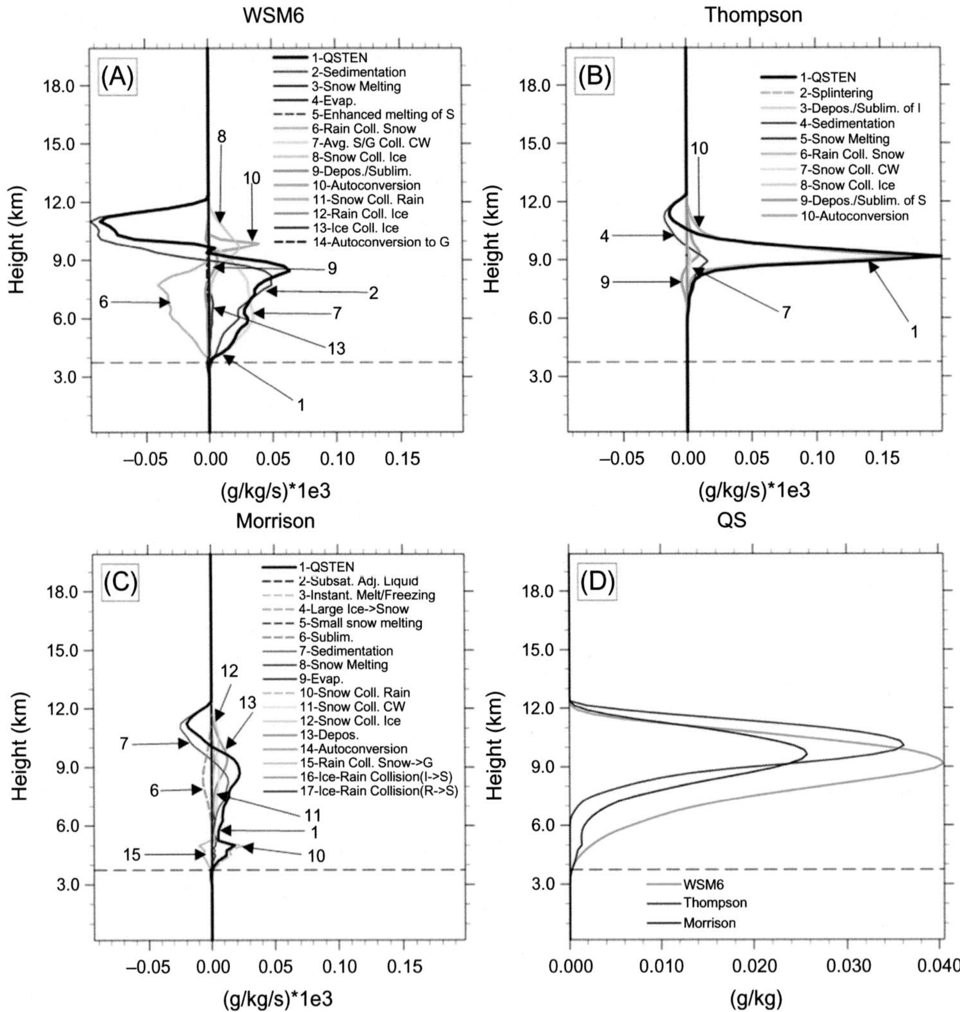

FIG. 10 The domain-averaged vertical profiles of snow mass mixing ratio averaged from 21 to 30min into the simulations for the (A) WSM6, (B) Thompson, and (C) Morrison runs. Panel (D) is the snow mass mixing ratio from the three runs (the *green line* is the WSM6 run, the *red line* is the Thompson run, and the *black line* is the Morrison run). The *black dashed line* is the 0°C line.

deposition, and ice-rain collisions. The main pathways for graupel reduction in the Morrison run are sublimation and melting, although at this time they have small magnitudes.

Overall, similar to the warm rain processes, all three schemes have similar prominent pathways to the production of frozen hydrometeors, in which heavier hydrometeorological particles prey on smaller ones. However, these pathways are quantitatively different in domain-averaged vertical magnitude distributions, indicating that uncertainties remain in how to accurately simulate population properties of the simulated frozen hydrometeors. We will further illustrate this in the following section.

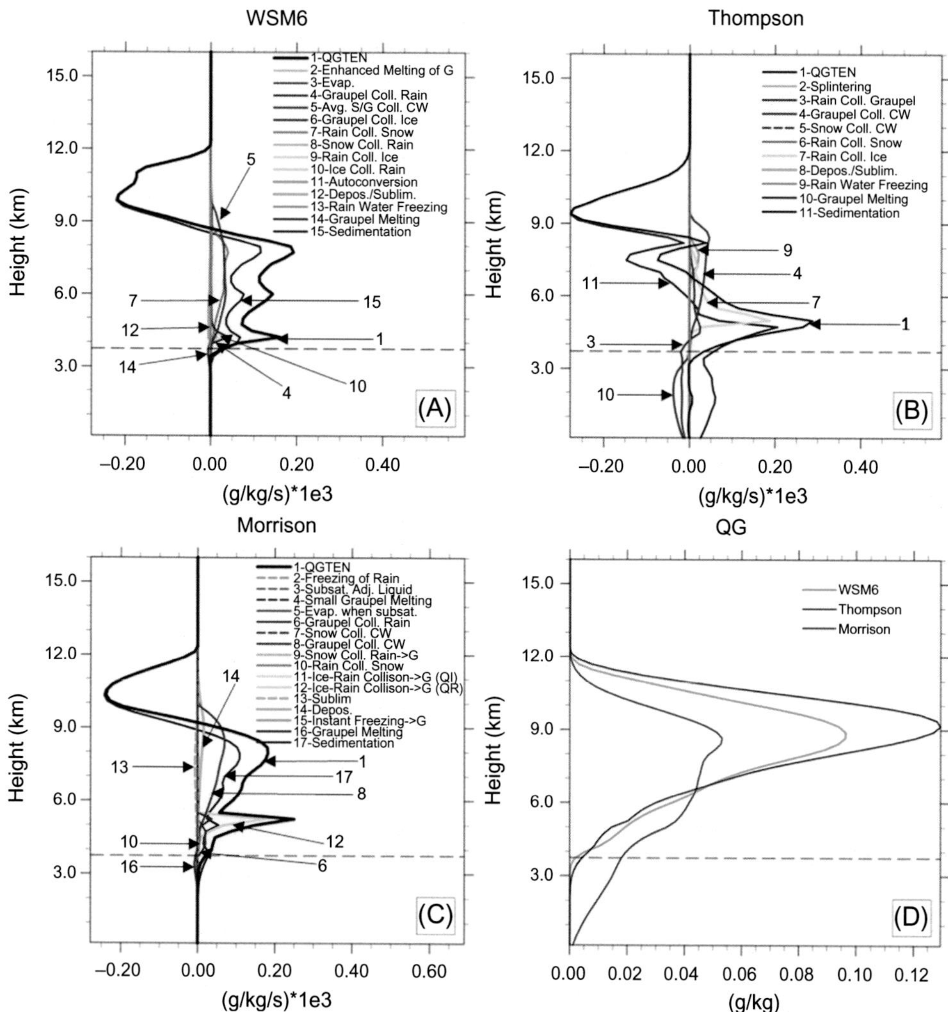

FIG. 11 The domain-averaged vertical profiles of graupel mass mixing ratio averaged from 21 to 30 min into the simulations for the (A) WSM6, (B) Thompson, and (C) Morrison runs. Panel (D) is the graupel mass mixing ratio from the three runs (the *green line* is the WSM6 run, the *red line* is the Thompson run, and the *black line* is the Morrison run). The *black dashed line* is the 0°C line.

8 Uncertainties in the simulated frozen hydrometeor particle properties and size distributions

Fig. 12 shows the cloud ice content and the particle number concentration of cloud ice vs MVD over the entire domain and the same time period as in Fig. 9. For a given mass, the MVD of cloud ice in the WSM6 run is larger than in the Thompson and Morrison runs. The Morrison run produces more mass and a greater range of the MVD of cloud ice than in the Thompson

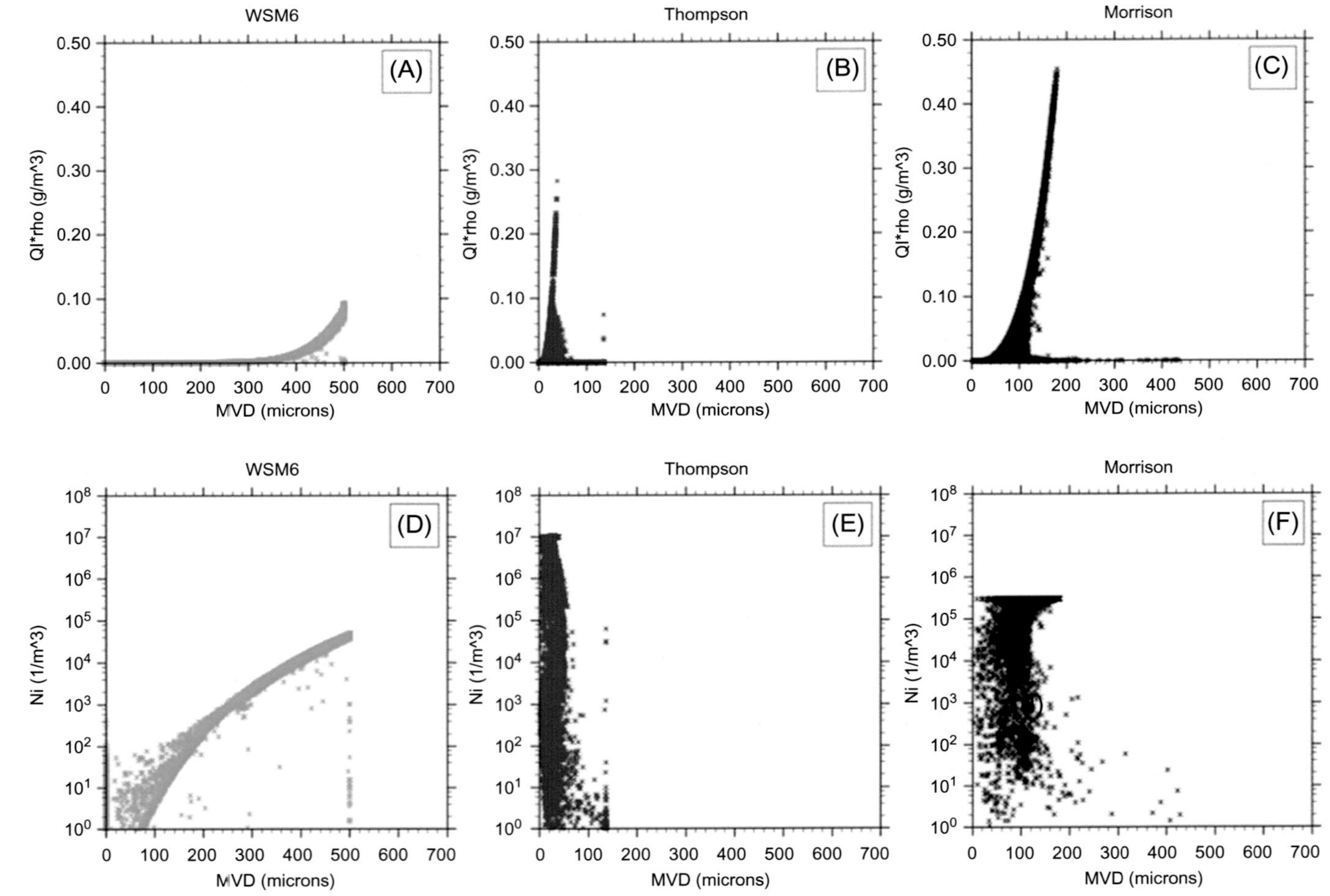

FIG. 12 The cloud ice content vs mean volume diameter (μm) during the 21–30 min time period from the (A) WSM6, (B) Thompson, and (C) Morrison runs. The cloud ice number concentration (m^{-3}) vs mean volume diameter (μm) during the 21–30 min time period from the (D) WSM6, (E) Thompson, and (F) Morrison runs.

run. All these indicate that the effective size of cloud ice particles (assumed to be bullet-shaped) in the WSM6 run is greater than in the other two runs. Among the three runs, the cloud ice particles produced in the Thompson run are the smallest in size while the total mass of cloud ice particles in the Morrison run is the greatest. Overall, there are more cloud ice particles with smaller MVD in the Thompson and Morrison runs than in the WSM6 run (see Fig. 12D–F).

Fig. 13 shows the snow content and number concentration vs MVD (note that for the Thompson run the MVD is calculated as the mass mean diameter because the density of snow is not constant, it varies with snow particle size) over the entire domain and the same time period as in Fig. 10. The mean size of snow particles in both the WSM6 and Morrison runs do not exceed 4 mm, with most of the snow particles in the Morrison run being smaller than 2 mm. In the Thompson run, the MVD of all the snow particles is less than 1 mm. Overall, there are many more snow particles of smaller sizes in both the Thompson and Morrison runs than the WSM6 run (Fig. 13A–C). The size of snow particles in both the WSM6 and Morrison runs does not exceed 4.0 mm because of the limit placed on the slope parameter for snow in both schemes. The size difference between the runs fundamentally results from the different assumptions in snow physical properties (size distribution parameters) and different size-dependent process parameterizations between individual schemes, despite the fact that the differences between the schemes lead to differences in dynamics, which in turn feedback to the microphysical differences. These differences in the size distribution of snow particles have two consequences. First, they lead to different sedimentation rates for the same mass mixing ratio due to the differences in size-dependent fall speed. Second, they affect all the size-dependent microphysics processes contributing to the changes in the total snow content such as deposition/sublimation, collection, and riming to form graupel.

The noticeable difference in the contributing terms to the mass mixing ratio tendency equation of graupel between the three runs is in sedimentation below the freezing level. This is an indication that the size of the graupel is different between the schemes. Fig. 14 shows the graupel content and number concentration vs MVD for the time period and area over which the profiles shown in Fig. 11 are averaged. It is clearly seen that the MVD for graupel is larger in the Thompson run than in either the WSM6 run or the Morrison run (Fig. 14A–C) and, overall, there are more graupel particles of smaller sizes in both the Morrison and WSM6 runs than the Thompson run (Fig. 14D–F). Consistent with the differences in the density and fall speed parameters of graupel between the schemes, graupel is bigger and falls faster in the Thompson run than in the other two runs. The comparison of sedimentation terms shown in Fig. 11 also confirms this as the Thompson run has more sedimentation from 4 to 6 km, just above the freezing level. Consequently, the graupel in the Thompson run reaches the ground, as is seen in Fig. 11D while the graupel in the other two runs does not.

It is important to emphasize, as a summary of what has been shown so far, that the simulated hydrometeor particles from the three schemes are different physically in size distributions. Further, the assumed density for frozen hydrometeors is not all the same between the schemes. Numerically, the differences in the simulated hydrometeor particle properties and size distributions not only represent the uncertainties in assumed bulk hydrometeor properties and size distributions, but also the uncertainties in the parameterization formulations. These differences are due to the fact that we do not completely understand and cannot easily observe all the microphysical processes of cloud and precipitation production in nature.

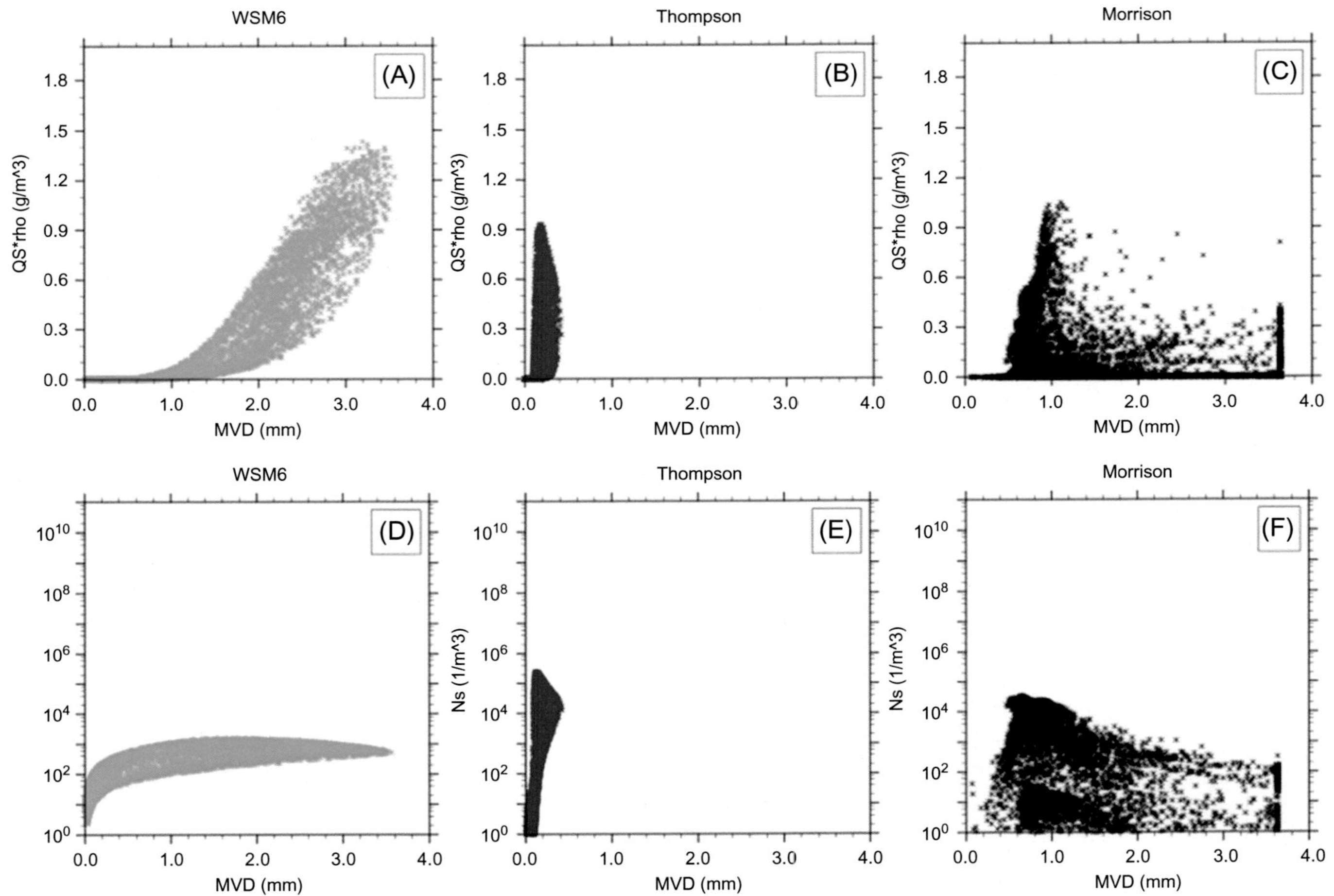

FIG. 13 The snow content vs mean volume diameter (μm) during the 21–30 min time period from the (A) WSM6, (B) Thompson, and (c) Morrison runs. The snow number concentration (m^{-3}) vs mean volume diameter (μm) during the 21–30 min time period from the (D) WSM6, (E) Thompson, and (F) Morrison runs.

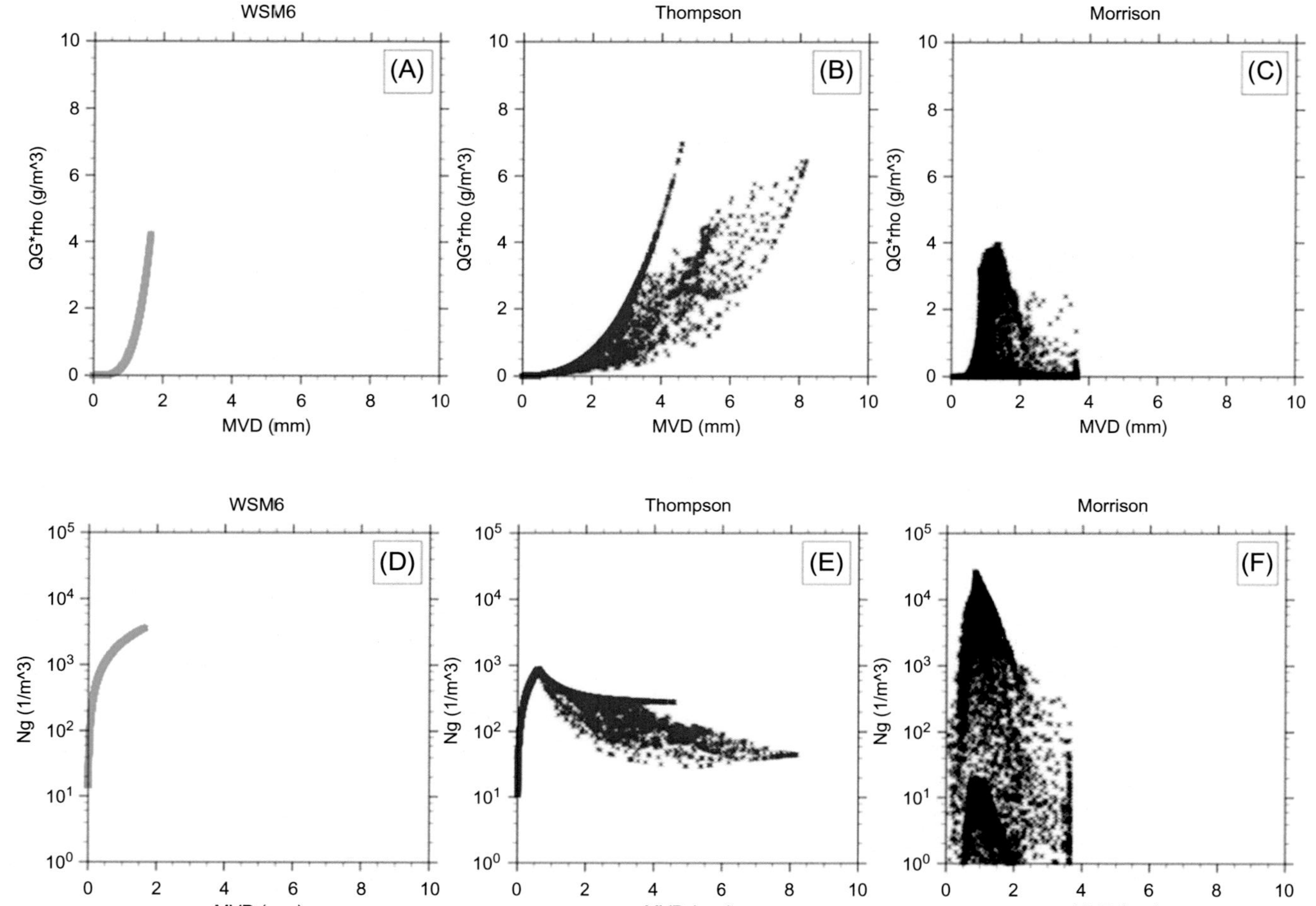

FIG. 14 The graupel content vs mean volume diameter (μm) during the 21–30 min time period from the (A) WSM6, (B) Thompson, and (C) Morrison runs. The graupel number concentration (m^{-3}) vs mean volume diameter (μm) during the 21–30 min time period from the (D) WSM6, (E) Thompson, and (F) Morrison runs.

Given the importance of the hydrometeor particle properties and size distributions in the parameterization of microphysical processes, realistic reproduction of observed values of the effective size, and mass production rate of precipitating hydrometeors should be used to reduce uncertainties in bulk microphysical schemes.

9 Uncertainties in the interaction of aerosols and clouds

The uncertainties shown above in the hydrometeor size distributions and process parameterizations will intrinsically lead to uncertainties in simulations of the interaction between aerosols and clouds. Results from another set of experiments with the idealized squall line case are used to illustrate this. This set of experiments uses an identical model configuration as before, except that aerosol-aware versions of the Thompson and the Morrison microphysics parameterization schemes are used. In both schemes, aerosol "awareness" is included through the use of the Thompson and Eidhammer (2014) aerosol scheme. This aerosol scheme is available in the community WRF code for the Thompson parameterization, and for these experiments, was implemented in the Morrison microphysics parameterization to allow a fair comparison of aerosol impact.

In the aerosol-aware schemes, the number mixing ratio of cloud water (N_c) is a prognostic variable; otherwise N_c is a constant user-prescribed value. The aerosol content is initialized with default profiles of "water-friendly" (hygroscopic) and "ice-friendly" (nonhygroscopic, primarily dust) aerosols. The nucleation rate of cloud droplets is found using a lookup table of activated aerosol fraction, derived from a parcel model by Feingold and Heymsfield (1992), with additional changes by Eidhammer et al. (2009). Cloud ice activation is a function of ice-friendly aerosol content, following DeMott et al. (2010). The freezing of water droplets is impacted by aerosols only by modifying the temperature at which supercooled liquid undergoes homogeneous freezing. Aerosol content varies in time, due to the effects of wet scavenging (by rain, snow, and graupel), deposition nucleation, freezing of aqueous aerosols (Koop et al., 2000), and an idealized ground-based source.

As in the first set of model runs, the experiments are initialized with a warm perturbation, and the response of the two parameterizations to the warm bubble begins to differ very quickly. At 4 min (not shown) into the simulation, both schemes have begun to produce cloud droplets via condensation. As previously seen, the condensation process in the two schemes leads to nearly identical average profiles of cloud water mixing ratio. Once some cloud droplets exist, the autoconversion of cloud droplets into rain and growth of raindrops via the collection of cloud droplets also begin. By 5 min into the simulations, the autoconversion and collection terms in the Thompson run have become larger than in the Morrison run (Fig. 15B and C) and by 10 min, the Thompson run contains much less cloud water and more rain water than the Morrison run (Fig. 15I and J). Since aerosol-activated condensation is described by a similar saturation adjustment scheme in both the schemes, it is no surprise that the profiles of the condensation terms from the two schemes (Fig. 15A and F) are nearly identical. The differences in the autoconversion and collection terms, however, are indicative of the uncertainty in the calculation of these terms.

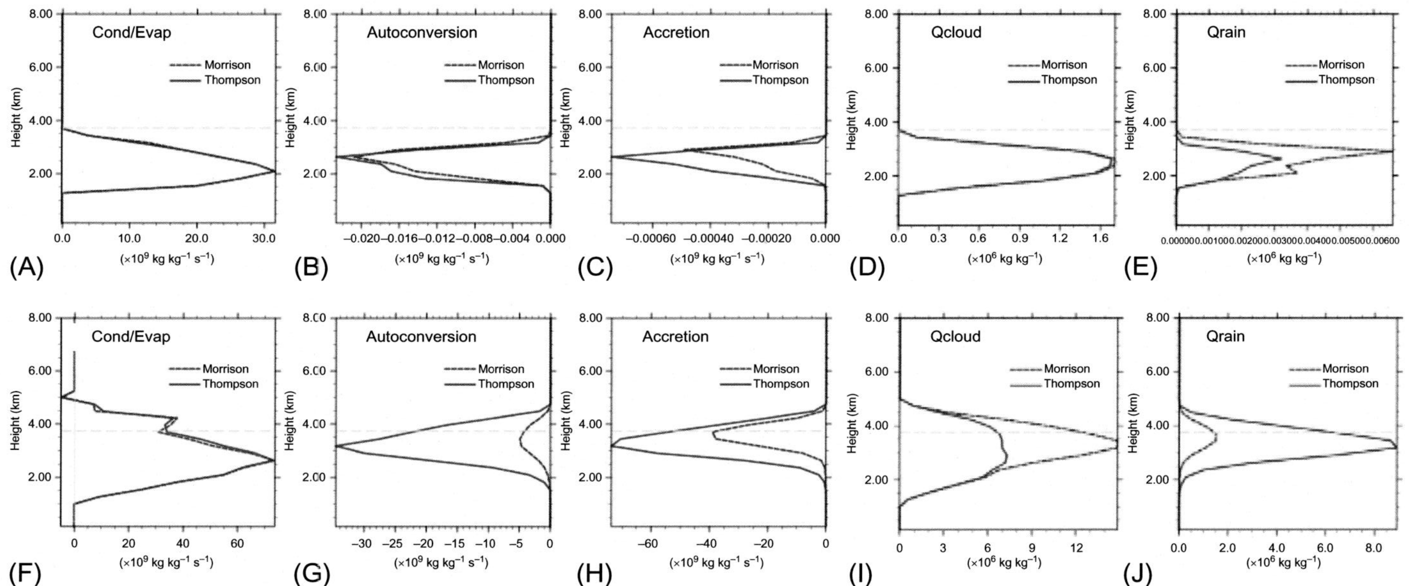

FIG. 15 Domain-average vertical profiles of the largest terms in the cloud water mixing ratio tendency equation for the Thompson (*blue solid lines*) and Morrison (*red dashed lines*) schemes at 5 min (*top row*: A–E) and 10 min (*bottom row*: F–J) into the simulation. The terms are: condensation/evaporation of cloud droplets (A and F), autoconversion of cloud droplets to rain (B and G), collection of cloud droplets by raindrops (C and H). Panels D and I show cloud water mixing ratio while (E and J) show rain water mixing ratio. Note that the horizontal axes differ in scale between panels.

For the parameterization of the aerosol-dependent autoconversion rate, the schemes use very different approaches. The Morrison scheme adopted an equation from Khairoutdinov and Kogan (2000), derived from aircraft measurements as well as from large-eddy simulations with explicit microphysics. This autoconversion rate is a function of cloud water mass mixing ratio and number concentration:

$$\left(\frac{\partial q_r}{\partial t}\right)_{autoconv} = 1350 q_c^{2.47} N_c^{-1.79}, \tag{5}$$

where q is mass mixing ratio (kg kg^{-1}), N is number concentration (cm^{-3}), and the subscripts r and c refer to rain and cloud water, respectively. In the Thompson scheme, the classical approach of Berry and Reinhardt (1974) is used, which depends upon the cloud water content and the characteristic diameters (D_b and D_f) of the assumed gamma distribution of drop sizes (Thompson et al., 2008).

$$\left(\frac{\partial q_r}{\partial t}\right)_{autoconv} = \frac{0.027 \rho q_c \left(0.0625 \times 10^{20} D_b^3 D_f - 0.4\right)}{\frac{3.72}{\rho q_c}\left(0.5 \times 10^6 D_b - 7.5\right)} \tag{6}$$

where ρ is the air density. This form of the autoconversion calculation has been widely adopted in other microphysics schemes as well (Thompson et al., 2008).

From a comparison of the above equations, it is clear that the rate of mass conversion from cloud to rain categories by autoconversion will vary with the choice of equation describing the process, and the underlying assumptions that govern the related number concentration and characteristic drop sizes. It should also be noted that for many parameterized microphysical processes, if a change in number concentration is also computed, it is often a function of the change in mass for that hydrometeor species.

It is important to point out that the Thompson and Morrison schemes also have slightly different definitions of cloud droplets and rain. Understanding this difference can shed more light onto the uncertainties illustrated by the differences between the two schemes. In the Thompson scheme, cloud drops may range in size from 1 to 100 μm while raindrops can be 50 μm to 5 mm in diameter. In the Morrison scheme, cloud drops are between 1 and 60 μm while raindrops are in the range of 20 μm to 2.8 mm in diameter. Besides the differences in formulation, the distinctions in size definition will also contribute to the divergence in autoconversion rates. However, an experiment using the Morrison scheme in which the cloud and raindrop size limits were modified to match those of the Thompson scheme showed almost no difference in the early time periods (not shown).

The accretion or collection term is also a primary contributor to the cloud and rainwater budgets at this time. This term describes the loss of cloud droplets due to raindrops colliding and coalescing with them, collecting the smaller drop into the larger one. The Morrison scheme computes the droplet accretion rate using the results from the study by Khairoutdinov and Kogan (2000), which is dependent only on cloud and rainwater content:

$$\left(\frac{\partial q_r}{\partial t}\right) = 67 (q_c q_r)^{1.15} \tag{7}$$

The Thompson formulation follows Verlinde et al. (1990), and more explicitly depends on raindrop diameter and fall speed characteristics, and the collection efficiency varies with the size distributions of both cloud and raindrops. As a consequence, as shown in Fig. 15, the Thompson scheme converts cloud water to rainwater more quickly than the Morrison scheme, due to the larger autoconversion and accretion rates. By the 10-min forecast time, the run using the Thompson parameterization has a peak in the average rain mixing ratio that is more than 5 times greater than in the run using the Morrison parameterization.

Several additional experiments were performed using the Thompson and Morrison aerosol-aware parameterizations, in which the aerosol load was either increased or decreased by 20%. The resulting structure of the simulated squall line is shown in Fig. 16. The modifications to the aerosol content had a small impact on the speed of the squall line advancement and the intensity of the cold pool, however, the impact of changing the parameterization scheme was much greater than the impact of changing the aerosol content by 20%.

It is important to note that since the same aerosol activation scheme is used in the aerosol-aware versions of the Thompson and the Morrison microphysics parameterization schemes, the different impact of aerosols on the simulated convection between the two schemes is due to their differences in the parameterized size-dependent processes, particularly in the parameterization of autoconversion and collection of cloud droplets by raindrops. These results illustrate the large uncertainty in cloud and precipitation responses to aerosols in simulations of convection. The implication for process studies of aerosol-convection interaction using

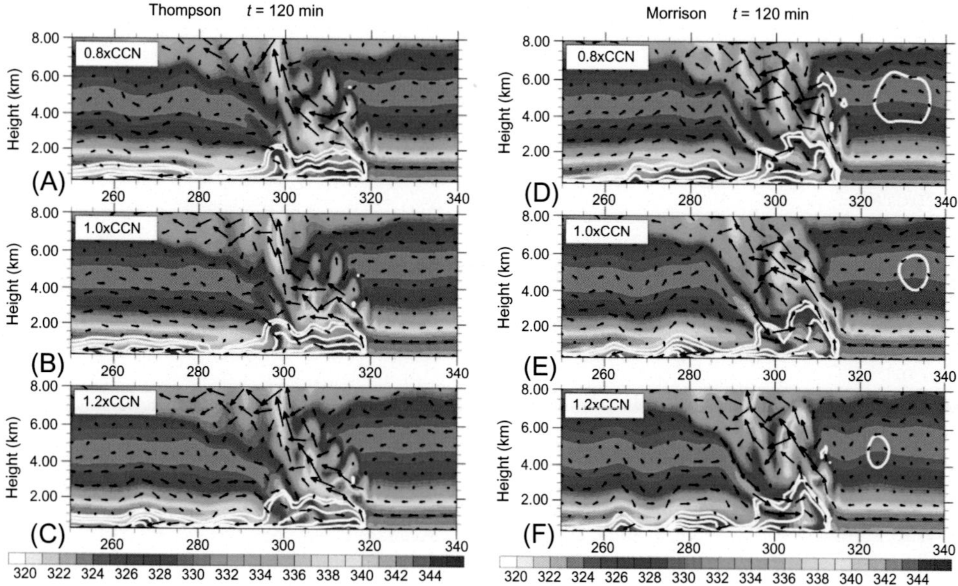

FIG. 16 Cross section of equivalent potential temperature (color-shaded in K), potential temperature perturbation (white lines, contour interval of 2 K beginning with −1 K), along cross-section winds at 120 min into the simulations with modified aerosol load. The *left column* is from the Thompson run and the right column is from the Morrison run. Panels (A and D) have a 20% decrease in aerosols, panels (B and E) have the default level of aerosols, and panels (C and F) have a 20% increase in aerosol load.

these two schemes is that unless there are reliable observations of hydrometeor population characteristics in terms of mass and size distributions to constrain the schemes to validate the simulations, caution must be applied to any quantitative interpretation of the results.

10 Summary

In this chapter, major uncertainties in many different microphysics parameterization schemes used in atmospheric models are illustrated using the WRF model simulations of an idealized squall line development. It has been shown that these uncertainties exist in parameterized microphysical processes, in particular, the mass-size relations of hydrometeor particles (e.g., gamma or exponential distributions), the fall speed calculation, the activation of cloud condensation nuclei into drops, homogeneous, and heterogeneous ice nucleation. Great uncertainties exist as to how hydrometeor shape parameters affect the accuracy of microphysical process parameterizations and how to effectively parameterize microphysical processes responsible for frozen hydrometeor production. It is also important to keep in mind that the three schemes illustrated in this chapter result from the development based on fits of meteorological simulations to observational data. The specific parameterizations may depend on specific observations and regions of simulation and forecast applications. This further generates uncertainty even in the choice of an optimal microphysics scheme for deep convection simulations. All these uncertainties could be diminished if microphysics parameterization schemes could be developed based on the same well-established theory and representative observations of clouds in various conditions and locations around the world. Unfortunately, great gaps remain in both theoretical understanding and observations in both nature and laboratories. We will continue facing and having to deal with the uncertainties in microphysics parameterizations in the foreseeable future.

References

Anthes, R., Warner, T.T., 1978. Development of hydrodynamic models suitable for air pollution and other mesometerological studies. Mon. Weather Rev. 106, 1045–1078. https://doi.org/10.1175/1520-0493(1978)106<1045:DOHMSF>2.0.CO;2.

Anthes, R.A., Hsie, E.-Y., Kuo, Y.-H., 1987. Description of the Penn State/NCAR Mesoscale Model version 4 (MM4). NCAR Tech. Note NCAR/TN-282+STR, 66 pp. [Available from MMM Division, NCAR, P.O. Box 3000, Boulder, CO 80307].

Beheng, K.D., 2010. The evolution of raindrop spectra: a review of basic microphysical essentials. In: Rainfall: State of the Science. Geophys. Monogr., vol. 191. Amer. Geophys. Union, pp. 29–48.

Berry, E.X., Reinhardt, R.L., 1974. An analysis of cloud drop growth by collection: part II. Single initial distributions. J. Atmos. Sci. 31, 1825–1831.

Chen, S.H., Sun, W.Y., 2002. A one-dimensional time dependent cloud model. J. Meteorol. Soc. Jpn. 80, 99–118.

Cotton, W.R., Anthes, R.A., 1989. Storm and Cloud Dynamics. Acadamic Press 883 pp.

DeMott, P.J., et al., 2010. Predicting global atmospheric ice nuclei distributions and their impacts on climate. Proc. Natl. Acad. Sci. U. S. A. 107, 11217–11222. https://doi.org/10.1073/pnas.0910818107.

Dudhia, J., 1989. Numerical study of convection observed during the Winter Monsoon Experiment using a mesoscale two-dimensional model. J. Atmos. Sci. 46, 3077–3107. https://doi.org/10.1175/1520-0469(1989)046<3077:NSOCOD>2.0.CO;2.

Eidhammer, T., DeMott, P.J., Kreidenweis, S.M., 2009. A comparison of heterogeneous ice nucleation parameterizations using a parcel model framework. J. Geophys. Res.. 114, D06202. https://doi.org/10.1029/2008JD011095.

Feingold, G., Heymsfield, A.J., 1992. Parameterizations of condensational growth of droplets for use in general circulation models. J. Atmos. Sci. 49, 2325–2342. https://doi.org/10.1175/1520-0469(1992)049,2325:POCGOD.2.0.CO;2.

Hong, S.-Y., Lim, J.-O.J., 2006. The WRF single-moment 6-class microphysics scheme (WSM6). J. Korean Meteorol. Soc. 42, 129–151.

Hong, S.-Y., Dudhia, J., Chen, S.-H., 2004. A revised approach to ice microphysical processes for the bulk parameterization of clouds and precipitation. Mon. Weather Rev. 132, 103–119.

Hsie, E.-Y., Anthes, R.A., Keyser, D., 1984. Numerical simulation of frontogenesis in a moist atmosphere. J. Atmos. Sci. 41, 2581–2594. https://doi.org/10.1175/1520-0469(1984)041<2581:NSOFIA>2.0.CO;2.

Ikawa, M., Saito, K., 1991. Description of a Nonhydrostatic Model Developed at the Forecast Research Department of the MRI. Meteorological Research Institute Tech. Rep. 28, 238 pp. [Available online at http://www.mri-jma.go.jp/Publish/Technical/DATA/VOL_28/28_en.html.].

Jackson, E.A., 1991. Perspectives of Nonlinear Dynamics. vol. 2. Cambridge University Press, p. 633.

Kessler, E., 1969. On the distribution and continuity of water substance in atmospheric circulation. In: Meteor. Monogr., No. 32Amer. Meteor. Soc. 84 pp.

Khairoutdinov, M., Kogan, Y., 2000. A new cloud physics parameterization in a Large-Eddy simulation model of marine stratocumulus. Mon. Weather Rev. 128, 229–243.

Koop, T., Luo, B.P., Tsias, A., Peter, T., 2000. Water activity as the determinant for homogeneous ice nucleation in aqueous solutions. Nature 406, 611–614. https://doi.org/10.1038/35020537.

Koren, I., Feingold, G., 2011. Aerosol–cloud–precipitation system as a predator-prey problem. Proc. Natl. Acad. Sci. U. S. A. 108(30). https://doi.org/10.1073/pnas.1101777108.

Lafore, J., Moncrieff, M.W., 1989. A numerical investigation of the organization and interaction of the convective and stratiform regions of tropical squall lines. J. Atmos. Sci. 46, 521–544.

Lin, Y.L., Farley, R.D., Orville, H.D., 1983. Bulk parameterization of the snow field in a cloud model. J. Appl. Meteorol. Climatol. 22, 1065–1092. https://doi.org/10.1175/1520-0450(1983)022<1065:BPOTSF>2.0.CO;2.

Morrison, H., Shupe, M.D., Curry, J.A., 2003. Modeling clouds observed at SHEBA using a bulk microphysics parameterization implemented into a single-column model. J. Geophys. Res. 108, 4255. https://doi.org/10.1029/2002JD002229.

Morrison, H., Curry, J.A., Khvorostyanov, V.I., 2005. A new double-moment microphysics parameterization for application in cloud and climate models. Part I: description. J. Atmos. Sci. 62, 1665–1677.

Morrison, H., Thompson, G., Tatarskii, V., 2009. Impact of cloud microphysics on the development of trailing stratiform precipitation in a simulated squall line: comparison of one- and two-moment schemes. Mon. Weather Rev. 137, 991–1007.

Murakami, M., 1990. Numerical modeling of dynamical and microphysical evolution of an isolated convective cloud. The 19 July 1981 CCOPE cloud. J. Meteorol. Soc. Jpn. 68, 107–128.

Pruppacher, H.R., Klett, J.D., 1997. Microphysics of Clouds and Precipitation. Kluwer Academic 954 pp.

Ramkrishna, D., 2000. Population Balances: Theory and Applications to Particulate Systems in Engineering. Academic Press, San Diego, CA, p. 355.

Reisner, J., Rasmussen, R.M., Bruintjes, R.T., 1998. Explicit forecasting of supercooled liquid water in winter storms using the MM5 mesoscale model. Q. J. Roy. Meteorol. Soc. 124, 1071–1107.

Rutledge, S.A., Hobbs, P.V., 1983. The mesoscale and microscale structure and organization of clouds and precipitation in midlatitude cyclones. VIII: a model for the "seeder-feeder" process in warm-frontal rainbands. J. Atmos. Sci. 40, 1185–1206.

Seifert, A., 2011. Uncertainty and complexity in cloud microphysics. In: Proceedings of the ECMWF Workshop on Model Uncertainty, 20–24 June 2011. https://www.ecmwf.int/sites/default/files/elibrary/2011/14857-uncertainty-and-complexity-cloud-microphysics.pdf.

Skamarock, W.C., et al., 2008. A Description of the Advanced Research WRF version 3. NCAR Tech. Rep. TN-4751STR, 113 pp.

Smoluchowski, M.V., 1916. Drei Vorträge über Diffusion, Brown'sche Molekularbewegung und Koagulation von Kolloidteilchen. Phys. Z. XVII, 557–571 585–599.

Stevens, B., Feingold, G., 2009. Untangling aerosol effects on clouds and precipitation in a buffered system. Nature 461, 607–613.

Straka, J.M., 2009. Cloud and Precipitation Microphysics: Principles and Parameterizations. Cambridge University Press, p. 392.

Thompson, G., Eidhammer, T., 2014. A study of aerosol impacts on clouds and precipitation development in a large winter cyclone. J. Atmos. Sci. 71, 3636–3658. https://doi.org/10.1175/JAS-D-13-0305.1.

Thompson, G., Rasmussen, R.M., Manning, K., 2004. Explicit forecasts of winter precipitation using an improved bulk microphysics scheme. Part I: description and sensitivity analysis. Mon. Weather Rev. 132, 519–542.

Thompson, G., Field, P.R., Rasmussen, R.M., Hall, W.D., 2008. Explicit forecasts of winter precipitation using an improved bulk microphysics scheme. Part II: implementation of a new snow parameterization. Mon. Weather Rev. 136, 5095–5115.

Verlinde, J., Flatau, P.J., Cotton, W.R., 1990. Analytical solutions to the collection growth equation: comparison with approximate methods and application to cloud microphysics parameterization scheme. J. Atmos. Sci. 47, 2871–2880.

Wacker, U., 1995. Competition of precipitation particles in a model with parameterized cloud microphysics. J. Atmos. Sci. 52, 2577–2589.

Weisman, M.L., Klemp, J.B., Rotunno, R., 1988. Structure and evolution of numerically simulated squalllines. J. Atmos. Sci. 45, 1990–2013.

CHAPTER

11

Mesoscale orographic flows

Haraldur Ólafsson[a] *and Hálfdán Ágústsson*[b,c,d]

[a]School of Atmospheric Sciences, Science Institute, and Faculty of Physical Sciences, University of Iceland, Veðurfélagið and the Icelandic Meteorological Office, Reykjavík, Iceland [b]Kjeller vindteknikk Ltd., Lillestrøm, Norway [c]Belgingur Ltd., Reykjavík, Iceland [d]School of Atmospheric Sciences, Science Institute, University of Iceland, Reykjavík, Iceland

1 Introduction

Mountains, and hills are a source of enormous variability in atmospheric flow at a multitude of scales. Variations in direction and speed of the flow, both in time and space, are associated with variations in all parameters that are generally associated with weather forecasting, including wind, temperature, cloudiness, and precipitation. In the vicinity of mountains, climate, weather, and in particular weather extremes may be highly influenced by the topography. In this chapter, a comprehensive, but short review of the impact of mountains on the atmospheric flow will be given and aspects of forecasting the different orographic wind patterns will be discussed.

2 Patterns of mountain flows

Atmospheric flow is described by the momentum equation

$$\underset{\text{I}}{\frac{\partial \mathbf{V}}{\partial t}} + \underset{\text{II}}{\mathbf{V}\cdot\nabla\mathbf{V}} = \underset{\text{III}}{-\frac{1}{\rho}\nabla P} \underset{\text{IV}}{-f\mathbf{k}x\mathbf{V}} - \underset{\text{V}}{g} \underset{\text{VI}}{-\frac{\partial(u'w')}{\partial z}} \underset{\text{VII}}{-\frac{\partial(v'w')}{\partial z}} \tag{1}$$

where $\mathbf{V}$ is the wind vector, with components u, v, and w, P is pressure, f is the Coriolis parameter, and g is gravity. Term I represents changes of the wind field in time, term II represents acceleration associated with spatial variability of the wind, term III is the pressure gradient force, term IV is the Coriolis force, term V is the gravity and terms VI and VII represent the turbulent vertical transfer of horizontal momentum.

Uncertainties in Numerical Weather Prediction
https://doi.org/10.1016/B978-0-12-815491-5.00011-2

In order to facilitate the analysis of orographic flows, two nondimensional numbers are defined, the Rossby number, and the nondimensional mountain height, sometimes referred to as the inverse Froude number. Scale analysis of terms II and IV in Eq. (1) yields respectively U^2/L and fU. The ratio of these two terms is the Rossby number

$$Ro = U/fL$$

A popular interpretation of the Rossby number is a measure of to what extent the flow is geostrophic. If the Rossby number is far below unity, term IV dominates term II and the flow is close to geostrophic. Another way of formulating an interpretation is whether it is the Coriolis term or the nonlinear terms that act to balance a pressure gradient in frictionless and stationary flow.

Another dimensionless number, referring to energy and associated with the gravity term (V) and the nonlinear terms (II) in Eq. (1) is the nondimensional mountain height or the inverse Froude number

$$Fr^{-1} = Nh/U$$

where N is the Brunt-Vaisala frequency or the static stability of the airmass, h represents the vertical distance a particle ascends, or the height of the obstacle on which the flow is impinging, and U is the flow speed. The Nh/U may be interpreted as the ratio of the potential energy associated with the stratification of the flow ascending the mountain, and the kinetic energy of the flow. Nh/U may also be interpreted as a measure of the internal gravity wave speed to the speed of the flow itself.

Patterns of atmospheric flow in the vicinity of a mountain may be classified according to the values of the above dimensionless numbers, Ro and Fr^{-1}. The diagram in Fig. 1 shows these patterns schematically. In the upper left corner (high Ro, low Nh/U), there is typically an airmass with low static stability impinging on low obstacles and there is a negligible

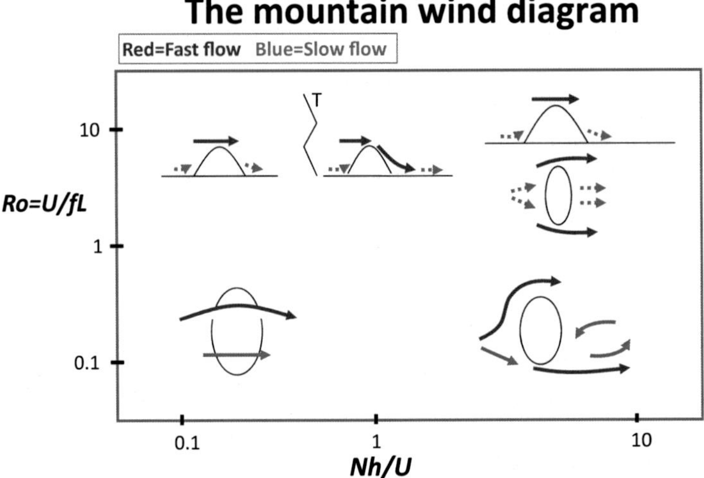

FIG. 1 A diagram of the main patterns of orographic flows, classified according to the values of the dimensionless numbers Ro and Nh/U.

impact of the Coriolis force upon the flow pattern. This is often referred to as flow over hills. Such flows are characterized by relatively fast flow on the top of the hills, but weaker flow at lower elevations, in between hills. The flow pattern may be sensitive to surface friction and the shape of the terrain, particularly on the downwind side, where a boundary-layer separation and wake formation may occur. Simulating such a pattern is hard and a true challenge for the parameterization of the numerical models. A review of flow over hills is given in, for instance, Belcher and Hunt (1998).

In the upper middle part of the diagram, there is a celebrated zone of amplified gravity waves and high orographic drag. The flow upstream of the mountain is in general not blocked, except perhaps at the bottom of the boundary-layer. Above the mountain, there are amplified gravity waves that may trail downstream and to some extent laterally. The flow in the gravity wave above the mountain is fast on its way downwards, but slow on its way upwards. Accordingly, the phase lines of the waves tilt upstream with height. The greater the amplitude of the waves are, the stronger is the acceleration in the downward flow. At the surface of the earth, this flow is often referred to as a downslope windstorm. A downslope windstorm may also occur where the waves are not very amplified. In this case, they are considered to break below a critical layer, above which the waves do not penetrate (e.g. Smith, 1985). Downslope acceleration and subsequent wake formation further downstream may also be described within the shallow water framework, where the flow moves from a subcritical to supercritical state over the mountain, accelerates over the downslopes, after which it decelerates suddenly in a hydraulic jump. The downslope acceleration does indeed occur in ideal, vertically nonstructured flow (e.g., Durran, 1990; Ólafsson and Bougeault, 1996), but the downslope acceleration is known to be sensitive to the vertical profile of the flow. Strong wave development is often associated with a stable layer or an inversion and a moderate change in the strength or the position of the inversion may have large impacts on the gravity waves and the downslope flow. Amplification of the waves and strong downslope flow may be enhanced by a positive vertical wind shear, while trapping of the wave energy at low levels, and strong downslope winds may also be favored by a negative vertical wind shear or a directional wind shear. The extent of the downslope windstorms, and the downstream flow pattern in general, is quite variable, ranging from smooth wavy field to well mixed turbulent flow. Hertenstein and Kuettner (2005) have proposed a diagram to classify the downstream flow where the key variables are the vertical wind shear, and the inversion strength. Ágústsson and Ólafsson (2010, 2014) have described two types of strong downslope flow an extended windstorm with no waves downstream, and a windstorm with a short downstream extension, and a wavy downstream flow field. The above papers include references to many articles dealing with different aspects of downslope acceleration.

In short, there is no universal theory of downslope windstorms in vertically nonuniform stratified flow, but in general, strong winds, and statically stable layers, particularly close to mountain top level, favor downslope acceleration. An extensive description of waves and other aspects of stratified and vertically structured flows over obstacles is given in Baines (1995).

In the upper right corner of the diagram in Fig. 1, there is blocked flow with corner/gap winds, and a wake, shown schematically in Fig. 2, and in a high-resolution simulation of real flow in Fig. 3. This is typical for high mountain ranges, and for moderate to low mountain ranges in weak winds. In blocked flows, there is a large area of weak winds upstream of

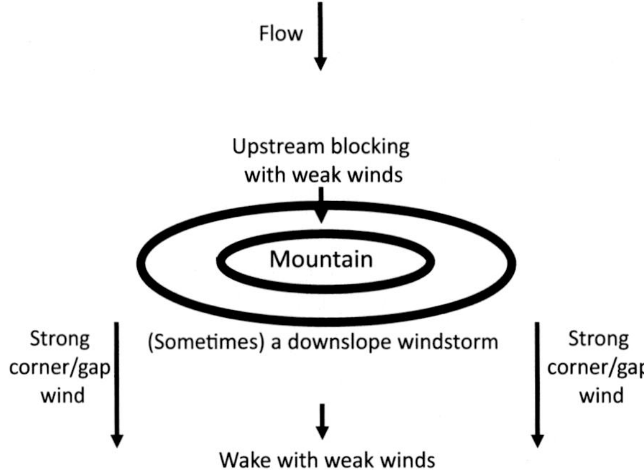

FIG. 2 A schematic of the main patterns of mesoscale orographic flows.

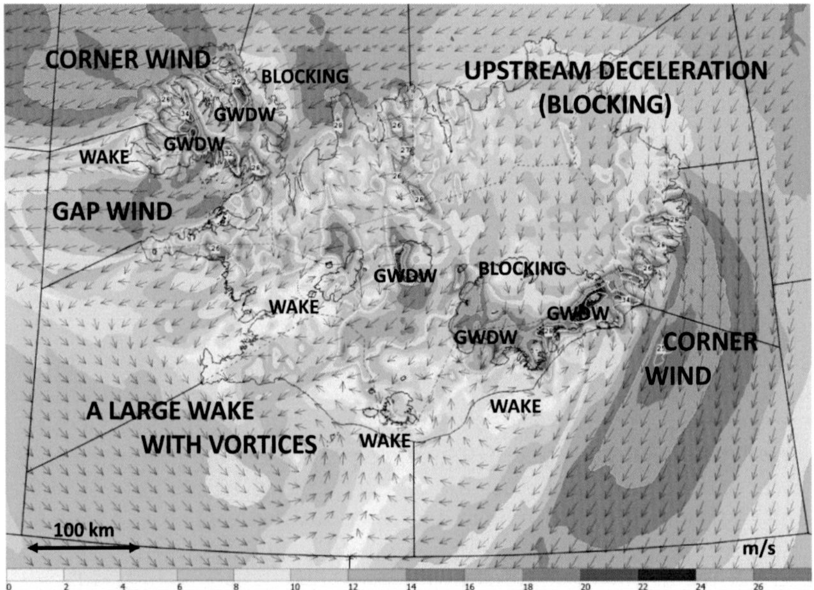

FIG. 3 Wind speed at 10 m (m/s) and identification of some of the main elements of mesoscale orographic flows in a case of northeasterly flow over Iceland on 29 June 2020 at 06 UTC. "GWDW" stands for gravity wave and a downslope windstorm. The flow is simulated in real-time by the numerical model Harmonie operated by the Icelandic Meteorological Office and the Danish Meteorological Institute at a horizontal resolution of 2.5 km.

the mountain, called a blocking, and a speed-up where the flow leaves the blocking. There is relatively fast flow at the mountain edges, from which jets may extend large distances. These jets are often referred to as corner winds or gap winds. Barrier winds may blow along the mountain, away from the blocking. A simple way of looking at the dynamics of the blocking is to consider conservation of energy along a streamline approaching the mountain, referring to terms II, III, and V in Eq. (1). This gives, after integration

$$0=\frac{1}{2}\rho V^2+P+\rho gz$$

here, the kinetic energy of the flow is redistributed to increased pressure, and increased potential. This happens as the flow meets the mountain, decelerates, ascends, and the pressure increases, relative to the pressure at the same elevation further upstream of the mountain. Traditionally, the term blocking refers to stagnant flow, i.e., that the surface flow is unable to overcome the potential barrier of the mountain, and stagnates. The onset of blocking has been dealt with by many authors, such as Smith (1989) and Smith and Grønås (1993). There may, however, be considerable upstream deceleration on the upstream side, but no blocking, and yet the region of decelerated flow is often referred to as an upstream blocking. On the downstream side of the mountain, there is relatively low pressure, and in flow described by linear theory (which, strictly speaking, refers to the low values of the Nh/U), the upstream, and downstream pressure anomalies are symmetric, but with opposite signs (Smith, 1989). The associated horizontal pressure gradient drives the corner or gap winds, and such winds have been investigated, and described by many authors, such as Ólafsson and Bougeault (1996), Zängl (2002), Gaberšek and Durran (2004, 2006), and others.

The orographic wake is an extended region of reduced wind speed downstream of a mountain. The dissipation of the kinetic energy of the flow of the wake may have taken place in breaking gravity waves or a hydraulic jump over the downslopes of the mountain, or elsewhere at the edges of the wake, referring to the last terms in Eq. (1), extended to all three dimensions. There may be a wake whether or not there is a downslope windstorm, and breaking waves. In fact, there are large, and extended wakes in flows with very high Nh/U with very little wave activity, and most of the flow diverted around the mountain, and not over the mountain. Dynamic aspects of mountain wakes, and how they relate to the ambient flow are discussed thoroughly in Smith et al. (1997), Rotunno et al. (1999), and Epifanio and Rotunno (2005), and aspects of the dissipation of kinetic energy, and the generation of potential vorticity associated with a mountain wake in stratified flows are presented comprehensively in Schär and Smith (1993), and idealized flows with bottom friction are addressed by Grubišić et al. (1995).

On the left side of the lower part of the diagram in Fig. 1, there is close to geostrophic flow over a large mountain ridge. This flow is characterized by a geopotential ridge above a wide, and large mountain ridge. The classic paper presenting such a solution is Queney (1948), and a popular textbook description of the ridge is derived from conservation of potential vorticity on a rotating earth. This flow pattern is on quite a large scale, and is, in general, well reproduced by numerical models.

To the right, in the lower part of the diagram, we have low Rossby number flows, which are at the same time blocked on the upstream side of the mountain, and with a downstream wake. A relatively large proportion of the low-level flow is diverted to the left as it meets the mountain, and on this side, there is acceleration. There is flow stagnation on the upstream right flank of the mountain (facing downstream), but acceleration on the right hand side of the wake (still facing downstream). Ólafsson (2000) and Petersen et al. (2005) presented a collection of solutions of such asymetric flows within the idealized framework. Classic examples of the upstream acceleration are the jet in southwesterly flows at Stad in W-Norway (Barstad and Grønås, 2005; Jonassen et al., 2012), the Cape Tobin jet at E-Greenland (Ólafsson et al.,

2009), and the northeasterly flows to the southeast of Greenland (Moore and Renfrew, 2005; Ólafsson and Ágústsson, 2009; Petersen et al., 2009). An example of a much celebrated jet to the right hand side of the wake (facing downstream) is the Greenland tip jet, extending hundreds of kilometers to the east from the southernmost tip of Greenland (Cape Farewell) (Doyle and Shapiro, 1999; Petersen et al., 2003). A popular presentation of the asymmetry of the flow field is geostrophic adjustment to the pressure anomalies generated by the mountain. The upstream high accelerates the flow that is deviated to the left of the mountain, while the downstream low accelerates the jet emanating from the right hand edge of the mountain (facing downstream).

3 Forecasting the orographic flows

A primary requirement for accurately reproducing patterns of orographic flows in a numerical weather prediction model must be that the model reproduces the topography accurately. Needless to say, this depends highly upon the horizontal resolution of the model; if the model resolution is too coarse to reproduce a mountain, the flow pattern associated with that mountain will inevitably also be missed by the model. Coarse-resolution models tend to underestimate the height of mountains, unless the width of the mountains is much greater than the distance between grid points of the model. In such cases, the maximum wind speed in corner winds, and the minimum wind speed in wakes, and blockings are systematically underestimated by the models. This underestimation is particularly strong if each of the two worlds, the real world, and the model world fall into two different regimes, i.e., blocked, and nonblocked flows. The intensity of corner winds, gap winds, wakes, and blockings may also be sensitive to the steepness of the slopes. Consequently, too gentle slopes in a numerical model tend to lead to an underestimation of the magnitude of the acceleration or the deceleration of the flow in the respective regions.

Even though all relevant topographic elements are well resolved, there are still uncertainties that may be characterized as systematic. Assessing these uncertainties is important for interpreting the model output for forecasting, and may be useful for systematic assessment of uncertainties in the context of postprocessing the model output. These uncertainties may, in general, be classified in the following categories: regions of spatial gradients, flow close to the limits of regime changes (e.g., blocked to nonblocked flow), gravity waves, and hydraulic jump-like flows, and features with high temporal variability.

Regions of strong horizontal, and vertical gradients are in general preferred locations for uncertainties, due to uncertain position of the edges, and to some extent to uncertainties in the magnitude of the jets. A small change in the direction of the wind upstream of a mountain will, in most cases, lead to a similar shift in the position of the orographic flow pattern. Certain locations, close to the edges of regions of weak or strong winds, will move from being inside of a region of strong winds to entering a region of weak winds or vice versa. A typical error of this kind is when a given location at the outer edge of a wake moves between being inside the corner wind, or inside the wake. Fig. 4 shows an example of a wake, and jets emanating from the mountains of SE-Iceland in northerly winds. The two panels in Fig. 4 are output from a model simulation with an interval of 1 h. There are negligible changes in the upstream winds, but inside the encircled area on the downstream side, the wind speed increases from 5 to

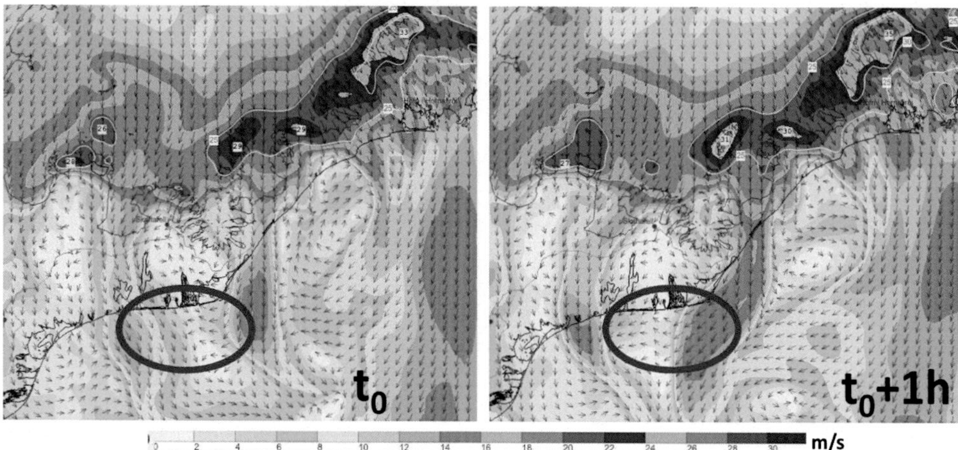

FIG. 4 Wind speed at 10m (m/s) in the wake of the mountains of SE-Iceland in northerly flow over Iceland on 5 June 2020 at 05 and 06 UTC. An area with large temporal variability in the wake is encircled. The flow is simulated in real-time by the numerical model Harmonie operated by the Icelandic Meteorological Office and the Danish Meteorological Institute at a horizontal resolution of 2.5km.

15m/s in the eastern part of the area, while in the western part of thearea, the wind speed is reduced from 13 to 3m/s. These specific changes in wind pattern occur over the ocean, and they are hardly ever verified, but in cases like this, the model is unlikely to be consistently correct in time, and space. In the case of Fig. 4, the jets emanate from lowerings in the topography, and they are modulated by gravity waves to an uncertain extent. The vertical extent of wakes is typically less than the height of the mountain that generates the wake, and in the case of a shallow wake, the relatively strong winds above the wake may penetrate down to the surface, either by turbulent mixing or by vertical transport of momentum in gravity waves. This penetration may be very hard for numerical models to reproduce accurately, regardless whether it is driven by gravity waves or turbulent mixing, and even if the models were very capable of reproducing both the mixing, and the gravity waves, both processes may be very sensitive to both wind, and static stability, and a small change or error in either parameter may lead to large changes or errors in the flow.

At the limits of flow regimes, small changes in the upstream flow or small modifications of model calculations, such as in the treatment of turbulent mixing, may trigger a shift from one flow pattern or regime to another. Blocked flow with stagnation on the upstream side of a mountain, and a wake on the downstream side may thus enter the "flow over" regime with only little upstream deceleration, and no wake. Fig. 5 illustrates such a change; first, there is an upstream blocking (encircled), but an hour later, the blocking has disappeared, and yet there are only minor changes in the upstream wind field, and the static stability.

A multitude of aspects of gravity waves, including uncertainties in their generation, amplification, propagation, and breaking have been addressed by a very large number of authors in the scientific literature. There is no universal theory of gravity waves in a nonuniformly layered flow, and their simulation is still a challenge to numerical models. The waves are sensitive to changes in the profile of wind, and stability for at least certain values of both profiles at certain levels, depending on the mountain on which the flow impinges, and the properties

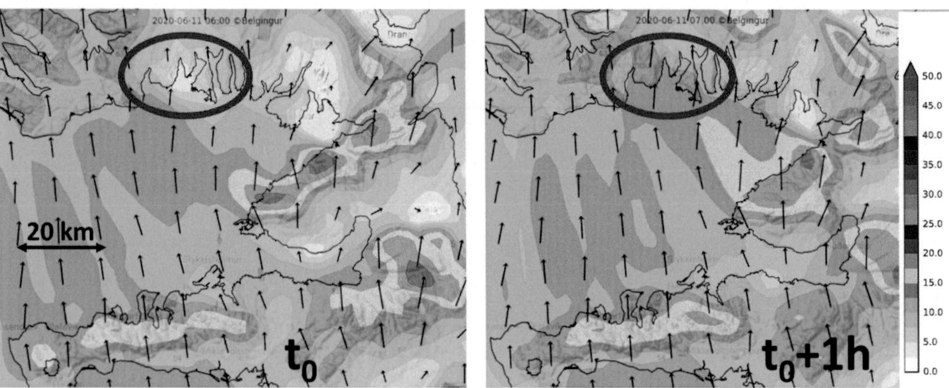

FIG. 5 Wind speed at 10 m (m/s) in southerly flow over Iceland on 11 June 2020 at 06 and 07 UTC. An area with large temporal variability in the blocking upstream of the mountains of NW-Iceland is encircled. The flow is simulated in real-time by the numerical model WRF, operated by Belgingur Ltd. at a horizontal resolution of 3 km.

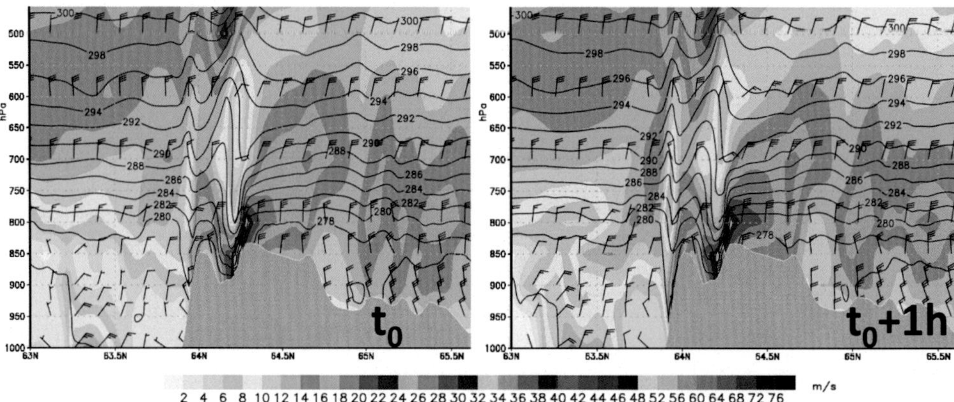

FIG. 6 Wind speed (m/s) and isentropes (K) in a S-N section across E-Iceland in northerly flow on 5 June 2020 at 15 and 16 UTC. The flow is simulated in real-time by the numerical model Harmonie operated by the Icelandic Meteorological Office and the Danish Meteorological Institute at a horizontal resolution of 2.5 km.

of the surface of the earth below the waves (e.g., Jonassen et al., 2014). They may also be sensitive to model details such as treatment of dissipation (Doyle et al., 2000) or even moisture (Rögnvaldsson et al., 2011). Even though many case studies appear to be able to reproduce the surface winds, and even parts of the wave pattern above, often after considerable tuning of tunable parameters of the numerical model, forecasting waves, and the associated downslope winds, remains a challenge. Fig. 6 illustrates the large spatial and temporal variability of the flow associated with gravity waves. In only 1 h, the downslope surface winds increase dramatically and the wave pattern above has changed very much. Yet, there are no clear signs of changes in the upstream vertical profiles of wind and temperature Reinecke and Durran (2009) expressed this uncertainty in the following words concluding a study based on cases from the T-REX field campaign: "neither case suggests that much confidence should be placed in the intensity of downslope winds forecast 12 or more hours in advance." Fig. 7, which shows the performance operational high-resolution numerical forecasts with lead time of 24h, based on data from 2017 to 2019 is in line with this; there is very little predictive skill for cases of downslope winds, where either the predicted or the observed winds are above 15m/s.

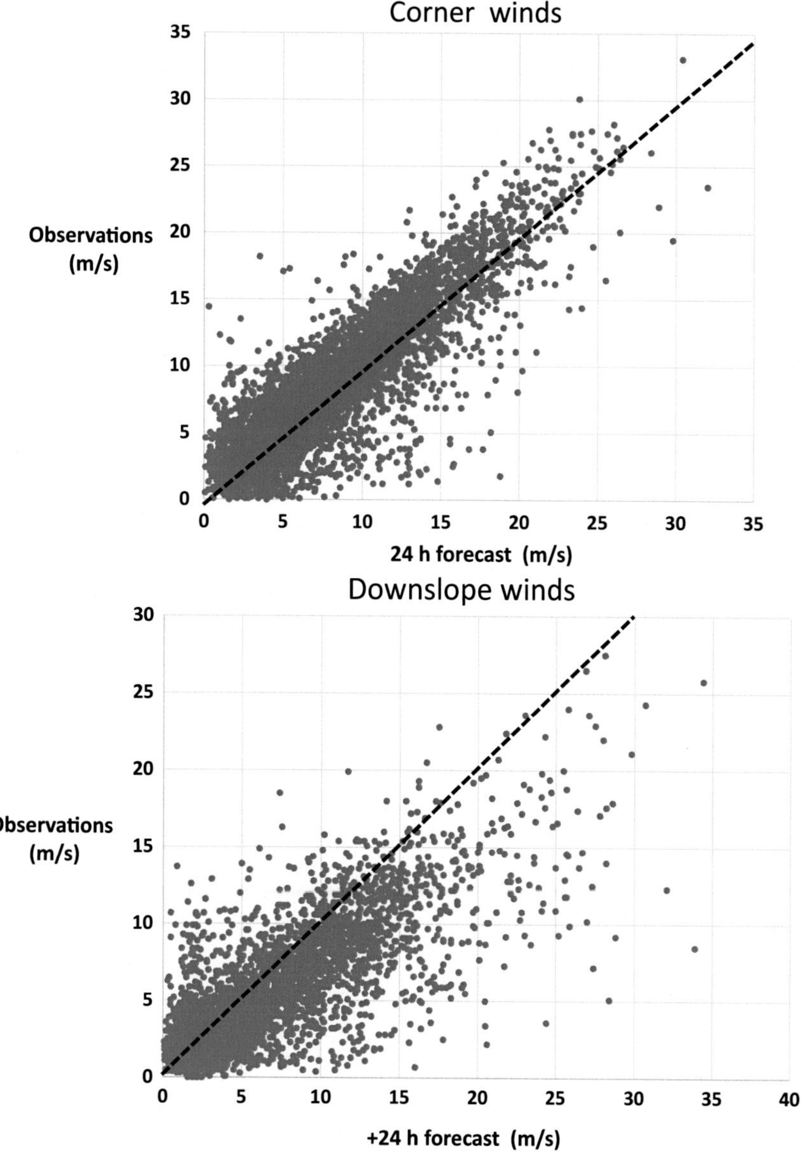

FIG. 7 Observed and forecasted (+24h) wind speed at Fagurhólsmýri, SE-Iceland in winds from the NE (Corner winds) and at Kvísker, SE-Iceland in winds from the W or N (Downslope winds). The numerical forecast is from the operational suite of the model Harmonie at a horizontal resolution of 2.5km with initial and boundary conditions from the ECMWF.

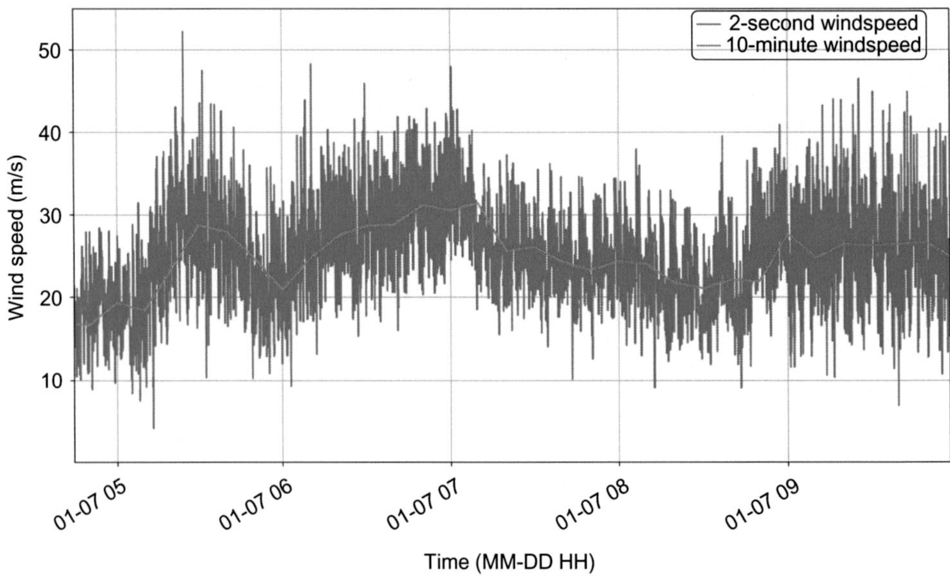

FIG. 8 Observations of surface wind speed at Kjalarnes, SW-Iceland on 7 January 2011 during a downslope windstorm. Data provided by the Icelandic Road Administration (Vegagerðin).

Temporal variability poses considerable uncertainty upon the forecasting of small-scale orographic flows, particularly on gravity waves, and downslope windstorms. High temporal variability of the winds is an integral part of gravity waves, and it is considered to be particularly important if the waves are breaking, as they go through phases of buildup, and breaking with large impacts on the surface wind speed. Turbulence is also associated with high temporal variability. Fig. 8 shows about 5 hours of instantaneous (2 s) wind speed at the foothills of Mt. Esja in SW-Iceland during a downslope windstorm The temporal variability is large on various time scales. There is definitely much turbulence, but one may assume that a part of the variability in Fig. 8, on the time scale of minutes to hours, is associated with gravity waves, either building up or breaking, and moving in space. Within the framework of shallow water, this may be expressed as variability associated with hydraulic jump-like flow features.

4 Future improvements

General improvements in the assessment of initial, and boundary conditions as well as more sophisticated treatment of elements such as turbulence, and dissipation in numerical models can be expected to lead to improvements of the representation of the above details of mesoscale orographic flows. To the extent that increased resolution leads to improvements in the representation of the topography, increased resolution may be expected to give better point forecasts. Apart from model issues such as representation of turbulence, and other boundary-layer processes, increased resolution tends to sharpen horizontal wind gradients, and may lead to fewer but bigger errors in regions with oscillating or uncertain position of the gradients.

In view of the sensitivity of gravity waves to the vertical profile of the atmosphere, improvements in vertical model resolution, and details in the initialization of features such as inversions, and vertical profile of winds will most likely lead to improvements in the forecasting of downslope windstorms.

References

Ágústsson, H., Ólafsson, H., 2010. The bimodal downslope windstorms at Kvísker. Meteorol. Atmos. Phys. 116, 27–42.

Ágústsson, H., Ólafsson, H., 2014. Simulations of observed lee waves and rotor turbulence. Mon. Weather Rev. 142, 832–849.

Baines, PG, 1995. Topographic Effects in Stratified Flows. Cambridge University Press, New York, pp. 1–482.

Barstad, I., Grønås, S., 2005. Southwesterly flows over southern Norway—mesoscale sensitivity to large-scale wind direction and speed. Tellus A 57 (2), 136–152.

Belcher, S.E., Hunt, J.C.R., 1998. Turbulent flow over hills and waves. Annu. Rev. Fluid Mech. 30, 507–538.

Doyle, J.D., Shapiro, M.A., 1999. Flow response to large-scale topography: the Greenland tip jet. Tellus A 5, 728–748.

Doyle, J.D., Durran, D.R., Chen, C., Colle, B.A., Georgelin, M., Grubisic, V., Hsu, W.R., Huang, C., Landau, D., Lin, Y.L., Poulos, G.S., Sun, W.Y., Weber, D.B., Wurtele, M.G., Xue, M., 2000. An intercomparison of model-predicted wave breaking for the 11 January 1972. Boulder windstorm. Mon. Weather Rev. 128 (3), 901–914.

Durran, D.R., 1990. Mountain waves and downslope winds. In: Atmospheric Processes over Complex Terrain, Meteorological Monographs, No. 45. American Metcorological Society, pp. 59–81.

Epifanio, C.C., Rotunno, R., 2005. The dynamics of orographic wake formation in flows with upstream blocking. J. Atmos. Sci. 62 (9), 3127–3150. https://doi.org/10.1175/JAS3523.1.

Gaberšek, S., Durran, D.R., 2004. Gap flows through idealized topography. Part I: forcing by large-scale winds in the nonrotating limit. J. Atmos. Sci. 61, 2846–2862.

Gaberšek, S., Durran, D.R., 2006. Gap flows through idealized topography. Part II: effects of rotation and surface friction. J. Atmos. Sci. 63, 2720–2739.

Grubišić, V., Smith, R.B., Schär, C., 1995. The effect of bottom friction on shallow-water flow past an isolated obstacle. J. Atmos. Sci. 52, 1985–2005.

Hertenstein, R.F., Kuettner, J.P., 2005. Rotor types associated with steep lee topography: influence of the wind profile. Tellus 57A (2), 117–135.

Jonassen, M.O., Ágústsson, H., Ólafsson, H., 2014. Impact of surface characteristics on flow over a mesoscale mountain. Q. J. R. Meteorol. Soc. 140, 2330–2341. https://doi.org/10.1002/qj.2302.

Jonassen, M.O., Ólafsson, H., Reuder, J., Olseth, J.A., 2012. Multi-scale variability of winds in the complex topography of southwestern Norway. Tellus. 64(11962)https://doi.org/10.3402/tellusa.v64i0.11962.

Moore, G.W.K., Renfrew, I.A., 2005. Tip jets and barrier winds: a QuikSCAT climatology of high wind speed events around Greenland. J. Clim. 18, 3713–3725.

Ólafsson, H., Bougeault, P., 1996. Nonlinear flow past an elliptic mountain ridge. J. Atmos. Sci. 53, 2465–2489.

Ólafsson, H., 2000. The impact of flow regimes on asymmetry of orographic drag at moderate and low Rossby numbers. Tellus 52 (4), 365–379.

Ólafsson, H., Ágústsson, H., 2009. Gravity wave breaking in easterly flow over Greenland and associated low level barrier- and reverse tip-jets. Meteorol. Atmos. Phys. 104, 191–197. https://doi.org/10.1007/s00703-009-0024-9.

Ólafsson, H., Ágústsson, H., Shapiro, M.A., Kristjánsson, J.E., Barstad, I., Dörnbrack, A., 2009. The Cape Tobin jet. In: Proc. Int. Conf. Alpine Meteorol., Raststatt, Germanypp. 224–225.

Petersen, G.N., Kristjánsson, J.E., Ólafsson, H., 2005. The effect of upstream wind direction on atmospheric flow in the vicinity of a large mountain. Q. J. R. Meteorol. Soc. 131, 1113–1128.

Petersen, G.N., Ólafsson, H., Kristjánsson, J.E., 2003. Flow in the lee of idealized mountains and Greenland. J. Atmos. Sci. 60, 2183–2195.

Petersen, G.N., Renfrew, I.A., Moore, G.W.K., 2009. An overview of barrier winds off southeastern Greenland during the Greenland flow distortion experiment. J. Atmos. Sci. 135, 1950–1967.

Queney, P., 1948. The problem of the airflow over mountains: a summary of theoretical studies. Bull. Am. Meteorol. Soc. 29, 16–26.

Reinecke, P.A., Durran, D.R., 2009. Initial condition sensitivities and the predictability of downslope winds. J. Atmos. Sci. 66 (11), 3401–3418.

Rögnvaldsson, Ó., Bao, J.W., Ágústsson, H., Ólafsson, H., 2011. Downslope windstorm in Iceland—WRF/MM5 model comparison. Atmos. Chem. Phys. 11, 103–120. https://doi.org/10.5194/acp-11-103-2011.

Rotunno, R., Grubišić, V., Smolarkiewicz, P.K., 1999. Vorticity and potential vorticity in mountain wakes. J. Atmos. Sci. 16, 2796–2810.

Schär, C., Smith, R.B., 1993. Shallow-water flow past isolated topograpy. 1. Vorticity production and wake formation. J. Atmos. Sci. 50, 1373–1400.

Smith, R.B., 1985. On severe downslope winds. J. Atmos. Sci. 43, 2597–2603.

Smith, R.B., 1989. Hydrostatic airflow over mountains. In: Advances in Geophysics, vol. 31. Academic Press, pp. 59–81.

Smith, R.B., Gleason, A.C., Gluhosky, P.A., Grubišić, V., 1997. The wake of St. Vincent. J. Atmos. Sci. 54, 606–623.

Smith, R.B., Grønås, S., 1993. Stagnation points and bifurcation in 3-D mountain airflow. Tellus 45, 28–43.

Zängl, G., 2002. Stratified flow over a mountain with a gap: linear theory and numerical simulations. Q. J. R. Meteorol. Soc. 128, 927–949.

CHAPTER

12

Numerical methods to identify model uncertainty

Harald Sodemann[a] *and Hanna Joos*[b]

[a]Geophysical Institute and Bjerknes Centre for Climate Research, University of Bergen, Bergen, Norway [b]Institute for Atmospheric and Climate Science, ETH Zürich, Zürich, Switzerland

1 Numerical tracers

Numerical weather prediction (NWP) is primarily concerned with predicting the state of the atmosphere for the key state variables temperature, horizontal and vertical wind, mass, pressure and humidity (Bjerknes, 1904). However, many more than those seven quantities can be present in the atmosphere and be relevant for weather processes at any point in time. These can be reactive or intert natural substances, such as mineral aerosol (dust), volatile organic compounds (VOCs), ozone, and CO_2, that each have individual atmospheric pathways and life times (Seinfeld and Pandis, 2016). A further example for natural tracers are stable isotopes of water, which are transported identical to the ordinary water cycle, while being split up according to their molecular mass, and being coupled to different phases through fractionation processes (Epstein, 1956). Besides such natural tracers, the use of artificial tracers can also be beneficial within a model framework. Artificial tracers of specific properties, that are inaccessible otherwise, such as air mass age or air mass origin, can support the interpretation of a model simulation (e.g., Boutle et al., 2011). Furthermore, the introduction of model variables that collect the contribution of specific processes, such as the condensational heat release, can support model development and validation (Section 2). Here we first review the numerical requirements for tracer advection (Section 1.1). Thereby, passive artificial and natural tracers are presented first, including stable isotope tracers to test a model's water cycle (Section 1.2). Then, active and reactive natural tracers are presented, including dust and chemical substances (Section 1.3). Throughout Section 1, we outline recent complementary and alternative developments, and discuss remaining challenges and uncertainties.

https://doi.org/10.1016/B978-0-12-815491-5.00012-4

1.1 Numerical requirements

Adding the advection of tracers increases the demands with respect to several properties of numerical schemes. In this section, we briefly consider how numerical requirements for tracer advection are enhanced, and exemplify this with several up-to-date numerical models. Additional details on the impact of numerical schemes on NWP uncertainty can be found in other chapters of this book.

In many cases, the most important property for a numerical scheme for tracer advection is arguably mass conservation. The emphasis of numerical weather forecasting models has naturally been on achieving the most reliable forecast product with the available computation resources, in particular in an operational context. This could lend conservation properties of a discretization scheme a lower priority than its efficiency. For example, semi-Lagrangian schemes allow for large time steps while maintaining stability, and are common across many operational NWP models. Yet, the computational advantage may come at the expense of violated mass conservation. In grid-scale models, higher-order schemes can be applied that reduce the discretization error, and that are designed to conserve mass. Furthermore, finite-volume methods combine long time steps of Semi-Lagrangian schemes with mass conservation. For chemical tracer variables, numerical artifacts such as negative quantities can require additional steps to maintain meaningful results, or prompt the implementation of schemes, which meet specific requirements, such as positive definiteness (Smolarkiewicz, 2005).

With many of the currently employed advection schemes, tracer quantities are already advected with mass-conservative, positive definite higher-order schemes. An example of this is the Bott scheme, used in the consortium for small-scale modeling (COSMO) model (Steppeler et al., 2003). The icosahedral nonhydrostatic (ICON) model, which also includes tracer variables, employs this scheme as well (Eckstein et al., 2018). In contrast, the climate high-resolution model (CHRM) (Majewski, 1991; Vidale et al., 2002), predecessor to the COSMO model, both the Leapfrog scheme and the Semi-Lagrangian scheme did not fulfill the requirements for a tracer quantity. In the implementation of a water tracer (see below, Sodemann et al., 2009), therefore, the advanced multidimensional positive-definite advection transport algorithm (MPDATA) scheme of Smolarkiewicz (2005) was implemented for the purpose of tracer advection. However, inconsistencies between the original leap-frog scheme and the MPDATA scheme lead to additional requirements for synchronization between the tracer and the original water quantity. Ultimately, it is therefore beneficial to use consistent schemes for tracer advection and the advection of the main quantities across a model, even though tradeoffs concerning the calculation efficiency need to be considered. Computational demand can thereby increase considerably, depending on the efficiency of the scheme, and the number of tracers to be advected.

An interesting aspect regarding the interplay of the model grid and tracer quantities is the interaction between flow properties and the preservation of gradients. Numerical diffusion of tracers is partly partly due to the grid spacing, i.e., the scale of preserved gradients, and partly due to additional diffusion operators that are employed to reduce the occurrence of nonlinear instabilities. For tracer quantities, there is a choice to be made between consistency with the main quantity (such as in the case of water tracers) and conservation of the gradients.

Rastigejev et al., 2010 demonstrated that in natural geophysical flow with convergence and divergence, gradients are more quickly reduced than expected from more idealized flow situations (Fig. 1). The reduction of gradients can in particular be a problem for the comparison with observations, and in case of reactive chemical transport simulations (Sodemann et al., 2011). Specific schemes have therefore been employed to preserve higher-order moment properties, such as the widely-used Prather (1986) scheme, which conserves second-order moments during advection. Semi-Lagrangian schemes, often employed in operational contexts due to their stability at large time steps, are prone to considerable diffusion originating from interpolation to the departure grid. Fully Lagrangian schemes may be most suited to preserve gradients, but have other numerical challenges, which is why they are not yet used in operational models (Bowman et al., 2014).

Parameterized processes have their own additional requirements when involved in tracer transport and modification. If a tracer is employed for a quantity that for example mimics the water cycle, all phase changes need to be implemented for that tracer. In that case, depending on the complexity of the parameterization schemes, this requires additional complexity of the model code to be added, e.g., when different hydrometeors or chemical reactions need to be implemented. Resulting challenges concern the maintenance of the model code over time, as well as additional resource demands during computation. Specific examples are given in the next sections. In particular, uncertainties related to parameterized processes can be identified with the methods described in the section on tendency diagnostics (Section 2).

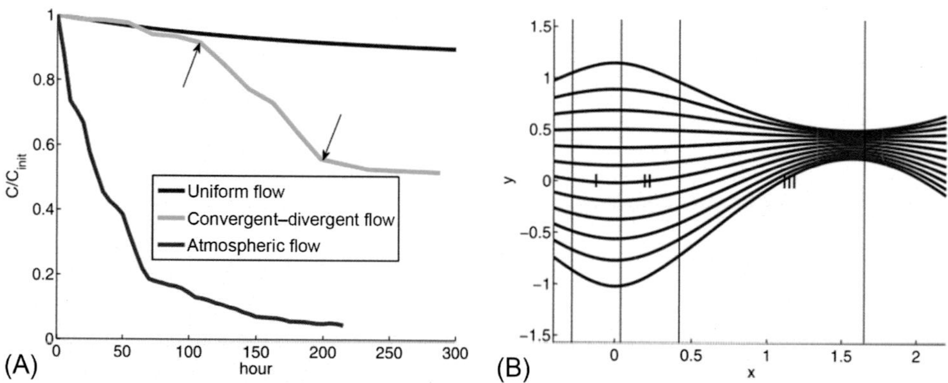

FIG. 1 (A) Maximum concentration of a plume as a function of time, as calculated by the global chemical transport model GEOS-Chem. An initially localized plume begins in China and propagates across the Pacific Ocean. The blue line represents the maximum plume concentration in a spatially uniform flow; the red line corresponds to an atmospheric flow. The green line represents a convergent-divergent flow, in which the Lyapunov exponent l is significant between times $125\,h < t < 200\,h$. The arrows denote the times at which the stretching turns on and off. (B) Streamlines of the model convergent-divergent flow in which a plume entering at the left boundary at $x \approx 0.28$ first expands (region I), then compresses (regions II and III), and then after the first constriction expands again. (Rastigejev et al., 2010, Figs. 1 and 4).

1.2 Passive tracers

Numerical tracers can be broadly classified into two categories, namely passive and active tracers. Passive tracers do not influence or feedback with the model dynamics. Examples for these are purely diagnostic, artificial tracers, that signify the age or origin of a substance. A further example for this category of tracers are stable water isotopes, as they specify additional information on a variable that is already existing as bulk quantity (i.e., all isotopologues of water) in the model dynamics. While passive tracers are calculated as a prognostic quantity, until now they mostly serve diagnostic purposes, e.g., to further understand the dynamics of a specific weather situation, or the behavior of parameterized processes. The following Sections exemplify passive tracer applications for artificial quantities that represent age and origin, and for natural (physical) quantities that add detail to other existing variables.

1.2.1 Airmass tracers

The most straightforward application of an airmass tracer is to mark or "tag" a specific airmass by origin. Therefore, this application of tracers is sometimes referred to as "tagging" (Koster et al., 1986). While airmass tracers, to some extent, correspond to long-lived atmospheric trace substances, such as SF_6, or certain PFCs, they are more accurately described as artificial tracers (with or without source and removal processes and background concentrations). One example is the tracing of boundary-layer air that is vented into the free atmosphere during cyclogenesis (Boutle et al., 2011). Fig. 2 shows a situation during a model run where a boundary-layer tracer allows to detect detrainment of boundary-layer air into the free troposphere. Further examples for tracer quantities are the latitudinal displacement of air under cyclogenesis, and airmass age. In the latter case, an initial age field is modified with a source term, increasing airmass age at every time step. Some numerical models are

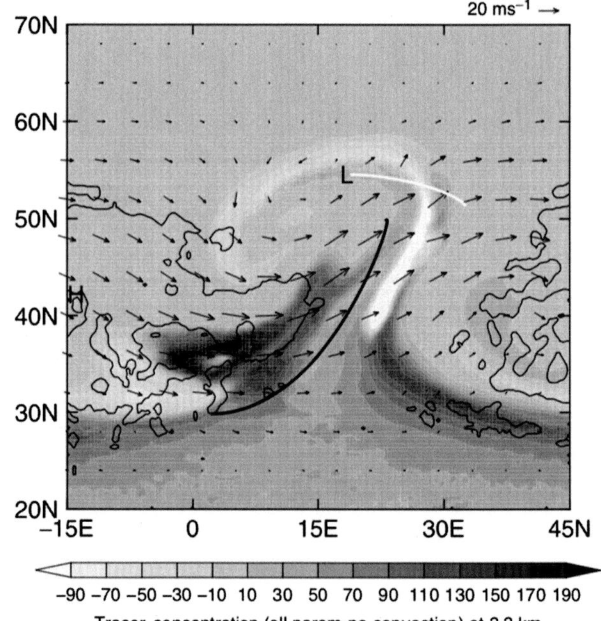

FIG. 2 Tracer concentration (arbitrary units) ventilated from the boundary layer by convection at 3.3 km (shading), with wind vectors at 3.3 km, at day 9. Cumulus-capped boundary layers are marked by thin black lines, with the warm/cold fronts shown by thick white/black lines. (Boutle et al., 2011, Fig. 4).

preequipped with generic tracer variables to allow for future testing and development, such as the application of research to operations at mesoscale (AROME) model (Seity et al., 2011). A versatile tracer module exists in the COSMO-ART model, enabling the use and definition of both passive and active tracers (Eckstein et al., 2018, see Section 1.3).

During the implementation of passive tracers, it is important to include all parameterized processes that involve airmass transport, such as moist convection and turbulence, to avoid erroneous results from an inconsistent tracer advection scheme (Sodemann, 2006). A particularly efficient, alternative solution to the advection of tracers within grid-scale models exists in the form of trajectory models and Lagrangian particle dispersion models, which can be coupled to the wind fields of NWP models both off-line (Stohl et al., 2005; Pisso et al., 2019) and on-line (Miltenberger et al., 2013). For such Lagrangian approaches, uncertainty may originate from the representation of dispersion processes, in particular if the boundary layer is characterized different stability regimes (Stohl and Thomson, 1999). A further source of error are inconsistencies between convection schemes in the model used to produce the boundary data, and the Lagrangian model (Forster et al., 2007).

1.2.2 Water vapor tracers

Water vapor was the last of the main prognostic variables added to operational forecast models in the 1970s. Still today, accurately predicting the advection of water vapor, as well as its liquid and solid phases, are challenging for current NWP models. Due to the small scales and phanse changes, many processes associated with water vapor require parameterization in models. Increasing model resolution, thereby, requires revision of how different processes are parameterized at a model and grid spacing. For example, prognostic precipitation was added once operational models became nonhydrostatic, reaching a horizontal grid-spacing where this was required (Steppeler et al., 2003). Parameterization of shallow and deep moist convection is still at least partly needed in operational NWP models. Cloud microphysical processes with mixed-phases and different hydrometeor categories add further complexity, and are still partly uncertain. Supercooled cloud water, for example, is a poorly constrained variable, but is known to play a key role in mixed-phase clouds.

Passive tracers in the water cycle have long been used to diagnose both model behavior and the coupling of processes in the Earth system. Koster et al. (1986) identified the evaporation sources for high-latitude precipitation from simulations with a GCM that included tagged water tracers. In contrast to airmass tracers, water tracers need to include all parameterized processes that affect the main water variable, including moist convection, cloud microphysics and precipitation, in what can be described as a *secondary water cycle* (Sodemann, 2006). Numaguti (1999) used such a secondary water cycle implementation in a GCM to investigate the age of precipitation, defined as the time since it last evaporated from the oceans. Such source-sink relationships allow to understand the coupling between ocean, land and atmosphere in Earth system models and the Earth system itself on different temporal and spatial scales. Age and origin of water vapor are thus the key pieces of information provided by water vapor tracers in numerical models. In conjunction with suitable observational constraints, this offers possibilities for reducing model uncertainty in an otherwise heavily parameterized part of NWP models.

Different tracer initialization methods allow to target the assessment of specific model characteristics in regional and global tracer-enabled models. Water vapor tracers can be released from specified areas, such as an ocean basin, or a region marked by particular surface

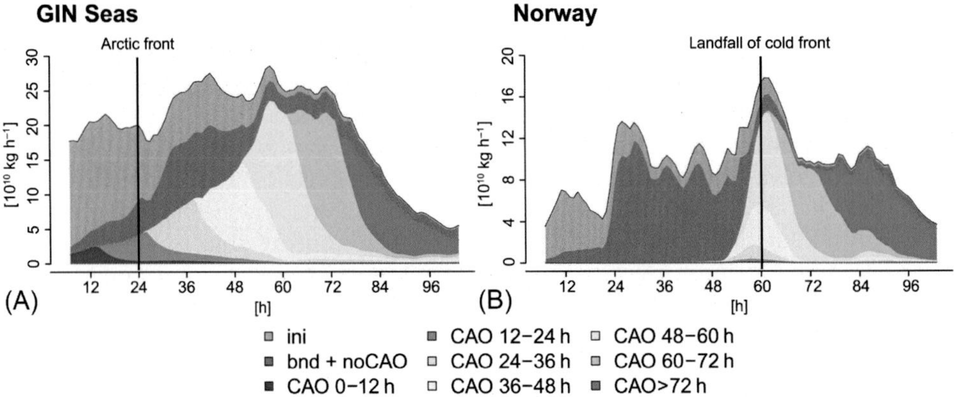

FIG. 3 Temporal evolution of tagged precipitation integrated over (A) the Greenland-Iceland-Norwegian (GIN) Seas and (B) Norway with contributions shown separately for tagged vapor initially present in the domain (light gray), entering the domain from the boundary or the surface outside of the CAO mask (dark gray), and associated with the CAO tracers that are released in 12-hourly intervals (color). Time axis is relative to the initial time of the simulation and the time when the CAO air mass reaches across the Arctic front and the cold front makes landfall are indicated in (A) and (B), respectively. *Data are from the COSMO 0.05° simulation (Papritz and Sodemann, 2018, Fig. 8).*

properties, such as SST, heat flux or sea level pressure (Numaguti, 1999; Sodemann et al., 2009; Sodemann and Stohl, 2013). In order to identify age properties, source areas can be activated during certain time periods only (Läderach, 2016). Papritz and Sodemann (2018) used such an approach to quantify the age of precipitation within a large cold-air outbreak in the Nordic Seas region (Fig. 3, color shading). In limited area models, tracer initialization methods can be used to mark and trace all water vapor entering at the boundaries of the domain (Fig. 3, dark gray shading). Furthermore, an initialization that marks all water vapor inside the model domain at an initial time with a specific tag, allows to assess the removal of the initial moisture during the time of a model simulation (Fig. 3, light gray shading, see also Sodemann et al., 2009; Winschall et al., 2012).

The fraction of initial and boundary tracers in the model domain can thereby serve as a measure of uncertainty for tracer approaches. Total tracer budgets can be obtained by comparing the sum of all tracers to the water variables from the primary water cycle. Over time, numerical deficiencies may accumulate, and lead to loss of the water inside the model domain, which can then be attributed to specific sources or processes. In some situations, where water vapor transitions through a boundary zone in a regional model, water vapor may also be erroneously marked twice if it becomes marked as boundary water (Sodemann, 2006). Numerical effects and diffusion properties become critical at longer times, and it is here that the conservative properties of the numerical schemes become most important. Such uncertainty can be efficiently attributed to individual parameterized processes in a model by means of diagnostic tendency output (Section 2).

1.2.3 Water vapor isotope tracers

While water vapor tracers are useful to detect model-specific characteristics and deficiencies, they predict artificial, and thus, nonobservable quantities. Ordinary water vapor measurements do not provide information about its age or origin. In situations where several poorly constrained processes interact, for example during evaporation, which is driven by low-level humidity gradients and boundary-layer turbulence, or in the case of precipitation, that involves several coupled microphysical processes (e.g., Stensrud, 2007), errors from different parameterizations can compensate each other, thereby providing the correct result for the wrong reason.

In this context, additional observable quantities are highly valuable for reducing model uncertainty. The stable isotope composition of water vapor and precipitation is a measurable quantity, and offers a way to identify and possibly reduce such uncertainty. Natural water occurs in several isotopologues. In addition to the most common water molecules, ${}^{1}H_2{}^{16}O$, there are additional variants, so-called isotopologues, that include the heavier molecules ${}^{18}O$, ${}^{17}O$ and ${}^{2}H$ instead of the lighter ${}^{16}O$ and ${}^{1}H$ (Epstein, 1956; Gat, 1996). Primarily, for natural waters the heavy isotopologues ${}^{2}H{}^{1}H\,{}^{16}O$ and ${}^{1}H_2{}^{18}O$ can be measured in addition to the common water molecule. Since their abundance is rare, on the order of less than 1 in 1000, sufficiently sensitive and precise measurement equipment is needed to quantify stable water isotopes (Galewsky et al., 2016). The advected water quantities are commonly reported as isotope ratios R, normalized on a delta-scale as deviation from a recognized standard, Vienna Standard Mean Ocean Water (VSMOW, e.g., Gat, 1996).

As the heavier isotopologues have a lower vapor pressure, they condense more readily and are less likely to evaporate. Therefore, when water changes phase in the atmosphere, evaporation and precipitation processes lead to the enrichment of the heavy isotopologues in the condensed phase and their preferential removal (depletion) from the vapor phase. Thereby, atmospheric water vapor, and rain formed from this vapor, become increasingly depleted compared to the average ocean water composition. This process is known as isotope fractionation. When water isotope fractionation is built into an atmospheric model, stable isotope measurements can be used as an additional constraint on the history of phase changes in an airmass, on the mixing of water vapor, and on the origin of water vapor from different source regions (e.g., Pfahl et al., 2012; Risi et al., 2008; Risi et al., 2012).

In a model system, the addition of stable water isotope tracers can be based on the implementation of a passive water vapor tracer as described above (Section 1.2.2). In practice, this implies additional water variables for each isotopologue, for example by adding another dimension to the water variables. Then, along the entire water cycle, temperature-dependent stable isotope fractionation processes are to be implemented during all phase transition processes (Joussaume et al., 1984). Thereby, one particular challenge are the small numbers compared to the main water field, due to the rarity of the stable isotopologues. Therefore, Pfahl et al. (2012) in their implementation scaled the isotope variables for advection with the mean isotope ratio of ocean water ${}^{18}R_{\mathrm{VSMOW}}$, such that concentrations instead of delta values are advected.

$$
{}^{18}q_x = {}^{18}q^*{}_x / \left({}^{18}R_{\mathrm{VSMOW}} \cdot {}^{18}j\right),
$$

where ${}^{18}q$ is the scaled specific humidity of the heavy isotope water in units of kg kg^{-1}, index x stands for the different water species, such as water vapor or cloud water, and ${}^{18}j$ is the ratio of the molecular weights of the heavy isotope and standard water. The scaling provides the same order of magnitude as the common water, thereby minimizing numerical inconsistencies between the fields. Accordingly, the approach improved results substantially over earlier implementations (Pfahl et al., 2012). The only requirement is then to perform a corresponding rescaling onto a delta-scale with respect to VSMOW during the treatment of actual phase changes and for working with output model fields:

$$\left(\delta^{18}O_x\right) = \left({}^{18}q_x/q^t{}_x - 1\right)\cdot 10^3$$

where the q_x^t stands for the total humidity of a water species x. A corresponding expression is formulated for the δ^2H.

1.2.4 *Challenges and developments*

Numerical accuracy, mass conservation and positive definiteness are even more important for an accurate result for stable isotope fractionation than for a water tracer by itself (Sturm et al., 2005). This is, in particular, important for additional isotope parameters, that are obtained from a combination of several isotopologues, such as the Deuterium excess (Dansgaard, 1964) and the ^{17}O-excess (Landais et al., 2008). In particular the ^{17}O-excess faces substantial numerical and computational challenges in order to match model results and observations in a meaningful way (Risi et al., 2013). In addition, model limitations are prevalent when working with isotope tracer implementations, also due to knowledge gaps in the physical understanding of mixed-phase cloud microphysics. One such example are fractionation processes during evaporation in strong-wind regimes (Bonne et al., 2019), and microphysical fractionation processes within clouds during super-saturation, and presence of super-cooled cloud water. Furthermore, isotope observations are still limited in terms of their spatial and temporal coverage. However, a large expansion of the available atmospheric in situ data is now ongoing due to the advent of laser-spectroscopic measurement capability (Galewsky et al., 2016; Kerstel et al., 2006). This observational development maintains the long-standing promise of a more well-founded water cycle in models due to the use of water isotopes (Risi et al., 2012), and motivates further development of this branch of NWP models.

One particular challenge regarding isotope tracers in models is to identify atmospheric situations where the stable isotopologues of water can contribute unique, new information. An example for the powerful capabilities to gain additional information from such isotope-enabled NWP models is presented in the study of Field et al., 2010. In a model simulation where the isotopic fractionation during the fall of the precipitation from cloud-base to the surface was deactivated, a clear difference in the composition of precipitation resulted. Based on such sensitivity studies, comparison with observations can then provide indications on the strength of parameterized postcondensation exchange processes that are otherwise not accessible. Aemisegger et al. (2015) presented a study using COSMOiso that demonstrated the importance of the role of surface evaporation to reproduce the observed stable isotope composition in several isotopologues in parallel, using the Deuterium excess parameter. Observations only match to the simulation when surface evaporation is activated in the regional model (Fig. 4, green curves). The possibilities in such regional studies are to resolve more detailed processes, for example regarding convection. Global models that have a higher degree of parameterizations can potentially benefit strongly, also when different model components are coupled. Isotope tracers may thereby contribute to help understand and reduce the uncertainty of the interpretation of past climate changes (Lewis et al., 2010). A further promising

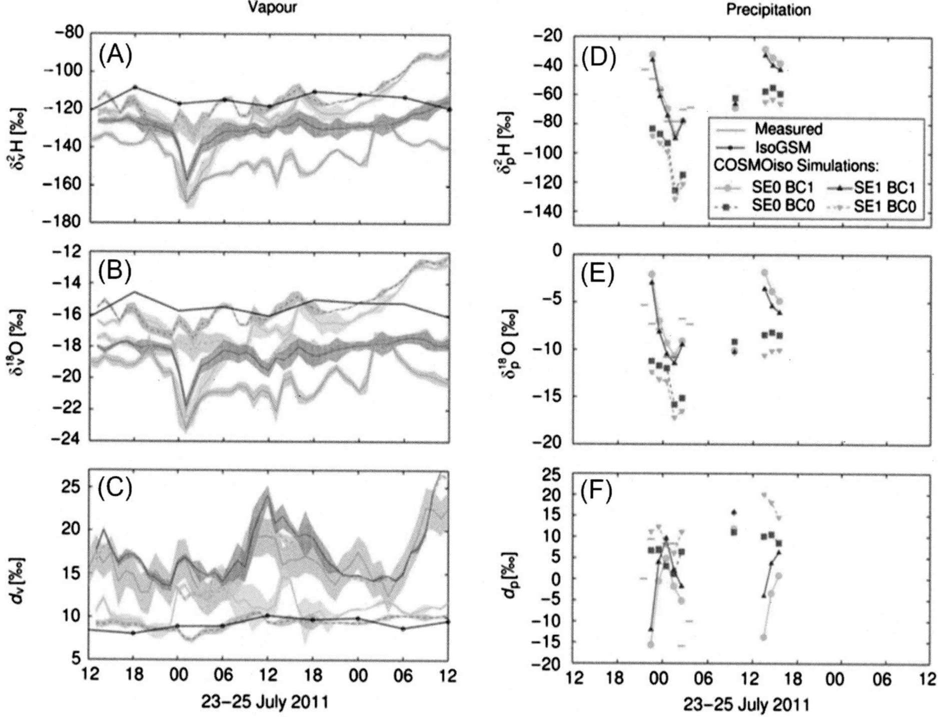

FIG. 4 Comparison between measured (orange) and modeled (blue and green) vapor (A)–(C) and precipitation (D)–(F) isotope signals for the cold front passage at 00UTC 24 July 2011. In panels (A)–(C), the measured hourly water vapor isotope signals are shown with a shaded band of the 5 s standard deviation from the measurements. The lowest model level water vapor isotope signals are taken from the COSMOiso simulations and interpolated bilinearly to the point of isotope measurement. The shaded area for the water vapor isotope simulations represents the standard deviation of the simulated isotope signals in a square of 56 km^2 around the isotope measurement station. In blue the simulations without fractionation during surface evaporation (SE0) are shown, in green the simulations with Craig-Gordon surface evaporation fractionation (SE1). Dashed lines show simulations without below cloud interaction (BC0), solid lines simulations with below-cloud interaction (BC1). The lowest model level water vapor isotope signals from IsoGSM used as boundary conditions are shown in black. Date ticks show hours in UTC. (Aemisegger et al., 2015, Fig. 3).

application is the use isotope tracers in data assimilation, mainly from satellite remote sensing instrumentation, to improve the accuracy of NWP model forecasts (Yoshimura et al., 2008).

1.3 Active tracers

Active tracers, in contrast to passive tracers, influence the evolution of the weather simulated by a forecast model. A simulation with active dust tracers, for example, can produce different results than would be obtained from a climatological dust distribution. By interaction with solar radiation, a dust tracer field can affect temperature and thus atmospheric stability. Another example for an active tracer is stratospheric ozone, which in addition can be chemically reactive. The main purpose of such active tracers is commonly to either obtain a more realistic forecast of the main model variables, i.e., a better weather forecast, or to predict the concentration field of a tracer variable itself. For example, harmful substances such as nitrous oxides from traffic emissions have negligible impact on the model forecast itself, while the pollutant concentration is coupled in complex ways to photochemistry and the concentration of emitted substances (Seinfeld and Pandis, 2016). Uncertainty of NWP models both affects and results from active tracers, due to feedbacks of parameterized, subgrid scale processes onto the model grid. This mutual connection provides a useful framework when considering the relation of active tracers to diagnostic tendency output (Section 2). In the following subsections, a brief overview and some major challenges of active tracer variables are given, using examples that impact weather forecasts, and such that are primarily an application of the weather forecast itself.

1.3.1 Natural and anthropogenic aerosols

Aerosols are particulate atmospheric constituents that can actively influence the forecast of an NWP model through several pathways. Apart from the general challenges associated with parameterizations of subgrid scale processes, this uncertainty originates from a lack in the understanding of several processes, ranging from the interaction of aerosols with radiation at different wavelengths, to cloud dynamics, and climatic changes (IPCC, 2013).

This section gives a brief overview of the sources of uncertainty involved in the implementation of an active aerosol cycle, using mineral aerosols as an example. Mineral aerosol, commonly also referred to as dust, provides several challenges when representing it as a tracer variable in a model. With its sources located at the ground, mobilization processes need to be represented by a source function for the mass flux of dust into the atmosphere. Semiempirical approaches on a limited data basis have long been used that include a power law relation between wind speed and soil properties, including vegetation, soil humidity and soil type (Marticorena and Bergametti, 1995). Recently, a mobilization approach based on brittle theory and saltation has been proposed, and applied with some success, thereby reducing the empirical aspect of dust emission schemes (Kok, 2011). The atmospheric dust burden is initially a direct result of this source function, where a major uncertainty lies in the aerosol size distribution, and its representation in a model. Over time, dry and wet deposition processes will remove part of the dust burden, and lead to changes in its size distribution. The varying efficiency of the dry and wet deposition processes is crucial in obtaining valid results. This removal efficiency is important, since the climatic impacts of long-lived dust components are substantial, with finer-mode mineral aerosol being a potentially important component of ice nucleation far away from its mobilization sources (Tegen and Fung, 1994).

Apart from such subgrid-scale uncertainty in aerosol simulations, there are challenges related to the meteorology and observational capacity in the main mobilization areas. Desert regions, in particular the Sahara as the largest global source of mineral dust, are sparsely populated and thus equally sparsely covered by meteorological and dust observations, apart from episodic field observation programs (e.g., Engelstaedter et al., 2015). In the Sahara, dust is often mobilized along intense gust fronts that form during excursions of precipitating midlatitude and tropical weather systems into the dry desert region (Knippertz et al., 2009). A key process for dust mobilization, the formation of submesoscale convective gust fronts, are not sufficiently resolved, even at the grid scale of current operational weather prediction models, nor by the observational network, and cause substantial uncertainty of the source function (Marsham et al., 2011).

Other natural aerosols that are included in some modeling systems are, for example, sea salt, pollen, black carbon, and volcanic aerosols. While they share many similarities with the life cycle of dust, some properties differ, such as the role in cloud processes, aerosol aging, and in particular, the emission processes. Volcanic aerosols are released during vertically and temporally complex eruption sequences (e.g., Eckhardt et al., 2008). Size distributions evolve rapidly, and additional reactive gases and water vapor initiate complex aging and deposition processes. Inverse methods appear as a promising pathway to constrain the large parameter space of the emission function of a volcanic eruption (Eckhardt et al., 2008). Pollen are relatively large natural aerosols, with important health implications. Challenges are again related to the correct simulation of the emission process, requiring accurate simulation of varied plant species, their spatial distribution, and their flowering process (Vogel et al., 2009). Observation of pollen are surprisingly limited, given their health implications to millions of people, mostly due to a hitherto tedious manual identification process. First robotic pollen measurement stations are now being developed and deployed, which hopefully allow for more verification data in the future.

1.3.2 *Reactive chemical tracers*

For matters of completeness, we briefly also consider implementations of reactive chemistry in models. Reactive chemistry has similar numerical challenges as other tracers regarding advection, yet with additional requirements and complexity added. The Prather (1986) scheme is one common choice in such applications. In addition to online methods, where the reactive chemistry is calculated at each time step, offline methods are available that use time-averaged or instantaneous wind fields and temperature, among other variables, for calculating advection and chemical reactions. In contrast to the previously discussed methods, the purpose of implementing reactive chemistry is, in most cases, an application of the NWP model, rather than a factor influencing the forecast itself. Examples of different fields of application are, for instance, urban air quality at different scales (Baklanov et al., 2014; Wolf-Grosse et al., 2017), radioactive emissions (Stohl et al., 2012), and tropospheric or stratospheric chemistry (Arnold et al., 2005). Thereby, reactive chemistry rarely impacts the model forecast quality on an NWP time scale, even though O_3, for example, is a strong absorber in the UV spectrum. For further details on this subject, we refer more specific literature available on this topic (e.g., Jacobson, 2005).

2 Tendency diagnostics

2.1 Purpose

The atmospheric circulation is strongly modified by processes that lead to changes in temperature and momentum. Consequently, the formation of clouds with its manifold microphysical processes and the associated release or consumption of latent heat, known as *diabatic processes*, has a strong impact on the atmospheric flow. Furthermore, momentum tendencies caused by turbulence, convection and gravity waves have an influence on the evolution of the atmospheric state. As these processes act on small scales, from micrometers to hundreds of meters, they are not resolved by common NWP models and are therefore parameterized. Thus, their effect is described by variables that are resolved by the model. As the representation of these small-scale processes with the help of parameterizations is subject to large uncertainties, a deficient representation can lead to erroneous forecasts (e.g., Leutbecher et al., 2017). In order to better understand the effect of various diabatic processes on the atmospheric flow, so-called *tendency diagnostics*, i.e., the output of parameterized tendencies from various processes, have been developed that allow for directly quantifying the modification of grid-scale variables through parameterizations.

2.2 Direct tendency framework

Direct tendency diagnostics have been widely used in order to assess the importance of microphysical processes, turbulence, convection and radiation on different weather systems. Igel and van den Heever (2014) used temperature tendency diagnostics in order to determine the importance of different microphysical heating rates for warm frontogenesis. In a high-resolution simulation of a springtime extratropical cyclone, they evaluated the contribution of each microphysical process to warm frontogenesis, the static stability, and finally the slope tendency equations. Thereby, Igel and van den Heever (2014) show that condensation and cloud droplet nucleation are the dominant latent heat releasing processes along the frontal surface. These processes are of similar importance for horizontal frontogenesis as the deformation and tilting terms at mid-levels. Diagnosing temperature tendencies and using them for the calculation of different warm frontal properties can thus lead to new insights into the importance of different microphysical processes for the characteristics of a warm front.

In a study using the COSMO model at high resolution, Miltenberger et al. (2016) analyzed temperature tendencies along trajectories representing Foehn flows over the Alps. In their study, the authors calculated trajectories online during the model run, based on the 3-dimensional wind fields at a 20s time resolution. Detailed analysis of the contribution of diabatic processes to the evolution of temperature along the flow shows that a major part of the air arriving in the foehn valley is warmed due to adiabatic processes. However, when precipitation occurs at the upstream side of the mountains, the air that descends into the foehn valley is influenced by cooling processes such as evaporation of rain, melting and sublimation (Miltenberger et al., 2016). The use of tendency diagnostics along trajectories representing the flow of interest can thus help to improve the understanding of complex interactions between dynamical and microphysical processes.

2.3 Potential vorticity framework

In other studies, the impact of temperature tendencies on the atmospheric flow is evaluated based on a potential vorticity (PV) framework. PV is defined as the product of absolute vorticity and static stability, and is conserved during adiabatic frictionless flow. If diabatic processes occur, PV is modified, which is then reflected in corresponding changes to the associated wind field. Therefore, the evaluation of the impact of temperature tendencies on PV sheds light on the modification of the atmospheric flow by diabatic processes, such as cloud microphysics.

Lamarque and Hess (1994) used a model forecast to diagnose the PV tendency due to advective, diabatic and diffusive processes based on tendencies in temperature and wind fields. Based on a method by Davis et al. (1993), Gray (2006) developed a more detailed method, which was then applied to the analysis of an extratropical cyclone (a case previously investigated in Stoelinga, 1996). With the Gray (2006) method, it is possible to disentangle the contribution of nonconservative processes from all parameterization schemes, accumulated in the form of tracer fields. Thereby, each PV-modifying process in the model is treated as a passive tracer advected by the resolved flow. At each model timestep the sinks and sources of PV due to nonconservative processes are diagnosed and added at each timestep. Additionally, the initial PV field is advected as passive tracer (Gray, 2006).

The Gray (2006) method allows to disentangle the processes leading to PV modification in the simulated weather system. In their setup, processes leading to PV modification due to latent heating or cooling (parametrization of convection, the cloud scheme, the boundary layer scheme, large-scale precipitation, rebalancing the cloud field after convection), radiative heating or cooling and mixing processes can be investigated separately. Gray (2006) applied her method in order to quantify the different PV-modifying processes in a model forecast and to investigate cross-tropopause transport. The method has later been extended and applied to different weather phenomena like convective storms (Chagnon and Gray, 2009), the diabatic modification of PV in extratropical cyclones (Chagnon et al., 2013; Chagnon and Gray, 2015; Martinez-Alvarado and Plant, 2014; Martinez-Alvarado et al., 2014b, Martinez-Alvarado et al., 2016) or the diabatic processes modifying the tropopause sharpness (Saffin et al., 2017).

Joos and Wernli (2012) presented a different method to diagnose PV tendencies, based on trajectory calculations. Thereby, the total diabatic modification of PV due to cloud formation is decomposed into the different contributions from single microphysical processes, such as condensation and evaporation of cloud liquid, freezing, melting, sublimation, depositional growth of ice and snow, evaporation of rain as well as longwave and shortwave radiative heating rates. Thereby, all microphysical heating rates are output with a high temporal resolution (every 15 min) and the associated change in PV is calculated separately for each process. The diabatic PV rates as well as heating rates are subsequently traced along selected trajectories that represent the atmospheric flow of interest. The Joos and Wernli (2012) method allows to investigate the diabatic history of each airparcel, thereby, allowing to separately quantify the importance of each process for the production or destruction of PV, and thus the atmospheric circulation. Their method has also been used in order to investigate the effect of differences in the parameterization of microphysical processes on the upper level flow (Joos and Forbes, 2016) on the formation of low-level PV anomalies in an idealized extratropical cyclone (Crezee et al., 2017) or the modification of an upper-level PV streamer by ice-phase processes (Hardy et al., 2017). An extended version of the Joos and Wernli (2012) method, which includes also the modification of PV due to friction, diffusion and turbulent processes, has been used in order to study the modification of PV near the tropopause (Spreitzer et al., 2019) and the formation of low-level positive PV anomalies in a real-case simulation of an extratropical cyclone (Attinger et al., 2019). A comparison of the PV-tracer method by Gray (2006) and the trajectory method by Joos and Wernli (2012) is presented in Martinez-Alvarado et al. (2014a).

2.4 Application examples

In this subsection, selected examples of the application of the methods introduced above are discussed in more detail. Chagnon et al., 2013 investigated the structure of diabatically modified PV in an extratropical cyclone by means of the PV tracer method described in Section 2.2. The tracers have been calculated in a model simulation of an extratropical cyclone with the Met Office Unified Model (UM). Each tracer $c_i(x,y,z,t)$ is set to zero at initial time, and then advected with the model's advection scheme. Each tracer can be modified by process i during every time step. The subsequent evolution of c_i may therefore be interpreted as a history of the total PV generated by process i (Chagnon et al., 2013). Additionally, an "advection-only" PV is calculated. This tracer is initialized with the full PV field and subsequently advected conservatively.

Fig. 5 shows PV tracers (in color) as well as the "advection-only" PV (gray line) along a vertical cross section through the tropopause fold, cold front and warm conveyor belt (WCB) of the simulated cyclone. Diabatic processes exhibit a dipole pattern across the tropopause. On the stratospheric side, PV has been produced diabatically (reddish colors), whereas on the eastern flank of the tropopause fold, the tropospheric side, PV is strongly reduced (bluish colors). The diabatic modification of PV therefore leads to a sharpening of the isentropic PV gradient across the tropopause. This sharpening of the PV gradient influences the wind speed and subsequently the advection of air along the jet. The positive part of the net diabatic PV dipole is primarily produced by the longwave radiation scheme. The negative diabatic PV is partly produced by the longwave radiation scheme, the large-scale cloud scheme and parameterized convection. The PV-tracer method thus clearly illustrates how diabatic processes alter the PV pattern, as well as the structure and exact location of the tropopause. With PV tracer methods it is possible to investigate, which processes are important for the modification of the upper-level flow, and provide guidance in which processes lead to deficiencies, and need improvement in weather forecast models.

In Joos and Wernli (2012), diabatic heating rates (DHR) and the associated diabatic PV rates (DPVR) are traced along trajectories representing the WCB, the most ascending airstream in extratropical cyclones. Due to its strong ascent, the WCB is subject of widespread

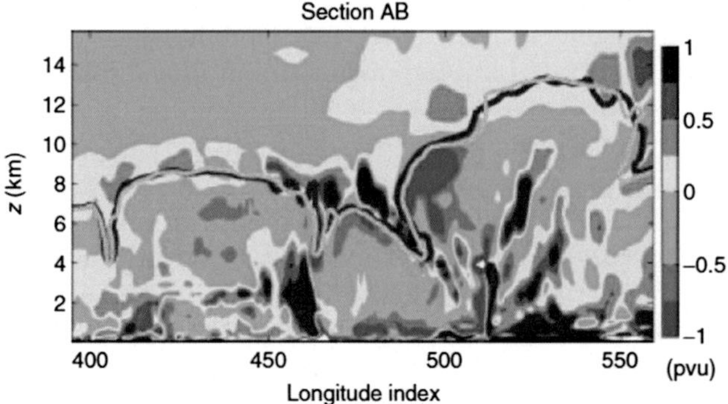

FIG. 5 Net diabatic PV (colored) in a vertical section through the cyclone. The solid black and gray lines depict the 2 PVU contour in the full PV and advection-only PV, respectively (Chagnon et al., 2013, Fig. 5C).

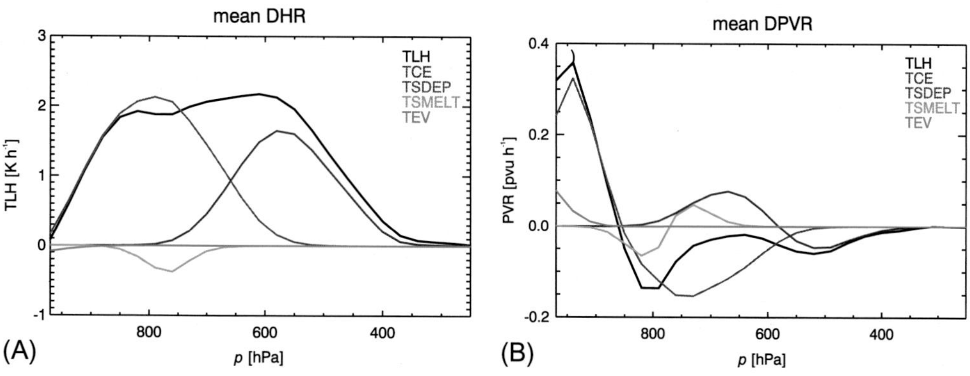

FIG. 6 Evolution of mean diabatic heating rates (DHR) (A) and diabatic PV rates (DPVR) (B) along a WCB. The black lines show the total DHR (DPVR), and colored lines show the contributions from the different microphysical processes to the total DHR (DPVR): condensation/evaporation (TCE), depositional growth of snow (TSDEP), melting of snow (TSMELT) and evaporation of rain (TEV) (Joos and Wernli, 2012, Fig. 10B, C).

cloud formation. The associated, extensive latent heating then further supports ascent of WCB airmasses. Fig. 6 depicts the most important microphysical heating and PV rates along a WCB associated with an extratropical cyclone in the North Atlantic.

The ascending air parcels are strongly heated by condensation of water vapor (purple line). An almost equal amount is, however, due to depositional growth of snow (red line) when reaching pressure altitudes in the troposphere of above 600 hPa. Furthermore, ascending air parcels are influenced by falling and melting snow (green line) that exerts a cooling effect. The falling snow is thereby produced by the ascending airstream itself. Each of the microphysical processes modifies the PV along the ascending airstream (Fig. 6B). The total DPVR (black line) exhibits positive values above 850 hPa and negative values below. Thus, PV is produced above 850 hPa and destroyed below. The contribution of each process to the total signal is depicted with the colored lines. The main contribution to the positive DPVR at low levels arises from condensation (purple line) and evaporation of rain (turquoise line). When the airparcels rise further, condensation starts to reduce PV. However, PV is produced due to depositional growth of snow, counteracting the influence of condensation. The total diabatic change of PV is the result of a complex interplay of a number of microphysical processes that partially act at the same height but with a different sign. The total PV modification and the associated changes in the wind field can only be understood and predicted correctly when taking into account the complex interplay of different processes. The method of tracing temperature tendencies and PV tendencies along relevant trajectories enables a detailed insight into processes modifying the atmospheric circulation.

Anomalously high or low PV values can have a strong impact on the atmospheric flow. In a study by Attinger et al. (2019), temperature and momentum tendencies due to all parameterized processes have been used in order to assess the importance of these processes for the formation of PV anomalies. Using a simulation of a maritime extratropical cyclone with the integrated forecasting sytem (IFS) from the European Centre for Midrange Weather Forecast (ECMWF), the DPVRs due to all temperature and momentum tendencies were output hourly (Attinger et al., 2019). Next, 24h backward trajectories were calculated from all PV anomalies in the lower troposphere that were associated with the cyclone's fronts and center. Along these trajectories, the accumulated change in PV was calculated separately for all processes. This made it possible to determine at every grid point how much and due to which process the PV had been modified. In Fig. 7, every colored point denotes whether PV has been increased (reddish colors) or decreased (bluish colors) during the last 24h along the backward trajectories.

The resulting PV modification due to all diabatic processes shows that PV has been produced in the previous 24h along trajectories ending at the cold, warm and bent-back front as well as in the cyclone center (Fig. 7A). However, PV has also been decreased in regions behind the cold front. This PV pattern results mainly from the contributions of three processes. PV production due to condensation is one of the most important processes (Fig. 7B). In addition, melting of snow also leads to a pronounced PV production along the bent-back front and the cyclone center (Fig. 7C). Sublimation of snow has two opposing effects (Fig. 7D). On the one hand it decreases PV behind the cold front, contributing to the observed PV dipole across the cold front. On the other hand, snow sublimation contributes to the strong PV production at the cyclone center. These results illustrate the benefit of tracing diabatic processes along trajectories, providing pathways to disentangle the complex interaction of processes that are responsible for the formation of dynamically relevant PV anomalies in extratropical cyclones.

2.5 Challenges and developments

The tendency and PV tracer diagnostics described above are not able to completely close the temperature or PV budget along tracers or trajectories. One reason for the presence of a residual in the budget is that the dynamical core of NWP models is generally not designed to conserve PV. Explicit or implicit numerical diffusion can produce changes in PV, as for example, has been shown for the Met Office UM by Saffin et al. (2016). Another limiting factor is that the processes leading to temperature tendencies are characterized by strong temporal fluctuations. Instantaneous temperature tendencies are therefore, not well represented for the tendency over an entire hour. Furthermore, trajectory positions are often calculated based on hourly wind fields leading to errors in the trajectory position.

The use of online trajectories, calculated with the same time resolution as the internal model timestep, (usually on the order of seconds) clearly reduces the position errors of trajectories (Miltenberger et al., 2013). Online trajectories then also allow to interpolate temperature tendencies to the trajectory position at every internal model timestep, providing more accurate temporal fluctuations. It will therefore be beneficial to future model development if online trajectories, which so far have been used in order to investigate precipitation formation in orographic flows and flow over complex topography (Miltenberger et al., 2015, 2016), as well as convection in extratropical cyclones (Oertel et al., 2019) become more widely used, and are implemented into more NWP models.

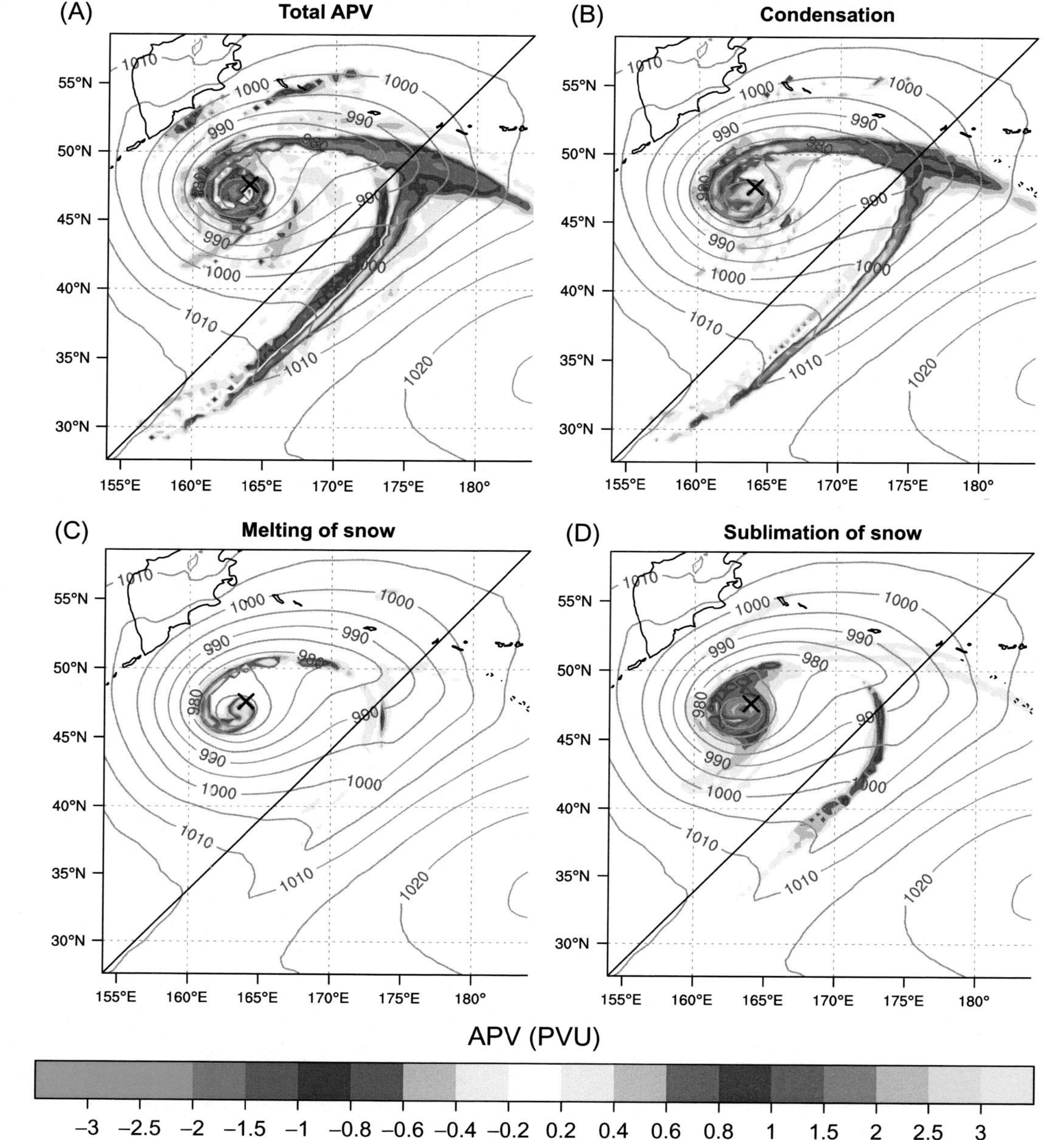

FIG. 7 Total accumulated PV (APV) due to different microphysical processes along 24h backward trajectories (Attinger et al., 2019, Fig. 7A,C,D, E).

References

Aemisegger, F., Spiegel, J.K., Pfahl, S., Sodemann, H., Eugster, W., Wernli, H., 2015. Isotope meteorology of cold front passages: a case study combining observations and modeling. Geophys. Res. Lett. 42 (13), 5652–5660. https://doi.org/10.1002/2015GL063988.

Arnold, S.R., Chipperfield, M.P., Blitz, M.A., 2005. A three- dimensional model study of the effect of new temperature-dependent quantum yields for acetone photolysis. J. Geophys. Res.. 110, D22305https://doi.org/10.1029/2005JD005998.

Attinger, R., Spreitzer, E., Boettcher, M., Forbes, R., Wernli, H., Joos, H., 2019. Quantifying the role of individual diabatic processes for the formation of PV anomalies in a North Pacific cyclone. Quart. J. Roy. Meteorol. Soc. 145, 2454–2476. https://doi.org/10.1002/qj.3573.

Baklanov, A., Schlünzen, K., Suppan, P., Baldasano, J., Brunner, D., Aksoyoglu, S., Carmichael, G., Douros, J., Flemming, J., Forkel, R., Galmarini, S., Gauss, M., Grell, G., Hirtl, M., Joffre, S., Jorba, O., Kaas, E., Kaasik, M., Kallos, G., Kong, X., Korsholm, U., Kurganskiy, A., Kushta, J., Lohmann, U., Mahura, A., Manders-Groot, A., Maurizi, A., Moussiopoulos, N., Rao, S.T., Savage, N., Seigneur, C., Sokhi, R.S., Solazzo, E., Solomos, S., Sørensen, B., Tsegas, G., Vignati, E., Vogel, B., Zhang, Y., 2014. Online coupled regional meteorology chemistry models in Europe: current status and prospects. Atmos. Chem. Phys. 14 (1), 317–398. https://doi.org/10.5194/acp-14-317-2014.

Bjerknes, V., 1904. The problem of weather prediction, considered from the viewpoints of mechanics and physics. Metz 21 (6), 1–7. https://doi.org/10.1127/0941-2948/2009/416.

Bonne, J.-L., Behrens, M., Meyer, H., Kipfstuhl, S., Rabe, B., Schönicke, L., Steen-Larsen, H.C., Werner, M., 2019. Resolving the controls of water vapour isotopes in the Atlantic sector. Nat. Commun. 1–10. https://doi.org/10.1038/s41467-019-09242-6.

Boutle, I.A., Belcher, S.E., Plant, R.S., 2011. Moisture transport in midlatitude cyclones. Q. J. Roy. Meteorol. Soc. 137, 360. https://doi.org/10.1002/qj.783.

Bowman, J.C., Yassaei, M.A., Basu, A., 2014. A fully Lagrangian advection scheme. J. Sci. Comput. 64 (1), 151–177. https://doi.org/10.1007/s10915-014-9928-8.

Chagnon, J.M., Gray, S.L., 2009. Horizontal potential vorticity dipoles on the convective storm scale. Quart. J. Roy. Meteor. Soc. 135 (643), 1392–1408. https://doi.org/10.1002/qj.468.

Chagnon, J.M., Gray, S.L., 2015. A diabatically generated potential vorticity structure near the extratropical tropopause in three simulated extratropical cyclones. Mon. Weather Rev. 143, 2337–2347. https://doi.org/10.1175/MWR-D-14-00092.1.

Chagnon, J.M., Gray, S.L., Methven, J., 2013. Diabatic processes modifying potential vorticity in a North Atlantic cyclone. Quart. J. Roy. Meteor. Soc. 139 (674), 1270–1282. https://doi.org/10.1002/qj.2037.

Crezee, B., Joos, H., Wernli, H., 2017. Diabatic potential vorticity anomalies related to clouds and precipitation in an idealized extratropical cyclone. J. Atmos. Sci. 74, 1403–1416. https://doi.org/10.1175/JAS-D-16-0260.1.

Dansgaard, W., 1964. Stable isotopes in precipitation. Tellus 16, 436–468.

Davis, C.A., Stoelinga, M.T., Kuo, Y.H., 1993. The integrated effect of condensation in numerical simulations of extratropical cyclogenesis. Mon. Weather Rev. 121, 2309–2330.

Eckhardt, S., Prata, A., Seibert, P., Stebel, K., Stohl, A., 2008. Estimation of the vertical profile of sulfur dioxide injection into the atmosphere by a volcanic eruption using satellite column measurements and inverse transport modeling. Atmos. Chem. Phys. 8 (14), 3881–3897.

Eckstein, J., Ruhnke, R., Pfahl, S., Christner, E., Diekmann, C., Dyroff, C., Reinert, D., Rieger, D., Schneider, M., Schröter, J., Zahn, A., Braesicke, P., 2018. From climatological to small-scale applications: simulating water isotopologues with ICON-ART-Iso (version 2.3). Geosci. Model Dev. 11 (12), 5113–5133. https://doi.org/10.5194/gmd-11-5113-2018.

Engelstaedter, S., Washington, R., Flamant, C., Parker, D.J., Allen, C.J.T., Todd, M.C., 2015. The Saharan heat low and moisture transport pathways in the Central Sahara-multiaircraft observations and Africa-LAM evaluation. J. Geophys. Res.-Atmos. 120 (10), 4417–4442. https://doi.org/10.1002/2015JD023123.

Epstein, S., 1956. Variations in the O18/O16 ratios of fresh water and ice. Nat. Acad. Sci, Nucl. Sci. Ser. Rep. No. 19, 20–25.

Field, R.D., Jones, D.B.A., Brown, D.P., 2010. Effects of postcondensation exchange on the isotopic composition of water in the atmosphere. J. Geophys. Res.. 115, D24305. https://doi.org/10.1029/2010JD014334.

Forster, C., Stohl, A., Seibert, P., 2007. Parameterization of convective transport in a Lagrangian particle dispersion model and its evaluation. J. Appl. Meteor. Climatol. 46 (4), 403–422. https://doi.org/10.1175/JAM2470.1.

Galewsky, J., Steen-Larsen, H.C., Field, R.D., Worden, J., Risi, C., Schneider, M., 2016. Stable isotopes in atmospheric water vapor and applications to the hydrologic cycle. Rev. Geophys. 54 (4), 809–865. https://doi.org/10.1002/2015RG000512.

Gat, J.R., 1996. Oxygen and hydrogen isotopes in the hydrologic cycle. Annu. Rev. Earth Planet. Sci. 24, 225–262.

Gray, S.L., 2006. Mechanisms of midlatitude cross-tropopause transport using a potential vorticity budget approach. J. Geophys. Res.. 111, D17113. https://doi.org/10.1029/2005JD006259.

Hardy, S., Schultz, D.M., Vaughan, G., 2017. Early evolution of the 23–26 September 2012 U.K. floods: Tropical storm Nadine and diabatic heating due to cloud microphysics. Mon. Weather Rev. 145, 543–563. https://doi.org/10.1175/MWR-D-16-0200.1.

Igel, A.L., van den Heever, S.C., 2014. The role of latent heating in warm frontogenesis. Quart. J. Roy. Meteor. Soc. 140 (678), 139–150. https://doi.org/10.1002/qj.2118.

IPCC, 2013. Climate Change 2013: The Physical Science Basis. In: Stocker, T.F., Qin, D., Plattner, G.-K., Tignor, M., Allen, S.K., Boschung, J., … Midgley, P.M. (Eds.), Contribution of Working Group I to the Fifth Assessment Report of the Intergovernmental Panel on Climate Change. Cambridge University Press, Cambridge, United Kingdom and New York, NY, USA 1535 pp.

Jacobson, M.Z., 2005. Fundamentals of Atmospheric Modeling, second ed Cambridge University Press, Cambridge, UK.

Joos, H., Forbes, R., 2016. Impact of different IFS microphysics on a warm conveyor belt and the downstream flow evolution. Q. J. Roy. Meteorol. Soc. 142, 2727–2739. https://doi.org/10.1002/qj.2863.

Joos, H., Wernli, H., 2012. Influence of microphysical processes on the potential vorticity development in a warm conveyor belt: a case study with the limited area model COSMO. Quart. J. Roy. Meteorol. Soc. 138, 407–418.

Joussaume, J., Sadourny, R., Jouzel, J., 1984. A general circulation model of water isotope cycles in the atmosphere. Nature 311, 24–29.

Kerstel, E.R.T., Iannone, R.Q., Chenevier, M., Kassi, S., Jost, H.J., Romanini, D., 2006. A water isotope (H-2, O-17, and O-18) spectrometer based on optical feedback cavity-enhanced absorption for in situ airborne applications. Appl Phys B-Lasers O 85 (2–3), 397–406. https://doi.org/10.1007/s00340-006-2356-1.

Knippertz, P., Trentmann, J., Seifert, A., 2009. High-resolution simulations of convective cold pools over the northwestern Sahara. J. Geophys. Res.. 114(D8), D08110. https://doi.org/10.1029/2008JD011271.

Kok, J.F., 2011. A scaling theory for the size distribution of emitted dust aerosols suggests climate models underestimate the size of the global dust cycle. PNAS 108 (3), 1016–1021.

Koster, R., Jouzel, J., Souzzo, R., Russell, G., Broecker, W., Rind, D., Eagleson, P., 1986. Global sources of local precipitation as determined by the NASA/GISS GCM. Geophys. Res. Lett. 13 (1), 121–124.

Läderach, A., 2016. Characteristic Scales of Atmospheric Moisture Transport (PhD dissertation). ETH Zürich.. https://doi.org/10.3929/ethz-a-010741025.

Lamarque, J.-F., Hess, P.G., 1994. Cross-tropopause mass exchange and potential vorticity budget in a simulated tropopause folding. J. Atmos. Sci. 51, 2246–2269. https://doi.org/10.1175/15200469(1994)051,2246:CTMEAP.2.0.CO;2.

Landais, A., Barkan, E., Luz, B., 2008. Record of δ18O and 17O- excess in ice from Vostok Antarctica during the last 150 000 years. Geophys. Res. Lett. 35, L02709–L02713.

Leutbecher, M., et al., 2017. Stochastic representations of model uncertainties at ECMWF: state of the art and future vision. Quart. J. Roy. Meteor. Soc. 143 (707), 2315–2339. https://doi.org/10.1002/qj.3094.

Lewis, S.C., LeGrande, A.N., Kelley, M., Schmidt, G.A., 2010. Water vapour source impacts on oxygen isotope variability in tropical precipitation during Heinrich events. Clim. Past 6, 325.

Majewski, D., 1991. The Europa-model of the Deutscher Wetterdienst. ECMWF Sem. Numer. Meth. Atmos. Models 2, 147–191.

Marsham, J.H., Knippertz, P., Dixon, N.S., Parker, D.J., Lister, G.M.S., 2011. The importance of the representation of deep convection for modeled dust-generating winds over West Africa during summer. Geophys. Res. Lett. 38(16) https://doi.org/10.1029/2011GL048368.

Marticorena, B., Bergametti, G., 1995. Modeling the atmospheric dust cycle: 1. Design of a soil-derived dust emission scheme. Journal of Geophysical Research: Atmospheres (1984–2012) 100 (D8), 16415–16430. https://doi.org/10.1029/95JD00690.

Martinez-Alvarado, O., Plant, R.S., 2014. Parametrized diabatic processes in numerical simulations of an extratropical cyclone. Quart. J. Roy. Meteor. Soc. 140 (682), 1742–1755. https://doi.org/10.1002/qj.2254.

Martinez-Alvarado, O., Joos, H., Chagnon, J., Boettcher, M., Gray, S.L., Plant, R.S., Methven, J., Wernli, H., 2014a. The dichotomous structure of the warm conveyor belt. Quart. J. Roy. Meteorol. Soc. 140, 1809–1824. https://doi.org/10.1002/qj.2276.

Martinez-Alvarado, O., Baker, L., Gray, S.L., Methven, J., Plant, R.S., 2014b. Distinguishing the cold conveyor belt and sting jet airstreams in an intense extratropical cyclone. Mon. Weather Rev. 142, 2571–2595. https://doi.org/10.1175/MWR-D-13-00348.1.

Martinez-Alvarado, O., Madonna, E., Gray, S.L., Joos, H., 2016. A route to systematic error in forecasts of Rossby waves. Quart. J. Roy. Meteor. Soc. 142 (694), 196–210. https://doi.org/10.1002/qj.2645.

Miltenberger, A.K., Pfahl, S., Wernli, H., 2013. An online trajectory module (version 1.0) for the nonhydrostatic numerical weather prediction model COSMO. Geosci. Model Dev. 6 (6), 1989–2004. https://doi.org/10.5194/gmd-6-1989-2013.

Miltenberger, A.K., Seifert, A., Joos, H., Wernli, H., 2015. A scaling relation for warm-phase orographic precipitation—a Lagrangian analysis for 2D mountains. Quart. J. Roy. Meteorol. Soc. 141, 2185–2198.

Miltenberger, A.K., Reynolds, S., Sprenger, M., 2016. Revisiting the latent heating contribution to foehn warming: Lagrangian analysis of two foehn events over the Swiss Alps. Quart. J. Roy. Meteor. Soc. 142 (698), 2194–2204. https://doi.org/10.1002/qj.2816.

Numaguti, A., 1999. Origin and recycling processes of precipitating water over the Eurasian continent: experiments using an atmospheric general circulation model. J. Geophys. Res. 104 (D2), 1957–1972.

Yoshimura, K., Kanamitsu, M., Noone, D., Oki, T., 2008. Historical isotope simulation using reanalysis atmospheric data. J. Geophys. Res. 113, D19108. https://doi.org/10.1029/2008JD010074.

Oertel, A., Boettcher, M., Joos, H., Sprenger, M., Konow, H., Hagen, M., Wernli, H., 2019. Convective activity in an extratropical cyclone and its warm conveyor belt—a case study combining observations and a convection-permitting model simulation. Q. J. R. Meteorol. Soc. 145, 1406–1426. https://doi.org/10.1002/qj.3500.

Papritz, L., Sodemann, H., 2018. Characterizing the local and intense water cycle during a cold air outbreak in the Nordic seas. Mon. Weather Rev. 146 (11), 3567–3588. https://doi.org/10.1175/MWR-D-18-0172.1.

Pfahl, S., Wernli, H., Yoshimura, K., 2012. The isotopic composition of precipitation from a winter storm—a case study with the limited-area model $COSMO_{iso}$. Atmos. Chem. Phys. 12 (3), 1629–1648. https://doi.org/10.5194/acp-12-1629-2012.

Pisso, I., Sollum, E., Grythe, H., Kristiansen, N., Cassiani, M., Eckhardt, S., Arnold, D., Morton, D., Thompson, R.L., Groot Zwaaftink, C.D., Evangeliou, N., Sodemann, H., Haimberger, L., Henne, S., Brunner, D., Burkhart, J.F., Fouilloux, A., Brioude, J., Philipp, A., Seibert, P., Stohl, A., 2019. The Lagrangian particle dispersion model FLEXPART version 10.3. Geosci. Model Dev. Discuss. 1–67. https://doi.org/10.5194/gmd-2018-333.

Prather, M., 1986. Numerical advection by conservation of 2nd-order moments. J. Geophys. Res. 91 (D6), 6671–6681.

Rastigejev, Y., Park, R., Brenner, M.P., Jacob, D.J., 2010. Resolving intercontinental pollution plumes in global models of atmospheric transport. J. Geophys. Res.. 115, D02302. https://doi.org/10.1029/2009JD012568.

Risi, C., Bony, S., Vimeux, F., 2008. Influence of convective processes on the isotopic composition (d18O and dD) of precipitation and water vapor in the tropics: 2. Physical interpretation of the amount effect. J. Geophys. Res. 113, D19306. https://doi.org/10.1029/2008JD009943.

Risi, C., Noone, D., Worden, J., Frankenberg, C., Stiller, G., Kiefer, M., Funke, B., Walker, K., Bernath, P., Schneider, M., Wunch, D., Sherlock, V., Deutscher, N., Griffith, D., Wennberg, P. O., Strong, K., Smale, D., Mahieu, E., Barthlott, S., Hase, F., García, O., Notholt, J., Warneke, T., Toon, G., Sayres, D., Bony, S., Lee, J., Brown, D., Uemura, R. and Sturm, C.: Process-evaluation of tropospheric humidity simulated by general circulation models using water vapor isotopologues: 1. Comparison between models and observations, J. Geophys. Res.-Atmos., 117 (D5), n/a–n/a, https://doi.org/10.1029/2011jd016621, 2012.

Risi, C., Landais, A., Winkler, R., Vimeux, F., 2013. Can we determine what controls the spatio-temporal distribution of d-excess and 17O-excess in precipitation using the LMDZ general circulation model? Clim. Past 9 (5), 2173–2193. https://doi.org/10.5194/cp-9-2173-2013.

Saffin, L., Methven, J., Gray, S.L., 2016. The non-conservation of potential vorticity by a dynamical core compared with the effects of parametrized physical processes. Quart. J. Roy. Meteor. Soc. 142, 1265–1275. https://doi.org/10.1002/qj.2729.

Saffin, L., Gray, S.L., Methven, J., Williams, K.D., 2017. Processes maintaining tropopause sharpness in numerical models. J. Geophys. Res. Atmos. 122, 9611–9627. https://doi.org/10.1002/2017JD026879.

Seinfeld, J.H., Pandis, S.N., 2016. Atmospheric Chemistry and Physics: From Air Pollution to Climate Change, third ed Wiley.

Seity, Y., Brousseau, P., Malardel, S., Hello, G., Bénard, P., Bouttier, F., Lac, C., Masson, V., 2011. The AROME-France convective-scale operational model. Mon. Weather Rev. 139 (3), 976–991. https://doi.org/10.1175/2010MWR3425.1.

Smolarkiewicz, P.K., 2005. Multidimensional positive defnite advection transport algorithm: an overview. Int. J. Numer. Meth. Fluids. 1–22.

Sodemann, H., 2006. Tropospheric Transport of Water Vapour: Lagrangian and Eulerian Perspectives.

Sodemann, H., Pommier, M., Arnold, S.R., Monks, S.A., Stebel, K., Burkhart, J.F., Hair, J.W., Diskin, G.S., Clerbaux, C., Coheur, P.-F., Hurtmans, D., Schlager, H., Blechschmidt, A.-M., Kristjánsson, J.E., Stohl, A., 2011. Episodes of cross-polar transport in the Arctic troposphere during July 2008 as seen from models, satellite, and aircraft observations. Atmos. Chem. Phys. 11, 3631. https://doi.org/10.5194/acp-11-3631-2011.

Sodemann, H., Stohl, A., 2013. Moisture origin and meridional transport in atmospheric Rivers and their association with multiple cyclones. Mon. Weather Rev. 141, 2850–2868. https://doi.org/10.1175/MWR-D-12-00256.s1.

Sodemann, H., Wernli, H., Schwierz, C., 2009. Sources of water vapour contributing to the Elbe flood in August 2002—a tagging study in a mesoscale model. Q. J. R. Meteorol. Soc. 135, 205–223. https://doi.org/10.1002/qj.374.

Spreitzer, E., Attinger, R., Boettcher, M., Forbes, R., Wernli, H., Joos, H., 2019. Modification of potential vorticity near the tropopause by non-conservative processes in the ECMWF model. J. Atmos. Sci. 76, 1709–1726. https://doi.org/10.1175/JAS-D-18-0295.1.

Stensrud, D.J., 2007. Parameterization Schemes. Cambridge University Press, Cambridge, UK.

Steppeler, J., Doms, G., Schättler, U., Bitzer, H., Gassmann, A., Damrath, U., Gregoric, G., 2003. Meso-gamma scale forecasts using the nonhydrostatic model LM. Meteorol. Atmos. Phys. 82 (1–4), 75–96.

Stoelinga, M.T., 1996. A potential vorticity-based study of the role of diabatic heating and friction in a numerically simulated baroclinic cyclone. Mon. Weather Rev. 124, 849–874.

Stohl, A., Seibert, P., Wotawa, G., Arnold, D., Burkhart, J.F., Eckhardt, S., Tapia, C., Vargas, A., Yasunari, T.J., 2012. Xenon-133 and caesium-137 releases into the atmosphere from the Fukushima Dai-ichi nuclear power plant: determination of the source term, atmospheric dispersion, and deposition. Atmos. Chem. Phys. 12, 2313–2343. https://doi.org/10.5194/acp-12-2313-2012.

Stohl, A., Thomson, D.J., 1999. A density correction for Lagrangian particle dispersion models. Bound.-Lay. Meteorol. 90 (1), 155–167. https://doi.org/10.1023/A:1001741110696.

Stohl, A., Forster, C., Frank, A., Seibert, P., Wotawa, G., 2005. Technical note: The Lagrangian particle dispersion model FLEXPART version 6.2. Atmos. Chem. Phys. 5 (9), 2461–2474. https://doi.org/10.5194/acp-5-2461-2005.

Sturm, K., Hoffmann, G., Langmann, B., Stichler, W., 2005. Simulation of d18O in precipitation by the regional circulation model REMOiso. Global Planet. Change 19, 3425–3444.

Tegen, I., Fung, I., 1994. Modeling of mineral dust in the atmosphere: sources, transport, and optical thickness. J. Geophys. Res. 99 (D11), 22897–22914.

Vidale, P.L., Lüthi, D., Heck, P., Schär, C., Frei, C., 2002. The Development of Climate HRM (ex-DWD EM). .

Vogel, B., Vogel, H., Baeumer, D., Bangert, M., Lundgren, K., Rinke, R., Stanelle, T., 2009. The comprehensive model system COSMO-ART - Radiative impact of aerosol on the state of the atmosphere on the regional scale. Atmos. Chem. Phys. 9 (22), 8661–8680. https://doi.org/10.5194/acp-9-8661-2009.

Winschall, A., Pfahl, S., Sodemann, H., Wernli, H., 2012. Impact of North Atlantic evaporation hot spots on southern alpine heavy precipitation events. Q. J. R. Mereorol. Soc. 138, 1245–1258. https://doi.org/10.1002/qj.987.

Wolf-Grosse, T., Esau, I., Reuder, J., 2017. Sensitivity of local air quality to the interplay between small- and large-scale circulations: a large-eddy simulation study. Atmos. Chem. Phys. 17 (11), 7261–7276. https://doi.org/10.5194/acp-17-7261-2017.

CHAPTER

13

Dynamic identification and tracking of errors in numerical simulations of the atmosphere

Haraldur Ólafsson

School of Atmospheric Sciences, Science Institute, and Faculty of Physical Sciences, University of Iceland, Veðurfélagið and the Icelandic Meteorological Office, Reykjavík, Iceland

1 Introduction

Errors in numerical weather prediction, and in simulations of the atmosphere in general, are either related to the numerical models themselves or the data used to drive the simulations, i.e., the initial or boundary conditions. Improvements in both the generation of initial conditions as well as aspects of the numerical calculations, including parameterizations, have been large in recent decades. The analysis has benefitted from new sources of remote sensing, including satellite-based observations, and progress in both the analysis, and the integration forward in time goes hand-in-hand with increased availability of computer power.

Although large synoptic scale errors in the short-to-medium range are not very common, they do still occur, not the least in connection with rapidly developing weather systems at mid-latitudes. It is not uncommon for forecasters to experience that a model may deliver variable results at different lead-times, but as the lead-times get shorter, the model simulations usually converge toward the correct solution. Sometimes, this convergence happens only 1–3 days before the time of forecast verification, and sometimes they do not converge at all. Mesoscale errors, commonly referred to by concepts such as shape of troughs, and position of fronts, are however quite common, even in short-range forecasts.

Many types of errors associated with the model themselves accumulate over time. Inaccuracies that result in small errors in a 12-h forecast may in other words be significant in a 10-day forecast. The state of the art of models is, however, so good that synoptic-scale disturbances do not generally suffer greatly from model errors within a time scale of a couple of days, and

Uncertainties in Numerical Weather Prediction
https://doi.org/10.1016/B978-0-12-815491-5.00013-6

even longer. Errors in the initial fields may, however, give great errors in forecasts with a lead time of only 1–3 days. They may also remain latent, and "explode" at a later stage in the simulation.

In this chapter, a way to analyze errors in short-to-medium range numerical simulations is presented. The method is based on comparison of simulations or forecasts with different lead times, identification of differences between the individual forecasts, and tracking of anomalies in the difference fields. This method may be useful in assessing sources of wrong forecasts, improving the observation network, and for assessing the uncertainty of the models in different flow situations. Furthermore, it gives a physical understanding of errors in simulations of the atmosphere, and weather forecasting.

2 Variables and functions, useful for error tracing in atmospheric flow

There are several conservative variables or functions in synoptic-scale atmospheric flow, in the sense that spatial anomalies in these variables or functions are materially conserved for up to several days. Other features of the flow fields may also live for many days, although they are not necessarily materially conserved. The mean sea level pressure (MSLP) and closed lows or troughs in the MSLP field are perhaps the most commonly used feature in assessing the atmospheric evolution on a time range of hours to days. As described by the hydrostatic equation, the surface pressure, or the sea-level pressure if the surface of the earth is at sea level, is a function of the integrated mass of a column of air above. At some levels, above a minimum in the mean sea level pressure field, the airmass is less dense than elsewhere, away from the minimum. There is, in other words, a pocket of relatively warm air somewhere above a surface low. The lifetime of synoptic-scale lows is typically a few days but may exceptionally be up to weeks.

The geopotential is a commonly used alternative to pressure at a given height. Traveling anomalies in the geopotential fields may be found at all levels of the atmosphere, but they are, in general, clearest, and best defined at low levels. A very useful approximation of the temporal evolution of the geopotential is presented by the quasigeostrophic geopotential tendency equation

$$\left(\nabla^2 + \frac{\partial}{\partial p}\left(\frac{f_0^2}{\sigma}\frac{\partial}{\partial p}\right)\right)\chi = -f_0\mathbf{V}_g\cdot\nabla\left(\frac{1}{f_0}\nabla^2\Phi + f\right) - \frac{\partial}{\partial p}\left(-\frac{f_0^2}{\sigma}\mathbf{V}_g\cdot\nabla\left(\frac{\partial\Phi}{\partial p}\right)\right)$$

where p is pressure, f is the Coriolis parameter, σ is a stability parameter, χ is the geopotential tendency, $\mathbf{V}_g$ is the geostrophic wind, and Φ is the geopotential.

The geopotential tendency equation relates changes in the geopotential to advection of vorticity, and warm or cold airmasses. Under many circumstances, both the vorticity, and in particular the potential temperature, are materially not far from being conserved in the atmosphere for up to several days. Consequently, tracking anomalies in maps showing differences between simulations in potential temperature or vorticity may be easier than tracking anomalies in the geopotential itself.

Another useful function is the potential temperature

$$\theta = T\left(\frac{P_0}{P}\right)^{R/c_p}$$

where θ is the potential temperature, T is temperature, P is pressure, P_0 is a reference pressure (often 1000 hPa), R is a constant, and c_p is the constant pressure specific heat. In the absence of diabatic processes, the potential temperature is a materially conserved quantity, and anomalies of potential temperature may often be traced for days in the atmosphere. The virtual potential temperature, and the equivalent potential temperature are related functions that take into account the presence of humidity in the atmosphere, and may also serve as a tracer.

In the absence of strong diabatic processes, anomalies of potential temperature in the atmosphere may live for several days. They may themselves have been created by diabatic processes, and by flow patterns such as local föhn, orographic jets or developing frontal systems.

Static stability, and vorticity can be combined into potential vorticity, which in the quasigeostrophic framework can be written as

$$q = \frac{1}{f_0}\nabla^2\Phi + f + \frac{\partial}{\partial p}\left(\frac{f_0}{\sigma}\frac{\partial\Phi}{\partial p}\right)$$

where q is the potential vorticity, and the other variables are the same as above.

The potential vorticity is better conserved than vorticity, as it takes into account the impact of the so-called stretching term in the vorticity equation. The material conservation of potential vorticity (and potential temperature) may be expressed as

$$\left(\frac{\partial}{\partial t} + \mathbf{V}_g \cdot \nabla\right) q = \frac{D_g q}{Dt} = 0$$

The potential vorticity is conserved in adiabatic flow, and anomalies of potential vorticity may be traced for several days in the atmospheric flow. If static stability is also constant, which sometimes is not far from being true, the vorticity is conserved. Assuming no change in time in the thermodynamic static stability may sound doubtful, but there are in fact many situations where vorticity anomalies are just as traceable as PV anomalies.

3 Precipitation and vertical velocity

Anomalies of precipitation may be analyzed, and traced in several ways, depending on the nature of the associated ascending motion. Orographic lifting is highly influenced by the speed of the wind impinging the mountain, through this simple equation:

$$w = hV/L$$

where w is the vertical velocity, h is the height of the mountain, V is the horizontal speed of the flow impinging the mountain, and L is the length scale of the mountain slope. The vertical velocity is, in other words, proportional to the speed of the horizontal wind, impinging the mountain, and the height of the mountain, as long as the impinging wind is strong enough to overcome the potential barrier of the mountain.

The wind speed is, to a large extent, dominated by the synoptic scale pressure gradient, and anomalies in the pressure field may be identified, and traced directly or through vorticity, potential vorticity or potential temperature. Anomalies of convective precipitation may often be linked to anomalies of potential temperature, which may be traced as explained previously. Finally, synoptic scale anomalies in the vertical velocity fields may be linked to the advection of temperature (geopotential thickness), or advection of vorticity at middle-to-upper tropospheric levels, through the quasigeostrophic omega equation

$$\sigma\nabla_H^2\omega + f^2\frac{\partial^2\omega}{\partial p^2} = f\frac{\partial}{\partial p}\left[\mathbf{V}_g\cdot\nabla_H\left(\zeta_g + f\right) - \nabla_H^2\left(\mathbf{V}_g\cdot\nabla_H\right)\frac{\partial\phi}{\partial p}\right]$$

where ω is the vertical velocity in pressure coordinates, ζ_g is the geostrophic vorticity, and the other parameters are as before.

In many cases, the anomalies in precipitation may be related to a combination of more than one of the above processes. It should also be kept in mind that all three processes of vertical velocity, direct orographic lifting due to impinging airflow, convection, and omega-equation lifting are derived from considerable simplifications of the atmospheric flow, simplifications that are not made in the state-of-the art numerical weather prediction tools, which are based on solving the full set of the primitive equations with advanced parameterizations of subgrid processes. Consequently, there are many anomalies of precipitation that may not have a clear connection to a well-defined anomaly in either vorticity, wind speed or potential temperature at some given height levels.

4 The method of error tracking

All the above functions can, and have been used successfully to trace anomalies in simulations of atmospheric flow, and thereby to analyze the origin of errors in cases of forecast failures.

Fig. 1 illustrates the method of detecting the dynamic origin of an error in a bad simulation. Two simulations are compared. One (*Good*) reproduces an event (a certain flow feature) at time t_0 while the other simulation (*Bad*) does not reproduce the same event in the same manner at t_0. The *Good* and the *Bad* forecasts are typically initialized with a 6–24 h interval. In most cases, the *Good* forecast would be the one with shorter lead time. In the subsequent paragraphs, the "difference field" refers to the difference between the values of the same fields in the two forecasts, *Good* and *Bad* at the same time. In an attempt to keep the presentation as clear as possible, the error tracking method will be described in the following steps (a)–(d)

(a) The field of interest of the *Bad* forecast is subtracted from the same field in the *Good* forecast. The field of interest would, in the case of a wrong wind forecast, be the field of wind speed or pressure, and in the case of a wrong temperature forecast, the field of

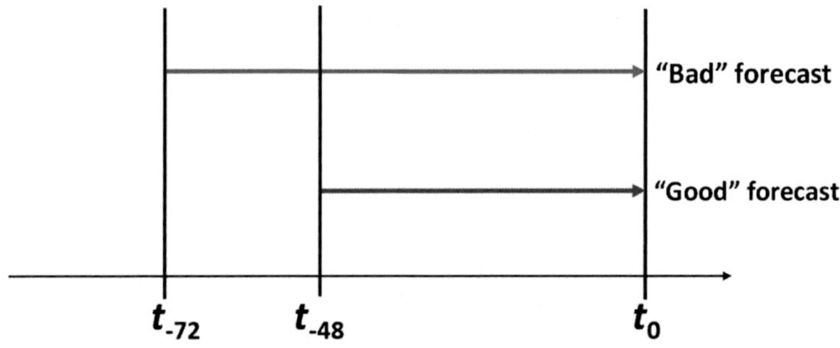

FIG. 1 A schematic of two forecasts (*Good* and *Bad*) with different lead times, compared at $t = t_0$.

temperature is of interest, as well as the fields of low-level winds in most cases. In the case of a wrong precipitation forecast, the primary field of interest is of course the precipitation field, but depending on the dynamics of the vertical velocity, fields of vertical velocity, horizontal velocity, pressure, temperature, vorticity or humidity would be of interest.

(b) The anomaly that appears in the differential field derived in (a) is related to an anomaly in a difference field of one of the conservative variables or the functions described in the preceding section of this chapter. A mean sea-level pressure anomaly in the differential field derived in (a) would typically be related to an anomaly in difference fields of vorticity, potential vorticity or in potential temperature at some tropospheric (or tropopause) level. An anomaly in vertical velocity may be related to an anomaly in the difference fields of the geopotential elsewhere in the flow field, by applying the previously discussed quasigeostrophic theory of rising motion or the simple theory of horizontal winds leading to vertical movements in orographic flows. An anomaly in the precipitation fields may also be related to an anomaly in humidity or potential temperature at low, middle or even upper tropospheric levels.

(c) The anomaly identified in the differential fields in (b) is traced back in time to the initial time of the forecast with the shortest lead time (usually the *Good* forecast).

(d) The airmass where the anomaly in (c) is situated at the initial time of the forecast with the shortest lead time is traced back to its location when the other forecast was initialized. At that location, the analysis of the *Bad* forecast was most likely wrong.

The best way to gain comprehension of the method of dynamic error tracking may be by regarding illustrative examples. In the following, such examples will be explained briefly with reference to the sources where they are explained in detail.

5 Wrong prediction of the surface pressure field

A comparison of two simulations by two different NWP models was made by Ólafsson (1998). Both models were operating in real-time, and they gave very different predictions of the surface pressure field over *E*-Iceland, and the surrounding waters (Fig. 2). With reference to the geopotential tendency equation, details of the pressure fields are related to anomalies in the mid-tropospheric vorticity field (Fig. 3). These anomalies are traced back in time, and they are found to be generated immediately downstream of Iceland and Greenland. It is concluded that the model that gives a bad forecast is representing the impact of orography on the atmospheric flow incorrectly in this specific case. Here, the errors could be traced visually in the vorticity field, while in other available fields the anomalies were not as clearly identifiable.

6 Wrong prediction of heavy precipitation

In a paper by Einarsson et al. (2004), two different predictions of precipitation in Norway are compared. The *Good* forecast predicted extreme precipitation, while the *Bad* one did not (Fig. 4). The precipitation event takes place in strong onshore northwesterly winds, impinging

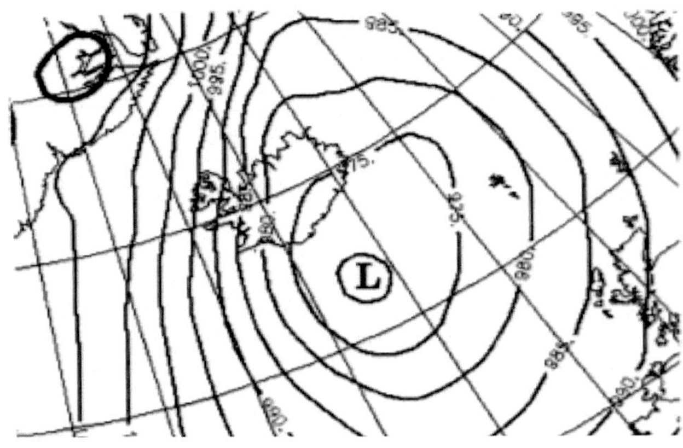

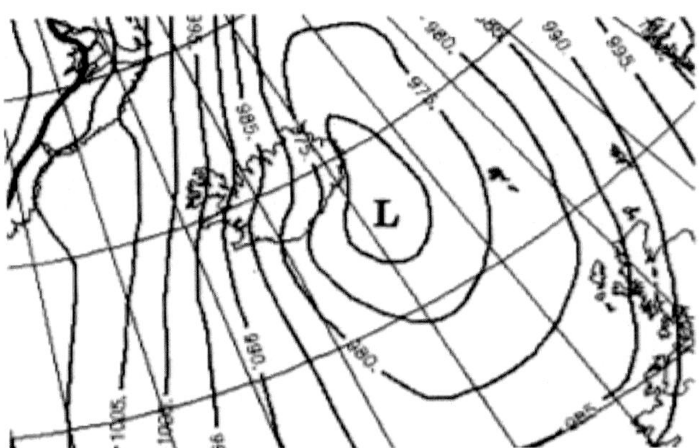

FIG. 2 Mean sea level pressure (hPa) in two different numerical forecasts. The upper panel is the *Bad* forecast, and the lower panel is the *Good* forecast. *From Ólafsson, H., 1998. Different predictions by two NWP models of the surface pressure field east of Iceland. Meteorol. Appl. 5, 253–261.*

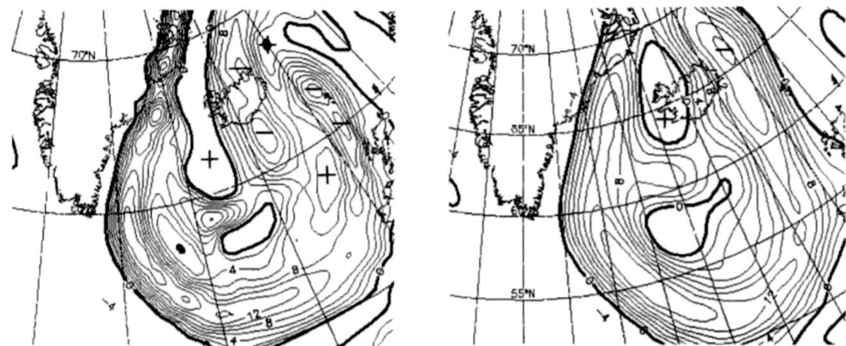

FIG. 3 Relative vorticity (s^{-1}) at 500 hPa at the same time as the mean sea level field in Fig. 1. The *Good* forecast is to the left, and the *Bad* forecast is to the right. *From Ólafsson, H., 1998. Different predictions by two NWP models of the surface pressure field east of Iceland. Meteorol. Appl. 5, 253–261.*

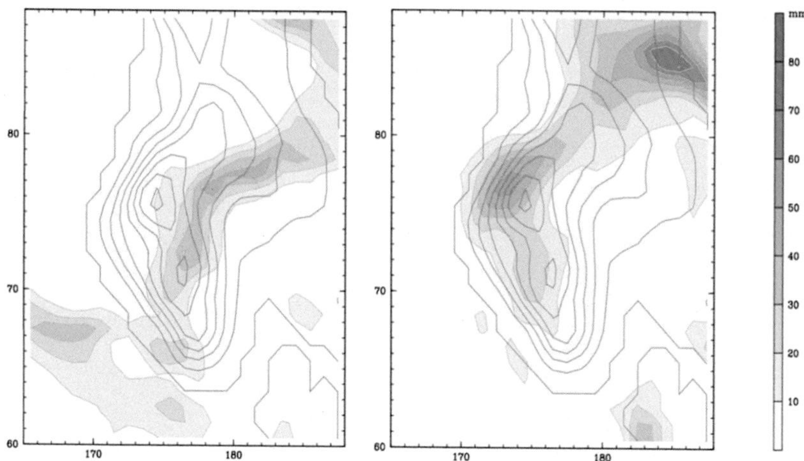

FIG. 4 Precipitation (mm/24 h) in S-Norway in simulations with two different lead times. The left panel is the *Bad* forecast and the right panel is the *Good* forecast. *From Einarsson, E.M., Ólafsson, H., Kristjánsson, J.E., 2004. Forecasting an extreme precipitation event in Norway. Proc. of the 11th AMS Conf. on Mountain Meteorol., Bartlett, New Hampshire, 2004.*

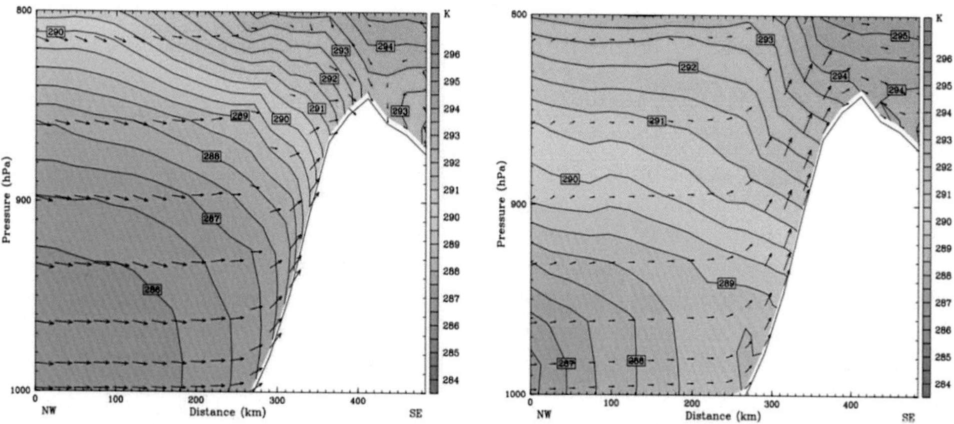

FIG. 5 Potential temperature (K) and wind barbs in a cross section from NW to SE (see Fig. 6). The left panel is the *Bad* forecast and the right panel is the *Good* forecast.

the mountains of Central Norway in the Møre region. The *Good* forecast has substantially stronger low-level winds impinging the mountains, than the *Bad* forecast does (Fig. 5). The stronger the low-level winds are, the stronger is the updraft, and the *Good* forecast has much stronger vertical velocity on the upstream side of the mountains than the *Bad* forecast does. The strong low-level winds in the *Good* forecast are associated with a strong sea-level pressure gradient, to which a low-level warm pocket of air on the low-pressure side of the

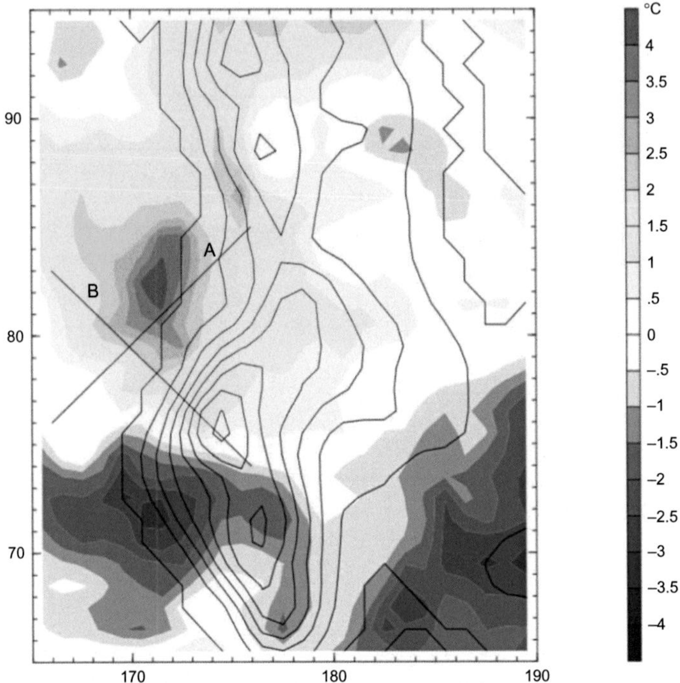

FIG. 6 Temperature difference (°C) between the two forecasts (*Good* minus *Bad*) at the time of the cross section in Fig. 5.

upstream winds (Fig. 6) contributes. The relatively warm air in the *Good* forecast, as well as the more correct position of the frontal system, is connected to more advection of air from the SE over Central-Norway, and the cyclone over S-Norway to be deeper in the *Good* forecast, than in the *Bad* forecast. The minimum pressure in the cyclone is more correct, and deeper in the *Good* than in the *Bad* simulation. This success in the *Good* simulation can be traced back to a more correct representation of the upper tropospheric trough, and the associated vorticity to the west of Ireland the analysis of the *Good* forecast than in the +24h forecast of the *Bad* simulation. Here, we do not have a clear answer to the question why the *Bad* forecast gave a poor representation of the upper-level vorticity in the +24h forecast, but the subsequent sequence of dynamical flow elements leading to enormous errors in local precipitation is well assessed.

A similar sequence of precipitation errors associated with the sensitivity of orographic precipitation to low-level winds was a study by Steensen et al. (2011). Here, the *Good* forecast gave extreme precipitation in Central Norway, while the *Bad* forecast gave much less precipitation. The *Good* forecast has winds, impinging the orography at low-levels, and these winds are stronger than in the *Bad* forecast, and the stronger winds, which are from the west can be related to lower mean sea level pressure in the *Good* forecast, to the north of the precipitation extreme (Fig. 7). In this case, the differences in the pressure of the two

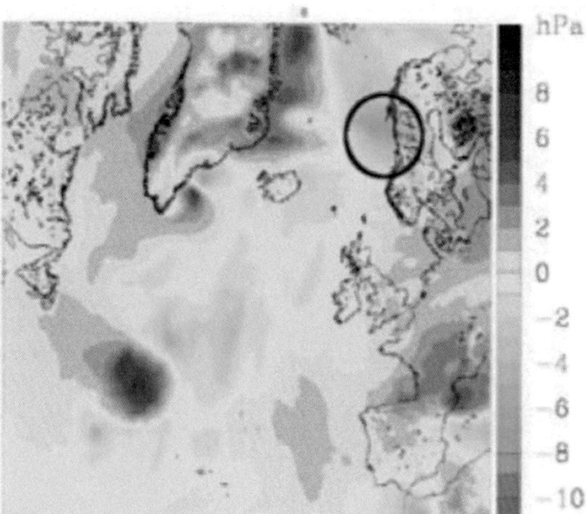

FIG. 7 Difference (*Good-Bad*) in the mean sea level pressure (hPa) at the time of comparison (t_0). The anomalously low pressure to the north of the extreme precipitation is encircled.

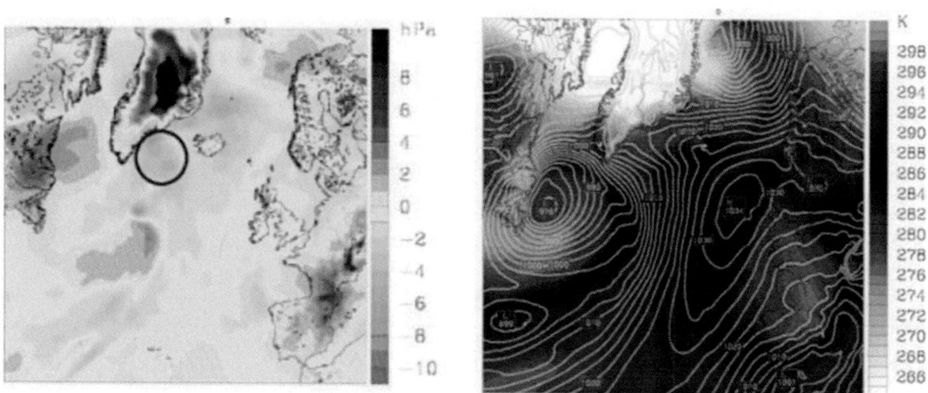

FIG. 8 Left panel: Difference (*Good-Bad*) in the mean sea level pressure (hPa) at the time of the initialization of the *Good* forecast. The pressure anomaly is encircled. Right panel: Potential temperature (K) at 850 hPa in the *Good* forecast.

forecasts formed a distinguishable anomaly that could be traced easily back to a region of strong warm advection southeast of Southeast-Greenland at the time of the initialization of the *Good* forecast (Fig. 8). The erroneous advection in the *Bad* forecast was directly from the cold front to the south of Greenland, 24 h earlier, indicating strong impact of a possibly small displacement of a cold front.

7 Wrong prediction of a rapidly deepening cyclone

Hagen (2008) applies dynamic error tracking on a case of a rapidly deepening cyclone in the vicinity of Iceland. Fig. 9 shows the mean sea level pressure in the two forecasts in question. The *Good* forecast has a deep cyclone at the North-East coast of Iceland, while the *Bad* forecast does not. Fig. 10 summarizes the dynamic sequence of the elements of the flow

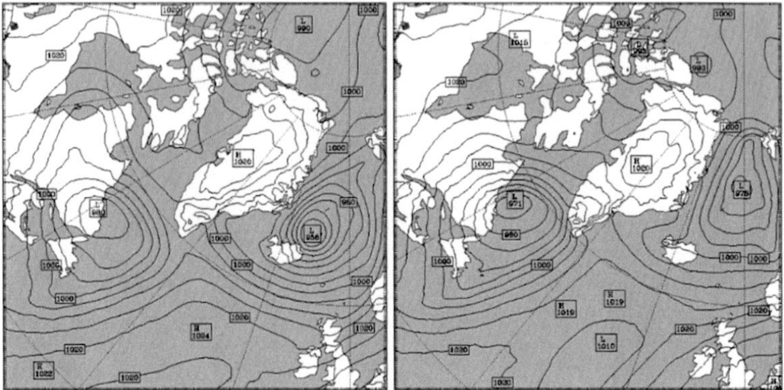

FIG. 9 Mean sea level pressure (hPa) in a *Good* forecast (left) and a *Bad* forecast (right). *From Hagen, B., 2008. Wind Extremes in the Nordic Seas: Dynamics and Forecasting. Thesis at the Geophysical Institute, University of Bergen, Norway.*

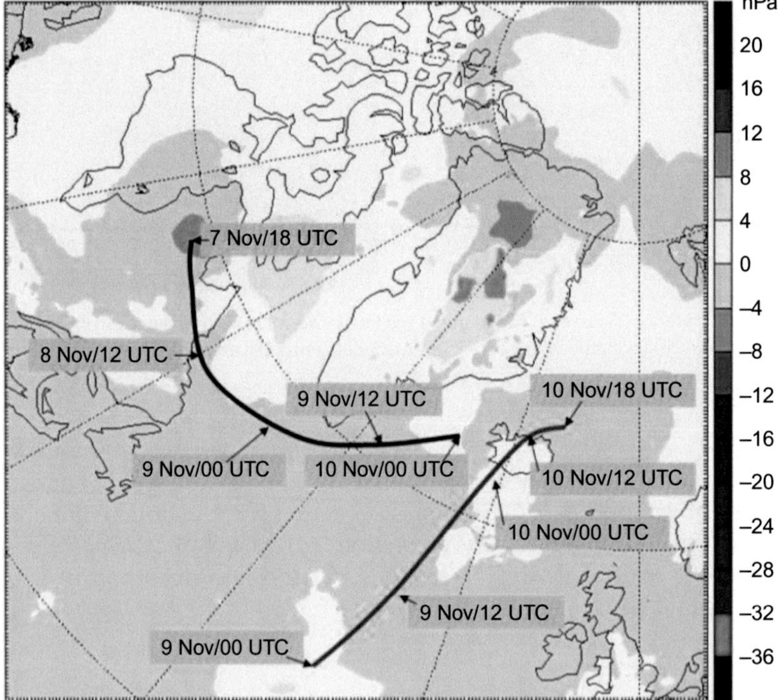

FIG. 10 The difference (*Good-Bad*) in mean sea level pressure (hPa) of the *Good* and the *Bad* forecasts and trajectories of the initial pressure anomaly *(blue)* and the explosive cyclone in the *Good* forecast *(red)*. *From Hagen, B., 2008. Wind Extremes in the Nordic Seas: Dynamics and Forecasting. Thesis at the Geophysical Institute, University of Bergen, Norway.*

leading up to the cyclone development, and the associated windstorm over Iceland. At the time of the initialization of the *Good* forecast, the *Good* forecast has lower mean sea level pressure than the *Bad* forecast, east of Hudson Bay. The maximum difference in the two forecasts is about 6 hPa. This pressure difference is advected to the east over the North-Atlantic where the *Good* forecast has more advection of cold air than the *Bad* forecast does, to the south of the anomaly of the pressure difference between the two forecasts. Subsequently, a classic baroclinic development takes place over the ocean, where the *Good* simulation, with more advection of cold air gives rapid cyclone development, while the *Bad* simulation does not. This is clearly illustrated in the equivalent potential temperature, and pressure fields in Fig. 11. The wrong pressure in the +24 h forecast of the *Bad* simulation appears to have been associated with wrong temperatures of the low-level airmasses. At the time of initialization of the *Bad* forecast, one radiosonde was missing from the regular network over Canada, leaving a large area void of regular upper-air observations.

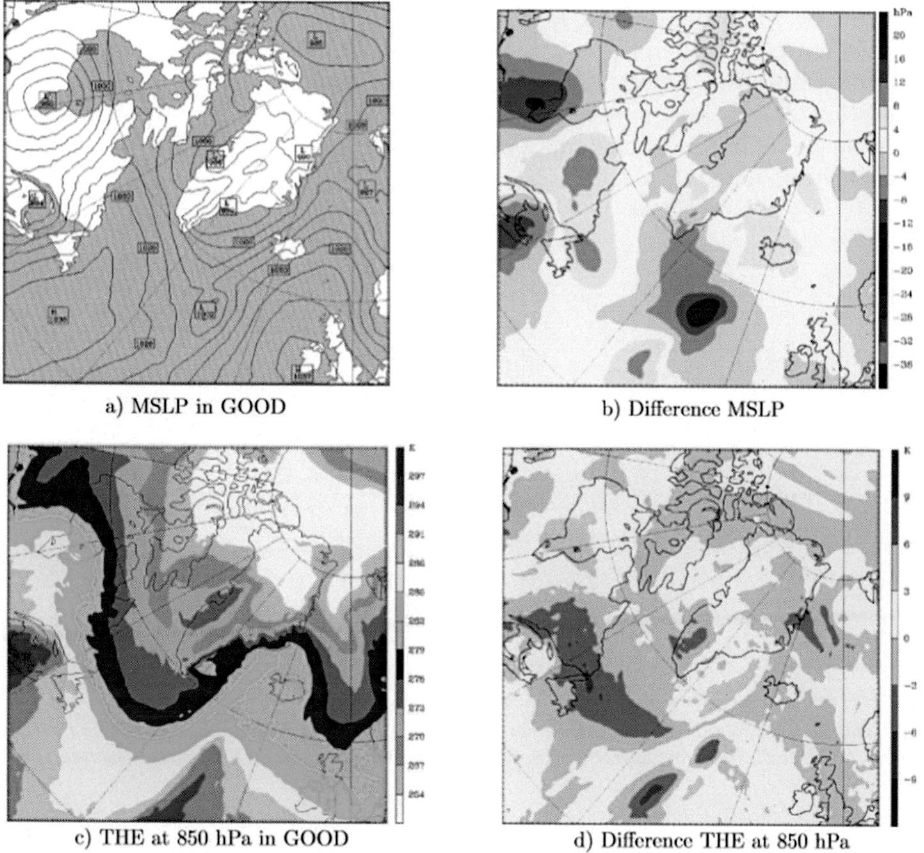

FIG. 11 (A) Mean sea level pressure (hPa), (B) difference in mean sea level pressure (*Good-Bad*), (C) equivalent potential temperature (K) at 850 hPa in the *Good* forecast, and (D) the difference in equivalent potential temperature (*Good-Bad*) at 850 hPa at the time when the cyclone starts to deepen rapidly.

8 Wrong prediction of the surface pressure field and mesoscale orographic impacts

A dynamic error-tracing case of mesoscale orographic impacts on the forecast quality is presented in Tveita (2008). Fig. 12 shows the mean sea level pressure in a *Good* and a Bad forecast. Both forecasts have a deep cyclone northeast of Iceland, and strong winds between Greenland and Iceland, while the shape of the pressure fields is different, so that the *Good* forecast has extreme wind at the north coast of Iceland, while the Bad forecast only has a moderate windstorm. The differences in the pressure fields of the two forecasts are related to variability in the low-level temperatures. The main anomaly in the temperature difference fields can be traced to the Cape Tobin region off the east coast of Greenland, at the outflow of extremely cold low-level airmasses over the sea-ice, blocked at the east coast of Greenland are advected over the sea in the Denmark Strait, between Greenland and Iceland. In this region, there are often very strong horizontal temperature gradients, and only a relatively small change in the low-level flow may lead to large changes in temperature. The initialization of the *Good* forecast consisted of such a small change from the +24h of the Bad forecast (Fig. 13), and the subsequent difference in the low-level temperatures in the two forecasts contributed to the difference in the mean sea level pressure shown in Fig. 12. This case illustrates that regions of strong horizontal gradients are prone to host sources of forecast errors, and that small changes in flow impinging major mountain ranges may in some cases lead to large subsequent changes in the atmospheric flow.

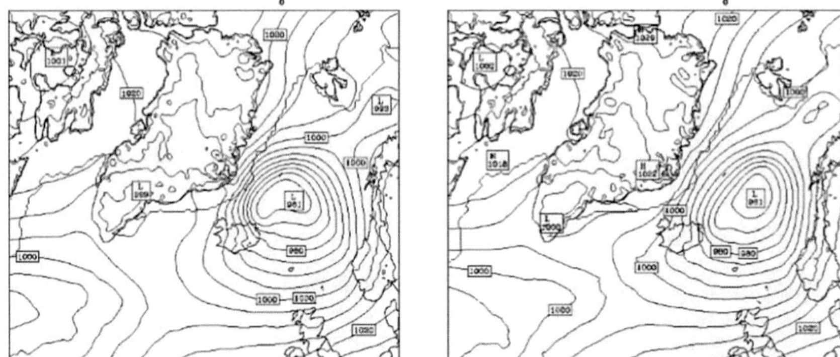

FIG. 12 Mean sea level pressure (hPa) in a *Good* simulation (left) and a *Bad* simulation (right). *From Tveita, B., 2008. Extreme Winds in the Nordic Seas: An Observational and Numerical Study. Thesis at the Geophysical Institute, University of Bergen, Norway.*

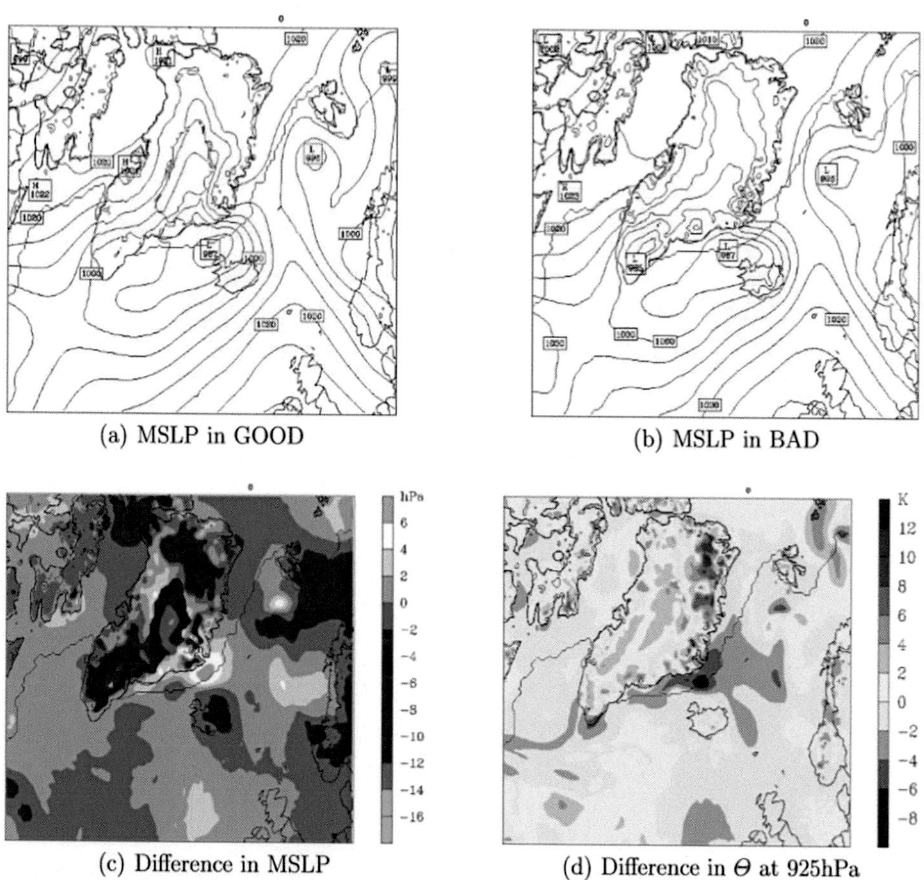

(a) MSLP in GOOD

(b) MSLP in BAD

(c) Difference in MSLP

(d) Difference in Θ at 925hPa

FIG. 13 Top panels: Mean sea level pressure (hPa) at the time of initialization of the *Good* forecast. The upper left panel (A) is the *Good* forecast and the upper right panel (B) is the *Bad* forecast. Bottom panels: (C) Difference (*Good-Bad*) in the mean sea level pressure (hPa) and (D) the temperature (K) at 850 hPa at the time of the initialization of the *Good* forecast.

References

Einarsson, E.M., Ólafsson, H., Kristjánsson, J.E., 2004. Forecasting an extreme precipitation event in Norway. In: Proc. of the 11th AMS Conf. on Mountain Meteorol., Bartlett, New Hampshire, 2004.

Hagen, B., 2008. Wind Extremes in the Nordic Seas: Dynamics and Forecasting. Thesis at the Geophysical Institute, University of Bergen, Norway.

Ólafsson, H., 1998. Different predictions by two NWP models of the surface pressure field east of Iceland. Meteorol. Appl. 5, 253–261.

Steensen, B.M., Óafsson, H., Jonassen, M.O., 2011. An extreme precipitation event in Central Norway. Tellus 63A, 675–686.

Tveita, B., 2008. Extreme Winds in the Nordic Seas: An Observational and Numerical Study. Thesis at the Geophysical Institute, University of Bergen, Norway.

Index

Note: Page numbers followed by *f* indicate figures, and *t* indicate tables.

T

U

V

W

Z

Printed in the United States
By Bookmasters